Outstanding Contributions to Logic

Volume 24

Outstanding Contributions to Logic puts focus on important advances in modern logical research. Each volume is devoted to a major contribution by an eminent logician. The series will cover contributions to logic broadly conceived, including philosophical and mathematical logic, logic in computer science, and the application of logic in linguistics, economics, psychology, and other specialized areas of study.

A typical volume of Outstanding Contributions to Logic contains:

- A short scientific autobiography by the logician to whom the volume is devoted
- The volume editor's introduction. This is a survey that puts the logician's contributions in context, discusses its importance and shows how it connects with related work by other scholars
- The main part of the book will consist of a series of chapters by different scholars that analyze, develop or constructively criticize the logician's work
- Response to the comments, by the logician to whom the volume is devoted
- A bibliography of the logician's publications Outstanding Contributions to Logic is published by Springer as part of the Studia Logica Library.

This book series, is also a sister series to Trends in Logic and Logic in Asia: Studia Logica Library. All books are published simultaneously in print and online. This book series is indexed in SCOPUS.

Proposals for new volumes are welcome. They should be sent to the editor-in-chief sven-ove.hansson@abe.kth.se

Alex Citkin · Ioannis M. Vandoulakis
Editors

V. A. Yankov on Non-Classical Logics, History and Philosophy of Mathematics

Springer

Editors
Alex Citkin
Metropolitan Telecommunications
New York, NY, USA

Ioannis M. Vandoulakis
Hellenic Open University
Athens, Greece

FernUniversität in Hagen
Hagen, Germany

ISSN 2211-2758 ISSN 2211-2766 (electronic)
Outstanding Contributions to Logic
ISBN 978-3-031-06845-4 ISBN 978-3-031-06843-0 (eBook)
https://doi.org/10.1007/978-3-031-06843-0

This Springer imprint is published by the registered company Springer Nature Switzerland AG
The registered company address is: Gewerbestrasse 11, 6330 Cham, Switzerland

Preface

This volume is dedicated to Vadim Yankov (Jankov[1]), the Russian logician, historian and philosopher of mathematics and political activist who was prosecuted in the former USSR.

In 1964, he defended his dissertation *Finite implicative structures and realizability of formulas of propositional logic* under the supervision of A. A. Markov. In the 1960s, Yankov published nine papers dedicated to non-classical propositional logics, predominantly to intermediate logics. Even today, these publications—more than fifty years later—still hold their place among the most quotable papers in logic. The reason for this is very simple: not only Yankov obtained significant results in propositional logic, but he also developed a machinery that has been successfully used to obtain new results up until our days.

Yankov studied the class of all intermediate logics, as well as some particular intermediate logic. He proved that the class of all intermediate logic ExtInt is not denumerable, and that there are intermediate logics lacking the finite model property, and he had exhibited such a logic. In addition, he proved that ExtInt contains infinite strongly ascending, strongly descending and independent (relative to set inclusion) subclasses of logics, each of which is defined by a formula on just two variables. Thus, it became apparent that ExtInt as a lattice has a quite a complex structure.

In 1953, G. Rose gave a negative answer to a hypothesis that the logic of realizability, introduced by S. Kleene in an attempt to give precise intuitionistic semantics to Int, does not coincide with Int. In 1963, Yankov constructed the infinite series of realizable formulas not belonging to Int.

In his 1968 paper, Yankov studied the logic of the weak law of excluded middle, and the logic defined relative to Int by a single axiom $\neg p \vee \neg\neg p$. Nowadays, this logic is often referred to as a Yankov (or Jankov) logic. In particular, Yankov has discovered that this logic has a very special place in ExtInt: it is the largest logic,

[1] In Russian, the last name is Янков. In the translations of papers of the 1960s by the American Mathematical Society, the last name was transliterated as "Jankov," while in the later translations, the last name is transliterated as "Yankov," which perhaps is more correct. In this volume, the reader will see both spellings.

a positive fragment of which coincides with the positive fragment of Int, while all extensions of the Yankov logic have distinct positive fragments.

In his seminal 1969 paper, Yankov described in detail the machinery mentioned above. The reader can find more on Yankov's achievements in intermediate logics in the exposition included in this volume.

However, not only Yankov's results in studying intermediate logics are important. His papers instigate the transition from matrix to algebraic semantics. Already in his 1963 papers, he started to use what is now known as Heyting or pseudo-Boolean algebras. At the same time, H. Rasiowa and R. Sikorski's book, *The Mathematics of Methamatematics*, was published, in which the pseudo-Boolean algebras were studied. Yankov made the Russian translation of this book (published in 1972), and it greatly influenced the researchers in the former Soviet Union. Besides, Yankov was one of the pioneers who studied not only intermediate logics—extensions of Int, but also extensions of positive and minimal logics and their fragments. It would not be an overstatement to say that Yankov is one of the most influential logicians of his time.

At the end of the 1960s and in the 1970s, Yankov got more involved in the political activities. In 1968, he joined other prominent mathematicians and co-signed the famous letter of the 99 Soviet mathematicians addressed to the Ministry of Health and the General Procurator of Moscow asking for the release of imprisoned Esenin-Vol'pin. As a consequence, Yankov lost his job at the Moscow Institute of Physics and Technology (MIPT), and most of the mathematicians who signed this letter faced severe troubles.

Since 1972, he started to publish abroad, for instance, in the dissident journal *Kontinent*, founded in 1974 by writer Vladimir Maximov that was printed in Paris and focused on the politics of the Soviet Union. In issue 18, he published the article "On the possible meaning of the Russian democratic movement." In 1981–1982 he wrote a "Letter to Russian workers on the Polish events," on the history and goals of the "Solidarity" trade union. Following these events, he was arrested in August 1982, and on January 21, 1983, the Moscow City Court sentenced him to four years in prison and three years in exile for anti-Soviet agitation and propaganda. He served his term in the Gulag labor camp, called "Dubravny Camp" in Mordovia, near Moscow, and exile in Buryatia in south-central Siberia. He was released in January 1987 and rehabilitated in 1991.

Despite his hard life in the Camp and the exile, Yankov started to study philosophy and the classic Greek language. The second editor was impressed when he visited him at home in Dolgoprudnyj, near Moscow, in 1990, and Yankov started to analyze the syntax of a passage from Plato's *Parmenides* in classic Greek. When he asked him where he studied classical Greek so competently, he was stunned Yankov's unexpected answer: "In prison"!

Thus, Yankov's philosophical concerns were shaped while he was imprisoned. His first, possibly philosophical publication was printed abroad in issue 43 (1985) of the journal *Kontinent*, entitled "Ethical-philosophical treatise," where he outlines his philosophical conception of existential history. A publication on the same theme in Russia was made possible only ten years later, in the journal *Voprosy Filosofii* (1998, 6).

After Yankov's acquaintance with the second editor's Ph.D. Thesis, he agreed to become a member of the Committee of Reviewers and then started to examine the history of Greek mathematics systematically but from a specific logical point of view. He was primarily concerned about the ontological aspects of Greek mathematical theories and the relevant ontological theories in pre-Socratic philosophy. He stated a hypothesis on the rise of mathematical proof in ancient Greece, which integrated into the broader context of his inquiry of the pre-Socratic philosophy.

This volume is a minimal appreciation to a mathematician and scholar who deserves our respect and admiration.

New York, USA Alex Citkin
Hagen, Germany Ioannis M. Vandoulakis

Contents

Contributors

Bezhanishvili Guram New Mexico State University, New Mexico, USA

Bezhanishvili Nick University of Amsterdam, Amsterdam, Netherlands

Citkin Alex Metropolitan Telecommunications, New York, USA

Denisova Tatiana Yu. Department of Philosophy and Law, Surgut State University, Surgut, Russia

Ghilardi Silvio Dipartimento di Matematica, Università degli Studi di Milano, Milano, Italy

Indrzejczak Andrzej Department of Logic, University of Łódź, Łódź, Poland

Muravitsky Alexei Louisiana Scholars' College, Northwestern State University, Natchitoches, LA, USA

Plisko Valery Faculty of Mechanics and Mathmetics, Moscow State University, Moscow, Russia

Suzuki Nobu-Yuki Faculty of Science, Department of Mathematics, Shizuoka University, Shizuoka, Japan

Vandoulakis Ioannis M. The Hellenic Open University, Athens, Greece and FernUniversität in Hagen, Hagen, Germany

Yankov Vadim A. Moscow, Russia

Chapter 1
Short Autobiography

Vadim A. Yankov

My full name is Vadim Anatol'evich Yankov. I was born on February 1st 1935, in Taganrog, Russia. During the Second World War, I was evacuated to Sverdlovsk. In 1952, I enrolled in the Department of Philosophy of Moscow State University. Faced with the "troubles" related to the ideologization in the humanities fields in the Soviet time, in 1953 I decided to transfer to the Department of Mechanics and Mathematics. In 1956, I was expelled from the University. The reasons given were my sharp criticism of the Komsomol, participation in a complaint against the conditions at the University students' cafeteria, and publication of an independent students' newspaper. Later, I was accepted into the University's distance remote program and obtained my diploma in 1959.

Since 1958, I have been employed in the Programming Department at the Steklov Institute of Mathematics (later, the Programming Department of the Institute of Mathematics of the Siberian Branch of the USSR Academy of Sciences). I worked in the research group developing one of the first programming languages, the ALPHA, an extension of ALGOL. After my graduation in 1959, I became a post-graduate student at the Department of Mathematics of the Moscow State University. Under the supervision of Andrey Markov, I prepared my thesis "Finite implicative structures

Editors' Note. In the book series "Outstanding Contributions to Logic" it is customary to include a scientific autobiography of the person the volume is dedicated to. Unfortunately, V. A. Yankov is not in a position to write his scientific autobiography, so we included a translation of his very brief and formal autobiography written a long time ago for the human resources department, and translated from Russian by Fiona Citkin. His scientific biography can be found in the overview papers by Citkin, Indrzejczak, Denisova, and Vandoulakis, which have been included in this volume.

V. A. Yankov (✉)
Volokolamskoe shosse 7a, apt.37, Moscow, Russia
e-mail: kirill_yankov@mail.ru

A. Citkin and I. M. Vandoulakis (eds.), *V. A. Yankov on Non-Classical Logics, History and Philosophy of Mathematics*, Outstanding Contributions to Logic 24,
https://doi.org/10.1007/978-3-031-06843-0_1

and realizability of formulas of propositional logic,"[1] for which I was awarded my PhD Degree in 1964. Since 1963, I have been an Assistant Lecturer at the Moscow Institute of Physics and Technology. In 1968, I was dismissed after co-signing the letter of my colleagues addressed to the Ministry of Health and the General Procurator of Moscow asking for the release of Aleksandr Esenin-Volpin.[2]

From 1968 to 1974, I worked as a Senior Lecturer at the Moscow Aviation Institute. Due to teaching overload during this period, my scientific achievement was substantially reduced.[3] Moreover, I was fired by the Institute's administration for my dissent views and discussions concerning the Soviet intervention in Czechoslovakia in 1968. During 1974–1982 I worked at the Enterprise Resource Planning Department of the Moscow Institute for Urban Economics.

From 1982 to 1987, I was imprisoned and exiled; the official reason for my arrest was anti-Soviet propaganda found in my publications on Soviet politics in foreign political journals.[4] During my confinement, I started studying Classic Greek by comparing Thucydides' works in the original and its Russian translation.

After my release in 1987, I worked in the Institute of Thermal Metallurgical Units and Technologies "STALPROEKT" until 1991.

Since 1991, I have been an Associate Professor at the Department of Mathematics, Logic and Intellectual Systems, Faculty of Theoretical and Applied and the Department of Logical and Mathematical Foundations of Humanitarian Knowledge, Institute of Linguistics of the Russian State University for the Humanities in Moscow. During this period, my research interests shifted to philosophy, history of philosophy and history of mathematics. I started lecturing regular courses on philosophy and the history of philosophy at the Russian State University of the Humanities, delivered a series of lectures in the Seminar of Philosophy of Mathematics of the Moscow

---

[1] *Editors' Note.* In Plisko's paper in this volume, the reader can find more information on Yankov's results in realizability.

[2] *Editors' Note.* The letter was signed by 99 prominent mathematicians. As a consequence, many of them had been compelled to leave their positions in academia. For more details, see Fuchs D.B. "On Soviet Mathematics of the 1950th and 1960th" in Golden Years of Soviet Mathematics, American Mathematical Society, 2007, p. 221.

[3] *Editors' Note.* In the early 1960s, Yankov published a series of papers dedicated to propositional logic, especially, intermediate propositional logics. In these papers he announced the results which were further developed and published with proofs in 1968–1969 (cf. the complete list of papers at the end of this volume). At this time, Yankov mentioned (in a letter to A. Citkin) that "the focus of my research interests has been shifted." More about Yankov's contribution to the theory of intermediate logics can be found in Citkin's expository paper in this volume.

[4] *Editors' Note.* During this period, Yankov published abroad some papers in which he criticized the Soviet regime. In November 1981 – January 1982 he published a "Letter to Russian workers about the Polish events" in which he expressed his support to the Polish workers that struggled for freedom. On the 9th of August 1982, when Yankov left his apartment to go to the office, he was arrested. In January 1983, he was sentenced to four years in prison and three years of exile. During the *Perestroika*, in January 1987, he was released, and then rehabilitated on the 30th of October 1991 from all charges against him.

State University and published papers in the history of mathematics. This activity culminated in the publication of my book *Interpretation of Early Greek Philosophy* in 2011.[5]

Complete Bibliography of Vadim Yankov

Янков В.А. О некотрых суперконструктивных исчислениях высказываний. ДАН СССР Сер. Мат. 151, 796–798 (1963) [Jankov, V. A. (1963). On certain superconstructive propositional calculi. *Soviet Mathematics Doklady*, *4*, 1103–1105.]

Янков В.А. О реализуемых формулах логики высказываний. ДАН СССР Сер. Мат. 151, 1035–1037 (1963) [Jankov, V. A. (1963). On the realizable formulae of propositional logic. *Soviet Mathematical Doklady*, *4*, 1146–1148.]

Янков В.А. О связи между выводимостью в интуиционистском исчислении высказываний и конечными импликативными структурами. ДАН СССР Сер. Мат. 151, 1293–1294 (1963) [Jankov, V. A. (1963). The relationship between deducibility in the intuitionistic propositional calculus and finite implicational structures. *Soviet Mathematics Doklady*, *4*, 1203–1204.]

Янков В.А.: О финитной общезначимости формул специального вида. ДАН СССР Сер. Мат. 174:302–304 (1967) [Jankov, V. A. (1967). Finite validity of formulas of special form. *Soviet Mathematics*, *8*(3), 648–650.]

Янков В.А.: Об исчислении слабого закона исключенного третьего. Изв. АН ССР Сер. Мат. 32б 1044–1051 (1968) [Jankov, V. A. (1969). Calculus of the weak law of the excluded middle. *Mathematics of the USSR-Izvestiya*, *2*(5), 997–1004.]

Янков В.А.: Построение последовательности сильно независимых суперинтуиционистских пропозициональных исчислений. ДАН СССР Сер. Мат. 181, 33–34 (1968) [Jankov, V. A. (1968). The construction of a sequence of strongly independent superintuitionistic propositional calculi. *Soviet Mathematics Doklady*, *9*, 806–807.]

Янков В.А.: О расширении интуиционистското пропозиционального исчисления до классического и минимального до интуиционистского . Изв. АН СССР Сер. Мат. 32, 208–211 (1968) [Jankov, V. A. (1968). On an extension of the intuitionistic propositional calculus to the classical one and of the minimal one to the intuitionistic one. *Mathematics of the USSR-Izvestiya*, *2*(1), 205–208.]

Янков В.А.: Три последовательности формул с двумя переменными в позитивной пропозициональной логике. Пзв. АН СССР Сер. Мат. 32, 880–883 (1968) [Jankov, V. A. (1968). Three sequences of formulas with two variables in positive propositional logic. *Mathematics of the USSR-Izvestiya*, *2*(4), 845–848.]

Янков В.А.: Конъюктивно неразложимые формулы в пропозициональных исчислениях. Izv Изв. АН СССР Сер. Мат., 33, 18–38 (1969) [Jankov, V. A. (1969). Conjunctively irresolvable formulae in propositional calculi. *Mathematics of the USSR, Izvestiya*, *3*, 17–35.]

Осипова В.А., Янков В.А. Математическая логика. 142 стр., МАИ (1974) [Osipova, V. A., Yankov, V. A. (1974). Mathematical logic. 142 p. Moscow institute for aviation, in Russian.]

Янков В.А.: Этико- философский трактат. Континент 43 271–301 (1974) [Yankov, V. A. Ethico-philosophical tractatus. *Kontinent*, *43*, 271–301, in Russian.]

Янков В.А.: Диалоговая теория доказательства для арифметики анализа и теории множеств. Изв. РАН Сер. Мат. 58(3) 140–168 (1994); [Yankov, V. A. (1995). Dialogue

[5] *Editors' Note*. The papers by Denisova and Vandoulakis in this volume have been devoted to Yankov's contribution to philosophy and history of mathematics.

theory of proofs for arithmetic, analysis and set theory. *Russian Academy of Science, Izvestiya Mathematics*, *44*(3), 571–600.]

Янков В.А.: Бесконечность и становление доказательства. Бесконечность в математике: философские и исторические аспекты, 20–24 Янус-К, Москва (1997) [Yankov, V. A. (1997). Infinity and establishment of proof. In: *Infinity in mathematics: Pholosophical and historical aspects*, 20–24, Janus-K, in Russian.]

Янков В.А.: Диалоговая интерпретация классического исчисления предикатов. Изв. РАН Сер. Мат. 61(1) 215–224 (1997) [Yankov, V. A. (1997). Dialogue interpretation of the classical predicate calculus. *Academy of Science, Izvestiya Mathematics*, *611*, 225–233.]

Янков В.А.: Становление доказательства в ранней греческой математике (гипотетическая реконструкция). Историко-Мат. Иссл. 37(2) 200–236 (1997) [Yankov, V. A., The establishment of proof in early Greek mathematics (hypothetical reconstruction). *Istor-Mat Issled*, *2*(37), 200–236, in Russian.]

Янков В.А.: Эскиз экзистенциальной истории. Вопосы филисофии, 6 3–28 (1998) [Yankov, V. A. (1998). A scetch of existential history. *Problems of Philosophy*, *6*, 3–28, in Russian.]

Янков В.А.: Типологические особенности арифметики древнего Египта и Месопотамии. Стили в математике: социокультурная философия математики, ред. А.Г. Барабашева, 265 – 269, Русская Христианская Гуманитарная Академия, Ст. Петербург (1999) [Yankov, V. A. (1999). Typological peculiarities of ancient Egypt and Mesopotamia arithmetics. In: A. G. Barabasheva (Ed.), *Styles in mathematics: Sociocultural philosophy of mathematics* (pp. 265–269). Russian Christian Humanitarian Academy, in Russian.]

Янков В.А.: Гиппас и рождение геометрии величин. Историко-Мат. Иссл. 40(5) 192–222 (2000) [Yankov, V. A., Hippas and the birth of the geometry of magnitude. *Istor-Mat Issled*, *40*(5), 192–222, 383, in Russian.]

Янков В.А.: Геометрия последователей Гиппаса. Историко-Мат. Иссл. 41(6) 285–319 (2001) [Yankov, V. A. (2001). The geometry of the followers of Hippas. *Istor-Mat Issled*, *41*6, 285–319, in Russian.]

Янков В.А.: Геометрия Анаксагора. Историко-Мат. Иссл. 43(8) 241–267 (2003) [Yankov, V. A., The geometry of Anaxagoras. *Istor-Mat Issled*, *43*8, 241–267, in Russian.]

Янков В.А.: Опыт и онтология математических обьектовю. Математика и опыт, pp 193 – 197, МГУ, Москва (2003) [Yankov, V. A. (2003). Experiment and ontology of mathematical objects. In: *Mathematics and experiment* (pp. 193–197). Moscow State University, Moscow, in Russian.]

Янков В.А.: Строение вещества в философии Анаксагора. Вопросы филосфии, 5,135 – 149 (2003) [Yankov, V. A. (2003). Structure of matter in philosophy of Anaxagoras. *Problems of Philosophy*, *5*, 135–149, in Russian.]

Янков В.А.: Истолквание ранней греческой философии. РГГУ, Москва, 853 с. (2011) [Yankov, V. A. (2011). Interpretation of early Greek philosophy. RGGU, Moscow, 853 p., in Russian.]

Янков В.А.: Диалоги и эссе. Кварта, 590 с. (2016) [Yankov, V. A. (2016). Dialogues and essays. Kvarta, 590 p., in Russian.]

Янков В.А.: Изераные лекции и статьи. Кварта, 414 с. (2022) [Yankov, V. A. (2022). Selected lectures and articles. Kvarta, 414 p. in Russian.]

Part I
Non-Classical Logics

Chapter 2
V. Yankov's Contributions to Propositional Logic

Alex Citkin

Abstract I give an exposition of the papers by Yankov published in the 1960s in which he studied positive and some intermediate propositional logics, and where he developed a technique that has successfully been used ever since.

Keywords Yankov's formula · Characteristic formula · Intermediate logic · Implicative lattice · Weak law of excluded middle · Yankov's logic · Positive logic · Logic of realizability · Heyting algebra

2020 Mathematics Subject Classification: Primary 03B55 · Secondary 06D20 · 06D75

2.1 Introduction

V. Yankov started his scientific career in early 1960s while writing his Ph.D. thesis under A. A. Markov's supervision. Yankov defended thesis "Finite implicative lattices and realizability of the formulas of propositional logic" in 1964. In 1963, he published three short papers Jankov (1963a, b, c) and later, in Jankov (1968a, b, c, d, 1969), he provided detailed proofs together with new results. All these papers are primarily concerned with studying *super-intuitionistic* (or super-constructive, as he called them) propositional logics, that is, logics extending the intuitionistic propositional logic Int. Throughout the present paper, the formulas are propositional formulas in the signature $\to, \wedge, \vee, \mathfrak{f}$, and as usual, $\neg p$ denotes $p \to \mathfrak{f}$ and $p \leftrightarrow q$ denotes $(p \to q) \wedge (q \to p)$; the logics are the sets of formulas closed under the rules Modus Ponens and substitution.

A. Citkin (✉)
Metropolitan Telecommunications, New York, USA
e-mail: acitkin@gmail.com

A. Citkin and I. M. Vandoulakis (eds.), *V. A. Yankov on Non-Classical Logics, History and Philosophy of Mathematics*, Outstanding Contributions to Logic 24,
https://doi.org/10.1007/978-3-031-06843-0_2

To put Yankov's achievements in a historical context, we need to recall that Int was introduced by Heyting (cf. Heyting 1930[1]), who defined it by a calculus denoted by IPC as an attempt to construct a propositional logic addressing Brouwer's critique of the law of excluded middle and complying with intuitionistic requirements. Soon after, Gödel (cf. Gödel 1932) observed that Int cannot be defined by any finite set of finite logical matrices and that there is a strongly descending (relative to set-inclusion) set of super-intuitionistic logics (*si-logics* for short); thus, the set of si-logics is infinite. Gödel also noted that IPC possesses the following property: for any formulas A, B, if $\mathsf{IPC} \vdash (A \vee B)$, then $\mathsf{IPC} \vdash A$, or $\mathsf{IPC} \vdash B$—the *disjunction property*, which was later proved by Gentzen.

Even though Int cannot be defined by any finite set of finite matrices, it turned out that it can be defined by an infinite set of finite matrices (cf. Jaśkowski 1936), in other words, Int enjoys the finite model property (f.m.p. for short). This led to a conjecture that every si-logic enjoys the f.m.p., which entails that every si-calculus is decidable.

At the time when Yankov started his research, there were three objectives in the area of si-logics: (a) to find a logic that has semantics suitable from the intuitionistic point of view, (b) to study the class of si-logics in more details, and (c) to construct a convenient algebraic semantics.

By the early 1960s the original conjecture that Int is the only si-logic enjoying the disjunction property and that the realizability semantics introduced by Kleene is adequate for Int were refuted: in Kreisel and Putnam (1957), it was shown that the logic of IPC endowed with axiom $(\neg p \to (q \vee r)) \to ((\neg p \to q) \vee (\neg p \to r))$ is strictly larger than Int, and in Rose (1953), a formula that is realizable but not derivable in IPC was given. Using the technique developed by Yankov, Wroński proved that in fact, there are continuum many si-logics enjoying the disjunction property (cf. Wroński 1973).

In Heyting (1941), Heyting suggested an algebraic semantics, and in 1940s, McKinsey and Tarski introduced an algebraic semantics based on topology. In his Ph.D. (Rieger 1949), which is not widely known even nowadays, Rieger essentially introduced what is called a "Heyting algebra," and in Rieger (1957), he constructed an infinite set of formulas on one variable that are mutually non-equivalent in IPC. It turned out (cf. Nishimura 1960) that every formula on one variable is equivalent in IPC to one of Rieger's formulas. We need to keep in mind that the book (Rasiowa and Sikorski 1963) was published only in 1963. In 1972, this book had been translated into Russian by Yankov, and it greatly influenced the studies in the area of si-logics.

By the 1960s, it also became apparent that the structure of the lattice of the si-logics is more complex than expected: in Umezawa (1959) it has was observed that the class of si-logics contains subsets of the order type of ω^{ω}; in addition, it contains infinite subsets consisting of incomparable relative to set-inclusion logics.

Generally speaking, there are two ways of defining a logic: semantically by logical matrices or algebras, and syntactically, by calculus. In any case, it is natural to ask whether two given logical matrices, or two given calculi define the same logic. More

[1] The first part was translated in Heyting (1998).

precisely, is there an algorithm that, given two finite logical matrices decides whether their logics coincide, and is there an algorithm that given two formulas A and B decides whether calculi $\mathsf{IPC} + A$ and $\mathsf{IPC} + B$ define the same logic? The positive answer to the first problem was given in Łoś (1949). But in Kuznetsov (1963), it was established that in a general case (in the case when one of the logics can be not s.i.), the problem of equivalence of two calculi is unsolvable. Note that if every si-logic enjoys the f.m.p., then every si-calculus would be decidable and consequently, the problem of equivalence of two calculi would be decidable as well.

In Jankov (1963a), Yankov considers four calculi:

(a) $\mathsf{CPC} = \mathsf{IPC} + (\neg\neg p \to p)$—the classical propositional calculus;
(b) $\mathsf{KC} = \mathsf{IPC} + (\neg p \vee \neg\neg p)$—the calculus of the weak law of excluded middle (nowadays the logic of KC is referred to as Yankov's logic);
(c) $\mathsf{BD}_2 = \mathsf{IPC} + ((\neg\neg p \wedge (p \to q) \wedge ((q \to p) \to p)) \to q)$;
(d) $\mathsf{SmC} = \mathsf{IPC} + (\neg p \vee \neg\neg p) + ((\neg\neg p \wedge (p \to q) \wedge ((q \to p) \to p)) \to q)$—the logic of SmC is referred to as Smetanich's logic and it can be also defined by $\mathsf{IPC} + ((p \to q) \vee (q \to r) \vee (r \to s))$

and he gives a criterion for a given formula to define it relative to IPC (cf. Sect. 2.7). In Jankov (1968a), Yankov studied the logic of KC, and he proved that it is the largest si-logic having the same positive fragment as Int. Moreover, in Jankov (1968d), Yankov showed that the positive logic, which is closely related to the logic of KC, contains infinite sets of mutually non-equivalent, strongly descending, and strongly ascending chains of formulas (cf. Sect. 2.6).

Independently, a criterion that determines by a given formula A whether $\mathsf{Int} + A$ defines Cl was found in Troelstra (1965). In Jankov (1968c), Yankov gave a proof of this criterion as well as a proof of a similar criterion for Johansson's logic (cf. Sect. 2.5).

In Jankov (1963b), Yankov constructed infinite sets of realizable formulas that are not derivable in IPC and that are not derivable from each other. Moreover, he presented the seven-element Heyting algebra in which all realizable formulas are valid (cf. Sect. 2.8).

Jankov (1963c) is perhaps the best-known Yankov's paper, and it is one of the most quoted papers even today. In this paper, Yankov established a close relation between syntax and algebraic semantics: with every finite subdirectly irreducible Heyting algebra $\mathbf{A}$ he associates a formula $X_{\mathbf{A}}$—a characteristic formula of $\mathbf{A}$, such that for every formula B, the refutability of B in $\mathbf{A}$ (i.e. $\mathbf{A} \not\models B$) is equivalent to $\mathsf{IPC} + B \vdash X_{\mathbf{A}}$. Jankov (1963c) is a short paper and does not contain proofs. The proofs and further results in this direction are given in Jankov (1969), and we discuss them in Sect. 2.3. Let us point out that characteristic formulas in a slightly different form were independently discovered in de Jongh (1968).

Applying the developed machinery of characteristic formulas, Yankov proved (cf. Jankov 1968b) that there are continuum many distinct si-logics, and that among them there are logics lacking the f.m.p. Because the logic without the f.m.p. presented by Yankov was not finitely axiomatizable, it left a hope that perhaps all si-calculi enjoy the f.m.p. (this conjecture was refuted in Kuznetsov and Gerčiu 1970.)

Let us start with the basic definitions used in Yankov's papers.

2.2 Classes of Logics and Their Respective Algebraic Semantics

2.2.1 Calculi and Their Logics

Propositional formulas are formulas built in a regular way from a denumerable set of propositional variables Var and connectives.

Consider the following six propositional calculi with axioms from the following formulas:

$$p \to (q \to p); \quad (p \to (q \to r)) \to ((p \to q) \to (p \to r)); \tag{I}$$
$$(p \land q) \to p; \quad (p \land q) \to q; \quad p \to (q \to (p \land q)); \tag{C}$$
$$p \to (p \lor q); \quad q \to (p \lor q); \quad (p \to r) \to ((q \to r) \to ((p \lor q) \to r)); \tag{D}$$
$$\mathfrak{f} \to p. \tag{N}$$

they have inference rules Modus Ponens and substitution:

Calculus	Connectives	Axioms	Description	Logic
IPC	$\to, \land, \lor, \mathfrak{f}$	I,C,D,N	intuitionistic	Int
MPC	$\to, \land, \lor, \mathfrak{f}$	I,C,D	minimal or Johansson's	Min
PPC	$\to, \land, \lor$	I,C,D	positive	Pos
IPC$^-$	$\to, \land, \mathfrak{f}$	I,C,N	$\{\to, \land, \mathfrak{f}\}$ – fragment of IPC	Int$^-$
MPC$^-$	$\to, \land, \mathfrak{f}$	I,C	$\{\to, \land, \mathfrak{f}\}$ – fragment of MPC	Min$^-$
PPC$^-$	$\to, \land$	I,C	$\{\to, \land, \}$ – fragment of PPC	Pos$^-$

If $\Sigma \subseteq \{\to, \land, \lor, \mathfrak{f}\}$, by a Σ-formula we understand a formula containing connectives only from Σ and in virtue of the Separation Theorem (cf., e.g., Kleene 1952, Theorem 49): for every $\Sigma \in \{\{\to, \land, \lor\}, \{\to, \land, \mathfrak{f}\}, \{\to, \land\}\}$, if A is a C-formula $\{\to, \land\}$-formula, $\mathsf{IPC} \vdash A$ if and only if $\mathsf{PPC} \vdash A$ or $\mathsf{IPC}^- \vdash A$, or $\mathsf{PPC}^- \vdash A$.

By a C-calculus we understand one of the six calculi under consideration, and a C-logic is a logic of the C-calculus. Accordingly, C-formulas are formulas in the signature of the C-calculus. For C-formulas A and B, by $A \overset{C}{\vdash} B$ we denote that formula B is derivable in the respective C-calculus extended by axiom B; that is, $C + A \overset{C}{\vdash} B$.

The relation between PPC and MPC (or between PPC$^-$ and MPC$^-$) is a bit more complex: for any formula $\{\to, \land, \lor, \mathfrak{f}\}$-formula A (or any $\{\to, \land, \mathfrak{f}\}$-formula A), $\mathsf{MPC} \vdash A$ (or $\mathsf{MPC}^- \vdash A$) if and only if $\mathsf{PPC} \vdash A'$ (or $\mathsf{PPC}^- \vdash A'$), where A' is a formula obtained from A by replacing all occurrences of $\mathfrak{f}$ with a propositional variable not occurring in A (cf., e.g., Odintsov 2008, Chap. 2). In virtue of the Separation Theorem, in the previous statement, PPC or PPC$^-$ can be replaced with IPC or IPC$^-$, respectively.

Figure 2.1 shows the relations between the introduced logics: a double edge depicts an extension of the logic without any extension of the language (e.g., $\mathsf{Min} \subset \mathsf{Int}$),

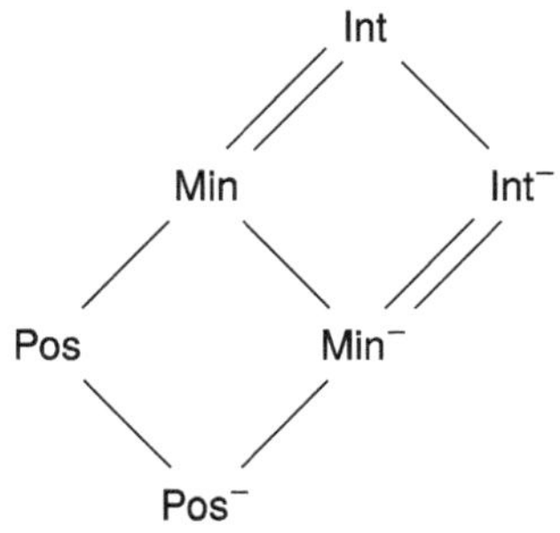

Fig. 2.1 Logics

while a single edge depicts an extension of the language but not of the class of theorems (e.g., if A is a $\{\to, \wedge, \neg\}$-formula, then $A \in \mathsf{Int}$ if and only if $A \in \mathsf{Int}^-$).

Let us observe that $((p \to \neg q) \to (q \to \neg p)) \in \mathsf{Min}^- \subseteq \mathsf{Min}$. Indeed, formula $(p \to (q \to r)) \to (q \to (p \to r))$ can be derived from the axioms (I). Hence, formula $(p \to (q \to \mathfrak{f})) \to (q \to (p \to \mathfrak{f}))$ is derivable too, that is, $(p \to \neg q) \to (q \to \neg p)$ is derivable in MPC^-.

We use **ExtInt**, **ExtMin**, **ExtPos**, **ExtInt**$^-$, **ExtMin**$^-$, **ExtPos**$^-$ to denote classes of logics extending, respectively, Int, Min, Pos, Int^-, Min^-, and Pos^-. Thus, **ExtInt** is a class of all si-logics.

2.2.2 Algebraic Semantics

As pointed out in the Introduction, the first Yankov papers were written before the book by Rasiowa and Sikorski (1963) was published, and the terminology used by Yankov in his early papers was, as he himself admitted in Jankov (1968b), misleading. What he then called an "implicative lattice"[2] he later called a "Brouwerian algebra," and then he finally settled with the term "pseudo-Boolean algebra". We use a commonly accepted terminology, which we clarify below.

2.2.2.1 Correspondences Between Logics and Classes of Algebras.

In a meet-semilattice $\mathbf{A} = (\mathsf{A}; \wedge)$ an element c is a *complement of element* a *relative to element* b if c is the greatest element of A such that $\mathsf{a} \wedge \mathsf{c} \leq \mathsf{b}$ (e.g. Rasiowa 1974a). If a semilattice $\mathbf{A}$ for any elements a and b contains a complement of a relative to b, we say that $\mathbf{A}$ is a *semilattice with relative pseudocomplementation*, and we denote the relative pseudocomplementation by $\to$.

[2] In some translations of the Yankov paper, this term was translated as "implicative structure" (e.g. Jankov 1963a).

Proposition 2.1 *Suppose that* $\mathbf{A}$ *is a meet-semilattice and* $\mathsf{a}, \mathsf{b}, \mathsf{c} \in \mathbf{A}$. *If* $\mathsf{a} \to \mathsf{b}$ *and* $\mathsf{a} \to \mathsf{c}$ *are defined in* $\mathbf{A}$, *then* $\mathsf{a} \to (\mathsf{b} \wedge \mathsf{c})$ *is defined as well and*

$$\mathsf{a} \to (\mathsf{b} \wedge \mathsf{c}) = (\mathsf{a} \to \mathsf{b}) \wedge (\mathsf{a} \to \mathsf{c}).$$

Proof Suppose that $\mathbf{A}$ is a meet-semilattice in which $\mathsf{a} \to \mathsf{b}$ and $\mathsf{a} \to \mathsf{c}$ are defined. We need to show that $(\mathsf{a} \to \mathsf{b}) \wedge (\mathsf{a} \to \mathsf{c})$ is the greatest element of $\mathbf{A}' := \{\mathsf{d} \in \mathbf{A} \mid \mathsf{a} \wedge \mathsf{d} \leq \mathsf{b} \wedge \mathsf{c}\}$.

First, we observe that $(\mathsf{a} \to \mathsf{b}) \wedge (\mathsf{a} \to \mathsf{c}) \in \mathbf{A}'$:

$$(\mathsf{a} \to \mathsf{b}) \wedge (\mathsf{a} \to \mathsf{c}) \wedge \mathsf{a} = (\mathsf{a} \wedge (\mathsf{a} \to \mathsf{b})) \wedge (\mathsf{a} \wedge (\mathsf{a} \to \mathsf{c})) \leq \mathsf{b} \wedge \mathsf{c},$$

because by the assumption, $\mathsf{a} \wedge (\mathsf{a} \to \mathsf{b}) \leq \mathsf{b}$ and $\mathsf{a} \wedge (\mathsf{a} \to \mathsf{c}) \leq \mathsf{c}$.

Next, we show that $(\mathsf{a} \to \mathsf{b}) \wedge (\mathsf{a} \to \mathsf{c})$ is the greatest element of $\mathbf{A}'$. Indeed, suppose that $\mathsf{d} \in \mathbf{A}'$. Then, $\mathsf{a} \wedge \mathsf{d} \leq \mathsf{b} \wedge \mathsf{c}$ and consequently,

$$\mathsf{a} \wedge \mathsf{d} \leq \mathsf{b} \text{ and } \mathsf{a} \wedge \mathsf{d} \leq \mathsf{c}.$$

Hence, by the definition of relative pseudocomplementation,

$$\mathsf{d} \leq \mathsf{a} \to \mathsf{b} \text{ and } \mathsf{d} \leq \mathsf{a} \to \mathsf{c},$$

which means that $\mathsf{d} \leq (\mathsf{a} \to \mathsf{b}) \wedge (\mathsf{a} \to \mathsf{c})$.

By an *implicative semilattice* we understand an algebra $(\mathsf{A}; \to, \wedge, \mathbf{1})$, where $(\mathsf{A}; \wedge)$ is a meet-semilattice with the greatest element $\mathbf{1}$ and $\to$ is a relative pseudocomplementation and accordingly, an algebra $(\mathsf{A}; \to, \wedge, \vee, \mathbf{1})$ is an *implicative lattice* if $(\mathsf{A}; \wedge \vee, \mathbf{1})$ is a lattice and $(\mathsf{A}; \to, \wedge, \mathbf{1})$ is an implicative semilattice (cf. Rasiowa 1974a). In implicative lattices, $\mathbf{0}$ denotes a constant (0-ary operation) that is the smallest element.

The logics described in the previous section have the following algebraic semantics:

Logic	Signature	Algebraic semantic	Denotation
Pos^-	$\{\to, \wedge, \mathbf{1}\}$	implicative semilattices	**BS**
Pos	$\{\to, \wedge, \vee, \mathbf{1}\}$	implicative lattices	**BA**
Min^-	$\{\to, \wedge, \mathfrak{f}, \mathbf{1}\}$	implicative semilattices with constant	**JS**
Min	$\{\to, \wedge, \vee, \mathfrak{f}, \mathbf{1}\}$	implicative semilattices with constant	**JA**
Int^-	$\{\to, \wedge, \mathbf{0}, \mathbf{1}\}$	bounded implicative semilattices	**HS**
Int	$\{\to, \wedge, \vee, \mathbf{0}, \mathbf{1}\}$	bounded implicative lattices	**HA**

As usual, in **JS** and **JA**, we let $\neg\mathsf{a} = \mathsf{a} \to \mathfrak{f}$, while in **HS** and **HA**, $\neg\mathsf{a} = \mathsf{a} \to \mathbf{0}$. Also, we use the following denotations: $\mathcal{L} := \{\mathsf{Pos}^-, \mathsf{Pos}, \mathsf{Min}^-, \mathsf{Min}, \mathsf{Int}^-, \mathsf{Int}\}$ and $\mathcal{A} := \{\mathbf{BS}, \mathbf{BA}, \mathbf{JS}, \mathbf{JA}, \mathbf{HS}, \mathbf{HA}\}$. For each $\boldsymbol{L} \in \mathcal{L}$, $\mathrm{Mod}(\boldsymbol{L})$ denotes the respective class of algebras. By a C-algebra we shell understand an algebra in the signature

$\Sigma \cup \{\mathbf{1}\}$, and we assume that Σ is always a signature of one of the six classes of logics under consideration.

Every class from $\mathcal{A}$ forms a variety. Moreover, **HS** and **HA** are subvarieties of, respectively, **JS** and **JA** defined by the identity $\mathfrak{f} \to x = \mathbf{1}$.

Remark 2.1 Let us observe that **BS** is a variety of all Brouwerian semilattices, and it was studied in detail in (cf. Köhler 1981); **BA** is a variety of all Brouwerian algebras (cf. Galatos et al. 2007); **JA** is a variety of all Johansson's algebras (j-algebras; cf. Odintsov 2008); and **HA** is a variety of all Heyting or pseudo-Boolean algebras (cf. Rasiowa and Sikorski 1963).

Let us recall the following properties of C-algebras.

Proposition 2.2 *The following holds:*

(a) every Brouwerian algebra forms a distributive lattice;
(b) every finite distributive lattice forms a Brouwerian algebra, and because it always contains the least element, it forms a Heyting algebra as well;
(c) every finite **BS***-algebra forms a Brouwerian algebra.*

(a) and (b) were observed in Rasiowa and Sikorski (1963) and Birkhoff (1948). (c) follows from the observation that in any finite **BS**-algebra **A**, for any two elements $\mathsf{a}, \mathsf{b} \in \mathbf{A}$, $\mathsf{a} \vee \mathsf{b}$ can be defined as a meet of $\{\mathsf{c} \in \mathbf{A} \mid \mathsf{a} \le \mathsf{c}, \mathsf{b} \le \mathsf{c}\}$.

As usual, given a formula A and a C-algebra, a map $\nu : Var \longrightarrow \mathbf{A}$ is called a *valuation* in **A**, and ν allows us to calculate a value of A in **A** by treating the connectives as operations of **A**. If $\nu(A) = \mathbf{1}$ for all valuations, we say that A is *valid* in **A**, in symbols, $\mathbf{A} \models A$. If for some valuation ν, $\nu(A) \neq \mathbf{1}$, we say that A is *refuted* in **A**, in symbols, $\mathbf{A} \not\models A$, in which case ν is called a *refuting valuation*. For a class of algebras $\mathbb{K}$, $\mathbb{K} \models A$ means that A is valid in every member of $\mathbb{K}$. Given a class of C-algebras $\mathbb{K}$, $\mathbb{K}_{fin}$ is a subclass of all finite members of $\mathbb{K}$.

For every logic $\boldsymbol{L} \in \mathcal{L}$, a respective class from $\mathcal{A}$ is denoted by $\mathrm{Mod}(\boldsymbol{L})$. A class of models $\mathbb{M}$ of logic $\boldsymbol{L}$ forms an *adequate algebraic semantics* of $\boldsymbol{L}$ if for each formula A, $A \in \boldsymbol{L}$ if and only if A is valid in all algebras from $\mathbb{M}$.

Proposition 2.3 *For every* $\boldsymbol{L} \in \mathcal{L}$ *class* $\mathrm{Mod}(\boldsymbol{L})$ *forms an adequate algebraic semantics. Moreover, each logic* $\boldsymbol{L} \in \mathcal{L}$ *enjoys the f.m.p.; that is,* $A \in \boldsymbol{L}$ *if and only if* $\mathrm{Mod}(\boldsymbol{L})_{fin} \models A$.

Proof The proofs of adequacy can be found in Rasiowa (1974a). The f.m.p. for Int follows from Jaśkowski (1936). The f.m.p. for $\mathsf{Int}^-, \mathsf{Pos}, \mathsf{Pos}^-$ follows from the f.m.p. for Int and the Separation Theorem.

As we mentioned earlier, for any formula A, $A \in \mathsf{Min}$ (or $A \in \mathsf{Min}^-$) if and only if $A^f \in \mathsf{Int}$ (or $A \in Int^-$), where A^f is a formula obtained from A by replacing every occurrence of $\mathfrak{f}$ with a new variable p. Because Int (and Int^-) enjoys the f.m.p., if $A \notin \mathsf{Min}$ (or $A \notin \mathsf{Min}^-$), there is a finite Heyting algebra **A** refuting A^f (finite **HS**-algebra refuting A^f). If ν is a refuting valuation, we can convert **A** into a **JA**-algebra (or into a **JS**-algebra) by regarding A as a Brouwerian algebra (or a Brouwerian semilattice) with $\mathfrak{f}$ being $\nu(A)$. It is clear that A is refuted in such a **JA**-algebra (**JS**-algebra).

2.2.2.2 Meet-Irreducible Elements

Let $\mathbf{A} = (\mathsf{A}; \wedge)$ be a meet-semilattice and $\mathsf{a} \in \mathbf{A}$. Element a is called *meet-irreducible*, if for every pair of elements b, c, $\mathsf{a} = \mathsf{b} \wedge \mathsf{c}$ entails that $\mathsf{a} = \mathsf{b}$ or $\mathsf{a} = \mathsf{c}$. And a is called *meet-prime* if $\mathsf{a} \leq \mathsf{b} \wedge \mathsf{c}$ entails that $\mathsf{a} = \mathsf{b}$ or $\mathsf{a} = \mathsf{c}$. For formulas where $\wedge$ is a conjunction, instead of meet-irreducible or meet-prime we say *conjunctively-irreducible* or *conjunctively-prime*.

If $\mathbf{A}$ is a semilattice, then elements a, b of $\mathbf{A}$ are *comparable* if $\mathsf{a} \leq \mathsf{b}$ or $\mathsf{b} \leq \mathsf{a}$, otherwise these elements are *incomparable*. A set of mutually incomparable elements is called an *antichain*. It is not hard to see that a meet of any finite set of elements is equal to a meet of a finite subset of mutually incomparable elements.

It is clear that every meet-prime element is meet-irreducible. In the distributive lattices, the converse holds as well.

The meet-irreducible elements play a role similar to that of prime numbers: every positive natural number is a product of primes. As usual, if a is an element of a semilattice, the representation $\mathsf{a} = \mathsf{a}_1 \wedge \cdots \wedge \mathsf{a}_n$ of a as a meet of finitely many meet-prime elements a_i, $i \in [1, n]$ is called a *finite decomposition* of a. This finite decomposition is *irredundant* if no factor can be omitted.

It is not hard to see that because the factors in a finite decomposition are meet-irreducible, the decomposition is irredundant if and only if the elements of its factors are mutually incomparable.

Proposition 2.4 *In any semilattice, if element* a *has a finite decomposition,* a *has a unique (up to an order of factors) irredundant finite decomposition. Thus, in finite semilattices, every element has a unique irredundant finite decomposition.*

Proof Indeed, if element a has two finite irredundant decompositions $\mathsf{a} = \mathsf{a}_1 \wedge \cdots \wedge \mathsf{a}_n$ and $\mathsf{a} = \mathsf{a}'_1 \wedge \cdots \wedge \mathsf{a}'_m$, then $\mathsf{a}_1 \wedge \cdots \wedge \mathsf{a}_n = \mathsf{a}'_1 \wedge \cdots \wedge \mathsf{a}'_m$ and

$$(\mathsf{a}_1 \wedge \cdots \wedge \mathsf{a}_n) \to (\mathsf{a}'_1 \wedge \cdots \wedge \mathsf{a}'_m) = \mathbf{1}.$$

Hence, for each $j \in [1, m]$,

$$(\mathsf{a}_1 \wedge \cdots \wedge \mathsf{a}_n) \to \mathsf{a}'_j = \mathbf{1}; \text{ that is, } (\mathsf{a}_1 \wedge \cdots \wedge \mathsf{a}_n) \leq \mathsf{a}'_j.$$

Because a'_j is meet-prime, $\mathsf{a}'_j \in \{\mathsf{a}_1, \ldots, \mathsf{a}_n\}$ and thus, $\{\mathsf{a}'_1, \ldots, \mathsf{a}'_m\} \subseteq \{\mathsf{a}_1, \ldots, \mathsf{a}_n\}$. By the same reason, $\{\mathsf{a}_1, \ldots, \mathsf{a}_n\} \subseteq \{\mathsf{a}'_1, \ldots, \mathsf{a}'_m\}$ and therefore, $\{\mathsf{a}_1, \ldots, \mathsf{a}_n\} = \{\mathsf{a}'_1, \ldots, \mathsf{a}'_m\}$.

Proposition 2.5 (Jankov 1969). *If a meet-semilattice* $\mathbf{A}$ *has a top element and all its elements have a finite irredundant decomposition, then* $\mathbf{A}$ *forms a Brouwerian semilattice.*

Proof We need to define on semilattice $\mathbf{A}$ a relative pseudocomplement $\to$. Because every element of $\mathbf{A}$ has a finite irredundant decomposition, for any two elements $\mathsf{a}, \mathsf{b} \in \mathbf{A}$ one can consider their finite irredundant decompositions $\mathsf{a} = \mathsf{a}_1 \wedge \cdots \wedge \mathsf{a}_n$

and $\mathsf{b} = \mathsf{b}_1 \wedge \cdots \wedge \mathsf{b}_m$. Now, we can define $\mathsf{a} \to \mathsf{c}$, where c is a meet-prime element, and then extend this definition by letting

$$\mathsf{a} \to (\mathsf{b}_1 \wedge \cdots \wedge \mathsf{b}_m) = (\mathsf{a} \to \mathsf{b}_1) \wedge \cdots \wedge (\mathsf{a} \to \mathsf{b}_m). \tag{2.1}$$

Proposition 2.1 ensures the correctness of such an extension.

Suppose $\mathsf{c} \in \mathbf{A}$ is meet-prime and $\mathsf{a} = \mathsf{a}_1 \wedge \cdots \wedge \mathsf{a}_n$ is a finite irredundant decomposition of a. Then we let

$$\mathsf{a} \to \mathsf{c} = \begin{cases} \mathbf{1}, & \text{if } \mathsf{a}_i \leq \mathsf{c} \text{ for some } i \in [1, n]; \\ \mathsf{c}, & \text{otherwise.} \end{cases}$$

Let us show that $\mathsf{a} \to \mathsf{c}$ is a pseudocomplement of a relative to c, that is, we need to show that $\mathsf{a} \to \mathsf{c}$ is the greatest element of $\mathbf{A}' := \{\mathsf{d} \in \mathbf{A} \mid \mathsf{a} \wedge \mathsf{d} \leq \mathsf{b}\}$.

Indeed, if $\mathsf{a}_i \leq \mathsf{c}$ for some $i \in [1, n]$, then

$$\mathbf{1} \wedge \mathsf{a} = \mathsf{a} = \mathsf{a}_1 \wedge \cdots \wedge \mathsf{a}_n \leq \mathsf{a}_i \leq \mathsf{c},$$

and obviously, $\mathbf{1}$ is the greatest of $\mathbf{A}'$.

Suppose now that $\mathsf{a}_i \not\leq \mathsf{c}$ for all $i \in [1, n]$. In this case, $\mathsf{a} \to \mathsf{c} = \mathsf{c}$, it is clear that $\mathsf{a} \wedge \mathsf{c} \leq \mathsf{c}$ (i.e., $\mathsf{a} \in \mathbf{A}'$), and we only need to verify that $\mathsf{d} \leq \mathsf{c}$ for every $\mathsf{d} \in \mathbf{A}'$.

Indeed, suppose that $\mathsf{a} \wedge \mathsf{d} \leq \mathsf{c}$; that is, $\mathsf{a}_1 \wedge \cdots \wedge \mathsf{a}_n \wedge \mathsf{d} \leq \mathsf{c}$. Then, $\mathsf{d} \leq \mathsf{c}$ because c is meet prime and $\mathsf{a}_i \not\leq \mathsf{c}$ for all $i \in [1, n]$.

Immediately from Propositions 2.5 and 2.2(c), we obtain the following statement.

Corollary 2.1 *Every finite meet-semilattice* $\mathbf{A}$ *with a top element in which every element has an irredundant finite decomposition forms a Brouwerian algebra. And because* $\mathbf{A}$ *is finite and has a bottom element,* $\mathbf{A}$ *is a Heyting algebra.*

2.2.3 *Lattices* Ded_C *and* $\mathsf{Lind}_{(C,k)}$

On the set of all C-formulas, relation $\overset{C}{\vdash}$ is a quasiorder and hence, the relation

$$A \overset{C}{\approx} B \quad \overset{\text{def}}{\Longleftrightarrow} \quad A \overset{C}{\vdash} B \text{ and } B \overset{C}{\vdash} A$$

is an equivalence relation. Moreover, the set of all C-formulas forms a semilattice relative to connecting formulas with $\wedge$. It is not hard to see that equivalence $\overset{C}{\approx}$ is a congruence and therefore, we can consider a quotient semilattice which is denoted by Ded_C.

For each $k > 0$, we consider the set of all formulas on variables $p_1, \ldots, p_k$. This set formulas a semilattice relative to connecting two given formulas with $\wedge$. It is not hard to see that relation

$$A \overset{C}{\sim} B \quad \overset{\text{def}}{\Longleftrightarrow} \quad \overset{C}{\vdash} A \leftrightarrow B$$

is a congruence, and by $\mathsf{Lind}_{(C,k)}$ we denote a quotient semilattice.

Theorem 2.1 (Jankov 1969) *For any C and $k > 0$, semilattices $\mathsf{Lind}_{(C,k)}$ and Ded_C are distributive lattices.*

Proof For $C \in \{\mathsf{PPC}, \mathsf{MPC}, \mathsf{IPC}\}$, it was observed in Rasiowa and Sikorski (1963). If $C \in \{\mathsf{PPC}^-, \mathsf{MPC}^-, \mathsf{IPC}^-\}$, by the Diego theorem (cf., e.g., Köhler 1981), lattice $\mathsf{Lind}_{(C,k)}$ is a finite implicative semilattice and, hence, a distributive lattice.

To convert Ded_C into a lattice we need to define a meet. Given two formulas A and B, we let

$$A \vee' B \quad = \quad (A \to p) \wedge ((B' \to p) \to p),$$

where formula B' is obtained from B by replacing the variables in such a way that formulas A and B have no variables in common, and p is a variable not occurring in formulas A and B'. If $C \in \{\mathsf{PPC}, \mathsf{MPC}, \mathsf{IPC}\}$, one can take

$$A \vee' B \quad = \quad A \vee B'.$$

A proof that Ded_C is indeed a distributive lattice can be found in Jankov (1969).

Meet-prime and meet-irreducible elements in $\mathsf{Lind}_{(C,k)}$ and Ded_C are called *conjunctively prime* and *conjunctively irreducible*, and because these lattices are distributive, every conjunctively irreducible formula is conjunctively prime and vice versa.

2.2.3.1 Congruences, Filters, Homomorphisms

Let us observe that every C-algebra $\mathbf{A}$ has a $\{\to, \wedge, \mathbf{1}\}$-reduct that is a Brouwerian semilattice, and therefore, any congruence on $\mathbf{A}$ is at the same time a congruence on its $\{\to, \wedge, \mathbf{1}\}$-reduct. It is remarkable that the converse is true too: every congruence on a $\{\to, \wedge, \mathbf{1}\}$-reduct can be lifted to the algebra.

Any congruence on a C-algebra $\mathbf{A}$ is uniquely defined by the set $\mathbf{1}/\theta := \{\mathsf{a} \in \mathbf{A} \mid (\mathsf{a}, \mathbf{1}) \in \theta\}$: indeed, it is not hard to see that $(\mathsf{b}, \mathsf{c}) \in \theta$ if and only if $(\mathsf{b} \leftrightarrow \mathsf{c}, \mathbf{1}) \in \theta$ (cf. Rasiowa 1974a). A set $\mathbf{1}/\theta$ forms a filter of $\mathbf{A}$: a subset $\mathsf{F} \subseteq \mathbf{A}$ is a *filter* if $\mathbf{1} \in \mathsf{F}$ and $\mathsf{a}, \mathsf{a} \to \mathsf{b} \in \mathsf{F}$ yields $\mathsf{b} \in \mathsf{F}$. The set of all filters of C-algebra $\mathbf{A}$ is denoted by $\mathsf{Flt}(\mathbf{A})$. It is not hard to see that a meet of an arbitrary system of filters is a filter and hence, $\mathsf{Flt}(\mathbf{A})$ forms a complete lattice. A set-join of two filters does not need to be a filter, but a join of any ascending chain of filters is a filter.

As we saw, every congruence is defined by a filter. The converse is true too: any filter F of a C-algebra $\mathbf{A}$ defines a congruence

$$(\mathsf{a},\mathsf{b}) \in \theta_{\mathsf{F}} \iff (\mathsf{a} \leftrightarrow \mathsf{b}) \in \mathsf{F}.$$

Moreover, the map $\mathsf{F} \longrightarrow \theta_{\mathsf{F}}$ is an isomorphism between complete lattices of filters and complete lattice of congruences (cf. Rasiowa 1974a). It is clear that any nontrivial C-algebra has at least two filters: $\{\mathbf{1}\}$ and the set of all elements of the algebra. The filter $\{\mathbf{1}\}$ is called *trivial*, and the filters that do not contain all the elements of the algebra are called *proper*. In what follows, by $\mathbf{A}/\mathsf{F}$ and a/F we understand $\mathbf{A}/\theta_{\mathsf{F}}$ and $\mathsf{c}/\theta_{\mathsf{F}}$.

If $\mathbf{A}$ is a C-algebra and $\mathsf{B} \subseteq \mathbf{A}$ is a subset of elements, there is the least filter $[\mathsf{B})$ of $\mathbf{A}$ containing B: $[\mathsf{B}) = \bigcap\{\mathsf{F} \in \mathsf{Flt}(A) \mid \mathsf{B} \subseteq \mathsf{F}\}$, and we write $[\mathsf{a})$ instead of $[\{\mathsf{a}\})$. The reader can easily verify that for any element a of a C-algebra $\mathbf{A}$, $[\mathsf{a}) = \{\mathsf{b} \in \mathbf{A} \mid \mathsf{a} \leq \mathsf{b}\}$.

Immediately from the definitions of a filter and a homomorphism, the following holds.

Proposition 2.6 *Suppose that* $\mathbf{A}$ *and* $\mathbf{B}$ *are* C*-algebras and* $\varphi : \mathbf{A} \longrightarrow \mathbf{B}$ *is a homomorphism of* $\mathbf{A}$ *onto* $\mathbf{B}$. *Then*

(a) If F *is a filter of* $\mathbf{A}$, *then* $\varphi(\mathsf{F})$ *is a filter of* $\mathbf{B}$;
(b) If F *is a filter of* $\mathbf{B}$, *then* $\varphi^{-1}(\mathsf{F})$ *is a filter of* $\mathbf{A}$.

A nontrivial algebra $\mathbf{A}$ is called *subdirectly irreducible* (s.i. for short) if the meet of all nontrivial filters is a nontrivial filter; or, in terms of congruences, the meet of all congruences that are distinct from the identity is distinct from the identity congruence.

Because every element a of a C-algebra $\mathbf{A}$ defines a filter $[\mathsf{a})$, the meet of all nontrivial filters of $\mathbf{A}$ coincides with $\bigcap\{[\mathsf{a}), \mathsf{a} \in \mathbf{A} \mid \mathsf{a} \neq \mathbf{1}\}$ and consequently, $\mathbf{A}$ is s.i. if and only if the set $\{\mathsf{a} \in \mathbf{A} \mid \mathsf{a} \neq \mathbf{1}\}$ contains the greatest element which is referred to as a *pretop* element or an *opremum* and is denoted by $\mathsf{m}_{\mathbf{A}}$.

Let us observe that immediately from the definition of a pretop element, if $\mathsf{m}_{\mathbf{A}}$ is a pretop element of a C-algebra $\mathbf{A}$ and F is a filter of $\mathbf{A}$, then, $\mathsf{m}_{\mathbf{A}} \in \mathsf{F}$ if and only if F is nontrivial. In terms of homomorphism, this can be stated in the following way.

Proposition 2.7 *Suppose that* $\mathbf{A}$ *is an s.i.* C*-algebra and* $\varphi : \mathbf{A} \longrightarrow \mathbf{B}$ *is a homomorphism of* $\mathbf{A}$ *into* C*-algebra* $\mathbf{B}$. *Then* φ *is an isomorphism if and only if* $\varphi(\mathsf{m}_{\mathbf{A}}) \neq \mathbf{1}_{\mathbf{B}}$.

The following simple proposition was observed in Jankov (1969) and it is very important in what follows.

Proposition 2.8 *Let* $\mathbf{A}$ *be a nontrivial* C*-algebra,* $\mathsf{a}, \mathsf{b} \in \mathbf{A}$ *and* $\mathsf{a} \not\leq \mathsf{b}$. *Then, there is a maximal (relative to* $\subseteq$*) filter* F *of* $\mathbf{A}$ *such that* $\mathsf{a} \in \mathsf{F}$ *and* $\mathsf{b} \notin \mathsf{F}$. *Furthermore,* $\mathbf{A}/\mathsf{F}$ *is an s.i.* C*-algebra with* b/F *being the pretop element.*

Proof First, let us observe that the condition $\mathsf{a} \not\leq \mathsf{b}$ is equivalent to $\mathsf{b} \notin [\mathsf{a})$. Thus, $\mathcal{F} := \{\mathsf{F} \in \mathsf{Flt}(\mathbf{A}) \mid \mathsf{a} \in \mathsf{F}, \mathsf{b} \notin \mathsf{F}\} \neq \varnothing$.

Next, we recall that the joins of ascending chains of filters are filters and therefore, $\mathcal{F}$ enjoys the ascending chain condition. Thus, by the Zorn Lemma, $\mathcal{F}$ contains a maximal element.

Let F be a maximal element of $\mathcal{F}$. We need to show that b/F is a pretop element of $\mathbf{A}/\mathsf{F}$.

Because $\mathsf{b} \notin \mathsf{F}$ (cf. the definition of $\mathcal{F}$), we know that $\mathsf{b}/\mathsf{F} \neq \mathbf{1}_{\mathbf{A}/\mathsf{F}}$.

Let $\varphi : \mathbf{A} \longrightarrow \mathbf{A}/\mathsf{F}$ be a natural homomorphism. By Proposition 2.6, for every filter F' of $\mathbf{A}/\mathsf{F}$, the preimage $\varphi^{-1}(\mathsf{F}')$ is a filter of $\mathbf{A}$. Because $\mathbf{1}_{\mathbf{A}/\mathsf{F}} \in \mathsf{F}'$,

$$\mathsf{F} = \varphi^{-1}(\mathbf{1}_{\mathbf{A}/\mathsf{F}}) \subseteq \varphi^{-1}(\mathsf{F}').$$

Hence, if $\mathsf{F}' \supsetneq \mathbf{1}_{\mathbf{A}}/\mathsf{F}$, then $\mathsf{b} \in \varphi^{-1}(\mathsf{F}')$ (because F is a maximal filter not containing b), and consequently, $\mathsf{b}/\mathsf{F} \in \mathsf{F}'$. Thus, b/F is in every nontrivial filter of $\mathbf{A}/\mathsf{F}$, which means that $\mathbf{A}/\mathsf{F}$ is s.i. and that b/F is a pretop element of $\mathbf{A}/\mathsf{F}$.

Corollary 2.2 *Suppose that $A \to B$ is a C-formula refuted in a C-algebra $\mathbf{A}$. Then there is an s.i. homomorphic image $\mathbf{B}$ of algebra $\mathbf{A}$ and a valuation ν in $\mathbf{B}$ such that*

$$\nu(A) = \mathbf{1}_{\mathbf{B}} \quad \textit{and} \quad \nu(B) = \mathsf{m}_{\mathbf{B}}.$$

Proof Suppose that ξ is a refuting valuation in $\mathbf{A}$; that is, $\xi(A \to B) \neq \mathbf{1}_{\mathbf{A}}$. Let $\xi(A) = \mathsf{a}$ and $\xi(B) = \mathsf{b}$. Then, $\mathsf{a} \not\leq \mathsf{b}$ and by Proposition 2.8, there is a filter F of $\mathbf{A}$ such that $\mathsf{a} \in \mathsf{F}$, $\mathsf{b} \notin \mathsf{F}$ and $\mathbf{A}/\mathsf{F}$ is subdirectly irreducible with b/F being a pretop element of $\mathbf{A}/\mathsf{F}$. Thus, one can take a natural homomorphism $\eta : \mathbf{A} \longrightarrow \mathbf{A}/\mathsf{F}$ and let $\nu = \eta \circ \xi$.

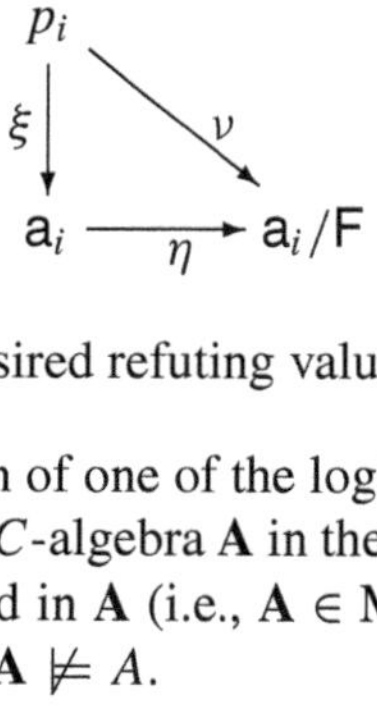

It is not hard to see that ν is a desired refuting valuation.

Suppose that $\boldsymbol{L}$ is an extension of one of the logics from $\mathcal{L}$ and A is a formula in the signature of $\boldsymbol{L}$. We say that a C-algebra $\mathbf{A}$ in the signature of $\boldsymbol{L}$ *separates A from $\boldsymbol{L}$* if all formulas from $\boldsymbol{L}$ are valid in $\mathbf{A}$ (i.e., $\mathbf{A} \in \mathrm{Mod}(\boldsymbol{L})$), while formula A is not valid in $\mathbf{A}$, that is, if $\mathbf{A} \models \boldsymbol{L}$ and $\mathbf{A} \not\models A$.

Corollary 2.3 *Suppose that $\mathbf{L}$ is a C-logic and A is a C-formula. If a C-algebra $\mathbf{A}$ separates formula A from $\mathbf{L}$, then there is an s.i. homomorphic image $\mathbf{B}$ of $\mathbf{A}$ and a valuation ν in $\mathbf{B}$ such that $\nu(A) = \mathsf{m}_{\mathbf{B}}$.*

Proof If formula A is invalid in $\mathbf{A}$, then there is a refuting valuation ξ in $\mathbf{A}$ such that $\xi(A) = \mathsf{a} < \mathbf{1}$. By Proposition 2.8, there is a maximal filter F of $\mathbf{A}$ such that $\mathsf{a} \notin \mathsf{F}$. Then, $\mathbf{B} := \mathbf{A}/\mathsf{F}$ is an s.i. algebra, and $\nu = \eta \circ \xi$, where ν is a natural homomorphism, is a desired refuting valuation.

Let us note that because $\mathbf{B}$ is a homomorphic image of $\mathbf{A}$, the finiteness of $\mathbf{A}$ yields the finiteness of $\mathbf{B}$.

Remark 2.2 In Jankov (1969), Corollary 2.3 (the Descent Theorem) is proved only for finite algebras. Yankov, being a disciple of Markov and sharing the constructivist view on mathematics, avoided using the Zorn Lemma which is necessary for proving Proposition 2.8 for infinite algebras.

2.3 Yankov's Characteristic Formulas

One of the biggest achievements of Yankov, apart from the particular results about si-logics, is the machinery that he had developed and used to establish these results. This machinery rests on the notion of a characteristic formula that he introduced in Jankov (1963c) and studied in detail in Jankov (1969).

2.3.1 Formulas and Homomorphisms

With each finite C-algebra $\mathbf{A}$ in the signature Σ we associate a formula $D_{\mathbf{A}}$ on variables $\{p_{\mathsf{a}}, \mathsf{a} \in \mathbf{A}\}$ in the following way: let $\Sigma_2 \subseteq \Sigma$ be a subset of all binary operation and $\Sigma_0 \subseteq \Sigma$ be a subset of nullary operations (constants); then

$$D_{\mathbf{A}} = \bigwedge_{\circ\in\Sigma_2} (p_{\mathsf{a}} \circ p_{\mathsf{b}} \leftrightarrow p_{\mathsf{a}\circ\mathsf{b}}) \wedge \bigwedge_{\mathfrak{c}\in\Sigma_0} (\mathfrak{c} \leftrightarrow p_{\mathfrak{c}}).$$

Example 2.1 Let $\mathbf{3} = (\{\mathsf{a}, \mathsf{b}, \mathbf{1}\}; \to, \wedge, \mathbf{1})$ be a Brouwerian semilattice, $\mathsf{a} \leq \mathsf{b} \leq \mathbf{1}$, and the operations are defined by the Cayley tables:

$\to$	a	b	1
a	1	1	1
b	a	1	1
1	a	b	1

$\wedge$	a	b	1
a	a	a	a
b	a	b	b
1	a	b	1

Then, in the Cayley tables, we replace the elements with the respective variables:

$\to$	p_{a}	p_{b}	$p_{\mathbf{1}}$
p_{a}	$p_{\mathbf{1}}$	$p_{\mathbf{1}}$	$p_{\mathbf{1}}$
p_{b}	p_{a}	$p_{\mathbf{1}}$	$p_{\mathbf{1}}$
$p_{\mathbf{1}}$	p_{a}	p_{b}	$p_{\mathbf{1}}$

$\wedge$	p_{a}	p_{b}	$p_{\mathbf{1}}$
p_{a}	p_{a}	p_{a}	p_{a}
p_{b}	p_{a}	p_{b}	p_{b}
$p_{\mathbf{1}}$	p_{a}	p_{b}	$p_{\mathbf{1}}$

and we express the above tables in the form of a formula:

$$\begin{aligned} D_3 = {} & (p_a \to p_a) \leftrightarrow p_1 \ \wedge (p_a \to p_b) \leftrightarrow p_1 \ \wedge (p_a \to p_1) \leftrightarrow p_1 \ \wedge \\ & (p_b \to p_a) \leftrightarrow p_a \ \wedge (p_b \to p_b) \leftrightarrow p_1 \ \wedge (p_b \to p_1) \leftrightarrow p_1 \ \wedge \\ & (p_1 \to p_a) \leftrightarrow p_a \ \wedge (p_1 \to p_b) \leftrightarrow p_b \ \wedge (p_1 \to p_1) \leftrightarrow p_1 \ \wedge \\ & (p_a \wedge p_a) \leftrightarrow p_a \ \wedge (p_a \wedge p_b) \leftrightarrow p_a \ \wedge (p_a \wedge p_1) \leftrightarrow p_a \ \wedge \\ & (p_b \wedge p_a) \leftrightarrow p_a \ \wedge (p_b \wedge p_b) \leftrightarrow p_b \ \wedge (p_b \wedge p_1) \leftrightarrow p_b \ \wedge \\ & (p_1 \wedge p_a) \leftrightarrow p_a \ \wedge (p_1 \wedge p_b) \leftrightarrow p_b \ \wedge (p_1 \wedge p_1) \leftrightarrow p_1 \ \wedge \\ & \mathbf{1} \leftrightarrow p_1. \end{aligned}$$

Let us note that formula D_3 is equivalent in Pos^- to a much simpler formula,

$$D' = ((p_b \to p_a) \to p_b) \wedge p_1.$$

The importance of formula $D_{\mathbf{A}}$ rests on the following observation.

Proposition 2.9 *Suppose that* **A** *and* **B** *are C-algebras. If for valuation* ν *in* **B**, $\nu(D_{\mathbf{A}}) = \mathbf{1_B}$, *then the map*

$$\eta : \mathsf{a} \mapsto \nu(p_a)$$

is a homomorphism.

Proof Indeed, for any $\mathsf{a}, \mathsf{b} \in \mathbf{A}$ and any operation $\circ$, formula $p_a \circ p_b \leftrightarrow p_{a \circ b}$ is a conjunct of $D_{\mathbf{A}}$ and hence, $\nu(p_a \circ p_b) = \nu(p_{a \circ b})$, because $\nu(D_{\mathbf{A}}) = \mathbf{1_b}$. Thus,

$$\eta(\mathsf{a} \circ \mathsf{b}) = \nu(p_{a \circ b}) = \nu(p_a \circ p_b) = \nu(p_a) \circ \nu(p_b) = \eta(p_a) \circ \eta(p_b).$$

It is not hard to see that η preserves the operations and therefore, η is a homomorphism.

Let us note that using any set of generators of a finite C-algebra **A**, one can construct a formula having properties similar to $D_{\mathbf{A}}$. Suppose that elements $\mathsf{g}_1, \ldots, \mathsf{g}_n$ generate algebra **A**. Then, each element $\mathsf{a} \in \mathbf{A}$ can be expressed via generators, that is, there is a formula $B_a(p_{g_1}, \ldots, p_{g_n})$ such that $\mathsf{a} = B_a(\mathsf{g}_1, \ldots, \mathsf{g}_n)$. If we substitute in $D_{\mathbf{A}}$ each variable p_a with formula B_a, we obtain a new formula $D'_{\mathbf{A}}(p_{g_1}, \ldots, p_{g_n})$, and this formula will posses the same property as formula $D_{\mathbf{A}}$. Because $D'_{\mathbf{A}}$ depends on the selection of formulas B_a, we use the notation $D_{\mathbf{A}}[B_{a_1}, \ldots, B_{a_m}]$, provided that $\mathsf{a}_1, \ldots, \mathsf{a}_m$ are all elements of **A**.

Proposition 2.10 *Suppose that* **A** *and* **B** *are C-algebras. If* ν *is a valuation in* **B** *and* $\nu(D_{\mathbf{A}}[B_{a_1}, \ldots, B_{a_m}]) = \mathbf{1_B}$, *then the map*

$$\eta : \mathsf{a} \mapsto \nu(B_a)$$

is a homomorphism.

Example 2.2 Let **3** be a three-element Heyting algebra with elements **0**, a, **1**. It is clear that **A** is generated by element a:

$$B_0(p_{\mathsf{a}}) = p_{\mathsf{a}} \wedge (p_{\mathsf{a}} \to \mathbf{0}), \quad B_{\mathsf{a}} = p_{\mathsf{a}}, \quad B_1 = (p_{\mathsf{a}} \to p_{\mathsf{a}}).$$

Formula $D_3[B_0(p_{\mathsf{a}}), B_{\mathsf{a}}(p_{\mathsf{a}}), B_1(p_{\mathsf{a}})]$ is equivalent in Int to the formula $(p_{\mathsf{a}} \to \mathbf{0}) \to \mathbf{0}$. It is not hard to verify that in any Heyting algebra $\mathbf{B}$, if element $\mathsf{b} \in \mathbf{B}$ satisfies condition $((\mathsf{b} \to \mathbf{0}) \to \mathbf{0}) = \mathbf{1}$ (i.e., $\neg\neg\mathsf{b} = \mathbf{1}$), then the map

$$\mathbf{0}_3 \mapsto \mathsf{b} \wedge (\mathsf{b} \to \mathbf{0}_{\mathbf{B}}), \quad \mathsf{a} \mapsto \mathsf{b}, \quad \mathbf{1}_3 \mapsto (\mathsf{b} \to \mathsf{b}),$$

that is, the map

$$\mathbf{0}_3 \mapsto \mathbf{0}_{\mathbf{B}}, \quad \mathsf{a} \mapsto \mathsf{b}, \quad \mathbf{1}_3 \mapsto \mathbf{1}_{\mathbf{B}},$$

is a homomorphism.

2.3.2 *Characteristic Formulas*

Now, we are in a position to define the Yankov characteristic formulas. These formulas are instrumental in studying different classes of logics. It also turned out that characteristic formulas, and only these formulas, are conjunctively indecomposable.

Definition 2.1 Suppose that $\mathbf{A}$ is a finite s.i. C-algebra (finite s.i. algebra, for short). Then the formula

$$X_{\mathbf{A}} := D_{\mathbf{A}} \to p_{\mathsf{m}_{\mathbf{A}}}$$

is a *Yankov (or characteristic) formula* of $\mathbf{A}$.

Let us observe that the valuation $\eta : p_{\mathsf{a}} \mapsto \mathsf{a}$ refutes $X_{\mathbf{A}}$, because clearly, $\eta(D_{\mathbf{A}}) = \mathbf{1}$, while $\eta(p_{\mathsf{m}_{\mathbf{A}}}) = \mathsf{m}_{\mathbf{A}} \neq \mathbf{1}$. That is,

$$\mathbf{A} \not\models X_{\mathbf{A}}. \tag{2.2}$$

Proposition 2.11 *Suppose that* $\mathbf{A}$ *is a finite s.i.* C*-algebra and* ν *is a refuting valuation of* $X_{\mathbf{A}}$ *in a* C*-algebra* $\mathbf{B}$ *such that* $\nu(D_{\mathbf{A}}) = \mathbf{1}_{\mathbf{B}}$*. Then, the map*

$$\varphi : \mathsf{a} \mapsto \nu(p_{\mathsf{a}})$$

is an isomorphism.

Proof Because $\nu(D_{\mathbf{A}}) = \mathbf{1}_{\mathbf{B}}$, by Proposition 2.9, φ ia a homomorphism. Because ν refutes $X_{\mathbf{A}}$, that is, ν refutes $D_{\mathbf{A}} \to p_{\mathsf{m}_{\mathbf{A}}}$, we know that $\nu(p_{\mathsf{m}_{\mathbf{A}}}) \neq \mathbf{1}_{\mathbf{B}}$ and consequently,

$$\varphi(\mathsf{m}_{\mathbf{A}}) = \nu(p_{\mathsf{m}_{\mathbf{A}}}) \neq \mathbf{1}_{\mathbf{B}}.$$

Thus, by Proposition 2.7, φ is an isomorphism.

Corollary 2.4 *If a characteristic formula of a finite s.i. C-algebra* $\mathbf{A}$ *is refuted in a C-algebra* $\mathbf{B}$*, then algebra* $\mathbf{A}$ *is embedded in a homomorphic image of algebra* $\mathbf{B}$.

The proof immediately follows from Corollary 2.2 and Proposition 2.11.

One of the most important properties of characteristic formula of C-algebra $\mathbf{A}$ is that $X_{\mathbf{A}}$ is the weakest formula refutable in $\mathbf{A}$. More precisely, the following holds.

Theorem 2.2 (Jankov 1969, Characteristic formula theorem) *A C-formula A is refutable in a finite s.i. C-algebra* $\mathbf{A}$ *if and only if* $A \overset{C}{\vdash} X_{\mathbf{A}}$.

Proof It is clear that if $A \overset{C}{\vdash} X_{\mathbf{A}}$, then A is refuted in $\mathbf{A}$, because $X_{\mathbf{A}}$ is refuted in $\mathbf{A}$.

To prove the converse statement, we will do the following:

(a) using a refuting valuation of A in $\mathbf{A}$, we will introduce a substitution σ such that formula $A' := \sigma(A)$ has the same variables as $X_{\mathbf{A}}$;

(b) we will prove that $\overset{C}{\vdash} A' \to X_{\mathbf{A}}$ by showing that formula $A' \to X_{\mathbf{A}}$ cannot be refuted in any C-algebra.

Indeed, because clearly $A \overset{C}{\vdash} A'$, (b) entails that $A \overset{C}{\vdash} X_{\mathbf{A}}$

(a) Assume that $\mathbf{A}$ is a k-element C-algebra, $\mathsf{a}_i, i \in [1, k]$ are all its elements, and that $q_1, \ldots, q_n$ are all variables occurring in A. Suppose that $\xi : q_i \mapsto \mathsf{a}_{j_i}$ is a refuting valuation of A in $\mathbf{A}$; that is,

$$\xi(A(q_1, \ldots, q_n)) = A(\xi(q_1), \ldots, \xi(q_n)) = A(\mathsf{a}_{j_1}, \ldots, \mathsf{a}_{j_n}) \neq \mathbf{1}_{\mathbf{A}}. \tag{2.3}$$

Let us consider formula A' obtained from A by a substitution $\sigma : q_i \mapsto p_{\mathsf{a}_{j_i}}$ and a valuation $\xi' : p_{\mathsf{a}_{j_i}} \mapsto \mathsf{a}_{j_i}, i \in [1, n]$, in $\mathbf{A}$:

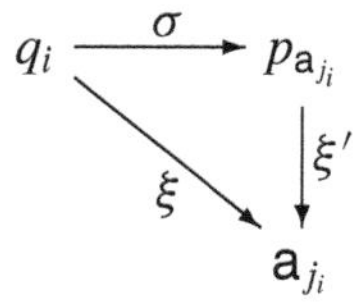

Let us note that A' contains variables only from $\{p_{\mathsf{a}_i}, i \in [1, k]\}$ but not necessarily all of them. To simplify notation and without losing generality, we can assume that A' is a formula in variables $\{p_{\mathsf{a}_i}, i \in [1, k]\}$ (if p_{a_i} does not occur in A', one simply can take $A' \wedge (p_{\mathsf{a}_i} \to p_{\mathsf{a}_i})$ instead of A' and let $\xi' : p_{\mathsf{a}_i} \mapsto \mathbf{1}_{\mathbf{A}}$).

Now, if we apply ξ' to A' and take into consideration (2.3), we get

$$A'(\mathsf{a}_{j_1}, \ldots, \mathsf{a}_{j_k}) = A(\xi'(p_1), \ldots, \xi'(p_n)) = A(\mathsf{a}_{j_1}, \ldots, \mathsf{a}_{j_n}) \neq \mathbf{1}_{\mathbf{A}}. \tag{2.4}$$

(b) For contradiction, assume that $\overset{C}{\nvdash} A' \to X_{\mathbf{A}}$. Thus, $\overset{C}{\nvdash} A' \to (D_{\mathbf{A}} \to p_{\mathsf{m}_{\mathbf{A}}})$ and therefore, $\overset{C}{\nvdash} (A' \wedge D_{\mathbf{A}}) \to p_{\mathsf{m}_{\mathbf{A}}}$ Then, there is a C-algebra in which formula

$(A' \wedge D_{\mathbf{A}}) \to p_{\mathsf{m}_{\mathbf{A}}}$ is refuted, and by Corollary 2.2, there is an s.i. C-algebra $\mathbf{B}$ and a valuation ν in $\mathbf{B}$ such that $\nu((A' \wedge D_{\mathbf{A}})) = \mathbf{1_B}$ and $\nu(p_{\mathsf{m}_{\mathbf{A}}}) = \mathsf{m}_{\mathbf{B}} \neq \mathbf{1_B}$; that is,

$$A'(\mathsf{b}_1, \ldots, \mathsf{b}_k) = \mathbf{1_B} \text{ and } D(\mathsf{b}_1, \ldots, \mathsf{b}_k) = \mathbf{1_B}, \tag{2.5}$$

where $\mathsf{b}_i = \nu(p_{j_i})$, $i \in [1, k]$. Let $\eta : \mathsf{a}_i \mapsto \mathsf{b}_i$:

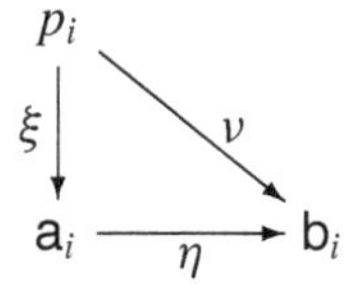

Then, because $D(\mathsf{b}_1, \ldots, \mathsf{b}_n) = \mathbf{1_B}$, η is a homomorphism and we can apply Proposition 2.9. Moreover, η is an isomorphism, because $\eta(\mathsf{m_A}) = \nu(p_{\mathsf{m}_{\mathbf{A}}}) = \mathsf{m}_{\mathbf{B}} \neq \mathbf{1}_{\mathsf{B}}$, and we can apply Proposition 2.7.

We have arrived at a contradiction: on the one hand, by (2.5), $A'(\mathsf{b}_1, \ldots, \mathsf{b}_n) = \mathbf{1_B}$, while on the other hand, by (2.4), $A'(\mathsf{a}_1, \ldots, \mathsf{a}_k) \neq \mathbf{1_A}$, and because η is an isomorphism,

$$\eta(A'(\mathsf{a}_1, \ldots 3, \mathsf{a}_k)) = A'(\eta(\mathsf{a}_1), \ldots, \eta(\mathsf{a}_n)) = A'(\mathsf{b}_1, \ldots, \mathsf{b}_k) \neq \mathbf{1_B}.$$

Example 2.3 Consider three-element Heyting algebra $\mathbf{3}$ from Example 2.2. Then, $X_{\mathbf{3}} = D(\mathbf{A}) \to p_{\mathsf{m}_{\mathbf{A}}}$. It is clear that $\mathsf{m}_{\mathbf{3}} = \mathsf{a}$, and from Example 2.2 we know that $D(\mathbf{3})$ is equivalent to $(p_{\mathsf{a}} \to \mathbf{0}) \to \mathbf{0}$. Therefore,

$$X_{\mathbf{3}} \text{ is equivalent in } \mathsf{Int} \text{ to } ((p_a \to \mathbf{0}) \to \mathbf{0}) \to p_a \text{ or to } \neg\neg p_{\mathsf{a}} \to p_{\mathsf{a}}.$$

2.3.3 Splitting

Suppose that $\mathbf{A}$ is a finite s.i. C-algebra and $X_{\mathbf{A}}$ is its characteristic formula. We already know from (2.2) that $\mathbf{A} \not\models X_{\mathbf{A}}$. But $X_{\mathbf{A}}$ possesses a much stronger property.

Proposition 2.12 *Suppose that* $\mathbf{A}$ *is a finite s.i. C-algebra and* $\mathbf{B}$ *is a C-algebra. Then,*

$$\mathbf{B} \not\models X_{\mathbf{A}} \iff \mathbf{A} \textit{ is embedded in a homomorphic image of } \mathbf{B}.$$

Proof If $\mathbf{A}$ is embedded in a homomorphic image of $\mathbf{B}$, then $\mathbf{B} \not\models X_{\mathbf{A}}$, because by (2.2), $\mathbf{A} \not\models X_{\mathbf{A}}$.

Conversely, suppose that $\mathbf{B} \not\models X_{\mathbf{A}}$; that is, $\mathbf{B} \not\models (D_{\mathbf{A}} \to p_{\mathsf{m}_{\mathbf{A}}})$. Then, we can use the same argument as in the proof of Theorem 2.2(b) and conclude that $\mathbf{A}$ is embedded in a homomorphic image of $\mathbf{B}$.

Let $\mathcal{A}$ be a class of finite s.i. C-algebras. We take $\overline{\mathcal{A}}$ to be a class of all finite s.i. C-algebras not belonging to $\mathcal{A}$. Denote by $\boldsymbol{L}(\mathcal{A})$ a logic of all formulas valid in each

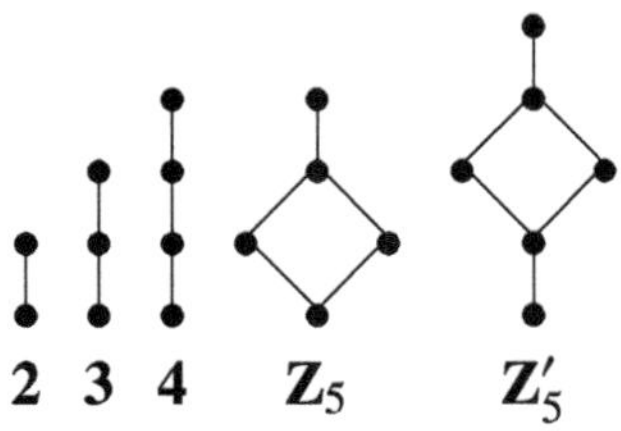

Fig. 2.2 Algebras

algebra from $\mathcal{A}$, and denote by $\overline{L}(\mathcal{A})$ a logic defined by characteristic formulas of algebras from $\mathcal{A}$ as additional axioms, that is, the logic defined by $C + \{X_{\mathbf{A}} \mid \mathbf{A} \in \mathcal{A}\}$. If $\mathcal{A}$ consists of a single algebra $\mathbf{A}$, we omit the curly brackets and write $L(\mathbf{A})$ and $\overline{L}(\mathbf{A})$.

Let us observe that if a two-element algebra is not in $\mathcal{A}$, logic $\overline{L}(\mathcal{A})$ is not trivial: no algebra $\mathbf{A}$ having more than two elements can be a subalgebra of a two-element algebra or its homomorphic image and hence, formula $X_{\mathbf{A}}$ is valid in a two-element algebra.

Corollary 2.5 *Suppose that $\mathcal{A}$ is a class of finite s.i. C-algebras. Then, logic $\overline{L}(\mathcal{A})$ is the smallest extension of C such that algebras from $\mathcal{A}$ are not its models.*

Proof We need to prove that for every C-logic L' for which $L \supsetneq L'$ is a proper extension of L', there is an algebra $\mathbf{A} \in \mathcal{A}$ that is a model for L'; that is, $\mathbf{A} \models A$ for every $A \in L'$.

For contradiction, assume that $L \supsetneq L'$ and for each algebra $\mathbf{A} \in \mathcal{A}$ there is a formula $A_{\mathbf{A}} \in L'$ such that $\mathbf{A} \not\models A_{\mathbf{A}}$. Then, by Theorem 2.2, $A_{\mathbf{A}} \overset{C}{\vdash} X_{\mathbf{A}}$. Hence, because $A_{\mathbf{A}} \in L'$ and L' is closed under Modus Ponens, $X_{\mathbf{A}} \in L'$, and subsequently, $L \subseteq L'$, because L is defined by $C + \{X_{\mathbf{A}} \mid \mathbf{A} \in \mathcal{A}\}$. Thus, we have arrived at a contradiction with the assumption that L is a proper extension of L'.

Example 2.4 If $\mathcal{A}$ consists of two algebras $\mathbf{Z}_5$ and $\mathbf{Z}'_5$ the Hasse diagrams of which are depicted in Fig. 2.2, then $\overline{L}(\mathcal{A})$ is Dummett's logic (cf. Idziak and Idziak 1988.)

If L is a C-logic, denote by L_{f} a class of all finite models of L, and by L_{fsi}—a class of all finite s.i. models of L. It should be clear that for any C-logics L and L', $L_{\mathsf{f}} = L'_{\mathsf{f}}$ if and only if $L_{\mathsf{fsi}} = L'_{\mathsf{fsi}}$. We say that two C-logics L and L' are finitely indistinguishable if $L_{\mathsf{f}} = L'_{\mathsf{f}}$ (in symbols, $L \approx_f L'$). Obviously, $\approx_f$ is an equivalence relation on the lattice $\mathbf{Ext}C$. Let us note that each $\approx_f$-equivalence class $[L]_f$ contains the largest element, namely a logic of all formulas valid in L_{f}. Moreover, by Corollary 2.5, $[L]_f$ contains the smallest element, namely $\overline{L}(\overline{\mathcal{A}})$—the logic defined relative to C by the characteristic formulas of all algebras from $\overline{\mathcal{A}}$, where $\mathcal{A} = L_{\mathsf{fsi}}$ Thus, if L is a C-logic and $\mathcal{A} = L_{\mathsf{fsi}}$, then $\approx_f$-equivalence class $[L]_f$ forms a segment

$$[L]_f = [\overline{L}(\overline{\mathcal{A}}), L(\mathcal{A})].$$

Let us point out that each $\approx_f$-equivalence class, contains a unique logic enjoying the f.m.p., namely, its largest logic, which is a logic defined by all finite models. Thus, if the cardinality of an $\approx_f$-equivalence class is distinct from one, this class contains logics lacking the f.m.p. (cf. Sect. 2.4 for examples). In fact (cf. Tomaszewski 2003, Theorem 4.8), there is an $\approx_f$-equivalent class of si-logics having continuum many members. Therefore, there are continuum many si-logics lacking the f.m.p.

The case in which class $\mathcal{A}$ consists of a single algebra plays a very special role.

Corollary 2.6 *Suppose that* $\mathbf{A}$ *is a finite s.i. C-algebra and* $X_{\mathbf{A}}$ *is its characteristic formula. Then, the logic* $\mathbf{L}$ *defined by* $C + X_{\mathbf{A}}$ *is the smallest extension of* C *for which* $\mathbf{A}$ *is not a model.*

Corollary 2.6 yields that for any logic $\boldsymbol{L} \in \mathbf{Ext}C$,

$$\text{either } \boldsymbol{L} \subseteq \boldsymbol{L}(\mathbf{A}), \text{ or } \boldsymbol{L} \supseteq \overline{\boldsymbol{L}}(\mathbf{A}).$$

Indeed, if $\mathbf{A}$ is a model of $\boldsymbol{L}$, then $\boldsymbol{L} \subseteq \boldsymbol{L}(\mathbf{A})$; otherwise, $\mathbf{A}$ is not a model of $\boldsymbol{L}$ and by Corollary 2.5, $\boldsymbol{L} \supseteq \overline{\boldsymbol{L}}(\mathbf{A})$.

Let us recall (cf., e.g., Kracht 1999; Galatos et al. 2007) that if $\boldsymbol{L}$ is a logic, a pair of its extension $(\boldsymbol{L}_1, \boldsymbol{L}_2)$ is a *splitting pair* of $\mathbf{Ext}\boldsymbol{L}$ if

$$\boldsymbol{L}_1 \nsubseteq \boldsymbol{L}_2, \text{ and for each } \boldsymbol{L}' \in \mathbf{Ext}\boldsymbol{L}, \text{ either } \boldsymbol{L}_1 \subseteq \boldsymbol{L}' \text{ or } \boldsymbol{L}' \subseteq \boldsymbol{L}_2,$$

and $\mathbf{A}$ is a *splitting algebra*, while $X_{\mathbf{A}}$ is a *splitting formula*.

Example 2.5 Consider Heyting algebra $\mathbf{3}$ from Example 2.3. Algebra $\mathbf{3}$ defines a splitting: for each logic $\boldsymbol{L} \in \mathbf{Ext}\mathsf{Int}$,

$$\text{either } \boldsymbol{L} \subseteq \boldsymbol{L}(\mathbf{3}) \text{ or } \boldsymbol{L} \supseteq \overline{\boldsymbol{L}}(\mathbf{3}),$$

and $\overline{\boldsymbol{L}}(\mathbf{3})$ is defined by $\mathsf{IPC} + X_3$. From Example 2.3, we know that formula X_3 is equivalent to formula $\neg\neg p_{\mathsf{a}} \to p_a$; that is, $\overline{\boldsymbol{L}}(\mathbf{3})$ is defined by $\mathsf{IPC} + \neg\neg p_{\mathsf{a}} \to p_a$ and therefore, $\overline{\boldsymbol{L}}(\mathbf{3}) = \mathsf{Cl}$. Thus, for any formula A refuted in $\mathbf{3}$, $\mathsf{Int} + A$ defines a logic extending Cl; that is, $\mathsf{Int} + A$ is Cl or a trivial logic.

Example 2.6 Let $\mathbf{n}$ denote a linearly ordered n-element Heyting algebra. Then, each nontrivial algebra $\mathbf{n}$ is s.i. and defines a splitting pair: for logic $\boldsymbol{L} \in \mathbf{Ext}\mathsf{Int}$,

$$\text{either } \boldsymbol{L} \subseteq \boldsymbol{L}(\mathbf{n}) \text{ or } \boldsymbol{L} \supseteq \overline{\boldsymbol{L}}(\mathbf{n}),$$

and $\overline{\boldsymbol{L}}(\mathbf{n})$ is defined by $\mathsf{IPC} + X_{\mathbf{n}}$. Logic $\boldsymbol{L}(\mathbf{n})$ is the smallest logic of the $n - 2$ slice introduced in Hosoi (1967).

2.3.4 Quasiorder

On the class of all finite s.i. C-algebras we introduce the following quasiorder: for any C-algebras $\mathbf{A}$ and $\mathbf{B}$,

$$\mathbf{A} \leq \mathbf{B} \overset{\text{def}}{\iff} X_{\mathbf{A}} \overset{C}{\vdash} X_{\mathbf{B}}.$$

The following theorem establishes the main properties of the introduced quasiorder.

Theorem 2.3 (Jankov 1963c, 1969) *Let $\mathbf{A}$ and $\mathbf{B}$ be finite s.i. C-algebras. The following conditions are equivalent:*

(a) $\mathbf{A} \leq \mathbf{B}$;
(b) $X_{\mathbf{A}}$ is refutable in $\mathbf{B}$;
(c) every formula refutable in $\mathbf{A}$ is refutable in $\mathbf{B}$;
(d) $\mathbf{A}$ is embedded in a homomorphic image of $\mathbf{B}$.

Proof (a) $\Rightarrow$ (b), because by (2.2), $\mathbf{B} \not\models X_{\mathbf{B}}$ and by the definition of quasiorder, $X_{\mathbf{A}} \overset{C}{\vdash} X_{\mathbf{B}}$.

(b) $\Rightarrow$ (c). If a formula A is refutable in $\mathbf{A}$, then by Theorem 2.2, $A \overset{C}{\vdash} X_{\mathbf{A}}$. By (b), $X_{\mathbf{A}}$ is refutable in $\mathbf{B}$ and then, by Theorem 2.2, $X_{\mathbf{A}} \overset{C}{\vdash} X_{\mathbf{B}}$. Hence, $A \overset{C}{\vdash} X_{\mathbf{B}}$ and consequently, A is refutable in $\mathbf{B}$, because $X_{\mathbf{B}}$ is refutable in $\mathbf{B}$.

(c) $\Rightarrow$ (d). Characteristic formula $X_{\mathbf{A}}$ is refutable in $\mathbf{A}$ and hence, by (c), formula $X_{\mathbf{A}}$ is refutable in $\mathbf{B}$. By Corollary 2.4, $\mathbf{A}$ is embedded in a homomorphic image of algebra $\mathbf{B}$.

(d) $\Rightarrow$ (a). Characteristic formula $X_{\mathbf{A}}$ is refutable in $\mathbf{A}$. Hence, if $\mathbf{A}$ is embedded in a homomorphic image $\mathbf{B}$, formula $X_{\mathbf{A}}$ is refutable in this homomorphic image and consequently, it is refutable in $\mathbf{B}$. Then, by Theorem 2.2, $X_{\mathbf{A}} \overset{C}{\vdash} X_{\mathbf{B}}$, which means that $\mathbf{A} \leq \mathbf{B}$.

Corollary 2.7 *Let $\mathbf{A}$ and $\mathbf{B}$ be finite s.i. C-algebras such that $\mathbf{A} \leq \mathbf{B}$ and $\mathbf{B} \leq \mathbf{A}$. Then, algebras $\mathbf{A}$ and $\mathbf{B}$ are isomorphic.*

Proof Indeed, by Theorem 2.3, $\mathbf{A} \leq \mathbf{B}$ entails that $\mathbf{A}$ is a subalgebra of a homomorphic image of $\mathbf{B}$ and hence, $\mathsf{card}(\mathbf{A}) \leq \mathsf{card}(\mathbf{B})$. Likewise, $\mathsf{card}(\mathbf{B}) \leq \mathsf{card}(\mathbf{A})$. Therefore, $\mathsf{card}(\mathbf{A}) = \mathsf{card}(\mathbf{B})$ and because $\mathbf{A}$ and $\mathbf{B}$ are homomorphic images of each other, their finiteness ensures that they are isomorphic.

The following corollaries are the immediate consequences of Theorem 2.3(d).

Corollary 2.8 *For any finite s.i. C-algebras $\mathbf{A}$ and $\mathbf{B}$, if $\mathbf{A} \leq \mathbf{B}$ and $\mathsf{card}(\mathbf{A}) = \mathsf{card}(\mathbf{B})$, then $\mathbf{A} \cong \mathbf{B}$.*

Let us observe that Corollary 2.7 entails that $\leq$ is a partial order and that by Corollary 2.8, any class $\mathcal{A}$ of finite s.i. C-algebras enjoys the descending chain condition. Hence, the following holds.

Corollary 2.9 *Let $\mathcal{A}$ be a class of finite s.i. C-algebras. Then $\mathcal{A}$ contains a subclass $\mathcal{A}^{(m)} \subseteq \mathcal{A}$ of pairwise nonisomorphic algebras that are minimal relative to $\leq$ such that*

$$\textit{for any algbera } \mathbf{A} \in \mathcal{A}, \textit{ there is an algebra } \mathbf{A}' \in \mathcal{A}^{(m)} \textit{ and } \mathbf{A}' \leq \mathbf{A}. \tag{2.6}$$

Proposition 2.13 *For any class of finite s.i. C-algebras $\mathcal{A}$,*

$$\overline{\boldsymbol{L}}(\mathcal{A}) = \overline{\boldsymbol{L}}(\mathcal{A}^{(m)}). \tag{2.7}$$

Proof Indeed, $\mathcal{A}^{(m)} \subseteq \mathcal{A}$ entails $\{X_{\mathbf{A}} \mid \mathbf{A} \in \mathcal{A}^{(m)}\} \subseteq \{X_{\mathbf{A}} \mid \mathbf{A} \in \mathcal{A}\}$ and subsequently, $\overline{\boldsymbol{L}}(\mathcal{A}^{(m)}) \subseteq \overline{\boldsymbol{L}}(\mathcal{A})$.

On the other hand, suppose that $\mathbf{A} \in \mathcal{A}$. Then, by (2.6), there is an algebra $\mathbf{A}' \in \mathcal{A}^{(m)}$ such that $\mathbf{A}' \leq \mathbf{A}$ and by definition, $X_{\mathbf{A}'} \overset{C}{\vdash} X_{\mathbf{A}}$. Thus, $X_{\mathbf{A}} \in \overline{\boldsymbol{L}}(\mathcal{A}^{(m)})$ for all $\mathbf{A} \in \mathcal{A}$, that is, $\overline{\boldsymbol{L}}(\mathcal{A}) \subseteq \overline{\boldsymbol{L}}(\mathcal{A}^{(m)})$.

2.4 Applications of Characteristic Formulas

In Jankov (1968b), the characteristic formulas were instrumental in proving that the cardinality of **ExtInt** is continuum and that there is an si.-logic lacking the f.m.p.

2.4.1 Antichains

Suppose that $\mathcal{A}$ is a class of finite s.i. C-algebras. We say that class $\mathcal{A}$ forms an *antichain* if for any $\mathbf{A}, \mathbf{B} \in \mathcal{A}$, algebras $\mathbf{A}$ and $\mathbf{B}$ are *incomparable*; that is, $\mathbf{A} \not\leq \mathbf{B}$ and $\mathbf{B} \not\leq \mathbf{A}$.

Let us observe that for any nonempty class of algebras $\mathcal{A}$, the subclass $\mathcal{A}^{(m)}$ forms an antichain.

Let C be a C-calculus and C be a set of formulas in the signature of C. Then C is said to be *strongly independent relative to* C if $C \setminus \{A\} \overset{C}{\nvdash} A$ for each formula $A \in C$. In other words, C is strongly independent relative to C if no formula from C can be derived in C from the rest of the formulas of C.

Let us observe that if C is a strongly independent set of C-formulas, then for any distinct subsets $C_1, C_2 \subseteq C$, the logics defined by C_1 and C_2 as sets of axioms, are distinct. Hence, if there is a countably infinite set C of strongly independent C-formulas, then the set of all extensions of the C-calculus is uncountable. This property of strongly independent sets was used in Jankov (1968b) for proving that the set of si-logics is not countable (cf. Sect. 2.5).

Antichains of finite s.i. C-algebras posses the following very important property.

Proposition 2.14 *Suppose that $\mathcal{A}$ is an antichain of finite s.i. C-algebras. Then the set $\{X_{\mathbf{A}} \mid \mathbf{A} \in \mathcal{A}\}$ is strongly independent.*

Proof For contradiction, suppose that for some $\mathbf{A} \in \mathcal{A}$,

$$\{X_{\mathbf{B}} \mid \mathbf{B} \in \mathcal{A} \setminus \{\mathbf{A}\}\} \vdash_C X_{\mathbf{A}}.$$

Recall that by (2.2), $\mathbf{A} \not\models A$ and hence, there is a $\mathbf{B} \in \mathcal{A} \setminus \{\mathbf{A}\}$ such that $\mathbf{A} \not\models X_{\mathbf{B}}$. Then, by Theorem 2.3, $\mathbf{B} \leq \mathbf{A}$, and we have arrived at a contradiction.

Corollary 2.10 *If there is an infinite antichain of finite s.i. C-algebras which are models of a given C-logic $\mathbf{L}$, then*

(a) the set of extensions of $\mathbf{L}$ is uncountable;
(b) there is an extension of $\mathbf{L}$ that cannot be defined by any C-calculus; that is, it cannot be defined by a finite set of axioms and the rules of substitution and Modus Ponens;
(c) there is a strongly ascending chain of C-logics.

In fact, if $\mathcal{A} = \{\mathbf{A}_i \mid i \geq 0\}$ is an infinite antichain of finite s.i. C-algebras, then logics $\boldsymbol{L}_k$ defined by $\{X_{\mathbf{A}_i} \mid i \in [1, k]\}$ form a strongly ascending chain, and consequently, logic $\overline{\boldsymbol{L}}(\mathcal{A})$ defined by $\{X_{\mathbf{A}} \mid \mathbf{A} \in \mathcal{A}\}$ cannot be defined by any C-calculus.

2.5 Extensions of C-Logics

In Jankov (1968b), it was observed that **ExtInt** is uncountable. To prove this claim, it is sufficient to present a countably infinite antichain of finite s.i. Heyting algebras.

Let $\mathcal{A}$ be a class of all finite s.i. Heyting algebras, generated by elements $\mathsf{a}, \mathsf{b}, \mathsf{c}$ and satisfying the following conditions:

$$\neg(\mathsf{a} \wedge \mathsf{b}) = \neg(\mathsf{b} \wedge \mathsf{c}) = \neg(\mathsf{c} \wedge \mathsf{a}) = \neg\neg\mathsf{a} \to \mathsf{a} = \neg\neg\mathsf{b} \to \mathsf{b} = \neg\neg(\mathsf{a} \vee \mathsf{b} \vee \mathsf{c}) = \mathbf{1} \tag{2.8}$$

$$\neg\mathsf{a} \vee \neg\mathsf{b} \vee (\neg\neg\mathsf{c} \to \mathsf{c}) = \mathsf{d}, \tag{2.9}$$

where d is a pretop element. Class $\mathcal{A}$ is not empty; moreover, it contains infinitely many members (cf. Fig. 2.3).

Conditions (2.8) and (2.9) yield that algebra is generated by three elements a, b, and c that are distinct from $\mathbf{0}$ such that elements a and b are *regular*, that is, $\neg\neg\mathsf{a} = \mathsf{a}$ and $\neg\neg\mathsf{b} = \mathsf{b}$, while element c is neither regular nor *dense*; that is $\neg\mathsf{c} \neq \mathbf{0}$.

The goal of this section is to prove the following theorem.

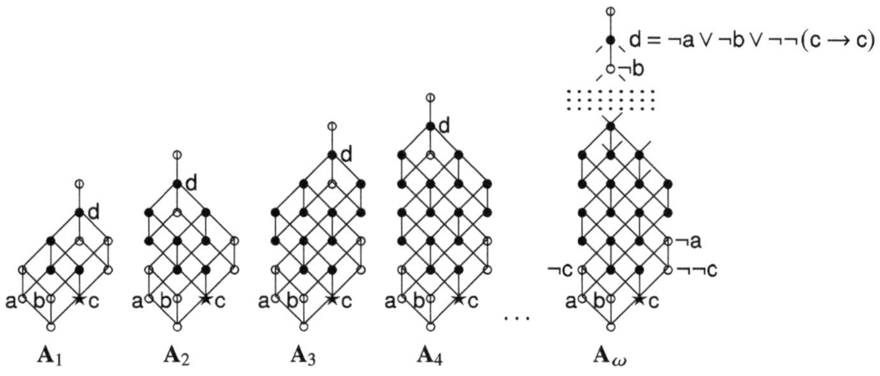

Fig. 2.3 Yankov's antichain

Theorem 2.4 *Logic* $\overline{\boldsymbol{L}}(\mathcal{A})$ *does not enjoy the finite model property.*

To show that $\overline{\boldsymbol{L}}(\mathcal{A})$ lacks the f.m.p., we will take the following formula:

$$A = \neg(p \wedge q) \wedge \neg(q \wedge r) \wedge \neg(p \wedge r) \wedge (\neg\neg p \to p) \wedge (\neg\neg q \to q) \wedge \neg\neg(p \vee q \vee r) \to \neg p \vee \neg q \vee (\neg\neg r \to r),$$

and we will prove the following two lemmas.

Lemma 2.1 $A \notin \overline{\boldsymbol{L}}(\mathcal{A})$.

Lemma 2.2 *A is valid in all finite models of* $\overline{\boldsymbol{L}}(\mathcal{A})$.

The proofs of Lemmas 2.1 and 2.2 can be found in Sect. 2.5.2.2, but first, we need to establish some properties of the algebras from $\mathcal{A}$ (cf. Sect. 2.5.1). In particular, we will prove (cf. Sect. 2.5.2.1) the following proposition, which has a very important corollaries on its own.

Proposition 2.15 *Algebras* $\{\mathbf{A}_i \mid i = 1, 2, \ldots\}$ *are minimal (relative to* $\leq$*) elements of* $\mathcal{A}$.

Corollary 2.11 *The class* $\{\mathbf{A}_i \mid i = 1, 2, \ldots\}$ *forms an antichain.*

Corollary 2.11 has three immediate corollaries, which at the time of the publication of Jankov (1968b) changed the view on the structure of **ExtInt**.

Corollary 2.12 *There are continuum many si-logics.*

Corollary 2.13 *There are si-logics that cannot be defined by an si-calculus.*

Corollary 2.14 *There exists a strictly ascending sequence* $\boldsymbol{L}_i, i > 0$, *of si-logics defined by si-calculi.*

Corollaries 2.12 and 2.13 follow immediately from Proposition 2.15 and Corollary 2.10. To prove Corollary 2.14, consider logics $\boldsymbol{L}_i$ defined by axioms $X_{\mathbf{A}_j}, j \in [1, i]$.

The rest of this section is dedicated to a proof of Proposition 2.15 and Theorem 2.4.

2.5.1 Properties of Algebras $\mathbf{A}_i$

In this section, $\mathbf{A}_i$ are algebras the diagrams of which are depicted in Fig. 2.3.

Proposition 2.16 *Each algebra $\mathbf{A}_i, i \in [1, \omega]$, contains precisely one set of three elements, namely $\{\mathsf{a}, \mathsf{b}, \mathsf{c}\}$, satisfying the following conditions:*

$$\neg(\mathsf{a} \wedge \mathsf{b}) = \neg(\mathsf{b} \wedge \mathsf{c}) = \neg(\mathsf{c} \wedge \mathsf{a}) = \neg\neg\mathsf{a} \to \mathsf{a} = \neg\neg\mathsf{b} \to \mathsf{b} = \mathbf{1} \quad (2.10)$$

$$\neg\mathsf{a} \neq \mathbf{1}, \quad \neg\mathsf{b} \neq \mathbf{1}, \quad \neg\neg\mathsf{c} \to \mathsf{c} \neq \mathbf{1}. \quad (2.11)$$

Proof It is not hard to see that in each $\mathbf{A}_i$, elements $\{\mathsf{a}, \mathsf{b}, \mathsf{c}\}$ satisfy conditions (2.10) and (2.11). Let us now show that there are no other elements satisfying these conditions.

It is clear that (2.11) yields that all elements $\mathsf{a}, \mathsf{b}, \mathsf{c}$ are distinct from $\mathbf{0}$, and $\mathsf{c} \neq \mathbf{1}$. Moreover, by (2.10), $\neg(\mathsf{a} \wedge \mathsf{c}) = \neg(\mathsf{b} \wedge \mathsf{c}) = \mathbf{1}$. Hence, $\mathsf{a} \wedge \mathsf{c} = \mathsf{b} \wedge \mathsf{c} = \mathbf{0}$ and therefore, $\mathsf{c} \wedge (\mathsf{a} \vee \mathsf{b}) = \mathbf{0}$. Hence, $\mathsf{a} \vee \mathsf{b} \leq \neg\mathsf{c}$ and consequently, $\neg\mathsf{c} \neq \mathbf{0}$.

Let us observe that in each algebra $\mathbf{A}_i$ there are precisely 8 elements for which $\neg\neg x = x$ holds:

$$\mathsf{A}_i^{(r)} := \{\mathbf{0}, \mathsf{a}, \mathsf{b}, \mathsf{a} \vee \mathsf{b}, \neg\mathsf{a}, \neg\mathsf{b}, \neg(\mathsf{a} \vee \mathsf{b}), \mathbf{1}\}.$$

Let us show that only elements a and b can potentially satisfy (2.10) and (2.11).

Indeed, we already know that we cannot use $\mathbf{0}$ and $\mathbf{1}$. In addition, we cannot use elements $\neg\mathsf{a}, \neg\mathsf{b}, \neg(\mathsf{a} \vee \mathsf{b})$, because for each $\mathsf{a}' \in \{\neg\mathsf{a}, \neg\mathsf{b}, \neg(\mathsf{a} \vee \mathsf{b})\}$ and for any $\mathsf{c} \in \mathbf{A}_i$, if $\mathsf{a}' \wedge \mathsf{c} = \mathbf{0}$, then $\mathsf{c} \leq \neg\mathsf{a}'$, that is, $\mathsf{c} \leq \mathsf{a} \vee \mathsf{b}$, while in algebra $\mathbf{A}_i$ all elements smaller then $\mathsf{a} \vee \mathsf{b}$ satisfy condition $\neg\neg\mathsf{c} = \mathsf{c}$.

This leaves us with elements a, b and $\mathsf{a} \vee \mathsf{b}$. But we cannot use element $\mathsf{a} \vee \mathsf{b}$, because neither $(\mathsf{a} \vee \mathsf{b}) \wedge \mathsf{a}$ nor $(\mathsf{a} \vee \mathsf{b}) \wedge \mathsf{b}$ is $\mathbf{0}$.

Next, we observe that in $\mathbf{A}_i$, there are just two elements c and $\neg\mathsf{a}$ whose intersection with a and b gives $\mathbf{0}$, but we cannot select $\neg\mathsf{a}$, because $\neg\neg\neg\mathsf{a} \to \neg\mathsf{a} = \mathbf{1}$, and this element would not satisfy (2.11). Thus, only elements a, b, and c satisfy conditions (2.10) and (2.11), and this observation completes the proof.

Next, we prove that in the homomorphic images of algebras $\mathbf{A}_i$, only images of elements a, b, and c may satisfy conditions (2.10) and (2.11).

Proposition 2.17 *Let algebra $\mathbf{A}_i, i \in [1, \omega]$, and $\varphi : \mathbf{A}_i \longrightarrow \mathbf{B}$ be a homomorphism onto algebra $\mathbf{B}$. If for some elements $\mathsf{a}', \mathsf{b}', \mathsf{c}' \in \mathbf{A}_i$, their images $\bar{\mathsf{a}}, \bar{\mathsf{b}}, \bar{\mathsf{c}}$ satisfy conditions* (2.10) *and* (2.11), *then elements $\mathsf{a} = \neg\neg\mathsf{a}', \mathsf{b} = \neg\neg\mathsf{b}'$, and $\mathsf{c} = \mathsf{c}' \wedge \neg\mathsf{a}' \wedge \neg\mathsf{b}'$ satisfy* (2.10) *and* (2.11).

Proof First, let us observe that $\varphi^{-1}(\mathbf{0}) = \{\mathbf{0}\}$; that is, $\mathbf{0}$ is the only element of $\mathbf{A}_i$ which φ sends to $\mathbf{0}$.

Indeed, assume for contradiction that there is an element $\mathsf{d}' \in \mathbf{A}_i$ such that $\mathbf{0} < \mathsf{d}'$ and $\varphi(\mathsf{d}') = \mathbf{0}$. Elements a, b, and c are the only atoms of $\mathbf{A}_i$ and therefore,

$\mathsf{a} \le \mathsf{d}'$, $\mathsf{b} \le \mathsf{d}'$, or $\mathsf{c} \le \mathsf{d}'$. Hence, $\varphi(\mathsf{a}) = \mathbf{0}$, $\varphi(\mathsf{b}) = \mathbf{0}$, or $\varphi(\mathsf{c}) = \mathbf{0}$ and therefore, $\varphi(\neg\mathsf{a}) = \mathbf{1}$, $\varphi(\neg\mathsf{b}) = \mathbf{1}$, or $\varphi(\neg\mathsf{c}) = \mathbf{1}$. Recall that $\varphi(\neg\mathsf{a}) = \neg\varphi(\mathsf{a}) = \neg\bar{\mathsf{a}}$ and by (2.11), $\neg\bar{\mathsf{a}} \neq \mathbf{1}$. Likewise, $\neg\bar{\mathsf{b}} = \varphi(\neg\mathsf{b}) \neq \mathbf{1}$. And if $\varphi(\neg\mathsf{c}) = \mathbf{1}$, then $\neg\bar{\mathsf{c}} = \mathbf{1}$ and consequently $\neg\neg\bar{\mathsf{c}} \to \bar{\mathsf{c}} = \mathbf{1}$, which contradicts (2.11). Thus, $\varphi^{-1}(\mathbf{0}) = \{\mathbf{0}\}$.

Next, let us show that $\mathsf{a} \wedge \mathsf{b} = \mathbf{0}$ and hence, $\neg(\mathsf{a} \wedge \mathsf{b}) = \mathbf{1}$. Indeed,

$$\varphi(\mathsf{a} \wedge \mathsf{b}) = \varphi(\neg\neg\mathsf{a}' \wedge \neg\neg\mathsf{b}') = \neg\neg\varphi(\mathsf{a}') \wedge \neg\neg\varphi(\mathsf{b}') = \varphi(\mathsf{a}') \wedge \varphi(\mathsf{b}') = \mathbf{0}.$$

Hence, $\mathsf{a} \wedge \mathsf{b} \in \varphi^{-1}(\mathbf{0})$ and therefore, $\mathsf{a} \wedge \mathsf{b} = \mathbf{0}$.

In addition, $\mathsf{a} \wedge \mathsf{c} = \neg\neg\mathsf{a}' \wedge (\mathsf{c} \wedge \neg\mathsf{a}' \wedge \neg\mathsf{b}') = \mathbf{0}$. Likewise, $\mathsf{b} \wedge \mathsf{c} = \mathbf{0}$.

$$\mathsf{a} \wedge \mathsf{c} = \neg\neg\mathsf{a}' \wedge (\mathsf{c} \wedge \neg\mathsf{a}' \wedge \neg\mathsf{b}') = \mathbf{0},\ \ \mathsf{b} \wedge \mathsf{c} = \mathbf{0},$$
$$\neg\neg\mathsf{a} \to \mathsf{a} = \neg\neg\neg\neg\mathsf{a}' \to \neg\neg\mathsf{a}' = \mathbf{1},\ \ \neg\neg\mathsf{b} \to \mathsf{b} = \mathbf{1}.$$

Thus, elements a, b, and c satisfy (2.10).

Next, we observe that by (2.11), $\neg\bar{\mathsf{a}} \neq \mathbf{1}$ and $\neg\bar{\mathsf{b}} \neq \mathbf{1}$, that is, $\bar{\mathsf{a}} > \mathbf{0}$ and $\bar{\mathsf{b}} > \mathbf{0}$. Hence,

$$\varphi(\mathsf{a}) = \varphi(\neg\neg\mathsf{a}') = \neg\neg\varphi(\mathsf{a}') = \neg\neg\bar{\mathsf{a}} \ge \bar{\mathsf{a}} > \mathbf{0}$$

and by the same reason, $\mathsf{b} \neq \mathbf{0}$. Thus, $\neg\mathsf{a} \neq \mathbf{1}$ and $\neg\mathsf{b} \neq \mathbf{1}$.

Now, let us show that $\neg\neg\mathsf{c} \to \mathsf{c} \neq \mathbf{1}$. That is, we need to demonstrate that

$$\neg\neg(\mathsf{c}' \wedge \neg\mathsf{a}' \wedge \neg\mathsf{b}') \to (\mathsf{c}' \wedge \neg\mathsf{a}' \wedge \neg\mathsf{b}') \neq \mathbf{1}.$$

To that end, we will show that

$$\varphi(\neg\neg(\mathsf{c}' \wedge \neg\mathsf{a}' \wedge \neg\mathsf{b}') \to (\mathsf{c}' \wedge \neg\mathsf{a}' \wedge \neg\bar{\mathsf{b}}) = \neg\neg(\bar{\mathsf{c}} \wedge \neg\bar{\mathsf{a}} \wedge \neg\bar{\mathsf{b}}) \to (\bar{\mathsf{c}} \wedge \neg\bar{\mathsf{a}} \wedge \neg\bar{\mathsf{b}}) \neq \mathbf{1}.$$

Indeed, recall that by (2.11), $\bar{\mathsf{a}} \wedge \bar{\mathsf{c}} = \mathbf{0}$ and hence, $\mathsf{c} \le \neg\bar{\mathsf{a}}$. Likewise, $\bar{\mathsf{c}} \le \neg\bar{\mathsf{b}}$ and hence,

$$\bar{\mathsf{c}} \le \neg\bar{\mathsf{a}} \wedge \neg\bar{\mathsf{b}} \text{ and consequently, } \bar{\mathsf{c}} \wedge \neg\bar{\mathsf{a}} \wedge \neg\bar{\mathsf{b}} = \bar{\mathsf{c}}.$$

Hence,

$$\neg\neg(\bar{\mathsf{c}} \wedge \neg\bar{\mathsf{a}} \wedge \neg\bar{\mathsf{b}}) \to (\bar{\mathsf{c}} \wedge \neg\bar{\mathsf{a}} \wedge \neg\bar{\mathsf{b}}) = \neg\neg\bar{\mathsf{c}} \to \bar{\mathsf{c}},$$

and by (2.11), $\neg\neg\bar{\mathsf{c}} \to \bar{\mathsf{c}} \neq \mathbf{1}$. This observation completes the proof.

Corollary 2.15 *Any homomorphic image of any algebra $\mathbf{A}_i$, $i \in [1, \omega]$, contains at most one set of elements satisfying conditions* (2.10) *and* (2.11).

Corollary 2.16 *None of the proper homomorphic images of algebras $\mathbf{A}_i$, $i \in [1, \omega]$, has elements satisfying conditions* (2.8) *and* (2.9).

Proof Suppose that $\mathbf{A}_i$ is an algebra the diagram of which is depicted in Fig. 2.3 and that $\mathsf{a}, \mathsf{b}, \mathsf{c} \in \mathbf{A}_i$ are elements satisfying conditions (2.8) and (2.9).

For contradiction, assume that $\varphi : \mathbf{A}_i \longrightarrow \mathbf{B}$ is a proper homomorphism of $\mathbf{A}_i$ onto $\mathbf{B}$ and that elements $\bar{\mathsf{a}}, \bar{\mathsf{b}}, \bar{\mathsf{c}} \in \mathbf{B}$ satisfy conditions (2.8) and (2.9). Then, these elements satisfy the weaker conditions (2.10) and (2.11). By Proposition 2.17, elements $\bar{\mathsf{a}}, \bar{\mathsf{b}}, \bar{\mathsf{c}}$ are images of some elements $\mathsf{a}', \mathsf{b}', \mathsf{c}' \in \mathbf{A}_i$ also satisfying condition (2.10) and (2.11). By Proposition 2.16, the set of elements of $\mathbf{A}_i$ satisfying (2.10) and (2.11) is unique; namely, it is $\{\mathsf{a}, \mathsf{b}, \mathsf{c}\}$. By (2.9), $\neg\mathsf{a} \vee \neg\mathsf{b} \vee (\neg\neg\mathsf{c} \to \mathsf{c})$ is a pretop element of $\mathbf{A}$ and hence, because φ is a proper homomorphism,

$$\varphi(\neg\mathsf{a} \vee \neg\mathsf{b} \vee (\neg\neg\mathsf{c} \to \mathsf{c})) = \neg\bar{\mathsf{a}} \vee \neg\bar{\mathsf{b}} \vee (\neg\neg\bar{\mathsf{c}} \to \bar{\mathsf{c}}) = \mathbf{1},$$

and we have arrived at a contradiction: elements $\bar{\mathsf{a}}, \bar{\mathsf{b}}$, and $\bar{\mathsf{c}}$ do not satisfy (2.9).

2.5.2 *Proofs of Lemmas*

2.5.2.1 Proof of Proposition 2.15

To prove Proposition 2.15, we need to show that no algebra $\mathbf{B} \in \mathcal{A}$ can be embedded in any homomorphic image of algebra $\mathbf{A}_i$, $i = 1, 2, \ldots$, as long as $\mathbf{B} \not\cong \mathbf{A}_i$.

From Corollary 2.16, we already know that none of the proper homomorphic images of algebras $\mathbf{A}_i$ contains elements satisfying conditions (2.8) and (2.9). Thus, no algebra from $\mathcal{A}$ can be embedded in a proper homomorphic image of any algebra $\mathbf{A}_i$.

Now, assume that $\mathbf{B} \in \mathcal{A}$ and $\varphi : \mathbf{B} \longrightarrow \mathbf{A}_i$ is an embedding. By the definition of $\mathcal{A}$, $\mathbf{B}$ is generated by some elements $\mathsf{a}, \mathsf{b}, \mathsf{c}$ satisfying conditions (2.8) and (2.9). Hence, because φ is an isomorphism, elements $\varphi(\mathsf{a}), \varphi(\mathsf{b}), \varphi(\mathsf{c})$ satisfy (2.8) and (2.9). By Proposition 2.16, there is a unique set of three elements that satisfy (2.10) and (2.11) and therefore, there is a unique set of three elements satisfying (2.8) and (2.9). By the definition of $\mathcal{A}$, this set generates algebra $\mathbf{A}_i$; that is, φ maps $\mathbf{B}$ onto $\mathbf{A}$ and thus φ is an isomorphism between $\mathbf{B}$ and $\mathbf{A}$.

2.5.2.2 Proof of Lemma 2.1

Syntactic proof (cf. Jankov 1968b). For contradiction, assume that $A \in \overline{\boldsymbol{L}}(\mathcal{A})$. Recall that by Proposition 2.13, $A \in \overline{\boldsymbol{L}}(\mathcal{A}^{(m)})$ and hence, for some minimal algebras $\mathbf{B}_i$, $i \in [1, n]$,

$$X_{\mathbf{B}_1}, \ldots, X_{\mathbf{B}_n} \vdash A.$$

On the other hand, by Proposition 2.5.2.1, $\{\mathbf{A}_i \mid i = 1, 2. \ldots\} \subseteq \overline{\boldsymbol{L}}(\mathcal{A}^{(m)})$. Class $\{\mathbf{A}_i \mid i = 1, 2. \ldots\}$ is infinite and thus, there is an $\mathbf{A}_k \notin \{\mathbf{B}_i, i \in [1, n]\}$. Observe that $\mathbf{A}_k \not\models A$: it is not hard to see that valuation $p \mapsto \mathsf{a}, q \mapsto \mathsf{b}, r \mapsto \mathsf{c}$ refutes A in every $\mathbf{A}_k$.

Hence, by Theorem 2.2, $A \vdash X_{\mathbf{A}_k}$ and therefore,

$$X_{\mathbf{B}_1}, \ldots, X_{\mathbf{B}_n} \vdash X_{\mathbf{A}_k}.$$

This contradicts Proposition 2.14, which states that the characteristic formulas of any antichain form a strongly independent set, and the subclass of all minimal algebras always forms an antichain.

Semantic proof. Observe that formula A is invalid in algebra $\mathbf{A}_\omega$, and let us prove that $\mathbf{A}_\omega$ is a model of $\overline{\boldsymbol{L}}(\mathcal{A})$. To that end, we prove that neither an algebra from $\mathcal{A}$ or its homomorphic image can be embedded into $\mathbf{A}_\omega$ and therefore, by Proposition 2.12, all formulas $X_{\mathbf{A}}$, $\mathbf{A} \in \mathcal{A}$, are valid in $\mathbf{A}_\omega$.

Indeed, by Proposition 2.17, not any algebra from $\mathcal{A}$ can be embedded in a proper homomorphic image of $\mathbf{A}_\omega$. In addition, by Proposition 2.16, $\mathbf{A}_\omega$ contains a unique set of three elements satisfying conditions (2.8) and (2.9), and these elements generate algebra $\mathbf{A}_\omega$. Thus, if algebra $\mathbf{A} \in \mathcal{A}$ was embedded in $\mathbf{A}_\omega$, its embedding would be a map onto $\mathbf{A}_\omega$, which is impossible, because $\mathbf{A}$ is finite, while $\mathbf{A}_\omega$ is infinite.

2.5.2.3 Proof of Lemma 2.2

We need to show that formula A is valid in all finite models of logic $\overline{\boldsymbol{L}}(\mathcal{A})$. To that end, we will show that every finite Heyting algebra $\mathbf{A}$ refuting A is not a model of $\overline{\boldsymbol{L}}(\mathcal{A})$, because there is a homomorphic image $\mathbf{B}$ of $\mathbf{A}$ in which one of the algebras from $\mathcal{A}$ is embedded. Because $\overline{\boldsymbol{L}}(\mathcal{A})$ is defined by characteristic formulas of algebras from $\mathcal{A}$, none of the members of $\mathcal{A}$ is a model of $\overline{\boldsymbol{L}}(\mathcal{A})$. Hence, if $\mathbf{A}' \in \mathcal{A}$ and $\mathbf{A}'$ is embedded in $\mathbf{B}$, algebra $\mathbf{B}$ and, consequently, algebra $\mathbf{A}$ are not models of $\overline{\boldsymbol{L}}(\mathcal{A})$.

Suppose that finite algebra $\mathbf{A}$ refutes formula

$$\begin{aligned}A =& \neg(p \wedge q) \wedge \neg(q \wedge r) \wedge \neg(p \wedge r) \wedge (\neg\neg p \to p) \wedge (\neg\neg q \to q) \wedge \neg\neg(p \vee q \vee r) \to \\ & \neg p \vee \neg q \vee (\neg\neg r \to r).\end{aligned}$$

Then, by Corollary 2.2, there is a homomorphic image $\mathbf{B}$ of algebra $\mathbf{A}$ and a valuation ν in $\mathbf{B}$ such that

$$\begin{aligned}&\nu(\neg(p \wedge q) \wedge \neg(q \wedge r) \wedge \neg(p \wedge r) \wedge (\neg\neg p \to p) \wedge (\neg\neg q \to q) \wedge \neg\neg(p \vee q \vee r)) = \mathbf{1_B} \\ &\nu(\neg p \vee \neg q \vee (\neg\neg r \to r)) = \mathsf{m}_\mathbf{B}.\end{aligned}$$

Let $\mathsf{a} = \nu(p)$, $\mathsf{b} = \nu(q)$, and $\mathsf{c} = \nu(r)$. Then, elements a, b, and c satisfy conditions (2.8) and (2.9) and therefore, these elements generate a subalgebra of $\mathbf{B}$ belonging to $\mathcal{A}$, and this observation completes the proof.

2.6 Calculus of the Weak Law of Excluded Middle

In Jankov (1968a), Yankov studied the logic of calculus $\mathsf{KC} := \mathsf{IPC} + \neg p \vee \neg\neg p$ which nowadays bears his name. Let us denote this logic by Yn.

A formula A is said to be *positive* if it contains only connectives $\wedge$, $\vee$ and $\rightarrow$. If $\boldsymbol{L}$ is an si-logic, $\boldsymbol{L}^+$ denotes a positive fragment of $\boldsymbol{L}$—the subset of all positive formulas from $\boldsymbol{L}$. We say that an si-logic $\boldsymbol{L}$ is a *p-conservative extension* of Int when $\boldsymbol{L}^+ = \mathsf{Int}^+$.

An s.i. calculus K *admits the derivable elimination of negation* if for any formula A there is a positive formula A^* such that $A \overset{\mathsf{K}\ \mathsf{K}}{\dashv\vdash} A^*$. If $\boldsymbol{L}$ is a logic of K, we say that $\boldsymbol{L}$ admits derivable elimination of negation. Given an si-logic $\boldsymbol{L}$, its extension $\boldsymbol{L}' \in \mathbf{Ext}\boldsymbol{L}$ is said to be *positively axiomatizable* relative to $\boldsymbol{L}$ just in case $\boldsymbol{L}'$ can be axiomatized relative to $\boldsymbol{L}$ by positive axioms.

The following simple proposition provides some different perspectives on the notion of derivable elimination of negation introduced in Jankov (1968a).

Proposition 2.18 *Suppose that* $\boldsymbol{L}$ *is an si-logic. Then, the following are equivalent:*

(a) $\boldsymbol{L}$ *admits derivable elimination of negation;*
(b) every extension of $\boldsymbol{L}$ *is positively axiomatizable relative to to* $\boldsymbol{L}$*;*
(c) any two distinct extensions of $\boldsymbol{L}$ *have distinct positive fragments.*

Proof (a) $\Longrightarrow$ (b) $\Longrightarrow$ (c) is straightforward.

(b) $\Longrightarrow$ (a). Suppose that $\boldsymbol{L}$ is defined by an s.i. calculus K. Then, for every formula A, consider logic $\boldsymbol{L}'$ defined by $\mathsf{K} + A$. If $\boldsymbol{L}' = \boldsymbol{L}$, that is, $A \in \boldsymbol{L}$, we have $A \overset{\mathsf{K}\ \mathsf{K}}{\dashv\vdash} (p \rightarrow p)$. If $\boldsymbol{L} \subsetneq \boldsymbol{L}'$, by assumption, there are positive formulas $B_i, i \in I$, such that $\boldsymbol{L}'$ is a logic of $\mathsf{K} + \{B_i, i \in I\}$. Thus, on the one hand, for every $i \in I$, $A \overset{\mathsf{K}}{\vdash} B_i$. On the other hand, $B_i, i \in I \overset{\mathsf{K}\ \mathsf{K}}{\dashv\vdash} A$, and consequently, there is a finite subset of formulas from $\{B_i, i \in I\}$, say, $B_1, \ldots, B_n$, such that $B_1, \ldots, B_n \overset{\mathsf{K}}{\vdash} A$. It is not hard to see that

$$A \overset{\mathsf{K}\ \mathsf{K}}{\dashv\vdash} \bigwedge_{i=1}^{n} B_i.$$

(c) $\Longrightarrow$ (b). Indeed, if $\boldsymbol{L}_1 \supseteq \boldsymbol{L}$, then $\boldsymbol{L}_1$ is a logic of $\mathsf{K} + \boldsymbol{L}_1^+$: the logics of $\mathsf{K} + \boldsymbol{L}_1^+$ and $\boldsymbol{L}_1$ cannot be distinct, because they have the same positive fragments and by (c) they must coincide.

Remark 2.3 Derivable elimination of negation is not the same as expressibility of negation. For instance, in IPC, $\neg p \dashv\vdash p \rightarrow q$, because $\vdash \neg p \rightarrow (p \rightarrow q)$ and a formula equivalent to $\neg p$ can be derived from $p \rightarrow q$ by substituting $p \wedge \neg p$ for q. At the same time, obviously, $\nvdash \neg p \leftrightarrow (p \rightarrow q)$. Similarly, $p \vee \neg p \dashv\vdash p \vee (p \rightarrow q)$, and Cl can be defined by $\mathsf{IPC} + p \vee (p \rightarrow q)$.

The goal of this section is to prove the following theorem.

Theorem 2.5 *The following holds:*

(a) Yn *is the greatest p-conservative extension of* Int;
(b) Yn *is a minimal logic admitting derivable elimination of negation.*

Corollary 2.17 *Logic* Yn *is a unique s.i. p-conservative extension of* Int *admitting derivable elimination of negation.*

Remark 2.4 Yn is a minimal logic admitting derivable elimination of negation, but it is not the smallest such logic: it was observed in Hosoi and Ono (1970) that all logics of the second slice are axiomatizable by implicative formulas. Hence, the smallest logic of the second slice has derivable elimination of negation. It is not hard to see that this logic is not an extension of Yn.

In Jankov (1968a), Yankov gave a syntactic proof of Theorem 2.5; we offer an alternative, semantic proof, and we start with studying the algebraic semantics of KC.

2.6.1 *Semantics of* KC

Let us start with a simple observation that any s.i. Heyting algebra $\mathbf{A}$ is a model for KC (that is, $\mathbf{A} \models (\neg p \vee \neg\neg p)$ if and only if each distinct from $\mathbf{0}$ element $\mathsf{a} \in \mathbf{A}$ is *dense*; that is, $\neg\neg\mathsf{a} = \mathbf{1}$ (or equivalently, $\neg\mathsf{a} = \mathbf{0}$). Thus, a class of all such algebras forms an adequate semantics for the Yankov logic, and we call these algebras the *Yankovean* algebras.

Let us recall some properties of dense elements that we need in the sequel. Suppose that $\mathbf{A}$ is a Heyting algebra and $\mathsf{a}, \mathsf{b} \in \mathbf{A}$. Then, it is clear that if $\mathsf{a} \leq \mathsf{b}$ and a is a dense element, then b is a dense element: $\mathsf{a} \leq \mathsf{b}$ implies $\neg\mathsf{b} \leq \neg\mathsf{a} = \mathbf{0}$. Moreover, if a and b are dense, so is $\mathsf{a} \wedge \mathsf{b}$: by Glivenko's Theorem $\neg\neg(\mathsf{a} \wedge \mathsf{b}) = \neg(\neg\mathsf{a} \vee \neg\mathsf{b}) = \neg(\mathbf{0} \vee \mathbf{0}) = \mathbf{1}$.

Theorem 2.6 (Jankov 1968a) *The class of all finite Yankovean algebras forms an adequate semantics for* KC.

Remark 2.5 In Jankov (1968a), Yankov offered a syntactic proof. We offer a semantic proof based on an idea used in McKinsey (1941).

Proof It is clear that $\mathbf{0} \oplus \mathbf{A} \models \neg p \vee \neg\neg p$ for all Heyting algebras $\mathbf{A}$, and we need to prove that for any formula A such that $\mathsf{KC} \vdash A$, there is a finite Yankovean algebra $\mathbf{B}$ in which A is refuted.

Suppose that $\mathsf{KC} \nvdash A$. Then, there is a Yankovean algebra $\mathbf{A}$ in which A is refuted. Let ν be a refuting valuation and $A_1, \ldots, A_n, \mathbf{1}$ be all the subformulas of A.

Consider a distributive sublattice $\mathbf{B}$ of $\mathbf{A}$ generated (as sublattice) by elements $\mathbf{0}, \nu(A_1), \ldots, \nu(A_n), \mathbf{1}$. Every finitely generated distributive lattice is finite (cf., e.g.,

Grätzer 2003), and any finite distributive lattice can be regarded as a Heyting algebra. Let us prove that (a) **B** is a Yankovean algebra, and (b) ν is a refuting valuation in **B**.

(a) First, let us note that the meets, the joins, and the partial orders in algebras **A** and **B** are the same. Hence, **B** has a pretop element: the join of all elements from **B** that are distinct from **1**, and therefore, it is an s.i. algebra. In addition, because algebra **A** is Yankovean, all its elements that are distinct from **0** are dense. Hence, as **B** is finite, the meet of all elements from **B** that are distinct from **0** is again a dense element and therefore, it is distinct from **0**. Thus, this meet is the smallest distinct from **0** element of **B** and therefore, all elements that are distinct from **0** are dense and **B** is a Yankovean algebra.

(b) Let us observe that if elements $\mathsf{a}, \mathsf{b} \in \mathbf{B}$, then $\mathsf{a} \wedge \mathsf{b}, \mathsf{a} \vee \mathsf{b} \in \mathbf{B}$, and $\neg\mathsf{a} \in \mathbf{B}$, because $\neg\mathsf{a} = \mathbf{1}$ if $\mathsf{g} = \mathbf{0}$ and $\neg\mathsf{a} = \mathbf{0}$ otherwise. In addition, if $\mathsf{a} \to \mathsf{b} \in \mathbf{B}$ and $\to'$ is an implication defined in **B**, then $\mathsf{a} \to' \mathsf{b} = \mathsf{a} \to \mathsf{b}$: by definition, $\mathsf{a} \to \mathsf{b}$ is the greatest element in $\{\mathsf{c} \in \mathbf{A} \mid \mathsf{a} \wedge \mathsf{c} \leq \mathsf{b}\}$, and because $\mathsf{a} \to \mathsf{b} \in \mathbf{B}$ and **A** and **B** have the same partial order, $\mathsf{a} \to \mathsf{b}$ is the greatest element in $\{\mathsf{c} \in \mathbf{B} \mid \mathsf{a} \wedge \mathsf{c} \leq \mathsf{b}\}$. Thus, because all elements are $\mathbf{0}, \nu(A_1), \ldots, \nu(A_n), \mathbf{1}$, all values of $\nu(A_1), \ldots, \nu(A_n)$ when ν is regarded as a valuation in **B** remain the same and therefore, ν refutes A in **B**.

Given a Heyting algebra **A**, one can adjoin a new bottom element and, in such a way, obtain a new Heyting algebra denoted by $\mathbf{0} \oplus \mathbf{A}$. For instance (cf. Fig. 2.2), $\mathbf{3} = \mathbf{0} \oplus \mathbf{2}$. It is not hard to see that $\mathbf{0} \oplus \mathbf{A}$ is a Yankovean algebra. On the other hand, any finite Yankovean algebra has the form $\mathbf{0} \oplus \mathbf{A}$, where **A** is a finite s.i. Heyting algebra.

Corollary 2.18 *The class of finite Yankovean algebras forms an adequate semantic for* KC.

Let us construct more adequate semantics for KC.

Observe that in any Heyting algebra **A**, the elements $\{\mathbf{0}\} \cup \{\mathsf{a} \in \mathbf{A} \mid \neg\neg\mathsf{a} = \mathbf{1}\}$ form a Heyting subalgebra of **A** denoted by $\mathbf{A}^{(d)}$. It is clear that if **A** is an s.i. algebra, then $\mathbf{A}^{d}$ is Yankovean. In the sequel, we use the following property of $\mathbf{A}^{(d)}$.

Proposition 2.19 *If* $\varphi : \mathbf{A} \longrightarrow \mathbf{B}$ *is a homomorphism of Heyting algebra* **A** *onto Heyting algebra* **B**, *then the restriction* $\widehat{\varphi}$ *of* φ *to* $\mathbf{A}^{(d)}$ *is a homomorphism of* $\mathbf{A}^{(d)}$ *onto* $\mathbf{B}^{(d)}$.

Proof It is clear that $\widehat{\varphi}(\mathbf{A}^{(d)})$ is a subalgebra of **B**. Moreover, because for any element of $\mathbf{A}^{(d)}$ that is distinct from **0**, $\neg\neg\mathsf{a} = \mathbf{1}$, it is clear that $\widehat{\varphi}(\neg\neg\mathsf{a}) = \neg\neg\widehat{\varphi}(\mathsf{a}) = \mathbf{1}$ and hence, $\widehat{\varphi}(\mathsf{a}) \in \mathbf{B}^{(d)}$; that is, $\varphi(\mathbf{A}^{(d)}) \subseteq \mathbf{B}^{(d)}$. Thus, we only need to show that $\widehat{\varphi}$ maps $\mathbf{A}^{(d)}$ onto $\mathbf{B}^{(d)}$.

Indeed, let us show that for any $\mathsf{b} \in \mathbf{B}^{(d)}$, the preimage $\widehat{\varphi}^{-1}(\mathsf{b})$ contains an element from $\mathbf{A}^{(d)}$.

Suppose that $\mathsf{b} \in \mathbf{B}^{(d)}$. If $\mathsf{b} = \mathbf{0}$, then trivially, $\mathbf{0} \in \widehat{\varphi}^{-1}(\mathsf{b})$. If $\mathsf{b} \neq \mathbf{0}$, then by the definition of $\mathbf{B}^{(d)}$, $\neg\neg\mathsf{b} = \mathbf{1}$, that is, $\neg\mathsf{b} = \mathbf{0}$, and consequently $\mathsf{b} \vee \neg\mathsf{b} = \mathsf{b}$. Hence, for any element $\mathsf{a} \in \widehat{\varphi}^{-1}(\mathsf{b})$, $\mathsf{a} \vee \neg\mathsf{a} \in \widehat{\varphi}^{-1}(\mathsf{b})$, and it is not hard to see that $\mathsf{a} \vee \neg\mathsf{a} \in \mathbf{A}^{(d)}$.

Theorem 2.7 *Suppose that Heyting algebras* $\{\mathbf{A}_i, i \in I\}$ *form an adequate semantics for* IPC. *Then algebras* $\{\mathbf{A}^{(d)}, i \in I\}$ *form an adequate semantics for* KC.

Proof It is clear that algebras $\mathbf{A}_i^{(d)}$ are models for KC, and we only need to prove that for any formula A not derivable in KC, there is an algebra $\mathbf{A}_i^{(d)}$ in which A is refuted. We already know that all finite Yankovean algebras form an adequate semantics for KC. Hence, it suffices to show that each Yankovean algebra can be embedded in a homomorphic image of some algebra $\mathbf{A}_i^{(d)}, i \in I$.

Let $\mathbf{B}$ be a finite Yankovean algebra and $X_{\mathbf{B}}$ be its characteristic formula. Then, by (2.2), $\mathbf{B} \not\models X_{\mathbf{B}}$ and consequently, IPC $\nvdash X_{\mathbf{B}}$, because algebras $\mathbf{A}_i$ $i \in I$, form an adequate semantics for IPC. For some $i \in I$, $\mathbf{A}_i \nvdash X_{\mathbf{B}}$ and by Proposition 2.12, $\mathbf{B}$ is embedded in a homomorphic image $\hat{\mathbf{A}}_i$ of algebra $\mathbf{A}_i$. Recall that $\mathbf{B}$ is Yankovean and all its elements that are distinct from $\mathbf{0}$ are dense. Clearly, embedding preserves density and hence, $\mathbf{B}$ is embedded in $\hat{\mathbf{A}}_i^{(d)}$. The observation that by Proposition 2.19 $\hat{\mathbf{A}}_i^{(d)}$ is a homomorphic image of $\mathbf{A}_i^{(d)}$ completes the proof.

Each Heyting algebra $\mathbf{A}$ can be adjoined with a new top element to obtain a new Heyting algebra that is denoted by $\mathbf{A} \oplus \mathbf{1}$. For instance, $\mathbf{3} = \mathbf{2} \oplus \mathbf{1}$, $\mathbf{4} = \mathbf{3} \oplus \mathbf{1}$, and $\mathbf{Z}_5 = \mathbf{2}^2 \oplus \mathbf{1}$ (cf. Fig. 2.2).

The following Heyting algebras are called *Jaśkowski matrices*, and they form an adequate semantics for IPC:

$$\mathbf{J}_0 = \mathbf{2}, \qquad \mathbf{J}_{k+1} = \mathbf{J}^k \oplus \mathbf{1}.$$

Corollary 2.19 *Algebras* $\mathbf{J}_k^{(d)}, k > 0$, *form an adequate semantics for* KC.

2.6.2 KC *from the Splitting Standpoint*

In what follows, algebra $\mathbf{Z}_5$, the Hasse diagram of which is depicted in Fig. 2.4, plays a very important role.

Proposition 2.20 *Suppose that* $\mathbf{A}$ *is a Heyting algebra. Then* $\mathbf{A} \not\models \neg p \vee \neg\neg p$ *if and only if algebra* $\mathbf{Z}_5$ *is a subalgebra of* $\mathbf{A}$.

Proof It should be clear that $\mathbf{Z}_5 \not\models \neg p \vee \neg\neg p$ (consider valuation $\nu(p) = \mathsf{a}$) and hence, any algebra $\mathbf{A}$ containing a subalgebra isomorphic to $\mathbf{Z}_5$ refutes $\neg p \vee \neg\neg p$.

Conversely, suppose that $\mathbf{A} \not\models \neg p \vee \neg\neg p$. Then, for some $\mathsf{a} \in \mathbf{A}$, $\neg\mathsf{a} \vee \neg\neg\mathsf{a} \neq \mathbf{1}$. It is not hard to verify that subset $\mathbf{0}, \neg\mathsf{a}, \neg\neg\mathsf{a}, \neg\mathsf{a} \vee \neg\neg\mathsf{a}, \mathbf{1}$ is closed under fundamental operations and, therefore, forms a subalgebra of $\mathbf{A}$. In addition, because $\neg\mathsf{a} \vee \neg\neg\mathsf{a} \neq \mathbf{1}$, all five elements of this subalgebra are distinct and thus, the subalgebra is isomorphic to $\mathbf{Z}_5$.

Proposition 2.20 entails that formula $\neg p \vee \neg\neg p$ is interderivable in IPC with characteristic formula $X_{\mathbf{Z}_5}$. Indeed, because $\mathbf{Z}_5 \not\models \neg p \vee \neg\neg p$, by Theorem 2.2,

$\mathbf{A} \vdash X_{\mathbf{Z}_5}$. On the other hand, for any algebra $\mathbf{A}$, if $\mathbf{A} \not\models \neg p \vee \neg\neg p$, then by Proposition 2.20, $\mathbf{A}$ has a subalgebra that is isomorphic to $\mathbf{Z}_5$ and by (2.2), $\mathbf{A} \not\models X_{\mathbf{Z}_5}$.

Corollary 2.20 Yn *coincides with* $\overline{\boldsymbol{L}}(\mathbf{Z}_5)$.

Thus Yn is the greatest logic for which $\mathbf{Z}_5$ is not a model.

2.6.3 Proof of Theorem 2.5

Proof of (a). Because Yn coincides with $\overline{\boldsymbol{L}}(\mathbf{Z}_5)$, for any si-logic $\boldsymbol{L}$, either $\boldsymbol{L} \subseteq \overline{\boldsymbol{L}}(\mathbf{Z}_5)$, or $\boldsymbol{L}(\mathbf{Z}_5) \subseteq \boldsymbol{L}$. Thus, to prove that Yn is the greatest p-conservative extension of Int, it suffices to show (i) that $\overline{\boldsymbol{L}}(\mathbf{Z}_5)$ is a p-conservative of extension of Int and (ii) that $\boldsymbol{L}(\mathbf{Z}_5)$ (and hence all it extensions) is not a p-conservative extension of Int.

(i) It is clear that any formula A and hence, any positive formula derivable in Int is derivable in KC. We need to show the converse: if a positive formula A is not derivable in Int, it is not derivable in KC.

Suppose that positive formula A is not derivable in Int. Then, it is refutable in a finite s.i. Heyting algebra $\mathbf{A}$. Consider algebra $\mathbf{0} \oplus \mathbf{A}$. Observe that operations $\wedge$, $\vee$, and $\to$ on elements of algebra $\mathbf{0} \oplus \mathbf{A}$ that are distinct from $\mathbf{0}$ coincide with the respective operations on $\mathbf{A}$. Hence, because A is a positive formula, the valuation refuting A in $\mathbf{A}$, refutes A in $\mathbf{0} \oplus \mathbf{A}$. Algebra $\mathbf{0} \oplus \mathbf{A}$ is Yankovean and, thus, it is a model for KC. Hence, formula A is not derivable in KC.

(ii) To prove lack of conservativity, let us observe that the following formula A,

$$((r \to (p \wedge q)) \wedge (((p \wedge q) \to r) \to r) \wedge ((p \to q) \to q) \wedge ((q \to p) \to p)) \to (p \vee q),$$

is valid in $\mathbf{Z}_5$ but refuted in $\mathbf{0} + \mathbf{Z}_5$ (cf. Fig. 2.4) by valuation $\nu(p) = \mathsf{a}$, $\nu(q) = \mathsf{b}$, $\nu(r) = \mathsf{c}$.

Remark 2.6 It is observed in Jankov (1968a) that formula A used in the proof of (ii) and formula

$$(((p \to q) \to q) \wedge ((q \to p) \to p)) \to (p \vee q)$$

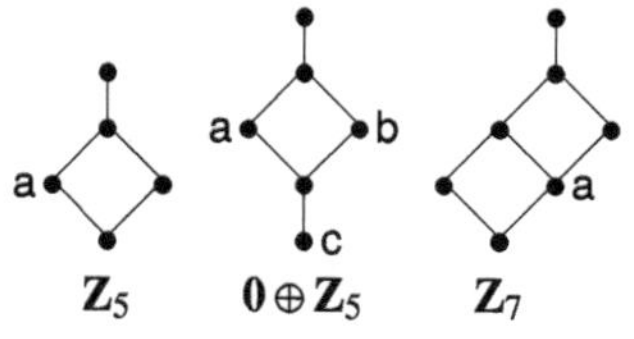

Fig. 2.4 Refuting algebras

are not equivalent in KC, but they are interderivable in KC. That is, even though the logic of KC is a p-conservative extension of Int, the relation of derivability in KC is stronger than that in IPC even for positive formulas.

Proof of (b). To prove (b) we need to show that (i) Yn admits the derivable elimination of negation and that (ii) if $\boldsymbol{L} \subsetneq \mathsf{Yn}$, then $\boldsymbol{L}$ does not admit the derivable elimination of negation. The latter follows immediately from Theorem 2.5 and Proposition 2.18: $\boldsymbol{L}$ and Yn are p-conservative extensions of Int and thus, they have the same positive fragments.

Let $\mathbf{A}$ and $\mathbf{B}$ be Heyting algebras. We say that $\mathbf{B}$ is a *p-subalgebra* of $\mathbf{A}$ when $\mathbf{B}$ is an implicative sublattice of $\mathbf{A}$. Let us note that if $\mathbf{A}$ is a model of some logic $\boldsymbol{L}$ and $\mathbf{B}$ is a p-subalgebra of $\mathbf{A}$, then $\mathbf{B}$ needs not to be a model of $\boldsymbol{L}$. For instance, consider algebras $\mathbf{Z}_5$ and $\mathbf{0} \oplus \mathbf{Z}_5$ from Fig. 2.4: $\mathbf{Z}_5$ is a p-subalgebra of $\mathbf{0} \oplus \mathbf{Z}_5$ (take elements $\mathsf{a}, \mathsf{b}, \mathsf{a} \wedge \mathsf{b}, \mathsf{a} \vee \mathsf{b}, \mathbf{1}$), formula $\neg p \vee \neg\neg p$ is valid in $\mathbf{0} \oplus \mathbf{Z}_5$, while it is not valid in $\mathbf{Z}_5$, because $\neg \mathsf{a} \vee \neg\neg \mathsf{a} < \mathbf{1}$.

Let us demonstrate that every extension $\boldsymbol{L}$ of Yn is positively axiomatizable relative to Yn.

For contradiction, assume that $\mathsf{Yn} \subseteq \boldsymbol{L}$ and that $\boldsymbol{L}$ is not positively axiomatizable relative to KC. Then, the logic $\boldsymbol{L}'$ defined relative to Yn by all positive formulas from $\boldsymbol{L}$ is distinct from $\boldsymbol{L}$; that is, $\mathsf{Yn} \subseteq \boldsymbol{L}' \subsetneq \boldsymbol{L}$. Let $A \in \boldsymbol{L} \setminus \boldsymbol{L}'$. Then, there is an s.i. Heyting algebra $\mathbf{A} \in \mathrm{Mod}(\boldsymbol{L}') \setminus \mathrm{Mod}(\boldsymbol{L})$ in which A is refuted by some valuation ν. We will construct a positive formula $X(\mathbf{A}, A, \nu)$ similar to a characteristic formula, and we will prove the following lemmas.

Lemma 2.3 *Formula* $X(\mathbf{A}, A, \nu)$ *is refuted in* $\mathbf{A}$.

Lemma 2.4 *Formula* $X(\mathbf{A}, A, \nu)$ *is valid in all algebras from* $\mathrm{Mod}(\boldsymbol{L})$.

Indeed, if $X(\mathbf{A}, A, \nu)$ is refuted in $\mathbf{A}$, and $\mathbf{A} \in \mathrm{Mod}(\boldsymbol{L}')$, then $X(\mathbf{A}, A, \nu) \notin \boldsymbol{L}'$.

On the other hand, if formula $X(\mathbf{A}, A, \nu)$ is valid in all algebras from $\mathrm{Mod}(\boldsymbol{L})$, then $X(\mathbf{A}) \in \boldsymbol{L}$. Recall that by definition, $\boldsymbol{L}$ and $\boldsymbol{L}'$ have the same positive formulas, and formula $X(\mathbf{A}, A, \nu)$ is positive. Hence, $X(\mathbf{A}) \in \boldsymbol{L}'$ and we have arrived at a contradiction and completed the proof.

Let us construct formula $X(\mathbf{A}, A, \nu)$.

Suppose $\mathbf{A}$ is a Heyting algebra, A is a formula, and valuation ν refutes A in $\mathbf{A}$. Then, we take A^{ν} to denote a formula obtained from A by substituting every variable q occurring in A with variable $p_{\nu(q)}$. It is clear that valuation $\nu' : p_{\nu(q)} \longrightarrow \nu(q)$ refutes A^{ν}. Let us also observe that because A^{ν} was obtained from A by substitution, A is refuted in every algebra in which A^{ν} is refuted.

If $\mathbf{A}$ is a Heyting algebra, and $\mathsf{B} \subseteq \mathbf{A}$ is a finite set of elements, by $\mathbf{A}^{+}[\mathsf{B}]$ we denote an implicative sublattice of $\mathbf{A}$ generated by elements B. Because B is finite, $\mathbf{A}^{+}[\mathsf{B}]$ contains the smallest element, namely $\bigwedge \mathsf{B}$ and therefore, $\mathbf{A}^{+}[\mathsf{B}]$ forms a Heyting algebra, which is denoted by $\mathbf{A}[\mathsf{B}]$. Note that $\mathbf{A}[\mathsf{B}]$ does not need to be a subalgebra of $\mathbf{A}$, because the bottom element of $\mathbf{A}^{+}[\mathsf{B}]$ may not coincide with the bottom element of $\mathbf{A}$.

Let $\mathbf{A}$ be a Heyting algebra and A be a formula refuted in $\mathbf{A}$ by valuation ν. Suppose that $\{A_1, \ldots, A_n\}$ is a set of all subformulas of A and suppose that $\mathsf{A}_{(A,\nu)} := \{\mathbf{0}, \mathbf{1}\} \cup \{\nu(A_i) \mid i \in [1, n]\}$. Let us observe that $\mathsf{A}_{(A,\nu)}$ contains all elements of $\mathbf{A}$ needed to compute the value of $\nu(A)$. It is not hard to see that $\mathsf{A}_{(A,\nu)} = \mathsf{A}_{(A^\nu,\nu')}$. Clearly, $\mathsf{A}_{(A,\nu)}$ does not need to be closed under fundamental operations, but $\mathbf{A}[\mathsf{A}_{(A,\nu)}]$ is a Heyting algebra, and the value of $\nu(A)$, or the value of $\nu'(A^\nu)$ for that matter, can be computed in the very same way as in $\mathbf{A}$. To simplify notation, we write $\mathbf{A}[A, \nu]$ instead of $\mathbf{A}[\mathsf{A}_{(A,\nu)}]$. Thus, ν refutes A in $\mathbf{A}[A, \nu]$ as long as it refutes A in $\mathbf{A}$.

If $\circ \in \{\wedge, \vee, \rightarrow\}$, by $\mathsf{A}^\circ_{(A,\nu)}$ we denote a set of all ordered pairs of elements of $\mathsf{A}_{(A,\nu)}$ for which $\circ$ is defined:

$$\mathsf{A}^\circ_{(A,\nu)} = \{(\mathsf{a}, \mathsf{b}) \mid \mathsf{a}, \mathsf{b}, \mathsf{a} \circ \mathsf{b} \in \mathsf{A}_{(A,\nu)}\}.$$

Consider formulas

$$D^+(\mathbf{A}, A, \nu) := \Big(\bigwedge_{\circ \in \{\wedge, \vee \rightarrow\}} \ \bigwedge_{(\mathsf{a},\mathsf{b}) \in \mathsf{A}^\circ_{(A,\nu)}} (p_{\mathsf{a}} \circ p_{\mathsf{b}} \leftrightarrow p_{\mathsf{a} \circ \mathsf{b}}) \Big),$$

and

$$X^+(\mathbf{A}, A, \nu) := D^+(\mathbf{A}, A, \nu) \rightarrow \bigvee_{\mathsf{a},\mathsf{b} \in \mathsf{A}_{(A,\nu)}, \mathsf{a} \neq \mathsf{b}} (p_{\mathsf{a}} \leftrightarrow p_{\mathsf{b}}).$$

Proof of Lemma 2.3.

Proof We will show that if $\mathbf{A}$ is an s.i. algebra, then $X(\mathbf{A}, A, \nu)$ is refuted in $\mathbf{A}$ by valuation ν'. Indeed,

$$\nu'(p_{\mathsf{a}} \circ p_{\mathsf{b}}) = \nu'(p_{\mathsf{a}}) \circ \nu'(p_{\mathsf{b}}) = \mathsf{a} \circ \mathsf{b} = \nu'(p_{\mathsf{a} \circ \mathsf{b}})$$

for all $(\mathsf{a}, \mathsf{b}) \in \mathsf{A}^\circ_{(A,\nu)}$ and all $\circ \in \{\wedge \vee, \rightarrow\}$ and therefore, $\nu'(D^+(\mathbf{A}, A, \nu)) = \mathbf{1}$.

On the other hand, $\nu'(p_{\mathsf{a}}) = \mathsf{a} \neq \mathsf{b} = \nu'(p_{\mathsf{b}})$; that is, ν' refutes every disjunct on the right-hand side of $X(\mathbf{A}, A, \nu)$, and therefore, ν' refutes whole disjunction, because $\mathbf{A}$ is s.i. and disjunction of two elements that are distinct of $\mathbf{1}$ is distinct from $\mathbf{1}$. Thus, ν' refutes $X(\mathbf{A}, A, \nu)$.

To prove Lemma 2.4, we will need the following property of $X(\mathbf{A}, A, \nu)$.

Proposition 2.21 *Let $\mathbf{A}$ be a Heyting algebra, and ν be a valuation refuting formula A in $\mathbf{A}$. Suppose that $\mathbf{B}$ is a Heyting algebra and η is a valuation refuting formula $X^+(\mathbf{A}, A, \nu)$ in $\mathbf{B}$ such that*

$$\eta(D^+(\mathbf{A}, A, \nu)) = \mathbf{1}_{\mathbf{B}}.$$

Then, η refutes A^ν in $\mathbf{B}[X^+, \eta]$ and therefore, A is refuted in $\mathbf{B}[X^+, \eta]$.

Proof Indeed, define a map $\xi : \mathsf{A}_{(A,\nu)} \longrightarrow \mathsf{B}_{(X^+,\eta)}$ by letting $\xi(\mathsf{a}) = \eta(p_a)$. Let $\bar{\mathsf{a}} = \eta(p_a)$ for every $\mathsf{a} \in \mathsf{A}_{(A,\nu)}$.

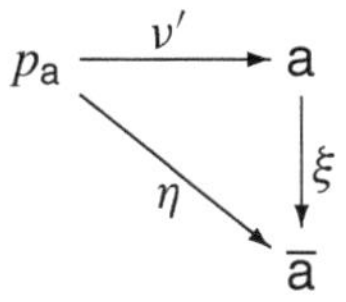

First, let us observe that $\eta(p_\mathbf{1}) = \mathbf{1_B}$. Indeed, by definition, $\mathbf{1} \in \mathbf{A}_{(A,\nu)}$, and $(\mathbf{1}, \mathbf{1}) \in \mathsf{A}^\circ_{(A,\nu)}$, because $\mathbf{1} \to \mathbf{1} = \mathbf{1} \in \mathsf{A}_{(A,\nu)}$. Hence, $(p_\mathbf{1} \to p_\mathbf{1}) \leftrightarrow p_\mathbf{1}$ is one of the conjuncts in $D^+(\mathbf{A}, A, \nu)$ and because $\eta(D^+(\mathbf{A}, A, \nu)) = \mathbf{1_B}$, we have $\eta((p_\mathbf{1} \to p_\mathbf{1}) \leftrightarrow p_\mathbf{1}) = \mathbf{1_B}$; that is, $\eta(p_\mathbf{1}) \to \eta(p_\mathbf{1}) = \eta(p_\mathbf{1})$ and $\eta(p_\mathbf{1}) = \mathbf{1_B}$.

Next, we observe that if $\mathsf{a}, \mathsf{b} \in \mathsf{A}_{(A,\nu)}$ and $\mathsf{a} \neq \mathsf{b}$, then $\bar{\mathsf{a}} \neq \bar{\mathsf{b}}$. Indeed, because η refutes $X^+(\mathbf{A}, A, \nu)$ and $\eta(D^+(\mathbf{A}, A, \nu)) = \mathbf{1_B}$,

$$\eta(\bigvee_{\mathsf{a},\mathsf{b} \in \mathsf{A}_{(A,\nu)}, \mathsf{a} \neq \mathsf{b}} (p_\mathsf{a} \leftrightarrow p_\mathsf{b})) \neq \mathbf{1_B},$$

and in particular, $\eta(p_\mathsf{a} \leftrightarrow p_\mathsf{b}) \neq \mathbf{1_B}$. Thus, $\eta(p_\mathsf{a}) \neq \eta(p_\mathsf{b})$; that is, $\bar{\mathsf{a}} \neq \bar{\mathsf{b}}$.

Lastly, we observe that if B, C and $B \circ C$ are subformulas of A and $\nu(B) = \mathsf{b}$ and $\nu(C) = \mathsf{c}$, then $\mathsf{b}, \mathsf{c}, \mathsf{b} \circ \mathsf{c} \in \mathsf{A}_{(A,\nu)}$. Moreover, $(\mathsf{b}, \mathsf{c}) \in \mathsf{A}^\circ_{(A,\nu)}$ and consequently, $(p_\mathsf{b} \circ p_\mathsf{c}) \leftrightarrow p_{(\mathsf{b} \circ \mathsf{c})}$ is one of the conjuncts of $D^+(\mathbf{A}, A, \nu)$ and by assumption,

$$\eta((p_\mathsf{b} \circ p_\mathsf{c}) \leftrightarrow p_{(\mathsf{b} \circ \mathsf{c})}) = \mathbf{1_B}.$$

Thus, $\eta(p_\mathsf{b}) \circ \eta(p_\mathsf{c}) = \eta(p_{\mathsf{b} \circ \mathsf{c}})$ and therefore, if $\nu(A) = \mathsf{a}$, then,

$$\eta(A^\nu) = \eta(p_\mathsf{a}) = \bar{\mathsf{a}}.$$

Recall that ν' refutes A^ν; that is $\nu'(A^\nu) \neq \mathbf{1}$ and hence, $\eta(A^\nu) \neq \eta(\mathbf{1}) = \mathbf{1_B}$.

Now, we can prove Lemma 2.4 and complete the proof of the theorem.

Proof of Lemma 2.4

Proof For contradiction, assume that formula $X(\mathbf{A}, A, \nu)$ is refuted in Heyting algebra $\mathbf{B} \in \mathrm{Mod}(\boldsymbol{L})$. By Corollary 2.2, we can assume that $\mathbf{B}$ is s.i. and that the refuting valuation η is such that

$$\eta(D^+(\mathbf{A}, A, \nu)) = \mathbf{1} \text{ and } \eta(\bigvee_{\mathsf{a},\mathsf{b} \in \mathsf{A}_{(A,\nu)}, \mathsf{a} \neq \mathsf{b}} (p_\mathsf{a} \leftrightarrow p_\mathsf{b})) \neq \mathbf{1_B}.$$

Thus, the condition of Proposition 2.21 is satisfied and hence, A is refuted in $\mathbf{B}[X^+, \eta]$. If we show that $\mathbf{B}[X^+, \eta] \in \mathrm{Mod}(\boldsymbol{L})$, we will arrive at a contradiction, because A was selected from $\boldsymbol{L} \setminus \boldsymbol{L}'$ and thus, A is valid in all models of $\boldsymbol{L}$.

Indeed, because $\mathbf{A} \in \mathrm{Mod}(\boldsymbol{L})$ and $\mathsf{KC} \subseteq \boldsymbol{L}$, $\mathbf{A}$ is a Yankovean algebra. Hence, for each element $\mathsf{a} \in \mathbf{A}$ that is distinct from $\mathbf{0}$, $\mathsf{a} \to \mathbf{0} = \mathbf{0}$ and $\mathbf{0} \to \mathsf{a} = \mathbf{1}$. Therefore, for each $\mathsf{a} \in \mathsf{A}_{(A,\nu)}$ that is distinct from $\mathbf{0}$, $\mathsf{a} \to \mathbf{0} = \mathbf{0}$ and $\mathbf{0} \to \mathsf{a} = \mathbf{1}$. Hence, $(p_{\mathsf{a}} \to p_{\mathbf{0}}) \leftrightarrow p_{\mathbf{0}}$ and $(p_{\mathbf{0}} \to p_{\mathsf{a}}) \leftrightarrow p_{\mathbf{1}}$ are conjuncts of $D^+(\mathbf{A}, A, \nu)$. By assumption, $\eta(D^+(\mathbf{A}, A, \nu)) = \mathbf{1}$ and subsequently, $\eta(p_{\mathsf{a}}) \to \eta(p_{\mathbf{0}}) = \eta(p_{\mathbf{0}})$ and $\eta(p_{\mathbf{0}}) \to \eta(p_{\mathsf{a}}) = \eta(p_{\mathbf{1}})$. The latter means that $\eta(p_{\mathbf{0}})$ is the smallest element of $\mathbf{B}[X^+, \eta]$. Let $\mathsf{c} = \eta(p_{\mathbf{0}})$. Then, for each distinct from c element $\mathsf{b} \in \mathbf{B}[X^+, \eta]$, $\mathsf{b} \wedge \mathsf{c} = \mathsf{c}$, $\mathsf{b} \vee \mathsf{c} = \mathsf{b}$, $\mathsf{b} \to \mathsf{c} = \mathsf{c}$ and $\mathsf{c} \to \mathsf{b} = \mathbf{1}$.

Recall that $\mathbf{B}$ is also a Yankovean algebra and therefore, for each element $\mathsf{b} \in \mathbf{B}$ that is distinct from $\mathbf{0}$, $\mathsf{b} \to \mathbf{0_B} = \mathbf{0_B}$ and $\mathbf{0_B} \to \mathsf{b} = \mathbf{1_B}$. Hence, the set of all elements of $\mathbf{B}[X^+, \eta]$ that are distinct from c together with $\mathbf{0_B}$ is closed under all fundamental operations and hence, it forms a subalgebra of $\mathbf{B}$. It is not hard to see that this subalgebra is isomorphic to $\mathbf{B}^+[X^+, \eta]$, and this entails that $\mathbf{B}^+[X^+, \eta]$ is isomorphic to a subalgebra of a model of $\boldsymbol{L}$ and therefore, $\mathbf{B}^+[X^+, \eta]$ is a model of $\mathrm{Mod}(\boldsymbol{L})$.

Remark 2.7 Formula $X(\mathbf{A}, A, \nu)$ is a characteristic formula of partial Heyting algebra. The reader can find more details about characteristic formulas of partial algebras in Tomaszewski (2003) and Citkin (2013).

2.7 Some Si-Calculi

Let us consider the following calculi.

(a) $\mathsf{CPC} = \mathsf{IPC} + (\neg\neg p \to p)$—the classical propositional calculus;

(b) $\mathsf{KC} = \mathsf{IPC} + (\neg p \vee \neg\neg p)$—the calculus of the weak law of excluded middle;

(c) $\mathsf{BD}_2 = \mathsf{IPC} + ((\neg\neg p \wedge (p \to q) \wedge ((q \to p) \to p)) \to q)$;

(d) $\mathsf{SmC} = \mathsf{IPC} + (\neg p \vee \neg\neg p) + ((\neg\neg p \wedge (p \to q) \wedge ((q \to p) \to p)) \to q)$.

Let us consider the algebras, whose Hasse diagrams are depicted in Fig. 2.2 and the following series of C-algebras defined inductively:

$$\mathbf{B}_0 = \mathbf{2}, \qquad \mathbf{B}_{k+1} = \mathbf{2}^k \oplus \mathbf{1};$$
$$\mathbf{J}_0 = \mathbf{2}, \qquad \mathbf{J}_{k+1} = \mathbf{J}^k \oplus \mathbf{1}.$$

Algebras $\mathbf{J}_k$ are referred to as Jaśkowski matrices. They were considered by S. Jaśkowski (cf. Jaśkowski 1975) and they form an adequate algebraic semantics for Int in the following sense:

$$\mathsf{Int} \vdash A \iff \mathbf{J}_k \models A \text{ for all } k > 0.$$

In Jankov (1963a), it was observed that algebras $\mathbf{J}_k^{(d)}$, $k > 1$, form an adequate semantics for KC:

$$\mathsf{KC} \vdash A \iff \mathbf{J}_k^{(d)} \models A \text{ for all } k > 0.$$

If C_1 and C_2 are two C-calculi, we write $C_1 = C_2$ to denote that C_1 and C_2 define the same logic. For instance, $\mathsf{Int} + (\neg\neg p \to p) = \mathsf{Int} + (p \vee \neg p)$, because both calculi define Cl.

As usual, if $\mathcal{A}$ is a class of C-algebras and C is a class of C-formulas, $\mathcal{A} \models C$ means that all formulas from C are valid in each algebra from $\mathcal{A}$, and $\mathcal{A} \not\models C$ denotes that at least one formula from C is invalid in some algebra from $\mathcal{A}$.

Theorem 2.8 (Jankov 1963a, Theorem 1) *Suppose that C is a set of formulas in the signature $\to, \wedge, \vee, \neg$. Then the following hold:*

(a) $\mathsf{IPC} + C = \boldsymbol{L}(\mathbf{2})$ *if and only if* $\mathbf{2} \models C$ *and* $\mathbf{3} \not\models C$*;*
(b) $\mathsf{IPC} + C = \boldsymbol{L}(\{\mathbf{J}_k^{(d)}, k \geq 0\})$ *if and only if* $\{\mathbf{J}_k^{(d)}, k \geq 0\} \models C$ *and* $\mathbf{Z}_5 \not\models C$*;*
(c) $\mathsf{IPC} + C = \boldsymbol{L}(\{\mathbf{B}_k, k \geq 0\})$ *if and only if* $\{\mathbf{B}_k, k \geq 0\} \models C$ *and* $\mathbf{4} \not\models C$*;*
(d) $\mathsf{IPC} + C = \boldsymbol{L}(\mathbf{3})$ *if and only if* $\mathbf{3} \models C$ *and* $\mathbf{4} \not\models C$ *and* $\mathbf{Z}_5 \not\models C$*.*

Proof In terms of splitting, we need to prove the following:

(a') $\boldsymbol{L}(\mathbf{2}) = \overline{\boldsymbol{L}}(\mathbf{3})$;
(b') $\boldsymbol{L}(\{\mathbf{J}_k^{(d)}, k \geq 0\}) = \overline{\boldsymbol{L}}(\mathbf{Z}_5)$;
(c') $\boldsymbol{L}(\{\mathbf{B}_k, k \geq 0\}) = \overline{\boldsymbol{L}}(\mathbf{4})$;
(d') $\boldsymbol{L}(\mathbf{3}) = \overline{\boldsymbol{L}}(\mathbf{4}, \mathbf{Z}_5))$.

(a') is trivial: the only s.i. Heyting algebra that does not contain $\mathbf{3}$ as a subalgebra is $\mathbf{2}$.

(b') was proven as Corollary 2.19.

(c') It is not hard to see that neither algebra $\mathbf{B}_k$ nor its homomorphic images or subalgebras contain a four-element chain subalgebra. On the other hand, if $\mathbf{B}$ is a finitely generated s.i. Heyting algebra such that $\mathbf{4}$ is not its subalgebra, then $\mathbf{B} \cong \mathbf{B}' \oplus \mathbf{1}$. Elements $\mathbf{0}$, $\mathbf{1}$, and the pretop element form a three-element chain algebra; hence, $\mathbf{B}'$ contains at most a two-element chain algebra and by (a'), $\mathbf{B}'$ is a Boolean algebra. Clearly, $\mathbf{B}'$ is a homomorphic image of $\mathbf{B}$ and hence, $\mathbf{B}'$ is finitely generated. Every finitely generated Boolean algebra is finite and therefore, $\mathbf{B}$ is isomorphic to one of the algebras $\mathbf{B}_k$.

(d') Let $\mathbf{A}$ be a finitely generated s.i. Heyting algebra that has no subalgebras isomorphic to $\mathbf{Z}_5$ and $\mathbf{4}$. Then by (b'), $\mathbf{A}$ is an s.i. Yankovean algebra and hence, $\mathbf{A} \cong \mathbf{0} \oplus \mathbf{A}' \oplus \mathbf{1}$. By (c'), $\mathbf{0} \oplus \mathbf{A}'$ is a Boolean algebra and therefore, $\mathbf{0} \oplus \mathbf{A}' \cong \mathbf{2}$. Thus, $\mathbf{A} \cong \mathbf{2} \oplus \mathbf{1} \cong \mathbf{3}$.

The above theorem can be rephrased as follows.

Theorem 2.9 (Jankov 1963a, Theorem 3) *Suppose that C is a set of formulas in the signature $\to, \wedge, \vee, \neg$. Then the following hold:*

(a) $\mathsf{IPC} + C = \mathsf{CPC}$ *if and only if* $\mathbf{2} \models C$ *and* $\mathbf{3} \not\models C$*;*
(b) $\mathsf{IPC} + C = \mathsf{KC}$ *if and only if* $\{\mathbf{J}_k^{(d)}, k \geq 0\} \models C$ *and* $\mathbf{Z}_5 \not\models C$*;*

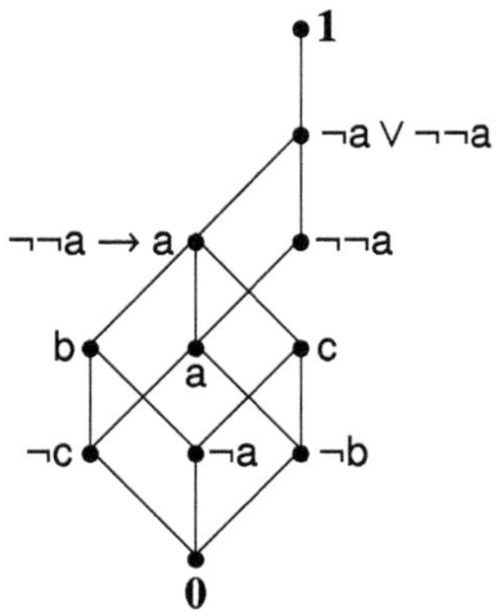

Fig. 2.5 Algebra refuting A

(c) $\mathsf{IPC} + C = \mathsf{BD}_2$ *if and only if* $\{\mathbf{B}_k, k \geq 0\} \models C$ *and* $\mathbf{4} \not\models C$;
(d) $\mathsf{IPC} + C = \mathsf{Sm}$ *if and only if* $\mathbf{3} \models C$ *and* $\mathbf{4} \not\models C$ *and* $\mathbf{Z}_5 \not\models C$.

Remark 2.8 It is noted in Jankov (1963a) that logic Sm can be defined by $\mathsf{IPC} + ((p \to q) \vee (q \to r) \vee (r \to s))$.

2.8 Realizable Formulas

In 1945, S. Kleene introduced a notion of realizability of intuitionistic formulas (cf. Kleene 1952). Formula A is said to be realizable when there is an algorithm that by each substitution of logical-arithmetical formulas gives a realization of the result. It was observed in Nelson (1947) that all formulas derivable in IPC are realizable; moreover, all formulas derivable in IPC from realizable formulas are realizable, while many classically valid formulas, $\neg\neg p \to p$, for instance, are not realizable. This observation gave the hope that the semantics of realizability is adequate for IPC. It turned out that this is not the case: in Rose (1953), it was proven that formula

$$C = ((\neg\neg A \to A) \to (\neg A \vee \neg\neg A)) \to (\neg A \vee \neg\neg A), \text{ where } A = (\neg p \vee \neg q)$$

is realizable not derivable in IPC. Indeed, formula C is refutable in the Heyting algebra whose Hasse diagram is depicted in Fig. 2.5 by substitution $\nu : p \mapsto \mathsf{b}$, $\nu : q \mapsto \mathsf{c}$, which entails $\nu : A \mapsto \mathsf{a}$.

In Jankov (1963b), Yankov constructed the following sequences of formulas: for each $n \geq 3$ and $i \in [1, n]$, let $\pi_n^i := \neg p_1 \wedge \cdots \wedge \neg p_{i-1} \wedge \neg p_{i+1} \wedge \cdots \wedge \neg p_n$ and

$$A_n := \bigwedge_{1 \leq k < m \leq n} \neg(p_k \wedge p_m) \wedge \bigwedge_{i=1}^{n-1} (\pi_{n-1}^i \to (p_i \vee p_n)) \to (p_n \vee \neg p_n)$$

and

$$B_n := \bigwedge_{1\leq i<j\leq n} \neg(p_i \wedge p_j) \wedge \bigwedge_{i=1}^{n-1}(\pi_{n-1}^{i} \to (p_i \vee p_j)) \to \bigvee_{i=1}^{n} p_i.$$

In addition, let

$$\rho := ((\neg\neg p \to (p \vee \neg p)) \wedge ((\neg\neg q \to q) \to (q \vee \neg q)) \wedge \neg(p \wedge q)) \to (p \vee \neg q).$$

Theorem 2.10 (Jankov 1963b, Theorem 1) *The following hold:*

(a) formulas A_3 and ρ are realizable and cannot be derived in IPC *from each other; thus, they are not derivable in* IPC*;*
(b) in IPC*, $A_3 \vdash C$ and $C \nvdash A_3$;*
(c) for any $n \geq 3$, formulas A_n and B_n are not derivable in IPC*; nevertheless, $A_3 \vdash A_n$ and $A_n \vdash B_n$ and hence, formulas A_n and B_n are realizable.*

Theorem 2.11 (Jankov 1963b, Theorem 2) *Every realizable formula is valid in algebra $\mathbf{Z}_7$ whose Hasse diagram is depicted in Fig. 2.4.*

Let us observe that formula $C' := ((\neg\neg p \to p) \to (\neg p \vee \neg\neg p)) \to (\neg p \vee \neg\neg p)$ (the skeleton of the Rose formula) is refuted in algebra $\mathbf{Z}_7$ by valuation $\nu : p \mapsto \mathsf{a}$. Hence, Theorem 2.11 entails that the skeleton $C' := ((\neg\neg p \to p) \to (\neg p \vee \neg\neg p)) \to (\neg p \vee \neg\neg p)$ of the Rose formula is not realizable.

On the other hand, formula C' is interderivable in Int with the characteristic formula $X_{\mathbf{Z}_7}$ of algebra $\mathbf{Z}_7$. Hence, if C' is not realizable, all realizable formulas are valid in $\mathbf{Z}_7$. Indeed, assume for contradiction that A is a realizable formula and is invalid in $\mathbf{Z}_7$. Then, by Theorem 2.2, in IPC, $A \vdash C'$ and therefore, C' should be realizable.

More information on the realizability of propositional formulas can be found in Plisko (2009).

2.9 Some Properties of Positive Logic

If A and B are positive formulas, let $A \leq B \overset{\text{def}}{\iff} \vdash A \to B$ in PPC. A set of formulas is *independent* if any two distinct formulas of this set are incomparable relative to $\leq$. In Jankov (1968d), Yankov constructed three infinite sequences of positive formulas on two variables: (a) independent, (b) strongly descending, and (c) strongly ascending. For the duration of this section, $\vdash$ means derivability in PPC (unless otherwise indicated).

2.9.1 Infinite Sequence of Independent Formulas

Consider the following sequence of positive formulas (cf. Jankov 1968d):

$$A_1 := p, \quad B_1 := q, \quad A_{k+1} := B_k \vee (B_k \to A_k), \quad B_{k+1} := A_k \vee (A_k \to B_k). \tag{2.12}$$

Let

$$C_k := (((A_k \to B_k) \to B_k) \wedge ((B_k \to A_k) \to A_k)) \to (A_k \vee B_k). \tag{2.13}$$

In the proofs, we use algebras $\mathbf{A}_i$, the Hasse diagrams of which are depicted in Fig. 2.6, and we use valuation

$$\nu : p \mapsto \mathsf{a}_1 \quad \nu : q \mapsto \mathsf{b}_1. \tag{2.14}$$

Let us observe that for all $1 \leq k \leq i$,

$$\nu(A_k) = \mathsf{a}_k, \quad \nu(B_k) = \mathsf{b}_k, \quad \text{and} \quad \nu(C_i) = \mathsf{c}_i.$$

Proposition 2.22 *Formulas $C_k, k > 0$, are independent in* PPC*; that is, for any $i \neq j$, $C_i \nvdash C_j$ and $C_j \nvdash C_i$.*

Proof First, let us observe that $\nvdash C_j \to C_i$ for any $i > j$, because valuation ν defined by (2.14) refutes formula $C_j \to C_i$.

Next, let us show that $\nvdash C_j \to C_i$ for any $i < j$. To this end, we will show that $\vdash (C_j \to C_i) \leftrightarrow C_i$ and consequently, $\nvdash (C_j \to C_i)$, because $\nvdash C_i$: it is refuted in $\mathbf{A}_i$ by valuation ν.

Let us prove $\vdash (C_j \to C_i) \leftrightarrow C_i$. By definition,

$$C_j \to C_i = C_j \to ((((A_i \to B_i) \to B_i) \wedge ((B_i \to A_i) \to A_i)) \to (A_i \vee B_i)),$$

and hence,

$$\vdash (C_j \to C_i) \leftrightarrow ((C_j \wedge ((A_i \to B_i) \to B_i) \wedge ((B_i \to A_i) \to A_i)) \to (A_i \vee B_i)). \tag{2.15}$$

Thus, by showing that

$$\vdash (((A_i \to B_i) \to B_i) \wedge ((B_i \to A_i) \to A_i)) \to C_j, \tag{2.16}$$

we will prove that (2.15) yields

$$\vdash (C_j \to C_i) \leftrightarrow ((((A_i \to B_i) \to B_i) \wedge ((B_i \to A_i) \to A_i)) \to (A_i \vee B_i)),$$

that is, that $\vdash (C_j \to C_i) \leftrightarrow C_i$.

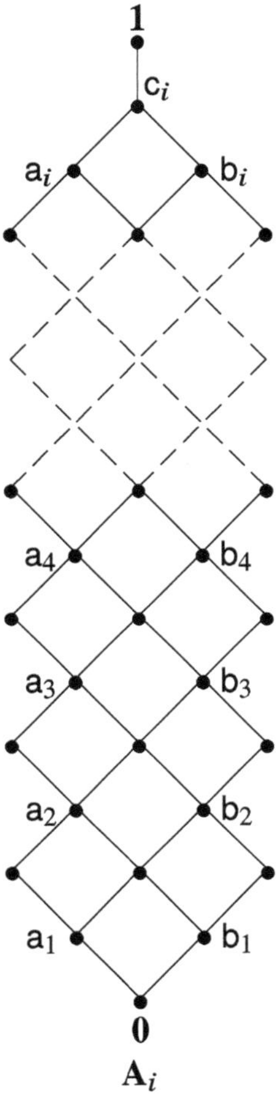

Fig. 2.6 Refuting algebra

To prove (2.16), we consider two cases: (a) $j = i + 1$ and (b) $j \geq i + 2$.

Case (a). Recall that $B_i \vdash A_i \to B_i$ and hence,

$$B_i,\ (A_i \to B_i) \to B_i,\ (B_i \to A_i) \to A_i \vdash (A_i \vee (A_i \to B_i)). \tag{2.17}$$

In addition, $B_i \to A_i,\ (B_i \to A_i) \to A_i \vdash A_i$ and hence,

$$B_i \to A_i,\ (A_i \to B_i) \to B_i,\ (B_i \to A_i) \to A_i \vdash (A_i \vee (A_i \to B_i)). \tag{2.18}$$

From (2.17) and (2.18),

$$B_i \vee (B_i \to A_i),\ (A_i \to B_i) \to B_i,\ (B_i \to A_i) \to A_i \ \vdash\ (A_i \vee (A_i \to B_i)), \tag{2.19}$$

and by the Deduction Theorem,

$$(A_i \to B_i) \to B_i,\ (B_i \to A_i) \to A_i \ \vdash\ (B_i \vee (B_i \to A_i)) \to (A_i \vee (A_i \to B_i)); \tag{2.20}$$

that is,

$$(A_i \to B_i) \to B_i,\ (B_i \to A_i) \to A_i \ \vdash\ A_{i+1} \to B_{i+1}. \tag{2.21}$$

Immediately from (2.21),

$$(A_i \to B_i) \to B_i,\ (B_i \to A_i) \to A_i \ \vdash\ ((A_{i+1} \to B_{i+1}) \to B_{i+1}) \to B_{i+1} \tag{2.22}$$

and consequently,

$$(A_i \to B_i) \to B_i,\ (B_i \to A_i) \to A_i \ \vdash\ ((A_{i+1} \to B_{i+1}) \to B_{i+1}) \to (A_{i+1} \vee B_{i+1}). \tag{2.23}$$

Hence,

$$\begin{gathered}(((A_i \to B_i) \to B_i) \wedge ((B_i \to A_i) \to A_i)) \ \vdash \\ ((((A_{i+1} \to B_{i+1}) \to B_{i+1}) \wedge ((B_{i+1} \to A_{i+1}) \to A_{i+1})) \to (A_{i+1} \vee B_{i+1}));\end{gathered} \tag{2.24}$$

that is,

$$(((A_i \to B_i) \to B_i) \wedge ((B_i \to A_i) \to A_i)) \vdash C_{i+1}. \tag{2.25}$$

Case (b). From (2.21),

$$(((A_i \to B_i) \to B_i) \wedge ((B_i \to A_i) \to A_i)) \vdash A_{i+1} \vee (A_{i+1} \to B_{i+1});$$

that is,

$$(((A_i \to B_i) \to B_i) \wedge ((B_i \to A_i) \to A_i)) \vdash B_{i+2}. \tag{2.26}$$

Using the definition of formulas A_j and B_j, by simple induction one can show that for any $j \geq i+2$,

$$(((A_i \to B_i) \to B_i) \wedge ((B_i \to A_i) \to A_i)) \vdash B_j \tag{2.27}$$

and subsequently,

$$\begin{gathered}(((A_i \to B_i) \to B_i) \wedge ((B_i \to A_i) \to A_i)) \ \vdash \\ (((A_j \to B_j) \to B_j) \wedge ((B_j \to A_j) \to A_j)) \to (A_j \vee B_j).\end{gathered}$$

Hence, by the definition of formula C_j,

$$(((A_i \to B_i) \to B_i) \wedge ((B_i \to A_i) \to A_i)) \vdash C_j.$$

2.9.2 Strongly Descending Infinite Sequence of Formulas

Consider the following sequence of formulas: for each $k > 0$,

$$D_k := \bigwedge_{i \leq k} C_i.$$

Let us prove that formulas D_i form a strongly descending (relative to $\leq$) sequence; that is, for any $0 < j < i$, $\vdash D_i \to D_j$, while $\nvdash D_j \to D_i$.

Proposition 2.23 *Formulas $D_k, k = 1, 2, \ldots$ form a strongly descending sequence.*

Proof Let $0 < j < i$. Then $\vdash D_i \to D_j$ trivially follows from the definition of D_k, and we only need to show that $\nvdash D_j \to D_i$.

Indeed, by the definition of formula D_j,

$$\vdash (D_j \to D_i) \leftrightarrow \bigwedge_{k \leq i}(D_j \to C_k),$$

and we will demonstrate that $\nvdash D_j \to C_i$ by showing that formula $D_i \to C_j$ is refuted in algebra $\mathbf{A}_i$ whose diagram is depicted in Fig. 2.6.

Let us consider valuation $\nu : p \mapsto \mathsf{a}$ and $\nu : q \mapsto \mathsf{b}$. Then, for all $j < i$, $\nu(C_j) = \mathbf{1}$ and hence, $\nu(\delta_j) = \mathbf{1}$, while $\nu(C_i) = \mathsf{c}_i < \mathbf{1}$ and hence, valuation ν refutes formula $D_j \to C_i$.

2.9.3 Strongly Ascending Infinite Sequence of Formulas

To construct a strongly ascending sequence of formulas of positive logic, one can use an observation from Wajsberg (1931) that a formula A is derivable in IPC if and only if a formula A^+ obtained from A by replacing any subformula of the form $\neg B$ with formula $B \to (p_1 \wedge \cdots \wedge p_n \wedge p_{n+1})$, where $p_1, \ldots, p_n$ is a list of all variables occurring in A, is derivable in PPC. Thus, if one takes any sequence $A_1, A_2, A_3, \ldots$ that is strongly ascending in IPC , the sequence $A_1^+, A_2^+, A_3^+, \ldots$ is strongly ascending in PPC. In particular, one can take sequence of formulas that are a strongly ascending in IPC on one variable constructed in Nishimura (1960) and obtain a desired sequence of formulas that is strongly ascending in PPC on two variables.

A proof that Int $\vdash A$ if and only if PPC $\vdash A^+$ can be done by simple induction. It appears that Yankov was not familiar with Wajsberg (1977) and his proof in Jankov (1968d) uses the same argument as the proof from Wajsberg (1931).

Let us also observe that the sequence $A_1, A_2, \ldots$ defined by (2.12) is strongly ascending. Indeed, it is clear that $\vdash A_k \rightarrow (B_k \vee (B_k \rightarrow A_k))$; that is , $\vdash A_k \rightarrow A_{k+1}$. On the other hand, $\nvdash A_{k+1} \rightarrow A_k$: the valuation ν refutes this formula.

By the Separation Theorem, all three sequences remain, respectively, independent, strongly ascending and strongly descending in IPC. Moreover, because IPC and KC have the same sets of derivable positive formulas, these three sequences retain their properties. And, if we replace q with $\neg(p \rightarrow p)$, we obtain three sequences of formulas that are independent, strongly descending, and strongly ascending in MPC.

2.10 Conclusions

In conclusion, let us point out that Yankov's results in intermediate logics not only changed the views on the lattice of intermediate logics but also instigated further research in this area. In 1971, in Kuznetsov (1971), it was observed that for any intermediate logic $\boldsymbol{L}$ distinct from Int, the segment [Int, $\boldsymbol{L}$] contains a continuum of logics. In the same year, using notion of a pre-true formula, which is a generalization of the notion of characteristic formula, Kuznetsov and Gerčiu presented a finitely axiomatizable intermediate logic without the f.m.p. (Kuznetsov and Gerčiu 1970). Using ideas from Jankov (1968b), Wroński proved that there are continuum many intermediate logics enjoying the disjunction property, among which are the logics lacking the f.m.p. (cf. Wroński 1973).

In Fine (1974), Fine introduced—for modal logics—formulas similar to Yankov's formulas, and he constructed a strongly ascending chain of logics extending S4.

In his Ph.D. Blok (1976), Blok linked the characteristic formulas with splitting, and studied the lattice of varieties of interior algebras. This line of research was continued by Routenberg in Rautenberg (1977, 1980), and his disciples (cf. Wolter 1993; Kracht 1999). Ever since, the splitting technique pioneered by Yankov is one of the main tools in the research of different classes on logics.

References

Birkhoff, G. (1948). *Lattice theory* (Vol. 25, revised ed.). American Mathematical Society Colloquium Publications. New York: American Mathematical Society.

Blok, W. (1976). *Varieties of interior algebras*. Ph.D thesis, University of Amsterdam.

Citkin, A. (2013). Characteristic formulas of partial Heyting algebras. *Logica Universalis, 7*(2), 167–193. https://doi.org/10.1007/s11787-012-0048-7.

Fine, K. (1974). An incomplete logic containing S4. *Theoria, 40*(1), 23–29.

Galatos, N., Jipsen, P., Kowalski, T., & Ono, H. (2007). *Residuated lattices: An algebraic glimpse at substructural logics*, Studies in logic and the foundations of mathematics (Vol. 151). Amsterdam: Elsevier B. V.

Gödel, K. (1932). Zum intuitionistischen Aussagekalkül. *Anzeiger Wien, 69*, 65–66.

Grätzer, G. (2003). *General lattice theory*. Basel: Birkhäuser.

Heyting, A. (1930). Die formalen Regeln der intuitionistischen Logik. I, II, III. *Sitzungsber Preuß Akad Wiss, Phys-Math Kl* 42–56, 57–71, 158–169.
Heyting, A. (1941). Untersuchungen über intuitionistische Algebra. *Verhdl. Akad. Wet. Amsterdam, Afd. Natuurk. I., Section, 18*, Nr. 2, 36 S.
Heyting, A. (1998). The formal rules of intuitionistic logic. In P. Mancocu (Ed.), *From Brouwer to Hilbert* (pp. 311–327). Oxford University Press.
Hosoi, T. (1967). On intermediate logics. I. *Journal of the Faculty of Science, University of Tokyo. Section I, 14*, 293–312.
Hosoi, T., & Ono, H. (1970). The intermediate logics on the second slice. *Journal of the Faculty of Science, University of Tokyo: Mathematics. Section IA, 17*, 457–461.
Idziak, K., & Idziak, P. (1988). Decidability problem for finite Heyting algebras. *The Journal of Symbolic Logic, 53*(3), 729–735. https://doi.org/10.2307/2274568.
Jankov, V. A. (1963a). On certain superconstructive propositional calculi. *Dokl Akad Nauk SSSR, 151*, 796–798. English translation in *Soviet Mathematics Doklady, 4*, 1103–1105.
Jankov, V. A. (1963b). On the realizable formulae of propositional logic. *Dokl Akad Nauk SSSR, 151*, 1035–1037. English translation in *Soviet Mathematics Doklady, 4*, 1146–1148.
Jankov, V. A. (1963c). On the relation between deducibility in intuitionistic propositional calculus and finite implicative structures. *Dokl Akad Nauk SSSR, 151*, 1293–1294. English translation in *Soviet Mathematics Doklady, 4*, 1203–1204.
Jankov, V. A. (1968a). Calculus of the weak law of the excluded middle. *Izv Akad Nauk SSSR Ser Mat, 32*, 1044–1051. English translation in *Mathematics of the USSR-Izvestiya, 2*(5), 997–1004.
Jankov, V. A. (1968b). The construction of a sequence of strongly independent superintuitionistic propositional calculi. *Dokl Akad Nauk SSSR, 181*, 33–34. English translation in *Soviet Mathematics Doklady, 9*, 806–807.
Jankov, V. A. (1968c). On an extension of the intuitionistic propositional calculus to the classical one and of the minimal one to the intuitionistic one. *Izv Akad Nauk SSSR Ser Mat, 32*, 208–211. English translation in *Mathematics of the USSR-Izvestiya, 2*(1), 205–208.
Jankov, V. A. (1968d). Three sequences of formulas with two variables in positive propositional logic. *Izv Akad Nauk SSSR Ser Mat, 32*, 880–883. English translation in *Mathematics of the USSR-Izvestiya, 2*(4), 845–848.
Jankov, V. A. (1969). Conjunctively irresolvable formulae in propositional calculi. *Izv Akad Nauk SSSR Ser Mat, 33*, 18–38. English translation in *Mathematics of the USSR-Izvestiya, 3*(1), 17–35.
Jaśkowski, S. (1936). Recherches sur le système de la logique intuitioniste. *Actual Science Industry, 393*, 58–61.
Jaśkowski, S. (1975). Investigations into the system of intuitionist logic. *Studia Logica, 34*(2), 117–120. Translated from the French by S. McCall (Actes Congr. Internat. Philos. Sci. (Paris, 1935), fasc. no. 6, pp. 58–61, Hermann, Paris, 1936), Stanisław Jaśkowski's achievements in mathematical logic (Proc. 20th Conf. History of Logic, Cracow, 1974).
de Jongh, D. (1968). *Investigations on intuitionistic propositional calculus*. Ph.D thesis, University of Wisconsin.
Kleene, S. C. (1952). *Introduction to metamathematics*. New York: D. Van Nostrand.
Köhler, P. (1981). Brouwerian semilattices. *Transactions of the American Mathematical Society, 268*(1), 103–126. https://doi.org/10.2307/1998339.
Kracht, M. (1999). *Tools and techniques in modal logic*, Studies in logic and the foundations of mathematics (Vol. 142). Amsterdam: North-Holland Publishing.
Kreisel, G., & Putnam, H. (1957). Eine Unableitbarkeitsbeweismethode für den intuitionistischen Aussagenkalkul. *Arch Math Logik Grundlagenforsch, 3*, 74–78.
Kuznetsov, A. V. (1963). Undecidability of the general problems of completeness, solvability and equivalence for propositional calculi. *Algebra i Logika Sem, 2*(4), 47–66.
Kuznetsov, A. V. (1971). Some properties of the lattice of varieties of pseudo-Boolean algebras. In: *XI All-Union Algebraic Colloquium. Abstracts* (pp. 255–256) (in Russian).
Kuznetsov, A. V., & Gerčiu, V. J. (1970). The superintuitionistic logics and finite approximability. *Dokl Akad Nauk SSSR, 195*, 1029–1032. (in Russian).

Łoś, J. (1949). On logical matrices. *Trav Soc Sci Lett Wrocław Ser B, 19*, 42.
McKinsey, J. C. C. (1941). A solution of the decision problem for the Lewis systems S2 and S4, with an application to topology. *The Journal of Symbolic Logic, 6*, 117–134.
Nelson, D. (1947). Recursive functions and intuitionistic number theory. *Transactions of the American Mathematical Society, 61*, 307–368. https://doi.org/10.2307/1990222.
Nishimura, I. (1962). On formulas of one variable in intuitionistic propositional calculus. *The Journal of Symbolic Logic, 25*, 327–331. https://doi.org/10.2307/2963526.
Odintsov, S. P. (2008). *Constructive negations and paraconsistency*, Trends in logic—studia logica library (Vol. 26). New York: Springer. https://doi.org/10.1007/978-1-4020-6867-6.
Plisko, V. (2009). A survey of propositional realizability logic. *Bull Symbolic Logic, 15*(1), 1–42. https://doi.org/10.2178/bsl/1231081768.
Rasiowa, H. (1974). *An algebraic approach to non-classical logics*, Studies in logic and the foundations of mathematics (Vol. 78). Amsterdam, New York: North-Holland Publishing, American Elsevier Publishing.
Rasiowa, H., & Sikorski, R. (1963). *The mathematics of metamathematics. Monografie Matematyczne, Tom* (Vol. 41). Warsaw: Państwowe Wydawnictwo Naukowe.
Rautenberg, W. (1977). The lattice of normal modal logics (Preliminary report). *The Bulletin of the Section of Logic, Polish Academy of Sciences, 6*, 193–201.
Rautenberg, W. (1980). Splitting lattices of logics. *Arch Math Logik Grundlag, 20*(3–4), 155–159. https://doi.org/10.1007/BF02021134.
Rieger, L. (1949). On the lattice theory of Brouwerian propositional logic. *Acta Fac Nat Univ Carol, Prague, 189*, 40.
Rieger, L. (1957). A remark on the so-called free closure algebras. *Czechoslovak Mathematical Journal, 7*(82), 16–20.
Rose, G. F. (1953). Propositional calculus and realizability. *Transactions of the American Mathematical Society, 75*, 1–19.
Tomaszewski, E. (2003). *On sufficiently rich sets of formulas*. Ph.D thesis, Institute of Philosophy, JagellonianUniversity, Krakov.
Troelstra, A. S. (1965). On intermediate propositional logics. *Nederl Akad Wetensch Proc Ser A 68=Indag Math, 27*, 141–152.
Umezawa, T. (1959). On intermediate many-valued logics. *The Journal of the Mathematical Society of Japan, 11*, 116–128.
Wajsberg, M. (1931). Untersuchungen über den Aussagenkalkül von A. *Heyting. Wiadom Mat, 46*, 45–101 (Translated in Wajsberg (1977)).
Wajsberg, M. (1977). On A. Heyting's propositional calculus. In S. Surma (Ed.), *Logical works* (pp. 132–171), Wydawnictwo Polskiej Akademii Nauk.
Wolter, F. (1993). *Lattices of modal logics*. Ph.D thesis, Freien Universität Berlin.
Wroński, A. (1973). Intermediate logics and the disjunction property. *Reports on Mathematical Logic, 1*, 39–51.

Chapter 3
Dialogues and Proofs; Yankov's Contribution to Proof Theory

Andrzej Indrzejczak

Abstract In the 1990s Yankov published two papers containing important contributions to proof theory and based on the application of a dialogical interpretation of proofs. In both cases the method is used for providing constructive proofs of important metalogical results concerning classical logic and fundamental mathematical theories. In the first paper it is shown that impredicative extensions of intuitionistic versions of arithmetic, analysis and set theory, enriched with suitable bar induction schemata, are sufficiently strong for proving the consistency of their classical counterparts. In the second paper the same method is applied to provide a constructive proof of the completeness theorem for classical logic. In both cases a version of a one-sided sequent calculus in Schütte-style is used and cut elimination is established in the second case. Although the obtained results are important, and the applied method is original and interesting, they have not received the attention they deserve from the wider community of researchers in proof theory. In this paper we briefly recall the content of both papers. We focus on essential features of Yankov's approach and provide comparisons with other results of similar character.

Keywords Proof theory · Dialogical interpretation of proofs · Consistency proofs · Bar induction · Arithmetic · Analysis · Set theory

3.1 Introduction

In the early 1990s, shortly after returning from internal exile in Siberia, Yankov published two papers containing important contributions to proof theory. More precisely the first of them Yankov (1994) presents constructive consistency proofs of classical arithmetic, analysis and set theory. The second one Yankov (1997) just shows the application of the general method developed in the first paper to obtain a constructive proof of the completeness of first order logic FOL. It is evident that both works were

A. Indrzejczak (✉)
Department of Logic, University of Łódź, Łódź, Poland
e-mail: andrzej.indrzejczak@filozof.uni.lodz.pl

A. Citkin and I. M. Vandoulakis (eds.), *V. A. Yankov on Non-Classical Logics, History and Philosophy of Mathematics*, Outstanding Contributions to Logic 24,
https://doi.org/10.1007/978-3-031-06843-0_3

prepared in extremely difficult times of his political imprisonment where he had no access to vast literature in ordinal proof theory. It explains why both papers have limited references. Only Spector's well known result on the consistency of analysis (Spector (1962)) is mentioned in addition to Shoenfield's textbook (Shoenfield (1967)) and some of Kreisel's papers on proof theory. Even Gentzen's works are not mentioned directly although it is evident that they are known to Yankov and provide some general background to his approach. What is sad is the fact that the reception of Yankov's work was sparse. There are no references to it in the works of leading experts in proof theory. In this paper we try to briefly characterise the results he obtained and put them in the wider context of proof-theoretic work which was conducted elsewhere.

In the next section we briefly recall the main trends and results obtained in proof theory after Gentzen's seminal works on the consistency of arithmetic. Then we characterise the main features of Yankov's approach as developed in these two papers. In particular, in Sect. 3.4 we present sequent calculi applied by Yankov and compare them with other ways of formalising mathematical theories. In Sect. 3.5 we describe his dialogue theory of proofs and compare it with other dialogical/game based approaches. Since Yankov's proof of consistency is based neither on cut elimination nor on any Gentzen-like reduction procedures but on direct justification of cut applications, it requires Brouwer's bar induction principle. In Sect. 3.6 bar induction will be generally explained and its specific forms introduced by Yankov briefly characterised. We close the presentation of his approach in Sect. 3.7 with a sketch of his proofs. The last section provides a summary and evaluation of his result in comparison to other approaches.

3.2 Consistency Proofs

One of the most important results in metamathematics which are known (at least by name) even to laymen are Gödel's two incompleteness theorems. In common opinion they have shown that an ambitious version of Hilbert's program is not realizable, at least in the original form. Gödel himself believed that his results do not destroy Hilbert's program decisively but some other outstanding scholars, like von Neumann, were convinced that the expectations for establishing a safe basis for arithmetic were buried. The strong tradition of proof theoretic research which soon started with Gentzen is a good evidence that Gödel's, more balanced, view was closer to truth. In fact, the result on the embeddability of classical arithmetic in intuitionistic arithmetic which was proved independently by Gödel and Gentzen may be seen as a first kind of consistency result for classical arithmetic in the sense that its consistency problem was reduced to the problem of the consistency of intuitionistic theory.

Gentzen recognized that the main problem with Hilbert's program was with its unclear explanation of the notion of finitist method[1] and started to investigate what kind of principles are sufficiently strong to provide a consistency proof for arithmetic, yet may be treated as sufficiently safe. He provided four proofs of consistency of suitable formalizations of Peano arithmetic PA. In fact, the first of them (unpublished after Bernays' criticism.) was based on the application of intuitionistically acceptable principles, namely the fan theorem which is a counterpart of König's lemma and a corollary of the bar induction theorem, which is the basis of Yankov's approach. Later Gentzen changed the strategy and based a proof of consistency on the principle of transfinite induction up to ϵ_0.

It is well known that Gentzen (1934) introduced special kinds of calculi enabling better analysis of proofs—natural deduction and sequent calculi. Frege and Hilbert-style axiomatic systems were formally correct and easy to describe but the construction of proofs was heavy and artificial, giving no insight as to how to discover a concrete proof and far from mathematical practice. Natural deduction was, as its name suggests, constructed as a formal system which corresponds to the actual practice of proof construction, in particular it enables an introduction of temporary assumptions for conditional and indirect proofs[2] instead of formalising the derivation of theses from axioms. Moreover, Gentzen was convinced that natural deduction calculi allow for better analysis of the structure of proofs, in particular by their transformation into special normal form. Sequent calculi were at first treated as a kind of technical device but soon started independent life in the realm of proof theory. The problem for Gentzen was that he was able to obtain normalizability of proofs only for intuitionistic logic whereas in sequent calculus the corresponding result, called by him Hauptsatz (cut elimination theorem), was established for classical and intuitionistic logic in an uniform way. Normal form theorems for classical and other logics were established in the 1960s by Prawitz (1965) (and independently by Raggio 1965) whereas Gentzen's original proof for intuitionism was rediscovered only recently by von Plato (2008). In the meantime sequent calculus became more and more popular due to the works of such logicians as Ketonen, Curry, Kleene, to mention just a few names. The earliest two proofs of the consistency of PA constructed by Gentzen (1936a) (and galley proofs added to the English translation of this paper in Szabo 1969) were established for a calculus which was a hybrid of natural deduction and sequent calculus (sequent natural deduction[3]). The third Gentzen (1938) (and fourth Gentzen 1943) one was provided for the sequent calculus version of PA.

Although it may be debatable whether Gentzen provided a properly finitist proof of the consistency of PA it is evident that he provided an important result which was also an improvement of Gödel's result. In fact, many mathematicians neglected Gödel's result claiming that his kind of unprovable statement is artificial and not

[1] Nowadays the view of Tait is often accepted according to which finitism is identified with the quantifier-free system of primitive recursive arithmetic PRA.

[2] In fact, a similar system of this kind independently introduced by Jaśkowski (1934) was called 'suppositional calculus'.

[3] More on the history and types of natural deduction and sequent calculi is in Indrzejczak (2010).

a 'real' arithmetic true sentence. Gentzen's principle of transfinite induction up to ϵ_0 is an example of a 'real' true and nonprovable arithmetic sentence. Moreover, he has shown in Gentzen (1943) that it is the best possible since his fourth proof shows that weaker principles (for ordinals below ϵ_0) are provable in his system. This way he opened the way for the foundation of ordinal proof theory developed by Takeuti, Schütte and his followers. It should be understood that what he obtained was not a trivial game where the ordinary induction principle is proved by means of stronger transfinite induction up to ϵ_0. His application of transfinite induction is restricted to recursive predicates; only so it has a constructive character in contrast to the nonrestricted induction principle which is a primitive rule of the system PA which is shown to be consistent.

Yet Gentzen believed that this was only the beginning and tried to prove the consistency of analysis.[4] His tragic death closed his research but other logicians continued his investigation.

Gentzen had a clear picture of three levels of infinitary involvement connected with arithmetic, analysis and set theory respectively. In his lecture (Gentzen (1936c)) elementary number theory is characterised as dealing only with the infinity of natural numbers which may be interpreted constructively in terms of potential infinity. Analysis introduces a higher level of commitments to infinity since we must deal with infinite sets of numbers. In case of set theory an unlimited freedom is allowed in dealing with infinity in the sense that it deals with infinite sets of infinite sets of any further order.

Gentzen's quest for a proof of the consistency of mathematics was continued in many ways and with the application of several ingenious techniques. We briefly point out only some chosen achievements, namely those which—in our opinion—may serve as the most direct points of reference to Yankov's results. More extensive information may be found in several textbooks and surveys, in particular: Takeuti (1987),[5] Pohlers (2009), Schwichtenberg (1977), Rathjen and Sieg (2018). One of the important features of Gentzen's original consistency proofs is the remarkable fact that despite differences all are based on some reduction steps rather than on full unrestricted cut elimination which yields consistency in a trivial way like in the case of pure classical and intuitionistic logic. In particular, in the third proof (the second one published in Gentzen 1938) no full cut elimination for a suitable sequent calculus is obtained due to the additional induction rule. Instead we have a reduction of proofs having some ordinal assigned, to proofs having lesser ordinals assigned. It opens the way for searching cut-free systems where consistency will be obtained as a simple corollary, similarly like in the case of pure logic, and even arithmetic without induction.[6]

---

[4] In the meantime he proved in a purely finitist way the consistency of simple type theory (Gentzen 1936b) but without the axiom of infinity so it is a rather weak result.

[5] The second edition is preferred since it contains also valuable appendices written by leading researchers in the field.

[6] Such a result was already present in Gentzen (1934).

Soon it appeared that the problems may be solved by extending the notion of a calculus in such a way that infinity is introduced in the realm of syntactic proofs. Independently, Novikov (1943) and Schütte (1977) showed how to avoid the difficulty with cut elimination in infinitary systems although the former work was rather unnoticed. In Schütte's approach a so called semiformal system with the ω-rule is introduced instead of a special rule of induction. It has the effect that cut elimination is a trivial extension of the ordinary proof but the cost is that the calculus is infinitary, i.e. proofs are well-founded trees with a denumerable number of branches. Additionally, Schütte introduced a simplified version of sequent calculus—the one-sided calculus which became very popular in proof theoretic investigations. Sequents are just collections (usually finite lists) of formulae and the number of rules is reduced. Such kind of calculus was also applied by Yankov and will be described below in detail. In fact, the level of infinity introduced to syntactical calculi by Schütte was earlier signalled by Hilbert and in fact even more economical infinitary systems are possible if we introduce also infinitary sequents. This was the solution proposed a little earlier by Novikov (1943) but wider known later due to Tait (1968). In such kind of calculi further simplifications follow since quantifiers are reduced to infinite conjunctions/disjunctions.

Different ways (avoiding the introduction of infinity to the realm of syntactic systems) of searching for cut elimination in enriched calculi were also investigated. Takeuti formulated a conjecture that cut is eliminable in the sequent calculi for full second-order logic Z2 (and logics of higher degrees) which, as is well known, is sufficient for the formulation of analysis. The troubles with Takeuti's conjecture was that rules for introduction of second-order quantifiers ($\forall$ into the antecedent or $\exists$ into the succedent) lack the subformula property since terms instantiated for second-order variables may be of any complexity, which destroys standard ways of proving this result. Proofs confirming Takeuti's conjecture were offered independently by Tait, Takahashi, Prawitz and Girard (see Rathjen 2018 for details). The problem with all these proofs is that they are not purely syntactical and highly non-constructive. Cut elimination is indirectly shown be means of applying the semantical resources due to Schütte. Problems with the lack of subformula property may be resolved by means of some stratification of formulae in the spirit of the ramified hierarchy of sets. In this way we enter into the field of predicative analysis.

In fact it is not necessary to be a conservative follower of Gentzen's syntactic approach to obtain similar results. In the meantime Spector (1962) proved the consistency of analysis with no use of sequent calculi or natural deduction but on the basis of Gödel's Dialectica interpretation and with the help of Brouwer's bar induction suitably generalised. It seems that his proof strongly influenced Yankov, especially concerning the role played by bar induction. We will return to this question in Sect. 3.6.

However, most of the outcomes concerning consistency results were developed in the spirit of Schütte's approach by the Munich school (Buchholz, Jäger, Pohlers) where several subsystems of analysis and set theory were investigated. In fact a development of so called ordinal proof analysis is strongly based on Gentzen's result (Gentzen 1943) where it is shown that in his sequent calculus for PA all induction principles for ordinals $< \epsilon_0$ are provable. In contemporary proof theory this approach was developed into a separate field of research which classifies systems by assigning to them the least ordinals such that suitable induction principles are not provable in them. The level of sophistication of these investigations is very high and was summarised by Rathjen (1999) as "tending to be at the limit of human tolerance".

This sort of analysis was also successfully applied to set theory, among others by Jäger, Arai, Rathjen. It is remarkable that most of the works on the consistency of set theory is concerned with different versions of Kripke-Platek theory KP not with ZF as in Yankov's paper. In KP the power set axiom is omitted and the axiom of separation and collection are restricted. Nonetheless this weak axiomatization is sufficient for the most of important applications of set theory and, moreover, it is amenable to ordinal analysis which makes it a favourite system for proof theoretic investigations. This branch of proof theory is sometimes called admissible proof theory since Barwise's admissible sets provide natural models for it.

Let me conclude this section with some remarks concerning the eliminability of axioms and cut as the main technique for obtaining the consistency of mathematical theories. It is known that cut is not fully eliminable in SC with axioms. Of course if axioms are atomic this is not a serious problem. Many approaches to arithmetic resolve the problem of axioms of PA in such a way but the principle of induction is not amenable to such treatment. Gentzen provided different forms of rules corresponding to the induction axiom (see below Sect. 3.4) but they did not allow for extending his proof of cut elimination. Negri and von Plato (2001) developed techniques for changing some types of axioms (like e.g. universal geometric formulae) into one-sided rules which allows for extending standard cut elimination proofs. But induction cannot be formalised in such a way. Of course any axiomatic sequent may be transformed into rules representing several schemata (see e.g. Indrzejczak 2018) but not necessarily fitting our purposes, in particular yielding rules which are suitable for proving cut elimination. Omega rules allow for elimination of induction principles but at the cost of introducing semi-formal (infinitary) calculi. Yankov's proof shows a different route.

I'm convinced that this short and very incomplete survey[7] is necessary to put Yankov's proposal in the proper context, to show that his ambitious but not recognized project belongs to a field of very active research developed before its publication and still actively continued, unfortunately without noting his interesting contribution.

3.3 Yankov's Approach

Both papers discussed below represent Yankov's conception of the so called 'dialogue' approach to solving metalogical problems. In Yankov (1994) it is used to prove the consistency of classical theories of Peano arithmetic, analysis and ZF set

[7] In particular research concerning bounded arithmetic (Parikh, Buss), reverse mathematics or predicative analysis were not mentioned since they are not related to Yankov's work.

theory, whereas in Yankov (1997) the completeness of FOL is demonstrated. In both cases proofs are carried out in a constructive way by means of the notion of the defensibility of a formula (sequent). This notion is introduced in the context of a dialogue (game) between two subjects called an opponent and a proponent. Each formula (sequent) is an object of discussion in which an opponent is trying to refute it and a proponent is trying to provide a defence. A formula (sequent) is defensible if for every opponent's attempt a proponent is able to find a true literal which is its subformula.

In case of consistency proofs the general strategy is simple. It is shown that the axioms of the theories under consideration are defensible and that every rule is defensibility-preserving from premisses to conclusions. Since no false formula is defensible it follows that the respective theories are consistent. The apparent simplicity of these proofs rests on the remarkable complexity of their concrete steps which involve the application of special versions of bar induction principles and arithmetization of the opponent's strategies, required for application of bar induction. Theories are formalised as one-sided sequent calculi with added axiomatic sequents. Proving defensibility-preservation for all rules except cut is relatively simple. Although Yankov in (1994) is sometimes referring to part of his proof as cut elimination it cannot be treated as such, at least in the standard sense of this word. The result is shown not by cut elimination but directly by showing that cut is also defensibility-preserving. It is this case which needs very complex argument and in particular bar induction in a variety of special forms must be applied to cover the cases of the quantifiers.

In the case of the completeness of FOL, soundness is just a corrolary of the consistency of arithmetic. In order to prove that FOL is complete again the dialogue interpretation is applied to a root-first proof search procedure. But since completeness is proved by means of constructive resources the usual strategy is not applicable. Thus instead of an indirect proof by construction of a model on the basis of an open infinite branch a direct proof is provided.

There are three main components of Yankov's method that are applied in both papers: (a) a (one-sided) cut-free sequent calculus used for the formalization of the respective theories; (b) the specific dialogical framework used for proving the defensibility of provable sequents, and (c) the application of specific bar induction principles required for the demonstration of the preservation of defensibility by cut. In the following sections we will briefly comment on all of them and sketch the proofs.

3.4 The Calculus

Yankov applies the one-sided sequent calculus invented by Novikov and Schütte and popularised by Tait and researchers from the Munich School (Buchholz, Pohlers) although his formalization of arithmetic is different. As we already mentioned this approach is usually based on the application of infinitary rules (so called semiformal systems), whereas Yankov's formalization is finitary and uses only additional axioms. The standard language of FOL is used with denumerably many variables and denumerably many predicate letters of any arity with constants: $\neg, \wedge, \vee, \forall, \exists$, but with formulae in negation normal form, i.e. with negations applied only to atomic formulae. Thus every occurrence of $\neg\varphi$ with φ nonatomic is to be treated as an abbreviation according to the rules: $\neg\neg\varphi := \varphi$, $\neg(\varphi \wedge \psi) := \neg\varphi \vee \neg\psi$, $\neg(\varphi \vee \psi) := \neg\varphi \wedge \neg\psi$, $\neg\forall x\varphi := \exists x\neg\varphi$, $\neg\exists x\varphi := \forall x\neg\varphi$. In the case of the languages for the theories considered only predicate and function symbols for recursive relations are permitted (no specification is provided which ones are preferred) so it is assumed that the law of excluded middle holds for atomic formulae. Also in case of analysis two sorts of variables are introduced, for numbers and functions from N to N (i.e. infinite sequences).

It is worth underlining that in the course of proof Yankov applies Smullyan's trick to the effect that free variables are substituted consecutively with the objects from the domain of the constructed model so in fact he is working only with closed formulae of a special kind. What objects are substituted for free variables depends on the theory under consideration. In case of PA these are natural numbers, for ZF these are sets. In analysis there are two domains for each sort of variables: numbers and infinite sequences of numbers.

Sequents are finite lists of formulae. Restriction to negative forms allows for a very economic calculus consisting of the following rules:

$$(W)\ \frac{\Gamma}{\Gamma, \varphi} \qquad (C)\ \frac{\Gamma, \varphi, \varphi}{\Gamma, \varphi} \qquad (P)\ \frac{\Gamma, \varphi, \psi, \Delta}{\Gamma, \psi, \varphi, \Delta} \qquad (Cut)\ \frac{\Gamma, \varphi \quad \Delta, \neg\varphi}{\Gamma, \Delta}$$

$$(\vee 1)\ \frac{\Gamma, \varphi}{\Gamma, \varphi \vee \psi} \qquad (\vee 2)\ \frac{\Gamma, \psi}{\Gamma, \varphi \vee \psi} \qquad (\wedge)\ \frac{\Gamma, \varphi \quad \Gamma, \psi}{\Gamma, \varphi \wedge \psi} \qquad (\forall)\ \frac{\Gamma, \varphi}{\Gamma, \forall x\varphi}\ (\exists)\ \frac{\Gamma, \varphi[x/t]}{\Gamma, \exists x\varphi}$$

with the usual eigenvariable condition for x being fresh in ($\forall$). In ($\exists$) t is an arbitrary term of the respective theory.

In the system for FOL the only axioms are of the form: $\varphi, \neg\varphi$, where φ is atomic. Such a formalization is well known to be adequate but Yankov provided a different kind of proof which we briefly describe in Sect. 3.7. In case of arithmetic, and other systems, this system must be enriched with axioms or rules. In general Yankov provides only axiomatic additions keeping the set of rules fixed. Moreover, in case of the theories considered the only logical axiom $\varphi, \neg\varphi$ is formulated without restriction to atomic formulae.

In case of PA additional axioms (or rules) are needed for identity, definitions of recursive operations and for induction. Yankov did not introduce any special set of axioms except for the induction axiom. In fact, if we just postulate that every true identity $t = s$ or its negation with closed terms is accepted as an axiom we do not even need special rules/axioms for identity since they are provable (see e.g. Mendelson 1964). Even logical axioms are provable so Yankov could safely dispense with them. Following Negri's and von Plato's approach (2001) we could as well express all such elementary nonlogical axioms by means of rules enabling cut elimination. In the framework of one-sided sequents they take the following form:

$$(true)\ \frac{\Gamma, t \neq s}{\Gamma}\ \text{for any true } t = s \qquad (false)\ \frac{\Gamma, t = s}{\Gamma}\ \text{for any false } t = s$$

But this is not necessary since for Yankov's proof of consistency cut elimination is not required. The only substantial axiom express the induction principle and has the following expected form:

$\neg\varphi(0) \vee \exists x(\varphi(x) \wedge \neg\varphi(sx)) \vee \forall x\varphi(x)$

Below for comparison's sake we recall two forms of Gentzen's rules (but again in one-sided sequent variation, not in his original version) and ω-rule of Schütte:

$$(Ind1)\ \frac{\Gamma, \varphi(0) \qquad \Gamma, \neg\varphi(a), \varphi(sa)}{\Gamma, \forall x\varphi} \qquad (Ind2)\ \frac{\Gamma, \neg\varphi(a), \varphi(sa)}{\Gamma, \neg\varphi(0), \forall x\varphi}$$

$$(\omega)\ \frac{\Gamma, \varphi(0) \qquad \Gamma, \varphi(1) \qquad \ldots}{\Gamma, \forall x\varphi}$$

with the usual eigenvariable condition of a being fresh variable (parameter). $(Ind1)$ was used for the first and second consistency proof based on natural deduction system in sequent form, whereas $(Ind2)$ was applied in the third consistency proof in the context of sequent calculus. One can easily show that the system with Yankov's axiom is equivalent to the one with any of the Gentzen's rules. In case of (ω) it can be shown that the induction principle is deducible but due to its infinitary character it cannot be derived in ordinary formalization of PA.

In case of analysis the language changes into a two-sorted one; in addition to ordinary individual variables denoting natural numbers we have variables denoting functions from N to N. Accordingly the rules for quantifiers are doubled. In fact we obtain a restricted form of a second-order system. The axiom of induction is preserved, additionally a principle of choice is added:

$\neg\forall x\exists y\varphi(x, y) \vee \exists f\forall x\varphi(x, f(x))$ where f is a functional variable.

The statement of ZF is more complicated since sequent counterparts of set-theoretic axioms of ZF are needed. We omit their formulation to save space.

3.5 The Dialogue Method

In all proofs Yankov consequently applies a method which he calls dialogue theory, or dialogue interpretation of proofs in respective systems. There are close affinities between Yankov's method and the dialogical method of Lorenzen (1961) as well as other approaches which are based on some kind of game involved in searching for a proof or a model (see e.g. Hintikka 1968). In fact, initially any attempt to compare mathematical activity to games was seen as odd within the intuitionistic approach. Hilbert was happy with such a comparison and considered it to be well-suited to the formalist stance in the philosophy of mathematics, but for Brouwer any such analogy was rather offensive.

Soon this approach found its way into the constructivist framework due to Lorenzen. Despite the similarities there are serious differences between the Lorenzen-style dialogical approach and Yankov's dialogues. It is common that in this kind of constructivist framework the roles of two game-players are assigned according to the kind of formula that is investigated. Thus, for example, we can have an $\forall$-player who is responsible for a choice in the case of a conjunctive or universally quantified formula, and an $\exists$-player who deals with disjunctions and existential formulae. The roles of both players may change during the game in the sense that any of them may either attack or defence some chosen formula at a given stage of the game.

There is an interesting paper of Coquand (1990) where such a Lorenzen-style approach is applied to Novikov's system (1943) and his notion of regular formulae. The latter may be seen as a sort of intuitionistic truth definition for classical infinitary one-sided sequent calculus. The main result in Novikov's paper is the original proof of admissibility of a special form of cut (or rather modus ponens). In Coquand's approach the concept of regularity obtains a natural reading in terms of a debate between two players. In this way a game-theoretic semantics of evidence for classical arithmetic is built, and a proof of admissibility of cut is obtained, which yields the consistency of the system. Coquand is unaware of the work of Yankov and despite applying some game-theoretic solution provides an approach which is significantly different. The strategies of players are characterised similarly to the Lorenzen approach, and the proof is based on the elimination of cut. Moreover, only classical arithmetic is considered although some remarks on a possible extension to analysis are formulated in the conclusion.

In Yankov's approach the strategies of both players are defined for all kinds of formulae but in a slightly different way. There is an opponent whose goal is to refute a formula and a proponent who tries to defend it. Moreover, an opponent is responsible for defining an (supposedly falsifying) interpretation, and she is not limited in any way in her choices. On the other hand, a proponent is only a proponent of the formula which is under consideration, but she is working with a model provided by the opponent. Moreover, a proponent is a constructive subject; her strategies are effectively computable functions of the opponent's strategies. In particular, it is assumed that she is capable of marking the steps of calculation with some system of ordinals of sufficiently constructive character. In PA, integers are sufficient. In analysis this system must allow for the enumeration of any collections of natural numbers and functions. In ZF these are simply the ordinals of the set-theoretic domain.

More formally, the opponent's tactics OT is an ordered pair $\langle g, f\rangle$ with g being an opponent's interpretation (OI), and f an opponent refutation tactics (ORT). OI depends on the theory which is being investigated. Thus in the case of arithmetic (but also first-order logic) it is simply an assignment of natural numbers to variables and extensions to predicates defined on numbers, so that all literals are definitely determined as true or false. In the case of analysis and ZF OI is determined accordingly on numbers/infinite sequences and on sets.

ORT is defined uniformly in all cases by induction on the length of a formula:

1. for a literal, a choice of any object from the domain of OI (natural number for analysis);
2. for $\varphi \wedge \psi$, a choice of φ or ψ and an ORT for the chosen subformula;
3. for $\varphi \vee \psi$, a pair consisting of an ORT for φ and for ψ;
4. for $\forall x\varphi$, a choice of an object o from the domain, and an ORT for $\varphi[x/o]$;
5. for $\exists x\varphi$, a function assigning to each object o an ORT for $\varphi[x/o]$.

Of course in analysis we have a division of cases for the quantifiers due to the two sorts of variables.

The proponent tactics PTs for some formula φ are constructive functionals on all possible OTs for φ with values being natural numbers for PA, their sequences for analysis and sets for ZF. Of course, to do this an arithmetization of OT in PA and ZF must be performed, in particular in ZF sequences of ordinals are assigned (for analysis it is not necessary).

Intuitively, every OT determines a list of subformulae of φ which are supposed to be refutable and then PT for this OT should determine a literal from this list which is true; the value of OT on this PT is just a number that is assigned to this literal (see item 1 in the definition of ORT). We omit a detailed description. A PT for φ is called a defence of φ if for each OT for φ a designated literal which is a subformula of φ which is true. Such a formula is called defensible. These notions are extended in a natural way to sequents. Let a sequent S be $\varphi_1, \ldots, \varphi_n$, then an OT for S is $\langle\langle g, f_1\rangle \ldots \langle g, f_n\rangle\rangle$ where g is some fixed OI and f_i is an ORT for φ_i. A PT for such a sequent is a computable function whose arguments are all possible OTs for S and whose values are pairs of natural numbers with the first item denoting a number of a formula in S and the second item denoting the designated number of its chosen subformula that is a literal. If PT for a sequent S designates a true literal that is a subformula of some φ_i, then S is defensible and this PT provides a defence for it.

3.6 Bar Induction

The last component of Yankov's method of proof is bar induction which appears in many specific forms suitable for the considered theories. Let us recall that bar induction was introduced by Brouwer as an intuitionistically acceptable counterpart of the ordinary induction principle. Informally bar induction is an inductive process which is running on sequences and is usually formulated in terms of well-founded trees (nodes as finite sequences) satisfying some conditions. It allows to "spread" any condition from the terminal nodes of a well-formed tree to all its nodes. The essential inductive step shows that if the condition holds for all immediate successors of some node (i.e. its one element extensions), then it holds for this node. More precisely, Brouwer's original bar induction may be stated as follows. The basic notion of a bar may be understood as a subset B of finite sequences of natural numbers such that for every $f : N \longrightarrow N$ there is n such that the sequence $s = \langle f(0), \ldots, f(n-1)\rangle$

is in B. It is sometimes said (in particular in Yankov 1994) that s secures B. Now, for given B and any predicate φ defined on sequences bar induction states that: (a) if it satisfies every sequence in B and (b) satisfies a sequence if it satisfies all its one-element extensions, then it satisfies the empty sequence (the root of a tree).

Bar induction, together with the principle of continuous choice and with the fan theorem may be seen as one of the fundamental Brouwerian principles related to his understanding of choice sequences. It is remarkable that bar induction although constructively acceptable is a very strong principle, stronger than transfinite induction considered by Gentzen. In fact the class of choice sequences assumed by bar induction is much wider than the class of general recursive functions. For example, bar induction added to a constructive version of ZF allows for proving its consistency (see Rathjen 2006) and added to classical logic yields the full second-order comprehension principle. Anyway in the context of intuitionism it is much weaker and we do not obtain the full force of the second-order system Z2. In fact, Brouwer's original bar principle is still too strong and leads to intuitionistically unwanted conclusions. To avoid them one must add some restrictive conditions on B; either that it is decidable for every finite sequence s that it belongs to B or not (decidable or detachable bar induction), or the stronger condition that every one-element extension of a sequence in B also belongs to B (monotonic bar induction).

It seems that Yankov's predilection to bar induction and corresponding bar recursion was influenced by Spector's proof of the consistency of analysis. This proof belongs to a significantly different tradition than all consistency proofs and other related proof theoretic works in Gentzen's tradition. On the basis of Gödel's dialectica interpretation Spector defined functionals by means of bar recursion. Yankov explicitly refers to his work. In particular, he developed the family of basic and auxiliary bar induction principles for all theories he considered. It is not possible to present all his considerations so we only briefly describe Yankov's original formulation of basic principles.

The central notion in the case of arithmetic is that of constructive functionals mapping sequences of natural numbers to natural numbers, and that of an initial segment. Let ϕ be such functional which plays the role of a bar and f an arbitrary function $N \longrightarrow N$ (sequence), then there is a finite initial segment $\kappa = \langle k_0, \ldots, k_n \rangle$ of f which secures ϕ in the sense that $\phi(f) = \phi(g)$ for any g that has this initial segment κ common with f. This is the basic form of the initial segment principle which is then enriched with formulations of two auxiliary forms. The latter are derivable from the basic one and are introduced only to facilitate the proof so we omit their formulation.

Now Yankov's original bar induction principle may be formulated for any constructive functional ϕ and predicate φ running over finite sequences of natural numbers (i.e. possible initial segments) in the following way:
Basis. $\varphi(\kappa)$ holds for any κ securing ϕ.
Step. If $\varphi(\kappa')$ holds for any single-element extension κ' of κ, then it holds for $\varphi(\kappa)$.
Conclusion: $\varphi(\curlywedge)$ holds, where $\curlywedge$ is the empty sequence.

Two auxiliary forms of bar induction are then introduced which facilitate the proof but are derivable from the basic form.

For analysis, two bar induction principles are formulated, one for natural numbers and one for functions of the type: $N \longrightarrow N$. Since they are similar we recall only the first. Let Φ run over functions defined on well-ordered sets of natural numbers and let $d(\Phi)$ denote the domain of Φ with its well ordering, then it may be formulated as follows:

Basis. $\varphi(\Phi)$ holds if $d(\Phi)$ encompasses all natural numbers.

Finite step. If $\varphi(\Phi')$ holds for any single-element extension Φ' of Φ (i.e. a new number is added to $d(\Phi)$), then it holds for $\varphi(\Phi)$.

Transfinite step. If $d(\Phi')$ is ordered by a limit ordinal and $\varphi(\Phi')$ holds, then there exists a Φ such that $d(\Phi)$ is a proper initial segment of $d(\Phi')$, Φ and Φ' coincide on $d(\Phi)$, and $\varphi(\Phi)$ holds.

Conclusion: $\varphi(\curlywedge)$ holds, where $\curlywedge$ is a function with empty domain.

It is interesting that in case of analysis the notion of constructive functional is not needed for suitable bar induction principles. Each ordinal sequence of functions is barred from above by functions which contain all the objects of the domain. Because of that the barring they introduce is called 'natural' by Yankov. Yankov acknowledges that his bar principles for analysis are not intuitionistically acceptable. However he believes that for an agent with somewhat extended constructive capacities dealing with potential infinity they may be as obvious as arithmetical bar induction in the finite case. In case of set theory the schema of introducing suitable bar induction principles and their assessment is similar (but with ordinals and their sequences) to the case of analysis and we omit presentation.

All bar induction principles have their corresponding bar recursion schemata.

As we mentioned in order to apply bar induction in consistency proofs an arithmetization of proof theoretic notions is required. In particular OTs are represented as "code sequences" (CS) and PTs as functionals on these sequences. We again leave out the details of the construction.

3.7 Proofs

As we mentioned above the structure of Yankov's consistency proof is simple. It is just shown that all axioms are defensible and all rules (including cut) are defensibility-preserving. Since inconsistent statements are obviously not defensible, consistency follows. Note that the strategy applied is different than in the consistency proofs briefly mentioned in Sect. 3.2. In general, in all approaches based on the application of some kind of sequent calculus the results are basically consequences of the subformula property of rules and because of that the position of cut is central. The most popular strategy works by showing that cut is eliminable; it was the work done by Gentzen (1934) for logic and arithmetic without induction and by Schütte, Tait and others for systems with infinitary rules. In general this strategy fails for systems with some axioms added but it may be refined by suitably restricting the class of axioms (e.g. basic, i.e. atomic sequents of Gentzen) which makes it possible to restrict the applications of cut to inessential ones and still obtain the result. In fact axioms of many specific forms may be introduced not only as basic sequents but also as corresponding rules working on only one side of a sequent and thus enabling full cut elimination. This is a strategy developed successfully by Negri and von Plato (2001). Unfortunately it does not work for PA. The induction principle either in the form of an additional (but not basic) sequent, or as a rule, is not amenable to these strategies. Thus (as in Gentzen 1936a and 1938) some additional kinds of reduction steps must be introduced to take control of the applications of cut and limit its use only to those which are not troublesome. Yankov does not try to eliminate or restrict the use of cut, instead he shows that it is also defensibility-preserving. Unsurprisingly this is the most difficult part of his consistency proofs. Without going into details we briefly sketch how it proceeds.

Showing that all logical and structural rules (except cut) are defensibility-preserving is easy. For example consider an application of ($\wedge$) and assume that both premisses are defensible. If an opponent is trying to disprove the left conjunct φ it is sufficient to apply a defense of the first premiss. Otherwise, a defense of the right premiss does the job.

In case of logical axioms $\varphi, \neg\varphi$ the proof goes by induction on the length of φ and applies a so called "mirror-chess" strategy.[8] In the basis a defense simply consists in chosing a true formula (both literals cannot be false), and this is possible since the proponent (being constructive) is assumed to be always in a position to compute the values of the involved terms. In the case of arithmetic it is obvious but it may be doubtful in the case of analysis and ZF. The induction step is easy. For example, in case of $\varphi := \psi \vee \chi$ if the opponent attacks ψ, then the proponent must choose $\neg\psi$ in $\neg\varphi := \neg\psi \wedge \neg\chi$ and follow a defense of $\psi, \neg\psi$ which holds by inductive hypothesis.

In case of nonlogical axioms we will illustrate the point with the case of the induction axiom. Any ORT f for it must consist of three components: f_1 for $\neg\varphi(0)$, f_2 for $\exists x(\varphi(x) \wedge \neg\varphi(sx))$ and f_3 for $\forall x\varphi(x)$. Consider f_3; if the chosen number is 0, then f_3' is an ORT for $\varphi(0)$ and the chess-mirror strategy wins against f_1, f_3'. If it is some $k \neq 0$, then f_3' is an ORT for $\varphi(k)$ and we consider f_2 to the effect that a sequence $\varphi(0) \wedge \neg\varphi(1), \ldots, \varphi(k-1) \wedge \neg\varphi(k)$ is obtained. Let n be the ordinal of the first conjunction in which the opponent chooses its first conjunct, i.e. $\varphi(n-1)$ for refutation. It cannot be 1 with some f_3'' since the proponent wins against f_1, f_3''. Continuing this way with other values of n the proponent eventually wins with $n = k$.

The last, and the most complex argument must be provided to show the defensibility-preservation of cut. It is again performed by induction on the length of the cut formula. Actually two proofs are given, indirect and direct but the whole proof is provided only for the latter which has the following form:

Suppose that two PTs ϕ_1 and ϕ_2 are given for both premisses of cut and f_1, f_2 are ORTs for Γ and Δ respectively. Then it is possible to construct ORTs f_3, f_4 for φ and $\neg\varphi$ such that ϕ_1, ϕ_2 are subject to the following conditions when applied to premisses of cut:

[8] Illustrated with an anecdote about two chess players who have a simultaneous match with a chess master and successively repeat his previous moves.

(a) either $\phi_1(f_1, f_3)$ designates a literal occuring in Γ;

(b) or $\phi_2(f_2, f_4)$ designates a literal occuring in Δ;

(c) or $\phi_1(f_1, f_3)$ designates a literal that is a subformula of φ and $\phi_2(f_2, f_4)$ designates a literal that is a subformula of $\neg\varphi$ and both literals are contradictory.

It is not difficult to observe that in case ϕ_1, ϕ_2 are defenses of premisses this assertion implies that the conclusion is also defensible since (c) is impossible in this case.

A proof is carried out by induction on the length of cut formula. In all cases the proof consists in constructing an extension of given strategies to determine a pair of contradictory literals that are subformulae of cut formulae. The basis is obvious since these are contradictory literals. The case of conjunction and its negation, i.e. the disjunction of the negated conjuncts, is also relatively easy. The main problem is with quantified formulae which additionally in case of analysis must be demonstrated twice for both kinds of bound variables. It is this point of the proof where the abilities of a constructive subject, i.e. a proponent, are needed for the construction of bar recursion. Basically a family of ORTs for $\exists x \neg\varphi$ is considered for all possible initial segments of respective sequences (depending on the theory under consideration). The construction is similar to the one provided by Spector in his proof of the consistency of analysis. In each case suitable bar induction principles are required to provide a verification of the recursion to the effect that the ternary disjunction (a), (b), (c) stated above holds and the recursive process terminates. Of course despite the general similarity of all proofs significant differences may be noticed. For example, in PA the computation process is uniquely determined whereas in case of analysis there is by contrast a search for suitable functions.

The general strategy of these proofs shows that Yankov's approach is in the middle between Gentzen-like approaches based on cut elimination or ordinal reduction and Spector's approach to consistency proof. It is also worth remarking his proof of the completeness of FOL since it is remarkably different from other proofs provided for cut-free formalizations of classical logic. The standard strategy is indirect in the sense that it is shown how to construct a proof-search tree for every unprovable sequent, with at least one, possibly infinite, open branch. Eventually such a branch provides sufficient resources for the construction of a falsifying model for the root-sequent. Similarly like in standard proofs Yankov defines some kind of fair procedure of systematic decomposition of compound formulae in the root-sequent, taking care of existentially quantified formulae since they may be invoked infinitely many times. It is carried out in terms of ORTs associated to each possible branch. But his aim is to provide an intuitionistic proof of the completeness of FOL and the general strategy sketched above is not feasible. In particular, instead of classical induction he applies bar induction and instead of König's lemma its intuitionistic counterpart, the fan theorem,[9] and all these replacements lead to changes in the proof which make it rather direct. It is shown that if a sequent is defensible, then all branches starting with them in a root-first proof search must finish with an axiomatic sequent,

[9] Classically the fan theorem is just the contrapositive of König's lemma, but intuitionistically they are not equivalent.

hence it is defensible. By construction cut is not used, so as a corrolary we obtain a non-constructive semantic proof of cut elimination for one-sided sequent calculus for FOL. Additionally, the proof is extended to the case of derivability of a sequent from an arbitrary set of premisses. To cover the case of infinite sets the notion of a sequent is also extended to cover infinite sequents. In this way Yankov has shown that his specific dialogic interpretation of proofs can work both ways in a uniform way—to show soundness of FOL and consistency of mathematical theories, and to show completeness of FOL.

3.8 Concluding Remarks

Yankov started his career as a programmer and his PhD as well as one of his early papers Yankov (1963) was devoted to Kleene's notion of realizability. His background is evident in his description of the constructive character of a proponent which is often presented in terms of computation processes. As he confesses his consistency proof for arithmetic was completed in 1968 whereas proofs for analysis and set theory were completed in 1991. For the contemporary reader two facts are very striking: the long period between the first (unpublished at the time of its discovery) proof and the final result, and very poor references. Both issues are certainly strongly connected with, not to say grounded in, the political and social activity of Yankov which had hard consequences for his life and career. It is not my role to focus on these aspects of Yankov's activity (but see the introduction to this volume) but it should be noted here that political imprisonment certainly had serious consequences for his scientific research. At the time when he was working on his proofs Yankov was not aware of the huge literature devoted to ordinal proof theory, including consistency proofs for a variety of subsystems of analysis and set theory. His background is in fact based on very limited access to works of some authors known in the 1960s. It is evident not only in case of his method and proofs but also in some other remarks he is making, for example concerning research on nonclassical implication (see the concluding remarks of Yankov 1994). Despite the modest links of his work with main achievements in the field, his results and the method applied are of great interest. Both papers are witnesses of the phenomena that science that is developed in isolation and outside main trends of research may lead to significantly different solutions than those obtained in the mainstream.

In particular, in contrast to research developed within ordinal proof theory Yankov did not care about establishing precise bounds on proofs. He developed his own view on the constructive capabilities of an agent solving problems, extended in such a way as to allow for establishing his results. He is aware that the bar induction principles he introduced are very strong and makes a remark that perhaps it is possible to find weaker tools enabling such proofs. Special bar principles for analysis and set theory are in fact justified in a way which can hardly be seen as constructive. One should notice at this point that Spector (1962) was also doubtful if the bar recursion applied in his proof is really constructive and Kreisel was even more critical in this respect.

Such attempts at proving the consistency of mathematical theories are sometimes criticised as being a pure formalist game, without any, or very little, foundational and epistemological value. In particular, Girard (2011) is very critical, not to say rude, in commenting on the achievements of the Munich school. It is obvious that such criticism may be also directed against Yankov's proofs. Anyway, it is a piece of very ambitious and ingenious work done in extremely hard circumstances and, as I pointed out above, in isolation. It certainly deserves high estimation and it is sad that so far it was not recognized by scholars working on proof theory and foundational problems.

We finish this short exposition of Yankov's achievements in proof theory with a citation from Yankov (1994) which best reflects his own view of what he obtained:

"David Hilbert identified the existence of mathematical objects with the consistency of the theory of these objects. Our arguments show that the verification of such consistency is connected with the possibility of including the implied collection of objects in some "constructive" context by conceiving the field of objects as the sphere of application of some constructive means by a constructive subject. Here the means must be powerful enough so that consistency can be proved, and, on the other hand, obvious enough so that we can evaluate their constructibility, that is, the means must be an extension of our finite constructive capabilities still close to us. I attempt to express this situation by the somewhat descriptive thesis that, in the final analysis, mathematical existence is the possibility of some constructive approach to us of mathematical objects."

References

Coquand, T. (1995). A semantics for classical arithmetic. *The Journal of Symbolic Logic, 60*(1), 325–337.

Gentzen, G. (1934). Untersuchungen über das Logische Schliessen. *Mathematische Zeitschrift, 39*, 176–210 and *39*, 405–431.

Gentzen, G. (1936a). Die Widerspruchsfreiheit der reinen Zahlentheorie'. *Mathematische Annalen, 112*, 493–565.

Gentzen, G. (1936b). Die Widerspruchsfreiheit der Stufenlogik. *Mathematische Zeitschrift, 3*, 357–366.

Gentzen, G. (1936c). Die Unendlichkeitsbegriff in der Mathematik, Semester-Berichte, Münster in: W., 9th Semester, Winter 1936–37, 65–80. *Mathematische Annalen, 112*, 493–565.

Gentzen, G. (1938). Neue Fassung des Widerspruchsfreiheitsbeweises für die reine Zahlentheorie, Forschungen zur Logik und zur Grundlegung der exakten Wissenschaften. *New Series, 4*, Leipzig 19–44.

Gentzen, G. (1943). Beweisbarkeit und Unbeweisbarkeit der Anfangsfällen der transfiniten Induktion in der reinen Zahlentheorie. *Mathematische Annalen, 120*, 140–161.

Girard, J. Y. (2011). *The blind spot*. Lectures on logic: European mathematical society.

Hintikka, J. (1968). Language-games for quantifiers. In *Studies in logical theory* (pp. 46–72). Blackwell.

Indrzejczak, A. (2010). *Natural deduction. Hybrid systems and modal logics*. Springer.

Indrzejczak, A. (2018). Rule-generation theorem and its applications. *The Bulletin of the Section of Logic, 47*(4), 265–282.

Jaśkowski, S. (1934). On the rules of suppositions in formal logic. *Studia Logica, 1*, 5–32.
Lorenzen, P. (1961). Ein dialogisches Konstruktivitätskriterium. In *Infinitistic methods* (pp. 193–200). Warszawa: PWN.
Mendelson, E. (1964). *Introduction to mathematical logic*. Chapman and Hall.
Negri, S., & von Plato, J. (2001). *Structural proof theory*. Cambridge: Cambridge University Press.
Novikov, P. S. (1943). On the consistency of certain logical calculus. *Matematiceskij Sbornik, 1254*, 230–260.
von Plato, J. (2008). Gentzen's proof of normalization for ND. *The Bulletin of Symbolic Logic, 14*(2), 240–257.
Pohlers, W. (2009). *Proof theory*. The first steps into impredicativity. Springer.
Prawitz, D. (1965). *Natural deduction*. Stockholm: Almqvist and Wiksell.
Raggio, A. (1965). Gentzen's Hauptsatz for the systems NI and NK. *Logique et Analyse, 8*, 91–100.
Rathjen, M. (1999). The realm of ordinal analysis. In B. S. Cooper & J. K. Truss (Eds.), *Sets and proofs* (pp. 219–280). Cambridge.
Rathjen, M. (2006). A note on bar induction in constructive set theory. *Mathematical Logic Quarterly, 52*, 253–258.
Rathjen, M., Sieg, W. (2018). *Proof theory. The Stanford encyclopedia of philosophy*, E. Zalta (Ed.). https://plato.stanford.edu/archives/fall2018/entries/proof-theory/.
Schütte, K. (1977). *Proof theory*. Berlin: Springer.
Schwichtenberg, H. (1977). Proof theory. In J. Barwise (Ed.), *Handbook of mathematical logic* (Vol. 1). Amsterdam: North-Holland.
Shoenfield, J. R. (1967). *Mathematical logic*. Addison-Wesley.
Spector, C. (1962). Provably recursive functionals in analysis: A consistency proof of analysis by an extension of principles formulated in current intuitionistic mathematics. *American Mathematical Society, Providence, 5*, 1–27.
Szabo, M. E. (1969). *The collected papers of Gerhard Gentzen*. Amsterdam: North-Holland.
Tait, W. W. (1968). Normal derivability in classical logic. In *The syntax and semantics of infinitary languages*, LNM (Vol. 72, pp. 204–236).
Takeuti, G. (1987). *Proof theory*. Amsterdam: North-Holland.
Yankov, V. A. (1963). On realizable formulas of propositional logic'. *Soviet Mathematics Doklady, 4*, 1035–1037.
Yankov, V. A. (1995). Dialogue theory of proofs for arithmetic, analysis and set theory. Russian Academy of Science *Izvestiya: Mathematics, 44*(3), 571–600. [Russian version 1994].
Yankov, V. A. (1997). Dialogue interpretation of the classical predicate calculus. *Izvestiya RAN: Seriya Matematicheskaya, 611*, 215–224.

Chapter 4
Jankov Formulas and Axiomatization Techniques for Intermediate Logics

Guram Bezhanishvili and Nick Bezhanishvili

Abstract We discuss some of Jankov's contributions to the study of intermediate logics, including the development of what have become known as Jankov formulas and a proof that there are continuum many intermediate logics. We also discuss how to generalize Jankov's technique to develop axiomatization methods for all intermediate logics. These considerations result in what we term subframe and stable canonical formulas. Subframe canonical formulas are obtained by working with the disjunction-free reduct of Heyting algebras and are the algebraic counterpart of Zakharyaschev's canonical formulas. On the other hand, stable canonical formulas are obtained by working with the implication-free reduct of Heyting algebras and are an alternative to subframe canonical formulas. We explain how to develop the standard and selective filtration methods algebraically to axiomatize intermediate logics by means of these formulas. Special cases of these formulas give rise to the classes of subframe and

In the English literature there are two competing spellings of Jankov's name: Jankov and Yankov. We follow the former because this is more common usage in the mathematical literature. The latter usage is more common in the philosophical literature, and is commonly used in this volume.

We are very happy to be able to contribute to this volume dedicated to V. A. Jankov. His work has been very influential for many generations of logicians, initially in the former Soviet Union, but eventually also abroad. In particular, it had a profound impact on our own research. While we have never met Professor Jankov in person, we have heard lots of interesting stories about him from our advisor Leo Esakia. Jankov is not only an outstanding logician, but also a role model citizen, who stood up against the Soviet regime. Because of this, he ended up in the Soviet political camps. A well-known Georgian dissident and human rights activist Levan Berdzenishvili spent several years there with Jankov. We refer to his memoirs (Berdzenishvili 2019) about the Soviet political camps of 1980s in general and about Jankov in particular. One chapter of the book "Vadim" (pp. 127–141) is dedicated to Jankov, in which he is characterized as follows: "I can say with certainty that in our political prison, Vadim Yankov, omnipotent and always ready to help, embodied in the pre-Internet era the combined capabilities of Google, Yahoo, and Wikipedia".

G. Bezhanishvili
New Mexico State University, New Mexico, USA
e-mail: guram@nmsu.edu

N. Bezhanishvili (✉)
University of Amsterdam, Amsterdam, Netherlands
e-mail: N.Bezhanishvili@uva.nl

A. Citkin and I. M. Vandoulakis (eds.), *V. A. Yankov on Non-Classical Logics, History and Philosophy of Mathematics*, Outstanding Contributions to Logic 24,
https://doi.org/10.1007/978-3-031-06843-0_4

stable intermediate logics, and the algebraic account of filtration techniques can be used to prove that they all posses the finite model property (fmp). The fmp results about subframe and cofinal subframe logics yield algebraic proofs of the results of Fine and Zakharyaschev. We conclude by discussing the operations of subframization and stabilization of intermediate logics that this approach gives rise to.

Keywords Intuitionistic logic · Intermediate logics · Splittings · Subframe logics · Axiomatization · Finite model property

2020 Mathematics Subject Classification Primary 03B55 · Secondary 06D20 · 06E15

4.1 Introduction

Intermediate logics are the logics that are situated between the intuitionistic propositional calculus IPC and the classical propositional calculus CPC. The study of intermediate logics was pioneered by Umezawa (1959). As was pointed out by Hosoi (1967), such a study may be viewed as a study of the classification of classically valid principles in terms of their interdeducibility in intuitionistic logic.

Jankov belongs to the first wave of researchers (alongside Dummett, Lemmon, Kuznetsov, Medvedev, Hosoi, de Jongh, Troelstra, and others) who obtained fundamental results in the study of intermediate logics. He is best known for developing algebra-based formulas, which he called characteristic formulas, but are now commonly known as Jankov formulas. This allowed him to obtain deep results about the complicated structure of the lattice of intermediate logics. His first paper on these formulas dates back to 1963 and is one of the early jewels in the study of intermediate logics (Jankov 1963).

From the modern perspective, Jankov formulas axiomatize splittings and their joins in the lattice of intermediate logics. But this came to light later, after the fundamental work of McKenzie (1972). In 1980s Blok and Pigozzi (1982) built on these results to develop a general theory of splittings in varieties with EDPC (equationally definable principal congruences). It should be pointed out that Jankov formulas were independently developed by de Jongh (1968). Because of this, Jankov formulas are also known as Jankov–de Jongh formulas (Bezhanishvili 2006, Remark 3.3.5). We point out that Jankov's technique was algebraic, while de Jongh mostly worked with Kripke frames.

Jankov (1968) utilized his formulas to develop a method for generating continuum many intermediate logics, thus refuting an earlier erroneous attempt of Troelstra (1965) to prove that there are only countably many intermediate logics. Jankov's method also allowed him to construct the first intermediate logic without the finite model property (fmp). These results had major impact on the study of lattices of intermediate and modal logics.

We can associate with any finite subdirectly irreducible Heyting algebra A the Jankov formula $\mathcal{J}(A)$. Then given an arbitrary Heyting algebra B, we can think of the validity of $\mathcal{J}(A)$ on B as forbidding A to be isomorphic to a subalgebra of a homomorphic image of B. This approach was adapted to modal logic by Rautenberg (1980) and was further refined by Kracht (1990) and Wolter (1993). An important result in this direction was obtained by Blok (1978), who characterized splitting modal logics and described the degree of Kripke incompleteness for extensions of the basic modal logic K.

Independently of Jankov, Fine (1974) developed similar formulas for the modal logic $\mathsf{S4}$ by utilizing its Kripke semantics. He associated a formula with each finite rooted $\mathsf{S4}$-frame $\mathfrak{F}$. The validity of such a formula on an $\mathsf{S4}$-frame $\mathfrak{G}$ forbids that $\mathfrak{F}$ is a p-morphic image of a generated subframe of $\mathfrak{G}$. Because of this, these formulas are sometimes called Jankov–Fine formulas in the modal logic literature (see Blackburn et al. 2001, p. 143; Chagrov and Zakharyaschev 1997, p. 332).

Fine (1985) undertook a different approach by "forbidding" p-morphic images of arbitrary (not necessarily generated) subframes. This has resulted in the theory of subframe logics, which was further generalized by Zakharyaschev (1996) to cofinal subframe logics. While Jankov and (cofinal) subframe formulas axiomatize large classes of logics, not every logic is axiomatized by them. This was addressed by Zakharyaschev (1989, 1992) who generalized these formulas to what he termed "canonical formulas" and proved that each intermediate logic and each extension of the modal logic $\mathsf{K4}$ is axiomatized by canonical formulas. Zakharyaschev's approach followed the path of Fine's and mainly utilized Kripke semantics. An algebraic approach to subframe and cofinal subframe logics via nuclei was developed for intermediate logics in Bezhanishvili and Ghilardi (2007) and was generalized to modal logics in Bezhanishvili et al. (2011).

In a series of papers (Bezhanishvili and Bezhanishvili 2009, 2011, 2012), we developed an algebraic treatment of Zakharyaschev's canonical formulas, as well as of subframe and cofinal subframe formulas. This was done for intermediate logics, as well as for extensions of $\mathsf{K4}$ (and even for extensions of weak $\mathsf{K4}$). A somewhat similar approach was undertaken independently and slightly earlier by Tomaszewski (2003). The key idea of this approach for intermediate logics is that the $\vee$-free reduct of each Heyting algebra is locally finite. This is a consequence of a celebrated result of Diego (1966) that the variety of (bounded) implicative semilattices is locally finite. Note that Heyting algebras have another locally finite reduct, which is even better known, namely the $\to$-free reduct. Indeed, it is a classic result that the variety of bounded distributive lattices is locally finite. Thus, it is possible to develop another kind of canonical formulas that also axiomatize all intermediate logics. This was done in Bezhanishvili and Bezhanishvili (2017) and generalized to modal logic in Bezhanishvili et al. (2016a).[1]

To distinguish between these two types of canonical formulas, we call the algebraic counterpart of Zakharyaschev's canonical formulas *subframe canonical for-*

[1] The results of Bezhanishvili and Bezhanishvili (2017) were obtained earlier than those in Bezhanishvili et al. (2016a). However, the latter appeared in print earlier than the former.

mulas. This is motivated by the fact that dually subframe canonical formulas forbid p-morphic images from subframes (see Sect. 4.5.1). On the other hand, we call the other kind *stable canonical formulas* because they forbid stable images of generated subframes (see Sect. 4.5.2). In special cases, both types of canonical formulas yield Jankov formulas. An additional special case for subframe canonical formulas gives rise to the subframe and cofinal subframe formulas of Fine (1985) and Zakharyaschev (1989, 1996). A similar special case for stable subframe formulas gives rise to new classes of stable and cofinal stable formulas studied in Bezhanishvili and Bezhanishvili (2017), Bezhanishvili et al. (2016b) for intermediate logics[2] and in Bezhanishvili et al. (2016a), Bezhanishvili et al. (2018a) for modal logics. Our aim is to provide a uniform account of this line of research.

In this paper we only concentrate on the theory of canonical formulas for intermediate logics, which is closer to Jankov's original motivation and interests. We plan to discuss the theory of canonical formulas for modal logics elsewhere. As a rule of thumb, we supply sketches of proofs only for several central results. For the rest, we provide relevant references, so that it is easy for the interested reader to look up the details.

The paper is organized as follows. In Sect. 4.2 we recall the basic facts about intermediate logics, review their algebraic and Kripke semantics, and outline Esakia duality for Heyting algebras. In Sect. 4.3 we overview the method of Jankov formulas and its main consequences, such as the Splitting Theorem and the cardinality of the lattice of intermediate logics. In Sect. 4.4 we extend the method of Jankov formulas to that of subframe and stable canonical formulas and show that these formulas axiomatize all intermediate logics. Section 4.5 provides a dual approach to subframe and stable canonical formulas. In Sect. 4.6 we review the theory of subframe and cofinal subframe logics and in Sect. 4.7 that of stable and cofinal stable logics. Finally, in Sect. 4.8 we discuss the operations of subframization and stabilization for intermediate logics and their characterization via subframe and stable formulas.

Acknowledgement. Special thanks go to Alex Citkin for putting this volume together and for his enormous patience with all our incurred delays. We are very grateful to Wes Holliday and Luca Carai for careful reading of the manuscript and for many useful suggestions.

4.2 Intermediate Logics and Their Semantics

In this preliminary section, to keep the paper self-contained, we briefly review intermediate logics and their algebraic and relational semantics.

---

[2] Again, the results of Bezhanishvili and Bezhanishvili (2017) were obtained earlier but appeared later than those in Bezhanishvili et al. (2016b).

4.2.1 Intermediate Logics

As we pointed out in the Introduction, a propositional logic L (in the language of IPC) is an *intermediate logic* if $\mathsf{IPC} \subseteq \mathsf{L} \subseteq \mathsf{CPC}$. Intermediate logics are also called superintuitionistic logics (Kuznetsov's terminology). To be more precise, a propositional logic L is a *superintuitionistic logic* (or *si-logic* for short) if $\mathsf{IPC} \subseteq \mathsf{L}$. Since CPC is the largest consistent si-logic, we have that intermediate logics are precisely the consistent si-logics.

We identify each intermediate logic L with the set of theorems of L. It is well known that the collection of all intermediate logics, ordered by inclusion, is a complete lattice, which we denote by Λ. The meet in Λ is set-theoretic intersection, while the join $\bigvee\{\mathsf{L}_i \mid i \in I\}$ is the least intermediate logic containing $\bigcup\{\mathsf{L}_i \mid i \in I\}$. Clearly IPC is the least element and CPC the largest element of Λ.

For an intermediate logic L and a formula φ, we denote by $L + \varphi$ the least intermediate logic containing $L \cup \{\varphi\}$. As usual, if φ is provable in L, we write $\mathsf{L} \vdash \varphi$. For $\mathsf{L}, \mathsf{M} \in \Lambda$, if $\mathsf{L} \subseteq \mathsf{M}$, then we say that M is an *extension* of L.

As we pointed out in the Introduction, Jankov (1968) proved that the cardinality of Λ is that of the continuum. Below we give a list of some well known intermediate logics (see, e.g., Chagrov and Zakharyaschev 1997, p. 112, Table 4.1).

1. $\mathsf{KC} = \mathsf{IPC} + (\neg p \vee \neg\neg p)$—the logic of the weak excluded middle.
2. $\mathsf{LC} = \mathsf{IPC} + (p \to q) \vee (q \to p)$—the Gödel–Dummett logic.
3. $\mathsf{KP} = \mathsf{IPC} + (\neg p \to q \vee r) \to (\neg p \to q) \vee (\neg p \to r)$—the Kreisel–Putnam logic.
4. $\mathsf{T}_n = \mathsf{IPC} + \mathsf{t}_n$ $(n \geq 1)$—the Gabbay–de Jongh logics—where

$$\mathsf{t}_n = \bigwedge_{i=0}^{n} ((p_i \to \bigvee_{i \neq j} p_j) \to \bigvee_{i \neq j} p_j) \to \bigvee_{i=0}^{n} p_i).$$

5. $\mathsf{BD}_n = \mathsf{IPC} + \mathsf{bd}_n$ $(n \geq 1)$, where

$$\begin{aligned} \mathsf{bd}_1 &= p_1 \vee \neg p_1, \\ \mathsf{bd}_{n+1} &= p_{n+1} \vee (p_{n+1} \to \mathsf{bd}_n). \end{aligned}$$

6. $\mathsf{LC}_n = \mathsf{LC} + \mathsf{bd}_n$ $(n \geq 1)$—the n-valued Gödel–Dummett logic.
7. $\mathsf{BW}_n = \mathsf{IPC} + \mathsf{bw}_n$ $(n \geq 1)$, where

$$\mathsf{bw}_n = \bigvee_{i=0}^{n} (p_i \to \bigvee_{j \neq i} p_j).$$

8. $\mathsf{BTW}_n = \mathsf{IPC} + \mathsf{btw}_n$ $(n \geq 1)$, where

$$\mathsf{btw}_n = \bigwedge_{0 \le i < j \le n}^{n} \neg(\neg p_i \wedge \neg p_j) \to \bigvee_{i=0}^{n} \Big(\neg p_i \to \bigvee_{j \ne i} \neg p_j\Big).$$

9. $\mathsf{BC}_n = \mathsf{IPC} + \mathsf{bc}_n$ ($n \ge 1$), where

$$\mathsf{bc}_n = p_0 \vee (p_0 \to p_1) \vee \cdots \vee (p_0 \wedge \cdots \wedge p_{n-1} \to p_n).$$

10. $\mathsf{ND}_n = \mathsf{IPC} + (\neg p \to \bigvee_{1 \le i \le n} \neg q_i) \to \bigvee_{1 \le i \le n} (\neg p \to \neg q_i)$ ($n \ge 2$)—Maksimova's logics.

4.2.2 Heyting Algebras

We next recall the algebraic semantics of intermediate logics. A *Heyting algebra* is a bounded distributive lattice A with an additional binary operation $\to$ satisfying

$$a \wedge x \le b \text{ iff } x \le a \to b$$

for all $a, b, x \in A$. It is well known (see, e.g., Rasiowa and Sikorski 1963, p. 124) that the class of Heyting algebras is equationally definable. For Heyting algebras A and B, a *Heyting homomorphism* is a bounded lattice homomorphism $h : A \to B$ such that $h(a \to b) = h(a) \to h(b)$ for $a, b \in A$.

Definition 4.1 Let Heyt be the category (and the corresponding equational class) of Heyting algebras and Heyting homomorphisms.

A *valuation* v on a Heyting algebra A is a map from the set of propositional variables to A. It is extended to all formulas in an obvious way. A formula φ is *valid* on A if $v(\varphi) = 1$ for every valuation v on A. If φ is valid on A we write $A \models \varphi$. For a class $\mathcal{K}$ of Heyting algebras we write $\mathcal{K} \models \varphi$ if $A \models \varphi$ for each $A \in \mathcal{K}$.

For a Heyting algebra A and a class $\mathcal{K}$ of Heyting algebras, let

$$\mathsf{L}(A) = \{\varphi \mid A \models \varphi\} \text{ and } \mathsf{L}(\mathcal{K}) = \bigcap \{\mathsf{L}(A) \mid A \in \mathcal{K}\}.$$

It is well known that if A is a nontrivial Heyting algebra and $\mathcal{K}$ is a nonempty class of nontrivial Heyting algebras, then $\mathsf{L}(A)$ and $\mathsf{L}(\mathcal{K})$ are intermediate logics. We call $\mathsf{L}(A)$ the *logic* of A and $\mathsf{L}(\mathcal{K})$ the *logic* of $\mathcal{K}$.

Definition 4.2 We say that an intermediate logic L is *sound and complete* with respect to a class $\mathcal{K}$ of Heyting algebras if $\mathsf{L} = \mathsf{L}(\mathcal{K})$; that is, $\mathsf{L} \vdash \varphi$ iff $\mathcal{K} \models \varphi$.

For a class $\mathcal{K}$ of Heyting algebras, we let $\mathbf{H}(\mathcal{K})$, $\mathbf{S}(\mathcal{K})$, $\mathbf{P}(\mathcal{K})$, and $\mathbf{I}(\mathcal{K})$ be the classes of homomorphic images, subalgebras, products, and isomorphic copies of algebras from $\mathcal{K}$. A *variety* is a class of algebras closed under **H**, **S**, and **P**. By Birkhoff's celebrated theorem (see, e.g., Burris and Sankappanavar 1981, Theorem 11.9), varieties are precisely the equationally definable classes of algebras.

By the well-known Lindenbaum algebra construction (see, e.g., Rasiowa and Sikorski 1963, Chap. VI), each intermediate logic L is sound and complete with respect to the variety of Heyting algebras

$$\mathcal{V}(\mathsf{L}) := \{A \in \mathsf{Heyt} \mid A \models \mathsf{L}\}.$$

This variety is often called the variety *corresponding to* L. We call $A \in \mathcal{V}(\mathsf{L})$ an L-*algebra*.

We recall that a Heyting algebra A is *subdirectly irreducible* if it has a least nontrivial congruence. It is well known (see, e.g., Balbes and Dwinger 1974, p. 179, Theorem 5) that A is subdirectly irreducible iff $A \setminus \{1\}$ has the largest element s, called the *second largest element* of A.

Remark 4.3 This result also originates with Jankov, who referred to these algebras as Gödelean (see Jankov 1963).

A Heyting algebra A is *finitely subdirectly irreducible* or *well-connected* if

$$a \vee b = 1 \Rightarrow a = 1 \text{ or } b = 1$$

for each $a, b \in A$. Obviously each subdirectly irreducible Heyting algebra is well-connected, but there exist infinite well-connected Heyting algebras that are not subdirectly irreducible. On the other hand, a finite Heyting algebra is subdirectly irreducible iff it is well-connected.

By another celebrated result of Birkhoff (see, e.g., Burris and Sankappanavar 1981, Theorem 8.6), each variety $\mathcal{V}$ is generated by subdirectly irreducible members of $\mathcal{V}$. Thus, each intermediate logic is complete with respect to the class of subdirectly irreducible algebras in $\mathcal{V}(\mathsf{L})$.

The next definition and theorem are well known and go back to Kuznetsov.

Definition 4.4 Let L be an intermediate logic.

1. Two formulas φ, ψ are L-*equivalent* if $\mathsf{L} \vdash \varphi \leftrightarrow \psi$.
2. L is *locally tabular* if for each natural number n, there are only finitely many non-L-equivalent formulas in n-variables.
3. L is *tabular* if L is the logic of a finite Heyting algebra.
4. L has the *finite model property* (fmp for short) if $\mathsf{L} \nvdash \varphi$ implies that there is a finite Heyting algebra A such that $A \models \mathsf{L}$ and $A \not\models \varphi$.
5. L has the *hereditary finite model property* (hfmp for short) if L and all its extensions have the fmp.

Theorem 4.5

(1) L *is locally tabular iff* $\mathcal{V}(\mathsf{L})$ *is locally finite (each finitely generated* $\mathcal{V}(\mathsf{L})$-*algebra is finite).*

(2) L *is tabular iff* $\mathcal{V}(\mathsf{L})$ *is generated by a finite algebra.*
(3) L *has the fmp iff* $\mathcal{V}(\mathsf{L})$ *is generated by the class of finite* $\mathcal{V}(\mathsf{L})$*-algebras.*
(4) L *has the hfmp iff each subvariety of* $\mathcal{V}(\mathsf{L})$ *is generated by the class of its finite algebras.*

The next definition also goes back to Kuznetsov (1974).

Definition 4.6

1. Let Λ_{t} be the subclass of Λ consisting of tabular intermediate logics.
2. Let Λ_{lt} be the subclass of Λ consisting of locally tabular intermediate logics.
3. Let Λ_{fmp} be the subclass of Λ consisting of intermediate logics with the fmp.
4. Let Λ_{hfmp} be the subclass of Λ consisting of intermediate logics with the hfmp.

We then have the following hierarchy of Kuznetsov (1974):

$$\Lambda_{\mathsf{t}} \subsetneq \Lambda_{\mathsf{lt}} \subsetneq \Lambda_{\mathsf{hfmp}} \subsetneq \Lambda_{\mathsf{fmp}} \subsetneq \Lambda.$$

4.2.3 Kripke Frames and Esakia Spaces

We now turn to Kripke semantics for intermediate logics. In this case Kripke frames are simply posets (partially ordered sets). We denote the partial order of a poset P by $\leq$. For $S \subseteq P$, the *downset* of S is the set

$${\downarrow}S = \{x \in P \mid \exists s \in S \text{ with } x \leq s\}.$$

The *upset* of S is defined dually and is denoted by ${\uparrow}S$. If S is a singleton set $\{x\}$, then we write ${\downarrow}x$ and ${\uparrow}x$ instead of ${\downarrow}\{x\}$ and ${\uparrow}\{x\}$.

We call $U \subseteq P$ an *upset* if ${\uparrow}U = U$ (that is, $x \in U$ and $x \leq y$ imply $y \in U$). A *downset* of P is defined dually. Also, we let $\max(U)$ and $\min(U)$ be the sets of maximal and minimal points of U.

Let $\mathsf{Up}(P)$ and $\mathsf{Do}(P)$ be the sets of upsets and downsets of X, respectively. It is well known that $(\mathsf{Up}(P), \cap, \cup, \to, \varnothing, P)$ is a Heyting algebra, where for each $U, V \in \mathrm{Up}(X)$, we have:

$$U \to V = \{x \in P \mid {\uparrow}x \cap U \subseteq V\} = P \setminus {\downarrow}(U \setminus V).$$

Similarly, $(\mathsf{Do}(P), \cap, \cup, \to, \varnothing, P)$ is a Heyting algebra, but we will mainly work with the Heyting algebra of upsets of X.

Each Heyting algebra A is isomorphic to a subalgebra of the Heyting algebra of upsets of some poset. We call this representation the *Kripke representation* of Heyting algebras. Let X_A be the set of prime filters of A, ordered by inclusion. Then X_A is a poset, known as the *spectrum* of A. Define the *Stone map* $\zeta : A \to \mathsf{Up}(X_A)$ by

$$\zeta(a) = \{x \in X_A \mid a \in x\}.$$

Then ζ is a Heyting algebra embedding, and we arrive at the following well-known theorem.

Theorem 4.7 (Kripke representation) *Each Heyting algebra is isomorphic to a subalgebra of* $\mathsf{Up}(X_A)$.

To recover the image of A in $\mathsf{Up}(X_A)$, we need to introduce a topology on X_A. We recall that a subset of a topological space X is *clopen* if it is both closed and open, and that X is *zero-dimensional* if clopen sets form a basis for X. A *Stone space* is a compact, Hausdorff, zero-dimensional space. By the celebrated Stone duality (Stone 1936), the category of Boolean algebras and Boolean homomorphisms is dually equivalent to the category of Stone spaces and continuous maps. In particular, each Boolean algebra A is represented as the Boolean algebra of clopens of a Stone space (namely, of the prime spectrum of A) which is unique up to homeomorphism.

Stone duality for Boolean algebras was generalized to Heyting algebras by Esakia (1974) (see also Esakia 2019).

Definition 4.8 An *Esakia space* is a Stone space X equipped with a *continuous* partial order $\leq$, meaning that the following two conditions are satisfied:

1. $\uparrow x$ is a closed set for each $x \in X$.
2. U clopen implies that $\downarrow U$ is clopen.

We recall that a map $f : P \to Q$ between two posets is a *p-morphism* if $\uparrow f(x) = f[\uparrow x]$ for each $x \in P$. For Esakia spaces X and Y, a map $f : X \to Y$ is an *Esakia morphism* if it is a continuous p-morphism.

Definition 4.9 Let Esa be the category of Esakia spaces and Esakia morphisms.

Theorem 4.10 (Esakia duality) Heyt *is dually equivalent to* Esa.

In particular, each Heyting algebra A is represented as the Heyting algebra of clopen upsets of the prime spectrum X_A of A, where the topology on X_A is defined by the basis

$$\{\zeta(a) \setminus \zeta(b) \mid a, b \in A\}.$$

We refer to this representation as the *Esakia representation* of Heyting algebras.

If we restrict Esakia duality to the finite case, we obtain that the category of finite Heyting algebras is dually equivalent to the category of finite posets. In particular, each finite Heyting algebra A is isomorphic to $\mathsf{Up}(X_A)$. We refer to this duality as *finite Esakia duality* (but point out that this finite duality has been known before Esakia; see, e.g., de Jongh and Troelstra 1966).

It follows from Esakia duality that onto Heyting homomorphisms dually correspond to one-to-one Esakia morphisms, and one-to-one Heyting homomorphisms to onto Esakia morphisms. In particular, homomorphic images of a Heyting algebra A correspond to closed upsets of X_A, while subalgebras of A to special quotients of X_A known as *Esakia quotients* (see, e.g., Bezhanishvili et al. 2010).

Definition 4.11 Let X be an Esakia space.

1. We call X *rooted* if there is $x \in X$, called the *root* of X, such that $X = \uparrow x$.
2. We call X *strongly rooted* if X is rooted and the singleton $\{x\}$ is clopen.

It is well known (see, e.g., Esakia 1979 or Bezhanishvili and Bezhanishvili 2008) that a Heyting algebra A is well-connected iff X_A is rooted, and that A is subdirectly irreducible iff X_A is strongly rooted.

We evaluate formulas in a poset P by evaluating them in the Heyting algebra $\mathsf{Up}(P)$, and we evaluate formulas in an Esakia space X by evaluating them in the Heyting algebra of clopen upsets of X. These clopen upsets are known as *definable upsets* of X, so such valuations are called *definable valuations*.

Since each intermediate logic L is complete with respect to Heyting algebras, it follows from Esakia duality that L is complete with respect to Esakia spaces (but not necessarily with respect to posets as it is known that there exist Kripke incomplete intermediate logics, Shehtman 1977).

For a class $\mathcal{K}$ of posets or Esakia spaces, let $\mathcal{K}^*$ be the corresponding class of Heyting algebras (of all upsets or definable upsets of members of $\mathcal{K}$). We then say that an intermediate logic L is the *logic* of $\mathcal{K}$ if L is the logic of $\mathcal{K}^*$.

Definition 4.12 Let P be a finite poset and $n \geq 1$.

1. The *length* of a chain in P is its cardinality.
2. The *depth* of P is $\leq n$, denoted $d(P) \leq n$, if all chains in P have length $\leq n$.
3. The *width* of $x \in P$ is $\leq n$ if the length of antichains in $\uparrow x$ is $\leq n$.
4. The *cofinal width* (or *top width*) of $x \in P$ is $\leq n$ if $|\max(\uparrow x)| \leq n$.
5. The *width* of P is $\leq n$, denoted $w(P) \leq n$, if the width of each $x \in P$ is $\leq n$.
6. The *cofinal width* (or *top width*) of P is $\leq n$, denoted $w_c(P) \leq n$, if the cofinal width of each $x \in P$ is $\leq n$.
7. The *branching* of P is $\leq n$, denoted $b(P) \leq n$, if each $x \in P$ has at most n distinct immediate successors.
8. The *divergence* of P is $\leq n$, denoted $div(P) \leq n$, if for each $x \in P$ and $Q \subseteq \uparrow x \cap \max(P)$ satisfying $|Q| \leq n$, there is $y \geq x$ with $\max(\uparrow y) = Q$.

The next theorem is well known (see, e.g., Chagrov and Zakharyaschev 1997).

Theorem 4.13 *Let* $n \geq 1$.

(1) KC *is the logic of all finite rooted posets that have a largest element.*
(2) LC *is the logic of all finite chains.*
(3) KP *is the logic of all finite rooted posets satisfying*

$$\forall x \forall y \forall z \big(x \leq y \wedge x \leq z \wedge \neg(y \leq z) \wedge \neg(z \leq y) \rightarrow \exists u (x \leq u \wedge u \leq y \wedge u \leq z) \wedge \forall v (u \leq v \rightarrow \exists w (v \leq w \wedge (y \leq w \vee z \leq w)))\big).$$

(4) T_n *is the logic of all finite rooted posets of branching* $\leq n$.
(5) BD_n *is the logic of all finite rooted posets of depth* $\leq n$.
(6) LC_n *is the logic of the chain of length* n.

(7) BW_n *is the logic of all finite rooted posets of width* $\leq n$.
(8) BTW_n *is the logic of all finite rooted posets of cofinal width* $\leq n$.
(9) BC_n *is the logic of all finite rooted posets of cardinality* $\leq n$.
(10) ND_n *is the logic of all finite rooted posets of divergence* $\leq n$.

Thus, each of these logics has the fmp. In fact, LC *as well as each* BD_n *is locally tabular, and each* LC_n *as well as each* BC_n *is tabular.*

4.3 Jankov Formulas

As we pointed out in the Introduction, Jankov first introduced his formulas in (Jankov 1963) under the name of characteristic formulas. They have since become a major tool in the study of intermediate and modal logics and are often referred to as Jankov formulas (Chagrov and Zakharyaschev 1997, p. 332), Jankov–de Jongh formulas (Bezhanishvili 2006, p. 59), or Jankov–Fine formulas (Blackburn et al. 2001, p. 143). In this paper we will refer to them as Jankov formulas. First results about Jankov formulas were announced in Jankov (1963). Proofs of these results together with further properties of Jankov formulas were given in Jankov (1969). Jankov (1968) utilized his formulas to prove that there are continuum many intermediate logics. He also gave the first example of an intermediate logic without the fmp.

4.3.1 Jankov Lemma

The basic idea of Jankov formulas is closely related to the method of diagrams in model theory (see, e.g., Chang and Keisler 1990, pp. 68–69). Let A be a finite Heyting algebra.[3] We can encode the structure of A in our propositional language by describing what is true and what is false in A. This way we obtain two finite sets of formulas, Γ and Δ, where p_a is a new variable for each $a \in A$:

$$\begin{aligned}\Gamma = &\{p_{a\wedge b} \leftrightarrow p_a \wedge p_b \mid a, b \in A\} \cup \\ &\{p_{a\vee b} \leftrightarrow p_a \vee p_b \mid a, b \in A\} \cup \\ &\{p_{a\to b} \leftrightarrow p_a \to p_b \mid a, b \in A\} \cup \\ &\{p_{\neg a} \leftrightarrow \neg p_a \mid a \in A\}\end{aligned}$$

and

$$\Delta = \{p_a \leftrightarrow p_b \mid a, b \in A \text{ with } a \neq b\}.$$

[3] While the assumption that A is finite is not essential, it suffices for our purposes.

Thus, Γ describes what is true and Δ what is false in A. We can then work with the multiple-conclusion rule Γ/Δ and prove that this rule is characteristic for A in the following sense:

Lemma 4.14 (Bezhanishvili et al. 2016b) *Let A be a finite Heyting algebra and B an arbitrary Heyting algebra. Then*[4]

$$B \not\models \Gamma/\Delta \text{ iff } A \in \mathbf{IS}(B).$$

However, at the time of Jankov, it was unusual to work with multiple-conclusion rules. Instead Jankov assumed that A is subdirectly irreducible. Then A has the second largest element s. Therefore, Δ can be replaced with p_s since everything that is falsified in A ends up underneath s. Thus, we arrive at the following notion of the Jankov formula of A:

Definition 4.15 Let A be a finite subdirectly irreducible Heyting algebra with the second largest element s. Then the *Jankov formula* of A is the formula

$$\mathcal{J}(A) = \bigwedge \Gamma \to p_s.$$

The defining property of Jankov formulas is presented in the following lemma, which we will refer to as the Jankov Lemma. Comparing the Jankov Lemma to Lemma 4.14, we see that the switch from the multiple-conclusion rule Γ/Δ to the formula $\mathcal{J}(A)$ requires on the one hand to assume that A is subdirectly irreducible and on the other hand to also work with homomorphic images and not only with isomorphic copies of subalgebras of B as in Lemma 4.14.

Lemma 4.16 (Jankov Lemma) *Let A be a finite subdirectly irreducible Heyting algebra and B an arbitrary Heyting algebra. Then*

$$B \not\models \mathcal{J}(A) \text{ iff } A \in \mathbf{SH}(B).$$

Proof (*Sketch*) First suppose that $A \in \mathbf{SH}(B)$. By evaluating each p_a as a, it is easy to see that A refutes $\mathcal{J}(A)$. Therefore, since $A \in \mathbf{SH}(B)$, we also have that $B \not\models \mathcal{J}(A)$.

Conversely, suppose that $B \not\models \mathcal{J}(A)$. By Wronski (1973, Lemma 1), there is a subdirectly irreducible homomorphic image C of B and a valuation v on C such that $v(\bigwedge \Gamma) = 1_C$ and $v(p_s) = s_C$, where s_C is the second largest element of C. Define $h : A \to C$ by setting $h(a) = v(p_a)$. That $v(\bigwedge \Gamma) = 1_C$ implies that h is a Heyting homomorphism, and that $h(s) = s_C$ yields that h is one-to-one. Thus, $A \in \mathbf{ISH}(B) \subseteq \mathbf{SH}(B)$.

Remark 4.17 Since the variety of Heyting algebras has the congruence extension property, we have $A \in \mathbf{SH}(B)$ iff $A \in \mathbf{HS}(B)$. Therefore, the conclusion of the Jankov Lemma is often formulated as follows:

[4] This lemma is closely related to Chang and Keisler (1990, Proposition 2.1.8).

$$B \not\models \mathcal{J}(A) \text{ iff } A \in \mathbf{HS}(B).$$

Since A is a finite subdirectly irreducible Heyting algebra, by finite Esakia duality, A is isomorphic to the algebra $\mathsf{Up}(P)$ of upsets of a finite rooted poset P. To simplify notation, instead of $\mathcal{J}(A)$ we will often write $\mathcal{J}(P)$. Thus, we obtain the following dual reading of the Jankov Lemma.

Lemma 4.18 *Let P be a finite rooted poset and X an Esakia space. Then $X \not\models \mathcal{J}(P)$ iff P is isomorphic to an Esakia quotient of a closed upset of X.*

4.3.2 Splitting Theorem

A very useful feature of Jankov formulas is that they axiomatize splittings in the lattice of intermediate logics. We recall that a pair (s, t) of elements of a lattice L *splits* L if L is the disjoint union of $\uparrow s$ and $\downarrow t$.

Definition 4.19 (Chagrov and Zakharyaschev 1997, Sect. 10.5) An intermediate logic L is a *splitting logic* if there is an intermediate logic M such that (L, M) splits the lattice Λ of intermediate logics.

The next theorem is due to Jankov (1963, 1969), although not in the language of splitting logics.

Theorem 4.20 (Splitting Theorem) *Let L be an intermediate logic. Then L is a splitting logic iff $\mathsf{L} = \mathsf{IPC} + \mathcal{J}(\mathsf{A})$ for some finite subdirectly irreducible Heyting algebra A.*

Proof (*Sketch*) First suppose that L is a splitting logic. Then there is an intermediate logic M such that (L, M) splits Λ. Since the variety Heyt of Heyting algebras is congruence-distributive and is generated by its finite members, a result of McKenzie (1972, Theorem 4.3) yields that M is the logic of a finite subdirectly irreducible Heyting algebra A. But then for an arbitrary Heyting algebra B we have $B \models \mathcal{J}(A)$ iff $B \models \mathsf{L}$. Thus, $\mathsf{L} = \mathsf{IPC} + \mathcal{J}(\mathsf{A})$.

For the converse, suppose that $\mathsf{L} = \mathsf{IPC} + \mathcal{J}(\mathsf{A})$ for some finite subdirectly irreducible Heyting algebra A. Let $\mathsf{M} = \mathsf{L}(A)$. We claim that (L, M) splits Λ. To see this, first note that as $A \not\models \mathcal{J}(A)$, we have $\mathcal{J}(A) \notin \mathsf{M}$. Therefore, $\mathsf{L} \neq \mathsf{M}$. Next let N be an intermediate logic such that $\mathsf{L} \not\subseteq \mathsf{N}$. Then $\mathcal{J}(A) \notin \mathsf{N}$. Thus, there is a Heyting algebra B such that $B \models \mathsf{N}$ and $B \not\models \mathcal{J}(A)$. By the Jankov Lemma, $A \in \mathbf{SH}(B)$. Therefore, $A \models \mathsf{N}$, and hence $\mathsf{N} \subseteq \mathsf{M}$.

The Splitting Theorem was generalized to varieties with EDPC by Blok and Pigozzi (1982).

Definition 4.21 Let L be an intermediate logic.

1. L is a *join-splitting logic* if L is a join in Λ of splitting logics.

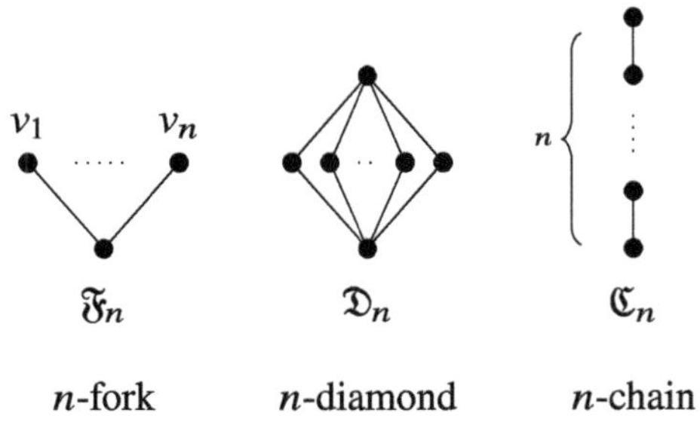

Fig. 4.1 The n-fork, n-diamond, and n-chain

2. L is *axiomatizable by Jankov formulas* if there is a set Ω of finite subdirectly irreducible Heyting algebras such that $\mathsf{L} = \mathsf{IPC} + \{\mathcal{J}(A) \mid A \in \Omega\}$.

As an immediate consequence of the Splitting Theorem we obtain:

Theorem 4.22 *An intermediate logic* L *is a join-splitting logic iff* L *is axiomatizable by Jankov formulas.*

To give examples of intermediate logics that are join-splitting, for each $n \geq 1$, let $\mathfrak{F}_n$ be the n-fork, $\mathfrak{D}_n$ the n-diamond, and $\mathfrak{C}_n$ the n-chain (see Fig. 4.1).

The next theorem is well-known (although finding an exact reference is a challenge).

Theorem 4.23

(1) $\mathsf{CPC} = \mathsf{IPC} + \mathcal{J}(\mathfrak{C}_2)$, *so* CPC *is a splitting logic.*
(2) $\mathsf{KC} = \mathsf{IPC} + \mathcal{J}(\mathfrak{F}_2)$, *so* KC *is a splitting logic.*
(3) $\mathsf{BD}_n = \mathsf{IPC} + \mathcal{J}(\mathfrak{C}_{n+1})$, *so each* BD_n *is a splitting logic.*
(4) $\mathsf{LC} = \mathsf{IPC} + \mathcal{J}(\mathfrak{F}_2) + \mathcal{J}(\mathfrak{D}_2)$, *so* LC *is a join-splitting logic.*
(5) $\mathsf{LC}_n = \mathsf{LC} + \mathcal{J}(\mathfrak{C}_{n+1})$, *so each* LC_n *is a join-splitting logic.*

This theorem shows that many well-known intermediate logics are indeed axiomatizable by Jankov formulas. However, not every intermediate logic is axiomatizable by Jankov formulas.

Theorem 4.24 (Chagrov and Zakharyaschev 1997, Proposition 9.50) BTW_3 *is not axiomatizable by Jankov formulas.*

We next give a criterion describing when an intermediate logic is axiomatizable by Jankov formulas. Define the following relation between Heyting algebras:

$$A \leq B \text{ iff } A \in \mathbf{SH}(B).$$

Remark 4.25 As follows from Remark 4.17, $A \leq B$ iff $A \in \mathbf{HS}(B)$.

If A is finite and subdirectly irreducible, then by the Jankov Lemma, $A \leq B$ iff $B \not\models \mathcal{J}(A)$. It was noted already by Jankov (1969) that $\leq$ is a quasi-order and that if A, B are finite and subdirectly irreducible, then $A \leq B$ and $B \leq A$ imply that A is isomorphic to B. Since the variety of Heyting algebras is congruence-distributive, this is also a consequence of Jónsson's Lemma (Jónsson 1968). The following result gives a criterion for an intermediate logic to be axiomatizable by Jankov formulas. For a proof see, e.g., Bezhanishvili (2006, Corollary 3.4.14).

Theorem 4.26 (Criterion of axiomatizability by Jankov formulas) *Let* L *and* M *be intermediate logics such that* $\mathsf{L} \subseteq \mathsf{M}$. *Then* M *is axiomatizable over* L *by Jankov formulas iff for every Heyting algebra* B *such that* $B \models \mathsf{L}$ *and* $B \not\models \mathsf{M}$ *there is a finite Heyting algebra* A *such that* $A \leq B$, $A \models \mathsf{L}$, *and* $A \not\models \mathsf{M}$.

As a consequence of this criterion, we obtain the following result about axiomatizability for extensions of a locally tabular intermediate logic.

Theorem 4.27 *Let* L *be a locally tabular intermediate logic. Then every extension of* L *is axiomatizable by Jankov formulas over* L.

Proof Let M be an extension of L and B a Heyting algebra such that $B \models \mathsf{L}$ and $B \not\models \mathsf{M}$. Then there is $\varphi(p_1, \ldots, p_n) \in \mathsf{M}$ such that $B \not\models \varphi(p_1, \ldots, p_n)$. Therefore, there is a valuation v on B refuting φ. Let A be the subalgebra of B generated by $\{v(p_1), \ldots, v(p_n)\}$. Then $A \leq B$ and A is finite since L is locally tabular. In addition, $A \models \mathsf{L}$ as A is a subalgebra of B and $A \not\models \varphi$ because $B \not\models \varphi$. Thus, $A \not\models \mathsf{M}$. By Theorem 4.26, M is axiomatizable over L by Jankov formulas.

Theorem 4.27 can be generalized in two directions. Firstly we have that every locally tabular intermediate logic is axiomatizable by Jankov formulas and every tabular intermediate logic is axiomatizable by finitely many Jankov formulas; see Bezhanishvili (2006, Theorems 3.4.24 and 3.4.27) (and also Citkin 1986; Tomaszewski 2003).

Theorem 4.28

(1) *Every locally tabular intermediate logic is axiomatizable by Jankov formulas.*
(2) *Every tabular intermediate logic is finitely axiomatizable by Jankov formulas.*

Secondly the assumption in Theorem 4.27 that L is locally tabular can be weakened to L having the hereditary finite model property.

Theorem 4.29 *Let* L *be an intermediate logic with the hereditary fmp. Then every extension of* L *is axiomatizable over* L *by Jankov formulas.*

Proof (*Sketch*) Let M be an extension of L. We let $\mathcal{X}$ be the set of all finite (non-isomorphic) subdirectly irreducible L-algebras A such that $A \not\models \mathsf{M}$ and consider

$$\mathsf{N} = \mathsf{L} + \{\mathcal{J}(A) \mid A \in \mathcal{X}\}.$$

Let B be a finite subdirectly irreducible Heyting algebra such that $B \models \mathsf{L}$. By definition of $\mathcal{X}$,

$$B \models \mathsf{N} \text{ iff } B \models \mathcal{J}(A) \text{ for each } A \in \mathcal{X} \text{ iff } B \models \mathsf{M}.$$

Since L has the hereditary fmp, both M and N have the fmp. Thus, $\mathsf{M} = \mathsf{N}$, and hence every extension of L is axiomatizable over L by Jankov formulas.

4.3.3 Cardinality of the Lattice of Intermediate Logics

Jankov formulas are also instrumental in determining cardinalities of different classes of intermediate logics. We call a set Ω of $\leq$-incomparable Heyting algebras an $\leq$-*antichain*. The next theorem is well known (see Jankov 1968 or Bezhanishvili 2006, Theorem 3.4.18).

Theorem 4.30 *Let Ω be a countably infinite $\leq$-antichain of finite subdirectly irreducible Heyting algebras. Then for $\Omega_1, \Omega_2 \subseteq \Omega$ with $\Omega_1 \neq \Omega_2$, we have*

(1) $\mathsf{IPC} + \{\mathcal{J}(A) \mid A \in \Omega_1\} \neq \mathsf{IPC} + \{\mathcal{J}(A) \mid A \in \Omega_2\}$.
(2) $\mathsf{L}(\Omega_1) \neq \mathsf{L}(\Omega_2)$.

Proof (1). Without loss of generality we may assume that $\Omega_1 \not\subseteq \Omega_2$. Therefore, there is $B \in \Omega_1$ with $B \notin \Omega_2$. Since $B \not\models \mathcal{J}(B)$, we have $B \not\models \mathsf{IPC} + \{\mathcal{J}(A) \mid A \in \Omega_1\}$. On the other hand, if $B \not\models \mathsf{IPC} + \{\mathcal{J}(A) \mid A \in \Omega_2\}$, then there is $A \in \Omega_2$ with $B \not\models \mathcal{J}(A)$. By the Jankov Lemma, $A \leq B$. However, since $A, B \in \Omega$, this contradicts the assumption that Ω is an $\leq$-antichain. Thus, $B \models \mathsf{IPC} + \{\mathcal{J}(A) \mid A \in \Omega_2\}$, and hence $\mathsf{IPC} + \{\mathcal{J}(A) \mid A \in \Omega_1\} \neq \mathsf{IPC} + \{\mathcal{J}(A) \mid A \in \Omega_2\}$.
(2). This is proved similarly to (1).

To construct countable $\leq$-antichains, it is more convenient to use finite Esakia duality and work with finite rooted posets. For this it is convenient to dualize the definition of $\leq$. Let P and Q be finite posets. We set

$$P \leq Q \text{ iff } P \text{ is isomorphic to an Esakia quotient of an upset of } Q.$$

The following lemma is an immediate consequence of finite Esakia duality.

Lemma 4.31 *Let P and Q be finite posets. Then*

$$P \leq Q \textit{ iff } \mathsf{Up}(P) \leq \mathsf{Up}(Q).$$

In the next lemma we describe two infinite $\leq$-antichains of finite rooted posets. The antichain Ω_1 is the dual version of Jankov's original antichain (Jankov 1968) of finite subdirectly irreducible Heyting algebras, while the antichain Ω_2 goes back to Kuznetsov (1974).

Lemma 4.32 *There exist countably infinite $\leq$-antichains of finite rooted posets.*

Proof (*Sketch*) We consider two countably infinite sets Ω_1 and Ω_2 of finite rooted posets. The set Ω_1 has infinite depth but width 3, while the set Ω_2 has infinite width but depth 3.

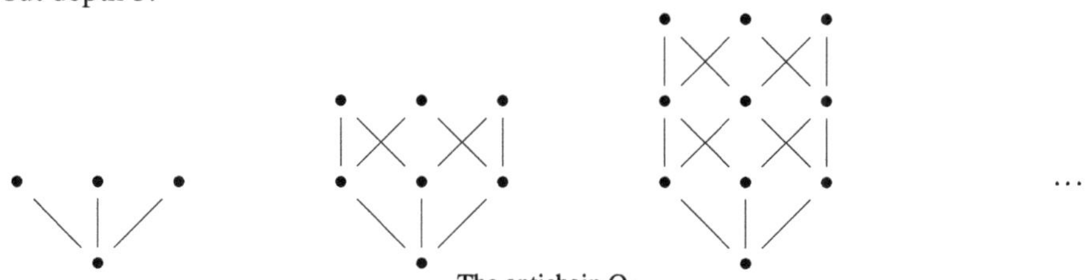

The antichain Ω_1

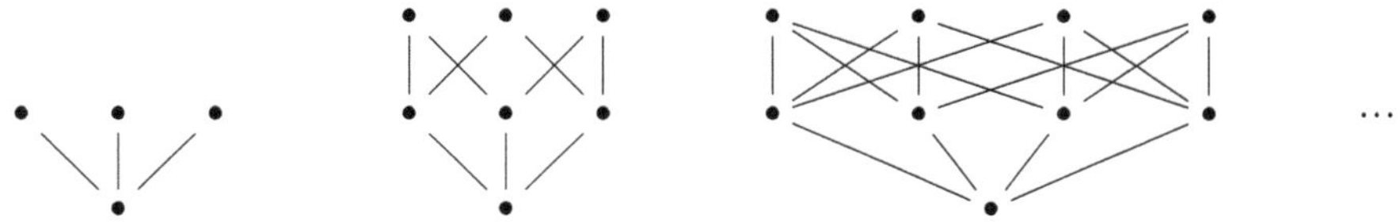

The antichain Ω_2

It is a somewhat tedious calculation to show that Ω_1 and Ω_2 are indeed $\leq$-antichains.

As an immediate consequence of Theorem 4.30 and Lemma 4.32 we obtain the cardinality bound for the lattice of intermediate logics.

Theorem 4.33 (Jankov 1968) *There are continuum many intermediate logics.*

In fact, Lemma 4.32 implies a stronger result that the cardinality of intermediate logics of width 3 is that of the continuum and that the cardinality of intermediate logics of depth 3 is that of the continuum.

Remark 4.34 We conclude this section by mentioning several more applications of Jankov formulas.

1. Jankov formulas play an important role for obtaining intermediate logics that lack the fmp and are Kripke incomplete. First intermediate logic without the fmp was constructed by Jankov (1968). Further examples of intermediate logics without the fmp were given by (Kuznetsov and Gerčiu 1970). In fact, there are continuum many intermediate logics without the fmp (see, e.g., Chagrov and Zakharyaschev 1997, Theorem 6.3; Bezhanishvili et al. 2008, Corollary 5.41). Litak (2002) used Jankov formulas to construct continuum many Kripke incomplete intermediate logics.
2. Wronski (1973) utilized Jankov formulas to construct continuum many intermediate logics with the disjunction property. (We recall that an intermediate logic L has the *disjunction property* if $\mathsf{L} \vdash \varphi \vee \psi$ implies $\mathsf{L} \vdash \varphi$ or $\mathsf{L} \vdash \psi$.)

3. Jankov formulas are essential in the study of HSC logics (hereditarily structurally complete intermediate logics). As was shown by Citkin (1978, 1987), there is a least HSC logic which is axiomatized by the Jankov formulas of five finite subdirectly irreducible Heyting algebras (see Bezhanishvili and Moraschini 2019 for a proof using Esakia duality). This was extended to extensions of K4 by Rybakov (1995).
4. A recent result (Bezhanishvili et al. 2022) shows that there is a largest variety $\mathcal{V}$ of Heyting algebras in which every profinite algebra is isomorphic to the profinite completion of some algebra in $\mathcal{V}$. Again, $\mathcal{V}$ is axiomatized by the Jankov formulas of four finite subdirectly irreducible Heyting algebras.
5. Jankov formulas are instrumental in the study of the refutation systems of Citkin (2013b) (these are formalisms that carry the information about what is not valid in a given logic).

4.4 Canonical Formulas

As follows from Theorem 4.24, Jankov formulas do not axiomatize all intermediate logics. However, by Theorem 4.28(1), they do axiomatize every locally tabular intermediate logic. By Theorem 4.5(1), these correspond to locally finite varieties of Heyting algebras. Although the variety Heyt of all Heyting algebras is not locally finite, both the $\vee$-free and $\to$-free reducts of Heyt generate locally finite varieties. Indeed, the $\vee$-free reducts of Heyt generate the variety of bounded implicative semilattices, which is locally finite by Diego's theorem (Diego 1966). Also, the $\to$-free reducts of Heyt generate the variety of bounded distributive lattices, which is well known to be locally finite (see, e.g., Grätzer 1978, p. 68, Theorem 1). On the one hand, these locally finite reducts can be used to prove the fmp of IPC and many other intermediate logics. On the other hand, they allow us to develop powerful methods of uniform axiomatization for all intermediate logics.

The key idea is to refine further the Jankov method discussed in the previous section. As we pointed out, the Jankov formula of a finite subdirectly irreducible Heyting algebra A encodes the structure of A in the full signature of Heyting algebras. The refinement of the method consists of encoding fully only the structure of locally finite reduct of A. Then the embedding of A into a homomorphic image of B discussed in the proof of the Jankov Lemma only preserves the operations of the reduct. Yet, the embedding may preserve the remaining operation ($\vee$ or $\to$ depending on which locally finite reduct we work with) only on some elements of A. These constitute what we call the "closed domain" of A. Thus, this new "generalized Jankov formula," which following Zakharyaschev (see Zakharyaschev 1989, 1992; Chagrov and Zakharyaschev 1997) we call the "canonical formula" of A, encodes fully the locally finite reduct of A that we work with, plus the remaining operation only partially, on the closed domain of A. Since we will mainly be working with two locally finite reducts of Heyting algebras, the $\vee$-free and $\to$-free reducts, we obtain two different types of canonical formulas. Based on the dual description of the homomorphisms involved (see Sect. 4.5), we call the canonical formulas associated with the $\vee$-free reduct "subframe canonical formulas," and the ones associated with the $\rightarrow$-free reduct "stable canonical formulas."

4.4.1 Subframe Canonical Formulas

In this section we survey the theory of subframe canonical formulas developed in Bezhanishvili and Bezhanishvili (2009) (under the name of $(\wedge, \rightarrow, 0)$-canonical formulas). These are the algebraic counterpart of Zakharyaschev's canonical formulas, and are a direct generalization of Jankov formulas.

In Sect. 4.3, with each finite subdirectly irreducible Heyting algebra A we associated the Jankov formula $\mathcal{J}(A)$ of A which encodes the structure of A in the full signature of Heyting algebras. The subframe canonical formula of A encodes the bounded implicative semilattice structure of A fully but the behavior of $\vee$ only partially on some specified subset $D \subseteq A^2$.

Definition 4.35 Let A be a finite subdirectly irreducible Heyting algebra, s the second largest element of A, and D a subset of A^2. For each $a \in A$ we introduce a new variable p_a and define the *subframe canonical formula* $\alpha(A, D)$ associated with A and D as

$$\begin{aligned}\alpha(A, D) = \Big(& \bigwedge\{p_{a\wedge b} \leftrightarrow p_a \wedge p_b \mid a, b \in A\} \wedge \\ & \bigwedge\{p_{a\rightarrow b} \leftrightarrow p_a \rightarrow p_b \mid a, b \in A\} \wedge \\ & \bigwedge\{p_{\neg a} \leftrightarrow \neg p_a \mid a \in A\} \wedge \\ & \bigwedge\{p_{a\vee b} \leftrightarrow p_a \vee p_b \mid (a, b) \in D\}\Big) \rightarrow p_s\end{aligned}$$

Remark 4.36 If $D = A^2$, then $\alpha(A, D) = \mathcal{J}(A)$.

Let A and B be Heyting algebras. We recall that a map $h : A \rightarrow B$ is an *implicative semilattice homomorphism* if

$$h(a \wedge b) = h(a) \wedge h(b) \text{ and } h(a \rightarrow b) = h(a) \rightarrow h(b)$$

for each $a, b \in A$. It is easy to see that implicative semilattice homomorphisms preserve the top element, but they may not preserve the bottom element. Thus, we call h *bounded* if $h(0) = 0$.

Definition 4.37 Let A, B be Heyting algebras, $D \subseteq A^2$, and $h : A \rightarrow B$ a bounded implicative semilattice homomorphism. We call D a $\vee$-*closed domain* of A and say that h satisfies the $\vee$-*closed domain condition* for D if $h(a \vee b) = h(a) \vee h(b)$ for $(a, b) \in D$.

To simplify notation, we abbreviate the $\vee$-closed domain condition by $\mathrm{CDC}_\vee$. An appropriate modification of the Jankov Lemma yields:

Lemma 4.38 (Subframe Jankov Lemma) *Let A be a finite subdirectly irreducible Heyting algebra, $D \subseteq A^2$, and B an arbitrary Heyting algebra. Then $B \not\models \alpha(A, D)$ iff there is a homomorphic image C of B and a bounded implicative semilattice embedding $h : A \rightarrowtail C$ satisfying $\mathrm{CDC}_\vee$ for D.*

Proof See (Bezhanishvili and Bezhanishvili 2009, Theorem 5.3).

Our second key tool for the desired uniform axiomatization of intermediate logics is what we call the Selective Filtration Lemma. The name is motivated by the fact that it provides an algebraic account of the Fine–Zakharyaschev method of selective filtration for intermediate logics (see, e.g., Chagrov and Zakharyaschev 1997, Theorem 9.34). For a detailed comparison of the algebraic and frame-theoretic methods of selective filtration we refer to Bezhanishvili and Bezhanishvili (2016). To formulate the Selective Filtration Lemma we require the following definition.

Definition 4.39 Let A, B be Heyting algebras with $A \subseteq B$. We say that A is a $(\wedge, \to)$-*subalgebra* of B if A is closed under $\wedge$ and $\to$, and we say that A is a $(\wedge, \to, 0)$-*subalgebra* of B if in addition $0 \in A$.

Lemma 4.40 (Selective Filtration Lemma) *Let B be a Heyting algebra such that $B \not\models \varphi$. Then there is a finite Heyting algebra A such that A is a $(\wedge, \to, 0)$-subalgebra of B and $A \not\models \varphi$. In addition, if B is subdirectly irreducible, then A can be chosen to be subdirectly irreducible as well.*

Proof Since $B \not\models \varphi$, there is a valuation v on B such that $v(\varphi) \neq 1_B$. Let $\mathrm{Sub}(\varphi)$ be the set of subformulas of φ and let A be the $(\wedge, \to, 0)$-subalgebra of B generated by $v[\mathrm{Sub}(\varphi)]$. If B is subdirectly irreducible, then it has the second largest element s, and we generate A by $\{s\} \cup v[\mathrm{Sub}(\varphi)]$. By Diego's theorem, A is finite. Therefore, A is a finite Heyting algebra, where

$$a \vee_A b = \bigwedge \{c \in A \mid a, b \leq c\}$$

for each $a, b \in A$. It is easy to see that $a \vee b \leq a \vee_A b$ and that $a \vee_A b = a \vee b$ whenever $a \vee b \in A$. Moreover, since for $a, b \in v[\mathrm{Sub}(\varphi)]$, if $a \vee b \in v[\mathrm{Sub}(\varphi)]$, then $a \vee_A b = a \vee b$, we see that the value of φ in A is the same as the value of φ in B. As $v(\varphi) \neq 1_B$, we conclude that $v(\varphi) \neq 1_A$. Thus, A is a finite Heyting algebra that is a $(\wedge, \to, 0)$-subalgebra of B and refutes φ. Finally, if B is subdirectly irreducible, then s is also the second largest element of A, so A is subdirectly irreducible as well.

Now suppose that $\mathsf{IPC} \nvdash \varphi$ and $n = |\mathrm{Sub}(\varphi)|$. Since the variety of bounded implicative semilattices is locally finite, there is a bound $c(\varphi)$ on the number of n-generated bounded implicative semilattices. Let $A_1, \ldots, A_{m(n)}$ be the list of finite subdirectly irreducible Heyting algebras such that $|A_i| \leq c(\varphi)$ and $A_i \not\models \varphi$.

For an algebra A refuting φ via a valuation v, let $\Theta = v[\mathrm{Sub}(\varphi)]$ and let

$$D^\vee = \{(a, b) \in \Theta^2 \mid a \vee b \in \Theta\}.$$

Consider a new list $(A_1, D_1^\vee), \ldots, (A_{k(n)}, D_{k(n)}^\vee)$, and note that in general $k(n)$ can be greater than $m(n)$ since each A_i may refute φ via different valuations.

We have the following characterization of refutability:

Theorem 4.41 (Bezhanishvili and Bezhanishvili 2009, Theorem 5.7 and Corollary 5.10) *Let B be a Heyting algebra.*

(1) *$B \not\models \varphi$ iff there is $i \leq k(n)$, a homomorphic image C of B, and a bounded implicative semilattice embedding $h : A_i \rightarrowtail C$ satisfying* CDC$_\vee$ *for $D_i^\vee$.*

(2) *$B \models \varphi$ iff $B \models \bigwedge_{i=1}^{k(n)} \alpha(A_i, D_i^\vee)$.*

Proof Since (2) follows from (1) and the Subframe Jankov Lemma, we only sketch the proof of (1). The right to left implication of (1) is straightforward. For the left to right implication, let $B \not\models \varphi$. By the Selective Filtration Lemma, there is a finite Heyting algebra A such that $A \not\models \varphi$ and A is a $(\wedge, \to, 0)$-subalgebra of B. If v is a valuation on B refuting φ, then as follows from the proof of the Selective Filtration Lemma, v restricts to a valuation on A refuting φ. Since by Birkhoff's theorem (see, e.g., Burris and Sankappanavar 1981, Theorem 8.6) A is isomorphic to a subdirect product of its subdirectly irreducible homomorphic images, there is a subdirectly irreducible homomorphic image A' of A such that $A' \not\models \varphi$. The valuation refuting φ on A' can be taken to be the composition $\pi \circ v$ where $\pi : A \twoheadrightarrow A'$ is the onto homomorphism. Because homomorphic images are determined by filters, A' is the quotient A/F by some filter $F \subseteq A$. Let G be the filter of B generated by F and let C be the quotient B/G. Then we have the following commutative diagram, and a direct verification shows that the embedding $A' \rightarrowtail C$ satisfies CDC$_\vee$ for $D^\vee$.

$$\begin{array}{ccc} A & \rightarrowtail & B \\ \downarrow & & \downarrow \\ A' & \rightarrowtail & C \end{array}$$

From this we conclude that the pair $(A', D^\vee)$ is one of the $(A_i, D_i^\vee)$ from the list, and the embedding of A' into a homomorphic image C of B satisfies CDC$_\vee$ for $D_i^\vee$.

Remark 4.42 The above sketch of the proof of Theorem 4.41(1) is simpler than the original proof given in Bezhanishvili and Bezhanishvili (2009, Theorem 5.7), where free algebras were used to obtain the list $(A_1, D_1^\vee), \ldots, (A_{k(n)}, D_{k(n)}^\vee)$.

As an immediate consequence, we arrive at the following uniform axiomatization of all intermediate logics by subframe canonical formulas.

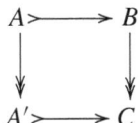

Theorem 4.43 (Bezhanishvili and Bezhanishvili 2009, Corollary 5.13) *Each intermediate logic* L *is axiomatizable by subframe canonical formulas. Moreover, if* L *is finitely axiomatizable, then* L *is axiomatizable by finitely many subframe canonical formulas.*

Proof Let $\mathsf{L} = \mathsf{IPC} + \{\varphi_i \mid i \in I\}$. Then $\mathsf{IPC} \nvdash \varphi_i$ for each $i \in I$. By Theorem 4.41, for each $i \in I$, there are $(A_{i1}, D_{i1}^{\vee}), \ldots, (A_{ik_i}, D_{ik_i}^{\vee})$ such that $\mathsf{IPC} + \varphi_i = \mathsf{IPC} + \bigwedge_{j=1}^{k_i} \alpha(A_{ij}, D_{ij}^{\vee})$. Thus, $\mathsf{L} = \mathsf{IPC} + \left\{\bigwedge_{j=1}^{k_i} \alpha(A_{ij}, D_{ij}^{\vee}) \mid i \in I\right\}$.

Remark 4.44

1. As we pointed out in the Introduction, canonical formulas were first introduced by Zakharyaschev (1989) where Theorem 4.43 was proved using relational semantics.
2. The notion of subframe canonical formulas can be generalized to that of multiple-conclusion subframe canonical rules along the lines of Lemma 4.14. This was done by Jeřábek (2009) whose approach was similar to that of Zakharyaschev (1989). In particular, Jeřábek proved that every intuitionistic multiple-conclusion consequence relation is axiomatizable by canonical rules. Jeřábek also gave an alternative proof of obtaining bases of admissible rules via these canonical rules and gave an alternative proof of Rybakov's decidability of the admissibility problem in IPC (Rybakov 1985).
3. An alternate approach to canonical formulas and rules using partial algebras was undertaken by Citkin (2013a, 2015).

4.4.2 Negation-Free Subframe Canonical Formulas

Definition 4.45 We call a propositional formula φ *negation-free* if φ does not contain $\neg$.

For those intermediate logics that are axiomatized by negation-free formulas, we can simplify subframe canonical formulas by dropping the conjunct

$$\bigwedge\{p_{\neg a} \leftrightarrow \neg p_a \mid a \in A\}$$

in the antecedent. The resulting formulas will be called negation-free subframe canonical formulas:

Definition 4.46 Let A be a finite subdirectly irreducible Heyting algebra, s the second largest element of A, and $D \subseteq A^2$ a $\vee$-closed domain of A. For each $a \in A$ we introduce a new variable p_a and define the *negation-free subframe canonical formula* $\beta(A, D)$ associated with A and D by

$$\beta(A, D) = \Big(\bigwedge \{p_{a \wedge b} \leftrightarrow p_a \wedge p_b \mid a, b \in A\} \wedge$$
$$\bigwedge \{p_{a \to b} \leftrightarrow p_a \to p_b \mid a, b \in A\} \wedge$$
$$\bigwedge \{p_{a \vee b} \leftrightarrow p_a \vee p_b \mid (a, b) \in D\} \Big) \to p_s.$$

We can then prove analogs of the results obtained in Sect. 4.4.1 and axiomatize each intermediate logic that is axiomatized by negation-free formulas by negation-free canonical formulas. The difference is that everywhere in Lemmas 4.38, 4.40 and Theorem 4.41 "bounded" needs to be dropped, and we need to work with not necessarily bounded implicative semilattice embeddings. Because of this, we only state the results without proofs.

Theorem 4.47 (Bezhanishvili and Bezhanishvili 2009, Corollaries 5.16 and 5.17) *Let φ be a negation-free formula such that $\mathsf{IPC} \nvdash \varphi$ and $n = |\mathrm{Sub}(\varphi)|$. Then there is a list $(A_1, D_1^{\vee}), \ldots, (A_{k(n)}, D_{k(n)}^{\vee})$ such that each A_i is a finite subdirectly irreducible Heyting algebra, $D_i \subseteq A_i^2$ is a $\vee$-closed domain of A_i, and for an arbitrary Heyting algebra B we have:*

(1) *$B \not\models \varphi$ iff there is $i \leq k(n)$, a homomorphic image C of B, and an implicative semilattice embedding $h : A_i \rightarrowtail C$ satisfying $\mathrm{CDC}_{\vee}$ for $D_i^{\vee}$.*

(2) *$B \models \varphi$ iff $B \models \bigwedge_{i=1}^{k(n)} \beta(A_i, D_i^{\vee})$.*

As a corollary, we obtain the following:

Corollary 4.48 (Bezhanishvili and Bezhanishvili 2009, Corollary 5.19) *Each intermediate logic L that is axiomatized by negation-free formulas is axiomatizable by negation-free canonical formulas. Moreover, if L is axiomatized by finitely many negation-free formulas, then L is axiomatizable by finitely many negation-free canonical formulas.*

Remark 4.49

1. Negation-free canonical formulas were first introduced by Zakharyaschev (1989) where Theorem 4.48 was proved using relational semantics.
2. The notion of negation-free subframe canonical formulas can be generalized to that of negation-free multiple-conclusion subframe canonical rules. This was done by Jeřábek (2009) who showed that every intuitionistic negation-free multiple-conclusion consequence relation is axiomatizable by negation-free multiple-conclusion canonical rules. Jeřábek's approach was similar to that of Zakharyaschev (1989).

4.4.3 Stable Canonical Formulas

In this section we survey the theory of stable canonical formulas of Bezhanishvili and Bezhanishvili (2017) (where they were called $(\wedge, \vee)$-canonical formulas). The theory is developed along the same lines as the theory of subframe canonical formulas, with the difference that stable canonical formulas require us to work with the $\to$-free reduct of Heyting algebras instead of the $\vee$-free reduct. We outline the similarities and differences between these two approaches.

We start by the following simple observation which will be useful throughout. Let A and B be Heyting algebras. If B is subdirectly irreducible and A is a subalgebra of B, then A does not have to be subdirectly irreducible. However, it is elementary to see that if B is well-connected and A is a bounded sublattice of B, then A is also well-connected. In particular, since a finite Heyting algebra is subdirectly irreducible iff it is well-connected, if B is well-connected and A is a finite bounded sublattice of B, then A is subdirectly irreducible.

We next define the stable canonical formula associated with a finite subdirectly irreducible Heyting algebra A and a subset D of A^2. This formula encodes the bounded lattice structure of A fully and the behavior of $\to$ partially, only on the elements of D.

Definition 4.50 Let A be a finite subdirectly irreducible Heyting algebra, s the second largest element of A, and $D \subseteq A^2$. For each $a \in A$ introduce a new variable p_a and set

$$\begin{aligned}\Gamma = &\{p_0 \leftrightarrow \bot\} \cup \{p_1 \leftrightarrow \top\} \cup \\ &\{p_{a \wedge b} \leftrightarrow p_a \wedge p_b \mid a, b \in A\} \cup \\ &\{p_{a \vee b} \leftrightarrow p_a \vee p_b \mid a, b \in A\} \cup \\ &\{p_{a \to b} \leftrightarrow p_a \to p_b \mid a, b \in D\}\end{aligned}$$

and

$$\Delta = \{p_a \leftrightarrow p_b \mid a, b \in A \text{ with } a \neq b\}.$$

Then define the *stable canonical formula* $\gamma(A, D)$ associated with A and D as

$$\gamma(A, D) = \bigwedge \Gamma \to \bigvee \Delta.$$

Remark 4.51 In Bezhanishvili and Bezhanishvili (2017, Definition 3.1) Δ is defined as $\{p_a \to p_b \mid a, b \in A \text{ with } a \not\leq b\}$.

Remark 4.52 Comparing $\gamma(A, D)$ and $\alpha(A, D)$, we see that the antecedent of $\gamma(A, D)$ encodes the bounded lattice structure of A and the implications in D, while the antecedent of $\alpha(A, D)$ encodes the bounded implicative semilattice structure of A and the joins in D.

The consequent of $\gamma(A, D)$ is more complicated than that of $\alpha(A, D)$. The intention in both cases is that the canonical formula is "pre-true" on the algebra. For

$\alpha(A, D)$, since the formula encodes implications of entire A, this can simply be expressed by introducing a variable for the second largest element s of A. For $\gamma(A, D)$ however we need a more complicated consequent because the formula encodes implications only from the designated subset D of A^2.

Remark 4.53 If $D = A^2$, then $\gamma(A, D)$ is equivalent to $\mathcal{J}(A)$ (see Bezhanishvili and Bezhanishvili 2017, Theorem 5.1).

Definition 4.54 Let A, B be Heyting algebras, $D \subseteq A^2$, and $h : A \to B$ a bounded lattice homomorphism. We call D a $\to$*-closed domain* of A and say that h satisfies the $\to$*-closed domain condition* for D if $h(a \to b) = h(a) \to h(b)$ for all $(a, b) \in D$.

We abbreviate the $\to$-closed domain condition by $\mathrm{CDC}_{\to}$. The next lemma is a version of the Jankov Lemma for stable canonical formulas.

Lemma 4.55 (Stable Jankov Lemma) *Let A be a finite subdirectly irreducible Heyting algebra, $D \subseteq A^2$ a $\to$-closed domain of A, and B a Heyting algebra. Then $B \not\models \gamma(A, D)$ iff there is a subdirectly irreducible homomorphic image C of B and a bounded lattice embedding $h : A \rightarrowtail C$ satisfying* $\mathrm{CDC}_{\to}$ *for D.*

Proof See (Bezhanishvili and Bezhanishvili 2017, Theorem 3.4).

Remark 4.56 The Stable Jankov Lemma plays the same role in the theory of stable canonical formulas as the Subframe Jankov Lemma in the theory of subframe canonical formulas, but it is weaker in that the C in the lemma is required to be subdirectly irreducible, while in the Subframe Jankov Lemma it is not. As is shown in Bezhanishvili and Bezhanishvili (2017, Remark 3.5), this assumption is necessary.

The second main ingredient for obtaining uniform axiomatization of intermediate logics by means of stable canonical formulas is the following Filtration Lemma, which goes back to McKinsey and Tarski (1946) (and for modal logics even further back to McKinsey 1941; McKinsey and Tarski 1944). The name is motivated by the fact that it provides an algebraic account of the method of standard filtration for intermediate logics (see, e.g., Chagrov and Zakharyaschev 1997, Sect. 5.3). For a detailed comparison of the algebraic and frame-theoretic methods of standard filtration we refer to Bezhanishvili and Bezhanishvili (2016).

Lemma 4.57 (Filtration Lemma) *Let B be a Heyting algebra such that $B \not\models \varphi$. Then there is a finite Heyting algebra A such that A is a bounded sublattice of B and $A \not\models \varphi$. In addition, if B is well-connected, then A is subdirectly irreducible.*

Proof Since $B \not\models \varphi$, there is a valuation v on B such that $v(\varphi) \neq 1_B$. Let A be the bounded sublattice of B generated by $v[\mathrm{Sub}(\varphi)]$. Since the variety of bounded distributive lattices is locally finite, A is finite. Therefore, A is a finite Heyting algebra, where

$$a \to_A b = \bigvee \{c \in A \mid a \wedge c \leq b\}$$

for each $a, b \in A$. Because $a \to b = \bigvee\{d \in B \mid a \wedge d \leq b\}$, it is easy to see that $a \to_A b \leq a \to b$ and that $a \to_A b = a \to b$ whenever $a \to b \in A$. Since for $a, b \in v[\mathrm{Sub}(\varphi)]$, if $a \to b \in v[\mathrm{Sub}(\varphi)]$, then $a \to_A b = a \to b$, we see that the value of φ in A is the same as the value of φ in B. As φ is refuted on B, we conclude that φ is refuted on A. Thus, A is a finite Heyting algebra that is a bounded sublattice of B and refutes φ. Finally, if B is well-connected, then so is A, and as A is finite, A is subdirectly irreducible.

Now suppose that $\mathsf{IPC} \nvdash \varphi$ and $n = |\mathrm{Sub}(\varphi)|$. Since the variety of bounded distributive lattices is locally finite, there is a bound $c(\varphi)$ on the number of n-generated bounded distributive lattices. Let $A_1, \ldots, A_{m(n)}$ be the list of all finite subdirectly irreducible Heyting algebras such that $|A_i| \leq c(\varphi)$ and $A_i \not\models \varphi$.

For an algebra A refuting φ via a valuation v, let $\Theta = v[\mathrm{Sub}(\varphi)]$ and let

$$D^{\to} = \{(a, b) \in \Theta^2 \mid a \to b \in \Theta\}.$$

Consider a new list $(A_1, D_1^{\to}), \ldots, (A_{k(n)}, D_{k(n)}^{\to})$, and note that in general $k(n)$ can be greater than $m(n)$ since each A_i may refute φ via different valuations.

The next theorem provides an alternative characterization of refutability to that given in Theorem 4.41. The proof is along similar lines of the proof of Theorem 4.41, but with appropriate adjustments since here we work with a different reduct of Heyting algebras.

Theorem 4.58 (Bezhanishvili and Bezhanishvili 2017, Theorem 3.7 and Corollary 3.9) *Let B be a subdirectly irreducible Heyting algebra.*

(1) *The following conditions are equivalent:*

(a) $B \not\models \varphi$.
(b) *There is $i \leq k(n)$ and a bounded lattice embedding $h : A_i \rightarrowtail B$ satisfying* $\mathrm{CDC}_{\to}$ *for $D_i^{\to}$.*
(c) *There is $i \leq k(n)$, a subdirectly irreducible homomorphic image C of B, and a bounded lattice embedding $h : A_i \rightarrowtail C$ satisfying* $\mathrm{CDC}_{\to}$ *for $D_i^{\to}$.*

(2) $B \models \varphi$ *iff* $B \models \bigwedge_{i=1}^{k(n)} \gamma(A_i, D_i^{\to})$.

Proof Since (2) follows from (1) and the Stable Jankov Lemma, we only sketch the proof of (1). The implications (1b)⇒(1c)⇒(1a) are straightforward. We prove the implication (1a)⇒(1b). Suppose that $B \not\models \varphi$. As B is subdirectly irreducible, it is well-connected. By the Filtration Lemma, there is a finite subdirectly irreducible Heyting algebra A such that $A \not\models \varphi$ and A is a bounded sublattice of B. Moreover, it follows from the proof of the Filtration Lemma that for each $a, b \in B$ such that $a \to b \in v[\mathrm{Sub}(\varphi)]$ we have $a \to_A b = a \to b$. From this we conclude that the pair $(A, D^{\to})$ is one of the $(A_i, D_i^{\to})$ from the list and the embedding of A into B satisfies $\mathrm{CDC}_{\to}$ for $D_i^{\to}$.

Remark 4.59 The above sketch of the proof of Theorem 4.58(1) is simpler than the original proof given in Bezhanishvili and Bezhanishvili (2017, Theorem 3.7), where free algebras were used to obtain the list $(A_1, D_1^{\to}), \ldots (A_{k(n)}, D_{k(n)}^{\to})$.

Remark 4.60 Theorem 4.58 plays the same role in the theory of stable canonical formulas as Theorem 4.41 in the theory of subframe canonical formulas, but it is weaker in that the B in the theorem is required to be subdirectly irreducible, while in Theorem 4.41 it is arbitrary.

As a consequence, we arrive at the following axiomatization of intermediate logics by means of stable canonical formulas, which is an alternative to Theorem 4.43. The proof is along the same lines as that of Theorem 4.43, the only difference being that we have to work with subdirectly irreducible Heyting algebras instead of arbitrary Heyting algebras. Since each variety of Heyting algebras is generated by its subdirectly irreducible members, the end result is the same.

Theorem 4.61 (Bezhanishvili and Bezhanishvili 2017, Corollary 3.10) *Each intermediate logic* L *is axiomatizable by stable canonical formulas. Moreover, if* L *is finitely axiomatizable, then* L *is axiomatizable by finitely many stable canonical formulas.*

Remark 4.62

1. The same way subframe canonical formulas can be generalized to subframe canonical rules (see Remark 4.44(2)), in Bezhanishvili et al. (2016b) stable canonical formulas were generalized to stable canonical rules and it was shown that every intuitionistic multiple-conclusion consequence relation is axiomatizable by stable canonical rules. These rules were used in Bezhanishvili et al. (2016) to give an alternative proof of the existence of bases of admissible rules and the decidability of the admissibility problem for IPC, thus providing an analogue of Jeřábek's result (Jeřábek 2009) via stable canonical rules.
2. Stable canonical formulas were generalized to substructural logics in Bezhanishvili et al. (2017).

4.5 Canonical Formulas Dually

In this section we discuss the dual reading of both subframe and stable canonical formulas. For subframe canonical formulas this requires a dual description of bounded implicative semilattice homomorphisms and for stable canonical formulas a dual description of bounded lattice homomorphisms. For the former we will work with the generalized Esakia duality of Bezhanishvili and Bezhanishvili (2009), and for the latter with Priestley duality for bounded distributive lattices (Priestley 1970, 1972).

4.5.1 Subframe Canonical Formulas Dually

As was shown in Bezhanishvili and Bezhanishvili (2009), implicative semilattice homomorphisms are dually described by means of special partial maps between Esakia spaces.

Definition 4.63 Let X and Y be Esakia spaces, $f : X \to Y$ a partial map, and $\mathrm{dom}(f)$ the domain of f. We call f a *partial Esakia morphism* if the following conditions are satisfied:

1. If $x, z \in \mathrm{dom}(f)$ and $x \le z$, then $f(x) \le f(z)$.
2. If $x \in \mathrm{dom}(f)$, $y \in Y$, and $f(x) \le y$, then there is $z \in \mathrm{dom}(f)$ such that $x \le z$ and $f(z) = y$.
3. $x \in \mathrm{dom}(f)$ iff there is $y \in Y$ such that $f[\uparrow x] = \uparrow y$.
4. $f[\uparrow x]$ is closed for each $x \in X$.
5. If U is a clopen upset of Y, then $X \setminus \downarrow f^{-1}(Y \setminus U)$ is a clopen upset of X.

Remark 4.64 If $\mathrm{dom}(f) = X$ and hence the partial Esakia morphism $f : X \to Y$ is total, then f is an Esakia morphism (see Lemma 4.66(2)).

The next result describes the topological properties of the domain of a partial Esakia morphism that will be used subsequently.

Lemma 4.65 *Let $f : X \to Y$ be a partial Esakia morphism.*

(1) $\mathrm{dom}(f)$ *is a closed subset of X.*
(2) *If Y is finite, then* $\mathrm{dom}(f)$ *is a clopen subset of X.*

Proof For a proof of (1) see Bezhanishvili and Bezhanishvili (2009, Lemma 3.7). For (2), in view of (1), it is sufficient to show that $\mathrm{dom}(f)$ is open. Let $x \in \mathrm{dom}(f)$. We set

$$\begin{aligned} D_1 &= Y \setminus \uparrow f(x), \\ D_2 &= Y \setminus (\uparrow f(x) \setminus \{f(x)\}), \\ U &= \downarrow f^{-1}(D_2) \setminus \downarrow f^{-1}(D_1). \end{aligned}$$

Since Y is finite and D_1, D_2 are downsets of Y, Definition 4.63(5) yields that $\downarrow f^{-1}(D_2)$ and $\downarrow f^{-1}(D_1)$ are clopen downsets of X. Therefore, U is clopen in X. Since $f(x) \in D_2$, we have $x \in \downarrow f^{-1}(D_2)$. Also, $f(x) \notin D_1$ and D_1 a downset of Y implies that $f^{-1}(D_1)$ is a downset of $\mathrm{dom}(f)$ by Definition 4.63(1). Thus, $x \notin \downarrow f^{-1}(D_1)$, and so $x \in U$. Therefore, it is sufficient to show that $U \subseteq \mathrm{dom}(f)$. Let $y \in U$. Then there is $z \in \mathrm{dom}(f)$ such that $y \le z$, $f(z) \notin (\uparrow f(x) \setminus \{f(x)\})$, and $f(z) \in \uparrow f(x)$. Thus, $f(z) = f(x)$. Since $y \le z$ and $z \in \mathrm{dom}(f)$, we have

$$\uparrow f(x) = \uparrow f(z) = f[\uparrow z] \subseteq f[\uparrow y],$$

where the second equality follows from Definition 4.63(1, 2). For the reverse inclusion, let $u \in \mathrm{dom}(f)$ and $y \le u$. Since $y \notin {\downarrow} f^{-1}(D_1)$, we have that $f(u) \in {\uparrow} f(x)$. Therefore, $f[{\uparrow} y] = {\uparrow} f(x)$, and we conclude by Definition 4.63(3) that $y \in \mathrm{dom}(f)$. Thus, $U \subseteq \mathrm{dom}(f)$, and hence $\mathrm{dom}(f)$ is clopen in X.

Lemma 4.66 *Let X, Y be Esakia spaces with Y finite and let $f : X \to Y$ be a partial Esakia morphism.*

(1) $\mathrm{dom}(f)$ *is an Esakia space in the induced topology and order.*
(2) f *restricted to* $\mathrm{dom}(f)$ *is an Esakia morphism.*

Proof (1). It is well known (see, e.g., Esakia 2019, Theorem 3.2.6) that a clopen subset of an Esakia space is an Esakia space in the induced topology and order. Thus, the result is immediate from Lemma 4.65(2).

(2). That f is a p-morphism follows from Definition 4.63(1, 2) and that f is continuous is proved in Bezhanishvili and Bezhanishvili (2009, Lemma 3.9).

Let A, B be Heyting algebras and X_A, X_B their Esakia spaces. Given an implicative semilattice homomorphism $h : A \to B$, define $h_* : X_B \to X_A$ by setting

$$\mathrm{dom}(h_*) = \{x \in X_B \mid h^{-1}(x) \in X_A\}$$

and for $x \in \mathrm{dom}(h_*)$ by putting $h_*(x) = h^{-1}(x)$.

Lemma 4.67 (Bezhanishvili and Bezhanishvili 2009, Theorem 3.14) $h_* : X_B \to X_A$ *is a partial Esakia morphism.*

Conversely, let X, Y be Esakia spaces and $f : X \to Y$ a partial Esakia morphism. Let X^*, Y^* be the Heyting algebras of clopen upsets of X, Y and define $f^* : Y^* \to X^*$ by

$$f^*(U) = X \setminus {\downarrow} f^{-1}(Y \setminus U)$$

for each $U \in Y^*$.

Lemma 4.68 (Bezhanishvili and Bezhanishvili 2009, Theorem 3.15) $f^* : Y^* \to X^*$ *is an implicative semilattice homomorphism.*

As was shown in Bezhanishvili and Bezhanishvili (2009, Theorem 3.27), this correspondence extends to a categorical duality between the category of Heyting algebras and implicative semilattice homomorphisms and the category of Esakia spaces and partial Esakia morphisms.

Definition 4.69 (Bezhanishvili and Bezhanishvili 2009, Definition 3.30) Let X and Y be Esakia spaces. We call a partial Esakia morphism $f : X \to Y$ *cofinal* if for each $x \in X$ there is $z \in \mathrm{dom}(f)$ such that $x \le z$.[5]

[5] In Bezhanishvili and Bezhanishvili (2009, Definition 3.30) these morphisms were called well partial Esakia morphisms.

By Bezhanishvili and Bezhanishvili (2009, Sect. 3.5), bounded implicative semilattice homomorphisms $h : A \to B$ dually correspond to cofinal partial Esakia morphisms $f : X_B \to X_A$.

We next connect the $\vee$-closed domain condition we discussed in Sect. 4.4.1 with Zakharyaschev's closed domain condition, which is one of the main tools in Zakharyaschev's frame-theoretic development of canonical formulas (Zakharyaschev 1989).

Let X, Y be Esakia spaces and $f : X \to Y$ a partial Esakia morphism. For $x \in X$ let $\min f[\uparrow x]$ be the set of minimal elements of $f[\uparrow x]$. Since $f[\uparrow x]$ is closed, $f[\uparrow x] \subseteq \uparrow \min f[\uparrow x]$ (see Esakia 2019, Theorem 3.2.1).

Definition 4.70 Let X, Y be Esakia spaces, $f : X \to Y$ a partial Esakia morphism, and $\mathfrak{D}$ a (possibly empty) set of antichains in Y. We say that f satisfies *Zakharyaschev's closed domain condition* (ZCDC for short) for $\mathfrak{D}$ if $x \notin \text{dom}(f)$ implies $\min f[\uparrow x] \notin \mathfrak{D}$.

Let A, B be Heyting algebras, $h : A \to B$ an implicative semilattice homomorphism, and $a, b \in A$. Let also X_A, X_B be the Esakia spaces of A, B, $f : X_B \to X_A$ the partial Esakia morphism corresponding to h, and $\zeta(a), \zeta(b)$ the clopen upsets of X_A corresponding to a, b. We let

$$\mathfrak{D}_{\zeta(a),\zeta(b)} = \{\text{antichains } \mathfrak{d} \text{ in } \zeta(a) \cup \zeta(b) \mid \mathfrak{d} \cap (\zeta(a) \setminus \zeta(b)) \neq \varnothing \\ \text{and } \mathfrak{d} \cap (\zeta(b) \setminus \zeta(a)) \neq \varnothing\}.$$

Lemma 4.71 (Bezhanishvili and Bezhanishvili 2009, Lemma 3.40) *Let A, B be Heyting algebras, $h : A \to B$ an implicative semilattice homomorphism, and $a, b \in A$. Let also X_A, X_B be the Esakia spaces of A, B and $f : X_B \to X_A$ the partial Esakia morphism corresponding to h. Then $h(a \vee b) = h(a) \vee h(b)$ iff f satisfies* ZCDC *for $\mathfrak{D}_{\zeta(a),\zeta(b)}$.*

Proof Using duality, it is sufficient to prove that for any clopen upsets U, V we have $f^*(U \cup V) = f^*(U) \cup f^*(V)$ iff f satisfies ZCDC for $\mathfrak{D}_{U,V}$.

$\Rightarrow$: Let $x \notin \text{dom}(f)$. If $\min f[\uparrow x] \in \mathfrak{D}_{U,V}$, then $f[\uparrow x] = \uparrow\min f[\uparrow x] \subseteq U \cup V$, but neither $f[\uparrow x] \subseteq U$ nor $f[\uparrow x] \subseteq V$. Therefore, $x \in f^*(U \cup V)$, but $x \notin f^*(U)$ and $x \notin f^*(V)$. This contradicts $f^*(U \cup V) = f^*(U) \cup f^*(V)$. Consequently, $\min f[\uparrow x] \notin \mathfrak{D}_{U,V}$, and so f satisfies ZCDC for $\mathfrak{D}_{U,V}$.

$\Leftarrow$: It is sufficient to show that $f^*(U \cup V) \subseteq f^*(U) \cup f^*(V)$ since the other inclusion always holds. Let $x \in f^*(U \cup V)$. Then $f[\uparrow x] \subseteq U \cup V$. We have that $x \in \text{dom}(f)$ or $x \notin \text{dom}(f)$. If $x \in \text{dom}(f)$, then $f[\uparrow x] = \uparrow f(x)$. Therefore, $f[\uparrow x] \subseteq U \cup V$ implies $\uparrow f(x) \subseteq U \cup V$, hence $\uparrow f(x) \subseteq U$ or $\uparrow f(x) \subseteq V$. Thus, $x \in f^*(U)$ or $x \in f^*(V)$, and so $x \in f^*(U) \cup f^*(V)$. On the other hand, if $x \notin \text{dom}(f)$, then as f satisfies ZCDC for $\mathfrak{D}_{U,V}$, we obtain that $\min f[\uparrow x] \notin \mathfrak{D}_{U,V}$. Therefore, $\min f[\uparrow x] \subseteq U$ or $\min f[\uparrow x] \subseteq V$. Thus, $f[\uparrow x] \subseteq \uparrow\min f[\uparrow x] \subseteq U$ or $f[\uparrow x] \subseteq \uparrow\min f[\uparrow x] \subseteq V$, which yields that $x \in f^*(U)$ or $x \in f^*(V)$. Consequently, $x \in f^*(U) \cup f^*(V)$, and so $f^*(U \cup V) \subseteq f^*(U) \cup f^*(V)$.

We are ready to give the dual reading of subframe canonical formulas of Sect. 4.4.1. Let A be a finite subdirectly irreducible Heyting algebra. By finite Esakia duality, its dual is a finite rooted poset P. Let $D \subseteq A^2$. We call the set $\mathfrak{D} = \{\mathfrak{D}_{\zeta(a),\zeta(b)} \mid (a, b) \in D\}$ *the set of antichains of P associated with D*. The following theorem is a consequence of the Subframe Jankov Lemma, Lemma 4.71, and generalized Esakia duality.

Theorem 4.72 (Bezhanishvili and Bezhanishvili 2009, Corollary 5.5) *Let A be a finite subdirectly irreducible Heyting algebra and P its dual finite rooted poset. Let $D \subseteq A^2$ and $\mathfrak{D}$ be the set of antichains of P associated with D. Then for each Esakia space X, we have $X \not\models \alpha(A, D)$ iff there is a closed upset Y of X and an onto cofinal partial Esakia morphism $f : Y \twoheadrightarrow P$ such that f satisfies* ZCDC *for $\mathfrak{D}$.*

From this we derive the following dual reading of Theorem 4.41.

Theorem 4.73 (Bezhanishvili and Bezhanishvili 2009, Corollaries 5.9 and 5.11) *Suppose* $\mathsf{IPC} \nvdash \varphi$ *and $(A_1, D_1^\vee), \ldots, (A_{k(n)}, D_{k(n)}^\vee)$ is the corresponding list of finite refutation patterns of φ. For each $i \leq k(n)$ let P_i be the dual finite rooted poset of A_i and $\mathfrak{D}_i$ the set of antichains of P_i associated with $D_i^\vee$. Then for an arbitrary Esakia space X, we have:*

(1) *$X \not\models \varphi$ iff there is $i \leq k(n)$, a closed upset Y of X, and an onto cofinal partial Esakia morphism $f : Y \twoheadrightarrow P_i$ satisfying* ZCDC *for $\mathfrak{D}_i$.*

(2) *$X \models \varphi$ iff $X \models \bigwedge_{i=1}^{k(n)} \alpha(A_i, D_i^\vee)$.*

We have a parallel situation with negation-free canonical formulas, the main difference being that "cofinal" has to be dropped from the consideration. We thus arrive at the following negation-free analogue of Theorem 4.73.

Theorem 4.74 (Bezhanishvili and Bezhanishvili 2009, Corollaries 5.16 and 5.17) *Suppose φ is a negation-free formula,* $\mathsf{IPC} \nvdash \varphi$, *and $(A_1, D_1^\vee), \ldots, (A_{k(n)}, D_{k(n)}^\vee)$ is the corresponding list of finite refutation patterns of φ. For each $i \leq k(n)$ let P_i be the dual finite rooted poset of A_i and $\mathfrak{D}_i$ the set of antichains of P_i associated with $D_i^\vee$. Then for an arbitrary Esakia space X, we have:*

(1) *$X \not\models \varphi$ iff there is $i \leq k(n)$, a closed upset Y of X, and an onto partial Esakia morphism $f : Y \twoheadrightarrow P_i$ satisfying* ZCDC *for $\mathfrak{D}_i$.*

(2) *$X \models \varphi$ iff $X \models \bigwedge_{i=1}^{k(n)} \beta(A_i, D_i^\vee)$.*

Remark 4.75 Zakharyschev's canonical formulas (Zakharyaschev 1989; Chagrov and Zakharyaschev 1997) are equivalent to subframe canonical formulas (see Bezhanishvili and Bezhanishvili 2009, Remark 5.6).

4.5.2 Stable Canonical Formulas Dually

Let A, B be Heyting algebras. We recall that a map $h : A \to B$ is a *lattice homomorphism* if

$$h(a \wedge b) = h(a) \wedge h(b) \text{ and } h(a \vee b) = h(a) \vee h(b)$$

for each $a, b \in A$. A lattice homomorphism $h : A \to B$ is *bounded* if $h(0) = 0$ and $h(1) = 1$. It is a consequence of Priestley duality for bounded distributive lattices (Priestley 1970, 1972) that bounded lattice homomorphisms $h : A \to B$ dually correspond to continuous order-preserving maps $f : X_B \to X_A$.

Definition 4.76 Let X, Y be Esakia spaces. We call a map $f : X \to Y$ a *stable morphism* if f is continuous and order-preserving.

Remark 4.77 The name "stable morphism" comes from modal logic, where it is used for continuous maps that preserve the relation (see Bezhanishvili et al. 2016a, 2018a; Ilin 2018). In Priestley duality for bounded distributive lattices these maps are known as Priestley morphisms.

Definition 4.78 (Bezhanishvili and Bezhanishvili 2017, Definition 4.1) Let X, Y be Esakia spaces and $f : X \to Y$ a stable morphism.

1. Let D be a clopen subset of Y. We say that f satisfies the *stable domain condition* (SDC for short) for D if

$$\uparrow f(x) \cap D \neq \varnothing \Rightarrow f[\uparrow x] \cap D \neq \varnothing.$$

2. Let $\mathfrak{D}$ be a collection of clopen subsets of Y. We say that $f : X \to Y$ satisfies the *stable domain condition* (SDC for short) for $\mathfrak{D}$ if f satisfies SDC for each $D \in \mathfrak{D}$.

Lemma 4.79 (Bezhanishvili and Bezhanishvili 2017, Lemma 4.3) *Let A, B be Heyting algebras, $h : A \to B$ a bounded lattice homomorphism, and $a, b \in A$. Let also X_A, X_B be the Esakia spaces of A, B, $f : X_B \to X_A$ the stable morphism corresponding to h, and $D_{\zeta(a),\zeta(b)} = \zeta(a) \setminus \zeta(b)$. Then $h(a \to b) = h(a) \to h(b)$ iff f satisfies* SDC *for $D_{\zeta(a),\zeta(b)}$.*

Proof Using duality it is sufficient to show that for any clopen upsets U, V we have $f^{-1}(U) \to f^{-1}(V) = f^{-1}(U \to V)$ iff f satisfies SDC for $D_{U,V}$.

$\Rightarrow$: Suppose that $\uparrow f(x) \cap D_{U,V} \neq \varnothing$. Then $\uparrow f(x) \cap U \nsubseteq V$. Therefore, $f(x) \notin U \to V$, so $x \notin f^{-1}(U \to V)$. Thus, $x \notin f^{-1}(U) \to f^{-1}(V)$, and so $\uparrow x \cap f^{-1}(U) \nsubseteq f^{-1}(V)$. This implies $f[\uparrow x] \cap U \nsubseteq V$, and hence $f[\uparrow x] \cap D_{U,V} \neq \varnothing$. Consequently, f satisfies SDC for $D_{U,V}$.

$\Leftarrow$: It is sufficient to show that $f^{-1}(U) \to f^{-1}(V) \subseteq f^{-1}(U \to V)$ since the other inclusion always holds. Suppose that $x \notin f^{-1}(U \to V)$. Then $f(x) \notin$

$U \to V$. Therefore, $\uparrow f(x) \cap U \nsubseteq V$, which means that $\uparrow f(x) \cap D_{U,V} \neq \varnothing$. Thus, $f[\uparrow x] \cap D_{U,V} \neq \varnothing$ by SDC for $D_{U,V}$. This means that $\uparrow x \cap (f^{-1}(U) \setminus f^{-1}(V)) \neq \varnothing$. Consequently, $\uparrow x \cap f^{-1}(U) \nsubseteq f^{-1}(V)$, implying that $x \notin f^{-1}(U) \to f^{-1}(V)$.

We recall from Sect. 4.2.3 that the Esakia dual of a subdirectly irreducible Heyting algebra is a strongly rooted Esakia space, the Esakia dual of a finite subdirectly irreducible Heyting algebra is a finite rooted poset, and the Esakia dual of a subdirectly irreducible homomorphic image of a Heyting algebra A is a strongly rooted closed upset of the Esakia dual of A. We also recall (Priestley 1970, Theorem 3) that bounded sublattices of A dually correspond to onto stable morphisms from the Esakia dual of A. Thus, Lemma 4.79 yields the following dual reading of the Stable Jankov Lemma and Theorem 4.58.

Theorem 4.80 (Bezhanishvili and Bezhanishvili 2017, Theorem 4.4)

(1) *Let A be a finite subdirectly irreducible Heyting algebra and P its dual finite rooted poset. For $D \subseteq A^2$, let $\mathfrak{D} = \{D_{\zeta(a),\zeta(b)} \mid (a,b) \in D\}$. Then for each Esakia space X, we have $X \not\models \gamma(A, D)$ iff there is a strongly rooted closed upset Y of X and an onto stable morphism $f : Y \twoheadrightarrow P$ such that f satisfies* SDC *for $\mathfrak{D}$.*
(2) *Suppose* $\mathsf{IPC} \nvdash \varphi$ *and $(A_1, D_1^{\to}), \ldots, (A_{k(n)}, D_{k(n)}^{\to})$ is the corresponding list of finite refutation patterns of φ. For each $i \le k(n)$, let P_i be the dual finite rooted poset of A_i and $\mathfrak{D}_i = \{D_{\zeta(a),\zeta(b)} \mid (a,b) \in D_i^{\to}\}$. Then for each strongly rooted Esakia space X, the following conditions are equivalent:*

(a) $X \not\models \varphi$.
(b) *There is $i \le k(n)$ and an onto stable morphism $f : Y \twoheadrightarrow P_i$ such that f satisfies* SDC *for $\mathfrak{D}_i$.*
(c) *There is $i \le k(n)$, a strongly rooted closed upset Y of X, and an onto stable morphism $f : Y \twoheadrightarrow P_i$ such that f satisfies* SDC *for $\mathfrak{D}_i$.*

(3) *For each strongly rooted Esakia space X, we have*

$$X \models \varphi \text{ iff } X \models \bigwedge_{i=1}^{k(n)} \gamma(A_i, D_i^{\to}).$$

Remark 4.81 When comparing the dual approaches to these two types of canonical formulas, we see that in the case of subframe canonical formulas we work with cofinal partial Esakia morphisms whose duals are bounded implicative semilattice homomorphisms, and Zakharyaschev's closed domain condition ZCDC provides means for the dual to also preserve $\vee$. On the other hand, in the case of stable canonical formulas we work with stable morphisms whose duals are bounded lattice homomorphisms, and the stable domain condition SDC provides means for the dual to also preserve $\to$. In the end, both approaches provide the same result, that all intermediate logics are axiomatizable either by subframe canonical formulas or by stable canonical formulas. However, both the algebra and geometry of the two approaches are different.

4.6 Subframe and Cofinal Subframe Formulas

As we saw in Remark 4.36, when the closed domain D of a subframe canonical formula $\alpha(A, D)$ is the entire A^2, then $\alpha(A, D)$ coincides with the Jankov formula $\mathcal{J}(A)$. Another extreme case is when $D = \varnothing$. In this case, we simply drop D and write $\alpha(A)$ or $\beta(A)$ depending on whether we work with $\alpha(A, D)$ (in the full signature of subframe canonical formulas) or with $\beta(A, D)$ (in the negation-free signature). As a result, we arrive at subframe formulas (when working with $\beta(A)$) or cofinal subframe formulas (when working with $\alpha(A)$), and the corresponding subframe and cofinal subframe logics. We briefly recall that subframe logics were first studied by Fine (1985) and cofinal subframe logics by Zakharyaschev (1996) for extensions of $\mathsf{K4}$. The study of subframe and cofinal subframe intermediate logics was initiated by Zakharyaschev (1989). Both Fine and Zakharyaschev utilized relational semantics. In this section we survey the theory of these logics utilizing algebraic semantics.

Definition 4.82 Let A be a finite subdirectly irreducible Heyting algebra.

1. We call $\alpha(A, \varnothing)$ the *cofinal subframe formula* of A and denote it by $\alpha(A)$.
2. We call $\beta(A, \varnothing)$ the *subframe formula* of A and denote it by $\beta(A)$.

The names "subframe formula" and "cofinal subframe formula" are justified by their connection to subframes and cofinal subframes of Esakia spaces discussed below (see Theorems 4.96 and 4.97).

Let A, B be Heyting algebras with A finite and subdirectly irreducible. As an immediate consequence of the Subframe Jankov Lemma we obtain that $B \not\models \alpha(A)$ iff there is a homomorphic image C of B and a bounded implicative semilattice embedding $h : A \rightarrowtail C$, and similarly for $\beta(A)$. However, this result can be improved by dropping homomorphic images from the consideration. For this we require the following lemma.

Lemma 4.83 *Let A, B, C be finite Heyting algebras and $h : B \twoheadrightarrow C$ an onto Heyting homomorphism.*

(1) *If $e : A \rightarrowtail C$ is an implicative semilattice embedding, then there is an implicative semilattice embedding $k : A \rightarrowtail B$ such that $h \circ k = e$.*
(2) *If in addition e is bounded, then so is k.*

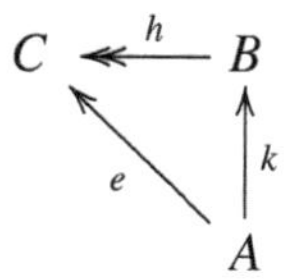

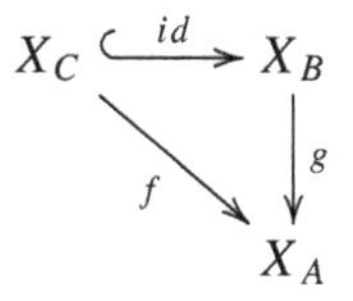

Proof (1). Let X_A, X_B, and X_C be the dual finite posets of A, B, and C. We identify A with the upsets of X_A, B with the upsets of X_B, and C with the upsets of X_C. Since C is a homomorphic image of B, we have that X_C is (isomorphic to) an upset of X_B. Also, since $e : A \to C$ is an implicative lattice embedding, there is an onto partial Esakia morphism $f : X_C \to X_A$ (see Bezhanishvili and Bezhanishvili 2009, Lemma 3.29). Viewing f also as a partial map $f : X_B \to X_A$, it is straightforward that f satisfies conditions (1), (2), and (5) of Definition 4.63. Therefore, by Chagrov and Zakharyaschev (1997, Theorem 9.7), $f^* : \mathsf{Up}(X_A) \to \mathsf{Up}(X_B)$ is an implicative semilattice embedding, and it is clear that $h \circ f^* = e$.

(2). Suppose in addition that e is bounded. Then $f : X_C \to X_A$ is cofinal, so $\max(X_C) \subseteq \mathrm{dom}(f)$. Let $y \in \max(X_A)$ and define a partial map $g : X_B \to X_A$ by setting $\mathrm{dom}(g) = \mathrm{dom}(f) \cup \max(X_B)$ and for $x \in \mathrm{dom}(g)$ letting

$$g(x) = \begin{cases} f(x) & \text{if } x \in \mathrm{dom}(f) \\ y & \text{if } x \in \max(X_B) \setminus \mathrm{dom}(f) \end{cases}$$

It is then straightforward that g satisfies conditions (1), (2), and (5) of Definition 4.63. Therefore, $g^* : \mathsf{Up}(X_A) \to \mathsf{Up}(X_B)$ is an implicative semilattice embedding. It follows from the definition of g that ${\downarrow}\mathrm{dom}(g) = X_B$ and $g|_{X_C} = f$. Thus, g^* is bounded (see Bezhanishvili and Bezhanishvili 2009, Lemma 3.32) and $h \circ g^* = f^*$.

Theorem 4.84

(1) *Let A be a finite subdirectly irreducible Heyting algebra and B an arbitrary Heyting algebra.*

(a) $B \not\models \beta(A)$ *iff there is an implicative semilattice embedding* $h : A \rightarrowtail B$.
(b) $B \not\models \alpha(A)$ *iff there is a bounded implicative semilattice embedding* $h : A \rightarrowtail B$.

(2) *Let A be a finite subdirectly irreducible Heyting algebra, P its dual finite rooted poset, and X an arbitrary Esakia space.*

(a) $X \not\models \beta(A)$ *iff there is an onto partial Esakia morphism* $f : X \twoheadrightarrow P$.
(b) $X \not\models \alpha(A)$ *iff there is an onto cofinal partial Esakia morphism* $f : X \twoheadrightarrow P$.

Proof Since (2) is the dual statement of (1), it is sufficient to prove (1). We first prove (1a). The right to left implication is follows from Theorem 4.47(1). For the left to right implication, suppose $B \not\models \beta(A)$. By the Selective Filtration Lemma, there is a finite $(\wedge, \to)$-subalgebra D of B such that $D \not\models \beta(A)$. By the Subframe Jankov Lemma, there is a homomorphic image C of D and an implicative semilattice embedding of A into C. By Lemma 4.83(1), there is an implicative semilattice embedding of A into D, and hence there is an implicative semilattice embedding of A into B. The proof of (1b) is similar but uses Lemma 4.83(2).

Remark 4.85 Theorem 4.84 improves (Bezhanishvili and Bezhanishvili 2009, Corollary 5.24) in that $B \not\models \beta(A)$ is equivalent to the existence of an implicative semilattice embedding of A directly into B, rather than a homomorphic image of B (and the same for $\alpha(A)$).

Definition 4.86 Let L be an intermediate logic.

1. We call L a *subframe logic* if there is a family $\{A_i \mid i \in I\}$ of finite subdirectly irreducible Heyting algebras such that $\mathsf{L} = \mathsf{IPC} + \{\beta(A_i) \mid i \in I\}$.
2. We call L a *cofinal subframe logic* if there is a family $\{A_i \mid i \in I\}$ of finite subdirectly irreducible Heyting algebras such that $\mathsf{L} = \mathsf{IPC} + \{\alpha(A_i) \mid i \in I\}$.
3. Let Λ_{Subf} be the set of subframe logics and Λ_{CSubf} the set of cofinal subframe logics.

Theorem 4.87 (Chagrov and Zakharyaschev 1997, Sect. 11.3)

(1) *Λ_{Subf} is a complete sublattice of Λ_{CSubf} and Λ_{CSubf} is a complete sublattice of Λ.*
(2) *The cardinalities of both Λ_{Subf} and $\Lambda_{\mathsf{CSubf}} \setminus \Lambda_{\mathsf{Subf}}$ are that of the continuum.*

Definition 4.88 Let φ be a propositional formula.

1. Call φ a *disjunction-free formula* if φ does not contain disjunction.
2. Call φ a *DN-free formula* if φ does not contain disjunction and negation.

Since $\alpha(A)$ encodes the bounded implicative semilattice structure and $\beta(A)$ the implicative semilattice structure of A, one would expect that subframe logics are exactly those intermediate logics that are axiomatizable by DN-free formulas and cofinal subframe logics are those that are axiomatizable by disjunction-free formulas. This indeed turns out to be the case, as was shown by Zakharyaschev (1989) using relational semantics. To give an algebraic proof and obtain other equivalent conditions for an intermediate logic to be a subframe or cofinal subframe logic, we introduce the following notation.

Definition 4.89

1. Let A, B be Heyting algebras with $A \subseteq B$. We say that A is a $(\wedge, \to)$*-subalgebra* of B if A is closed under $\wedge$ and $\to$, and we say that A is a $(\wedge, \to, 0)$*-subalgebra* of B if in addition $0 \in A$.
2. We say that a class $\mathcal{K}$ of Heyting algebras is *closed under* $(\wedge, \to)$*-subalgebras* if from $B \in \mathcal{K}$ and A being isomorphic to a $(\wedge, \to)$-subalgebra of B it follows that $A \in \mathcal{K}$.

3. We say that a class $\mathcal{K}$ of Heyting algebras is *closed under* $(\wedge, \to, 0)$*-subalgebras* if from $B \in \mathcal{K}$ and A being isomorphic to a $(\wedge, \to, 0)$-subalgebra of B it follows that $A \in \mathcal{K}$.

Theorem 4.90 *For an intermediate logic* L*, the following conditions are equivalent.*

(1) L *is a subframe logic.*
(2) L *is axiomatizable by DN-free formulas.*
(3) *The variety* $\mathcal{V}(\mathsf{L})$ *is closed under* $(\wedge, \to)$*-subalgebras.*
(4) *There is a class* $\mathcal{K}$ *of* L*-algebras closed under* $(\wedge, \to)$*-subalgebras that generates* $\mathcal{V}(\mathsf{L})$.

Proof (1)⇒(2). If L is a subframe logic, then L is axiomatizable by subframe formulas. But subframe formulas are DN-free formulas by definition. Thus, L is axiomatizable by DN-free formulas.

(2)⇒(3). Suppose $\mathsf{L} = \mathsf{IPC} + \{\varphi_i \mid i \in I\}$ where each φ_i is a DN-formula. Let A, B be Heyting algebras with $B \in \mathcal{V}(\mathsf{L})$ and A isomorphic to a $(\wedge, \to)$-subalgebra of B. From $B \in \mathcal{V}(\mathsf{L})$ it follows that each φ_i is valid on B. Since A is isomorphic to a $(\wedge, \to)$-subalgebra of B, each φ_i is also valid on A. Thus, $A \in \mathcal{V}(\mathsf{L})$.

(3)⇒(4). This is obvious.

(4)⇒(1). Let $\mathcal{X}$ be the set of all finite (non-isomorphic) subdirectly irreducible Heyting algebras such that $A \not\models \mathsf{L}$, and let

$$\mathsf{M} = \mathsf{IPC} + \{\beta(A) \mid A \in \mathcal{X}\}.$$

It is sufficient to show that $\mathsf{L} = \mathsf{M}$. Let B be a subdirectly irreducible Heyting algebra. It is enough to prove that $B \models \mathsf{L}$ iff $B \models \mathsf{M}$. First suppose that $B \not\models \mathsf{L}$. Then there is $\varphi \in \mathsf{L}$ such that $B \not\models \varphi$. By the Selective Filtration Lemma, there is a finite subdirectly irreducible Heyting algebra A such that $A \not\models \varphi$ and A is a $(\wedge, \to)$-subalgebra of B. By Theorem 4.84(1a), $B \not\models \beta(A)$. Therefore, $B \not\models \mathsf{M}$. Thus, $\mathsf{L} \subseteq \mathsf{M}$.

For the reverse inclusion, since L is the logic of $\mathcal{K}$, it is sufficient to show that if $B \in \mathcal{K}$, then $B \models \mathsf{M}$. If $B \not\models \mathsf{M}$, then $B \not\models \beta(A)$ for some $A \in \mathcal{X}$. By Theorem 4.84(1a), A is isomorphic to a $(\wedge, \to)$-subalgebra of B. Since $B \in \mathcal{K}$ and $\mathcal{K}$ is closed under $(\wedge, \to)$-subalgebras, $A \in \mathcal{K}$. Thus, $A \models \mathsf{L}$, a contradiction. Consequently, $B \models \mathsf{M}$, finishing the proof.

Theorem 4.90 directly generalizes to cofinal subframe logics.

Theorem 4.91 *For an intermediate logic* L*, the following conditions are equivalent.*

(1) L *is a cofinal subframe logic.*
(2) L *is axiomatizable by disjunction-free formulas.*
(3) *The variety* $\mathcal{V}(\mathsf{L})$ *is closed under* $(\wedge, \to, 0)$*-subalgebras.*
(4) *There is a class* $\mathcal{K}$ *of* L*-algebras closed under* $(\wedge, \to, 0)$*-subalgebras that generates* $\mathcal{V}(\mathsf{L})$.

Theorems 4.90 and 4.91 allow us to give a simple proof that each subframe and cofinal subframe logic has the fmp. This result for subframe modal logics above $\mathsf{K4}$ was first established by Fine (1985), for cofinal subframe intermediate logics by Zakharyaschev (1989), and for cofinal subframe modal logics above $\mathsf{K4}$ by Zakharyaschev (1996). Both Fine and Zakharyaschev used relational semantics. We will instead prove this result by utilizing the Selective Filtration Lemma. This is closely related to the work of McKay (1968).

Theorem 4.92

(1) *Each subframe logic has the fmp.*
(2) *Each cofinal subframe logic has the fmp.*

Proof Since $\Lambda_{\mathsf{Subf}} \subseteq \Lambda_{\mathsf{CSubf}}$, it is sufficient to prove (2). Let L be a cofinal subframe logic. By Theorem 4.91, L is axiomatized by a set of disjunction-free formulas $\{\chi_i \mid i \in I\}$. Suppose $\mathsf{L} \nvdash \varphi$. By algebraic completeness, there is an L-algebra B such that $B \not\models \varphi$. By the Selective Filtration Lemma, there is a finite $(\wedge, \to, 0)$-subalgebra A of B such that $A \not\models \varphi$. Since each χ_i is disjunction-free, we have that $B \models \chi_i$ implies $A \models \chi_i$ for each $i \in I$. Thus, A is an L-algebra, and hence L has the fmp.

We next justify the name "subframe logic" by connecting these logics to subframes of Esakia spaces.

Definition 4.93 (Chagrov and Zakharyaschev 1997, p. 289) Let X be an Esakia space. We call $Y \subseteq X$ a *subframe* of X if Y is an Esakia space in the induced topology and order and the partial identity map $X \to Y$ satisfies conditions (1), (2), and (5) of Definition 4.63.

The following is a convenient characterization of subframes of Esakia spaces.

Theorem 4.94 (Bezhanishvili and Ghilardi 2007, Lemma 2) *Let X be an Esakia space. Then $Y \subseteq X$ is a subframe of X iff Y is a closed subset of X and U a clopen subset of Y (in the induced topology) implies ${\downarrow}U$ is a clopen subset of X.*

We call a subframe Y of X *cofinal* if ${\downarrow}Y = X$. In Chagrov and Zakharyaschev (1997, p. 295) a weaker notion of cofinality was used, that ${\uparrow}Y \subseteq {\downarrow}Y$. See Bezhanishvili and Bezhanishvili (2009, Section 4) for the comparison of the two notions.

Recall that Esakia spaces X, Y are *isomorphic* in Esa if they are homeomorphic and order-isomorphic.

Definition 4.95 Let $\mathcal{K}$ be a class of Esakia spaces and X, Y Esakia spaces.

1. We call $\mathcal{K}$ *closed under subframes* if from $X \in \mathcal{K}$ and Y being isomorphic to a subframe of X it follows that $Y \in \mathcal{K}$.
2. We call $\mathcal{K}$ *closed under cofinal subframes* if from $X \in \mathcal{K}$ and Y being isomorphic to a cofinal subframe of X it follows that $Y \in \mathcal{K}$.

Dualizing Theorem 4.90 we obtain:

Theorem 4.96 *For an intermediate logic* L*, the following conditions are equivalent.*

(1) L *is a subframe logic.*
(2) L *is axiomatizable by DN-free formulas.*
(3) *The class of all Esakia spaces validating* L *is closed under subframes.*
(4) L *is sound and complete with respect to a class* $\mathcal{K}$ *of Esakia spaces that is closed under subframes.*

Proof (1)⇒(2). This is proved in Theorem 4.90.

(2)⇒(3). Let $X \models \mathsf{L}$ and Y be isomorphic to a subframe of X. Then $X^* \models \mathsf{L}$ and Y^* is isomorphic to a $(\wedge, \to)$-subalgebra of X^*. By Theorem 4.90, $Y^* \models \mathsf{L}$, and hence $Y \models \mathsf{L}$.

(3)⇒(4). This is straightforward.

(4)⇒(1). Let $\mathcal{K}^* = \{X^* \mid X \in \mathcal{K}\}$. We proceed as in the proof of Theorem 4.90 by showing that $\mathsf{L} = \mathsf{IPC} + \{\beta(A) \mid A \in \mathcal{X}\}$ (where we recall from the proof of Theorem 4.90 that $\mathcal{X}$ is the set of all finite (non-isomorphic) subdirectly irreducible Heyting algebras A such that $A \not\models \mathsf{L}$). Let B be a subdirectly irreducible Heyting algebra. That $B \not\models \mathsf{L}$ implies $B \not\models \{\beta(A) \mid A \in \mathcal{X}\}$ is proved as in Theorem 4.90. For the converse, since L is the logic of $\mathcal{K}$, it is sufficient to assume that $B = X^*$ for some $X \in \mathcal{K}$. If $B \not\models \beta(A)$ for some $A \in \mathcal{X}$, then by Theorem 4.84(1a), A is isomorphic to a $(\wedge, \to)$-subalgebra of B. Therefore, there is an onto partial Esakia morphism $f : X \to X_A$. Since X_A is finite, $\mathrm{dom}(f)$ is a clopen subset of X by Lemma 4.65(2). Therefore, it follows from Theorem 4.94 that $\mathrm{dom}(f)$ is a subframe of X. Thus, $\mathrm{dom}(f) \in \mathcal{K}$, so $\mathrm{dom}(f) \models \mathsf{L}$. But $f : \mathrm{dom}(f) \to X_A$ is an onto Esakia morphism by Lemma 4.66(2). Therefore, $X_A \models \mathsf{L}$, and hence $A \models \mathsf{L}$. The obtained contradiction proves that $B \models \{\beta(A) \mid A \in \mathcal{X}\}$, finishing the proof.

We also have the following dual version of Theorem 4.91, the proof of which is analogous to that of Theorem 4.96, and we skip it.

Theorem 4.97 *For an intermediate logic* L*, the following conditions are equivalent.*

(1) L *is a cofinal subframe logic.*
(2) L *is axiomatizable by disjunction-free formulas.*
(3) *The class of all Esakia spaces validating* L *is closed under cofinal subframes.*
(4) L *is sound and complete with respect to a class* $\mathcal{K}$ *of Esakia spaces that is closed under cofinal subframes.*

Remark 4.98 There are several other interesting characterizations of subframe and cofinal subframe logics.

1. In Bezhanishvili and Ghilardi (2007) it is shown that subframes of Esakia spaces correspond to nuclei on Heyting algebras and that cofinal subframes to dense nuclei. From this it follows that an intermediate logic L is a subframe logic iff its corresponding variety $\mathcal{V}(\mathsf{L})$ is a nuclear variety and that L is a cofinal subframe logic iff $\mathcal{V}(\mathsf{L})$ is a dense nuclear variety.
2. A different description of subframe and cofinal subframe formulas is given in Bezhanishvili (2006, Sect. 3.3.3) and Bezhanishvili and de Jongh (2018) (see also Ilin 2018), where it is shown that subframe formulas are equivalent to the NNIL-formulas of Visser et al. (1995).
3. Many important properties of logics (such as, e.g., canonicity and strong Kripke completeness) coincide for subframe and cofinal subframe logics (see, e.g., Chagrov and Zakharyaschev 1997, Theorems 11.26 and 11.28).

We finish this section with some examples of subframe and cofinal subframe logics. To simplify notation, we write $\beta(P)$ instead of $\beta(P^*)$ and $\alpha(P)$ instead of $\alpha(P^*)$. Then, recalling Fig. 4.1, we have:

Theorem 4.99 (Chagrov and Zakharyaschev 1997, p. 317, Table 9.7)

(1) $\mathsf{CPC} = \mathsf{LC} + \beta(\mathfrak{C}_2)$, *hence* CPC *is a subframe logic.*
(2) $\mathsf{LC} = \mathsf{IPC} + \beta(\mathfrak{F}_2)$, *hence* LC *is a subframe logic.*
(3) $\mathsf{BD}_n = \mathsf{IPC} + \beta(\mathfrak{C}_{n+1})$, *hence* BD_n *is a subframe logic.*
(4) $\mathsf{LC}_n = \mathsf{LC} + \beta(\mathfrak{C}_{n+1})$, *hence* LC_n *is a subframe logic.*
(5) $\mathsf{BW}_n = \mathsf{IPC} + \beta(\mathfrak{F}_{n+1})$, *hence* BW_n *is a subframe logic.*
(6) $\mathsf{KC} = \mathsf{IPC} + \alpha(\mathfrak{F}_2)$, *hence* KC *is a cofinal subframe logic.*
(7) $\mathsf{BTW}_n = \mathsf{IPC} + \alpha(\mathfrak{F}_{n+1})$, *hence* BTW_n *is a cofinal subframe logic.*

On the other hand, there exist intermediate logics that are not subframe (e.g., KC) and also ones that are not cofinal subframe (e.g., KP).

4.7 Stable Formulas

We can develop the theory of stable formulas which is parallel to that of subframe formulas. As we pointed out in Remark 4.53, if $D = A^2$, then $\gamma(A, D)$ is equivalent to the Jankov formula $\mathcal{J}(A)$. As with subframe formulas, we can consider the second extreme case when $D = \varnothing$. We call the resulting formulas stable formulas and denote them by $\gamma(A)$. Then the theory of (cofinal) stable formulas can be developed in parallel to the theory of (cofinal) subframe formulas. In Sect. 4.7.1 we survey the theory of stable formulas and the resulting stable logics, and in Sect. 4.7.2 that of cofinal stable formulas and the resulting cofinal stable logics. We also compare these new classes of logics to subframe and cofinal subframe logics.

4.7.1 Stable Formulas

Definition 4.100 Let A be a finite subdirectly irreducible Heyting algebra. We call $\gamma(A, \varnothing)$ the *stable formula* of A and denote it by $\gamma(A)$.

Let A, B be Heyting algebras with A finite and subdirectly irreducible. As an immediate consequence of the Stable Jankov Lemma we have $B \not\models \gamma(A)$ iff there is a subdirectly irreducible homomorphic image C of B and a bounded lattice embedding $h : A \rightarrowtail C$. In analogy with what happened in Sect. 4.6, we can improve this by dropping homomorphic images from consideration. For this we require the following lemma, which is an analogue of Lemma 4.83.

Lemma 4.101 (Bezhanishvili and Bezhanishvili 2017, Lemma 6.2) *Let A, B, C be finite Heyting algebras with A subdirectly irreducible, $h : B \twoheadrightarrow C$ an onto Heyting homomorphism, and $e : A \rightarrowtail C$ a bounded lattice embedding. Then there is a bounded lattice embedding $k : A \rightarrowtail B$ such that $h \circ k = e$.*

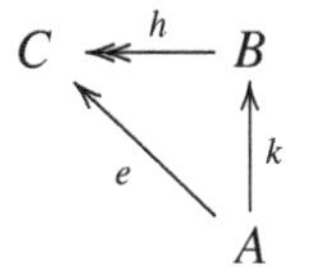

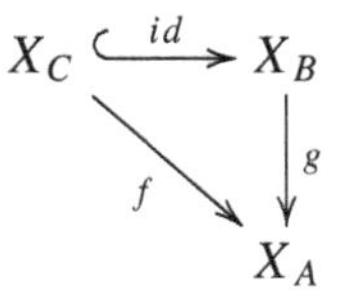

Proof Let X_A, X_B, and X_C be the dual finite posets of A, B, and C. We identify A, B, and C with the upsets of X_A, X_B, and X_C. Since A is subdirectly irreducible, X_A is rooted. As C is a homomorphic image of B, we have that X_C is (isomorphic to) an upset of X_B, and because $e : A \to C$ is a bounded lattice embedding, there is an onto stable map $f : X_C \twoheadrightarrow X_A$ such that $e = f^{-1}$.

Let x be the root of X_A. Define $g : X_B \to X_A$ by

$$g(y) = \begin{cases} f(y) & \text{if } y \in X_C \\ x & \text{otherwise} \end{cases}$$

Clearly g is a well-defined map extending f, and it is onto since f is onto. To see that g is stable, let $y, z \in X_B$ with $y \leq z$. First suppose that $y \in X_C$. Then $z \in X_C$ as X_C is an upset of X_B. Since f is stable, $f(y) \leq f(z)$. Therefore, by the definition of g, we have $g(y) \leq g(z)$. On the other hand, if $y \in X_B \setminus X_C$, then $g(y) = x$. As x

is the root of X_A, we have $x \leq u$ for each $u \in X_A$. Thus, $x \leq g(z)$ for each $z \in X_B$, which implies that $g(y) \leq g(z)$. Consequently, g is an onto stable map extending f.

From this we conclude that $g^{-1} : \mathsf{Up}(X_A) \to \mathsf{Up}(X_B)$ is a bounded lattice embedding such that $h \circ g^{-1} = f^{-1}$.

Theorem 4.102 (Bezhanishvili and Bezhanishvili 2017, Theorem 6.3) *Let A, B be subdirectly irreducible Heyting algebras with A finite. Then $B \not\models \gamma(A)$ iff there is a bounded lattice embedding of A into B.*

Proof The right to left implication follows directly from the Stable Jankov Lemma. For the left to right implication, suppose $B \not\models \gamma(A)$. By the Filtration Lemma, there is a finite bounded sublattice D of B such that $D \not\models \gamma(A)$. By the Stable Jankov Lemma, there is a subdirectly irreducible homomorphic image C of D and a bounded lattice embedding of A into C. By Lemma 4.101, there is a bounded lattice embedding of A into D, and hence a bounded lattice embedding of A into B.

The dual reading of Theorem 4.102 is as follows.

Theorem 4.103 (Bezhanishvili and Bezhanishvili 2017, Theorem 6.5) *Let A be a finite subdirectly irreducible Heyting algebra and P its dual finite rooted poset. For a strongly rooted Esakia space X, we have $X \not\models \gamma(A)$ iff there is an onto stable morphism $f : X \twoheadrightarrow P$.*

Definition 4.104

1. We call an intermediate logic L *stable* if there is a family $\{A_i \mid i \in I\}$ of finite subdirectly irreducible algebras such that $\mathsf{L} = \mathsf{IPC} + \{\gamma(A_i) \mid i \in I\}$.
2. We say that a class $\mathcal{K}$ of subdirectly irreducible Heyting algebras is *closed under bounded sublattices* if for any subdirectly irreducible Heyting algebras A and B, from $B \in \mathcal{K}$ and A being isomorphic to a bounded sublattice of B it follows that $A \in \mathcal{K}$.

In the next theorem, the equivalence of (1) and (2) is given in Bezhanishvili and Bezhanishvili (2017, Theorem 6.11). For further equivalent conditions for an intermediate logic to be stable see Bezhanishvili et al. (2016b, Theorem 5.3).

Theorem 4.105 *For an intermediate logic L, the following conditions are equivalent.*

(1) *L is a stable logic.*
(2) *The class $\mathcal{V}(\mathsf{L})_{\mathsf{si}}$ of subdirectly irreducible L-algebras is closed under bounded sublattices.*
(3) *There is a class $\mathcal{K}$ of subdirectly irreducible L-algebras that is closed under bounded sublattices and generates $\mathcal{V}(\mathsf{L})$.*

Proof (1)⇒(2). Let A, B be subdirectly irreducible Heyting algebras such that B is an L-algebra and A is isomorphic to a bounded sublattice of B. Suppose that A is not an L-algebra. Since L is stable, there is a finite subdirectly irreducible Heyting algebra C such that $\gamma(C) \in \mathsf{L}$ and $A \not\models \gamma(C)$. By Theorem 4.102, C is isomorphic to a bounded sublattice of A and hence to a bounded sublattice of B. Therefore, $B \not\models \gamma(C)$, a contradiction. Thus, A is an L-algebra.

(2)⇒(3). This is obvious.

(3)⇒(1). Let $\mathcal{X}$ be the set of finite (non-isomorphic) subdirectly irreducible Heyting algebras A such that $A \not\models \mathsf{L}$. Let

$$\mathsf{M} = \mathsf{IPC} + \{\gamma(A) \mid A \in \mathcal{X}\}.$$

It is sufficient to show that $\mathsf{L} = \mathsf{M}$. Let B be a subdirectly irreducible Heyting algebra. First suppose that $B \not\models \mathsf{L}$. Then there is $\varphi \in \mathsf{L}$ such that $B \not\models \varphi$. By the Filtration Lemma, there is a finite subdirectly irreducible Heyting algebra A such that $A \not\models \varphi$ and A is a bounded sublattice of B. By Theorem 4.102, $B \not\models \gamma(A)$. Therefore, $B \not\models \mathsf{M}$. Thus, $\mathsf{L} \subseteq \mathsf{M}$.

For the reverse inclusion, since L is the logic of $\mathcal{K}$, it is sufficient to show that if $B \in \mathcal{K}$, then $B \models \mathsf{M}$. If $B \not\models \mathsf{M}$, then $B \not\models \gamma(A)$ for some $A \in \mathcal{X}$. By Theorem 4.102, A is isomorphic to a bounded sublattice of B. Since $B \in \mathcal{K}$ and $\mathcal{K}$ is closed under bounded sublattices, $A \in \mathcal{K}$. Thus, $A \models \mathsf{L}$, a contradiction. Consequently, $B \models \mathsf{M}$, finishing the proof.

Remark 4.106 In Bezhanishvili and de Jongh (2018) a new class of formulas, called $\mathsf{ONNILLI}$, was described syntactically. It was shown that each formula in this class is preserved under stable images of posets and that for a finite subdirectly irreducible Heyting algebra A, the formula $\gamma(A)$ is equivalent to a formula in $\mathsf{ONNILLI}$. This provides another description of stable formulas.

Remark 4.107 Various subclasses of stable logics that are closed under MacNeille completions were studied in Bezhanishvili et al. (2018b). In Lauridsen (2019) a subclass of stable logics was identified and it was shown that an intermediate logic is axiomatizable by $\mathcal{P}_3$-formulas of the substructural hierarchy of Ciabattoni et al. (2012) iff it belongs to this subclass.

Definition 4.108 We say that a class $\mathcal{K}$ of strongly rooted Esakia spaces is *closed under stable images* if for any strongly rooted Esakia spaces X and Y, from $X \in \mathcal{K}$ and Y being an onto stable image of X it follows that $Y \in \mathcal{K}$.

We recall that for a class $\mathcal{K}$ of Esakia spaces, $\mathcal{K}^* = \{X^* \mid X \in \mathcal{K}\}$. It is easy to see that $\mathcal{K}^*$ is closed under bounded sublattices iff $\mathcal{K}$ is closed under stable images. Thus, as an immediate consequence of Theorem 4.105, we obtain:

Theorem 4.109 *For an intermediate logic* L, *the following conditions are equivalent.*

(1) L *is stable.*
(2) *The class of strongly rooted Esakia spaces of* L *is closed under stable images.*
(3) L *is sound and complete with respect to a class* $\mathcal{K}$ *of strongly rooted Esakia spaces that is closed under stable images.*

Theorem 4.109 can be thought of as a motivation for the name "stable logic." We next show that all stable logics have the fmp. This is an easy consequence of Theorem 4.105 and the Filtration Lemma.

Theorem 4.110 *Each stable logic has the fmp.*

Proof Let L be a stable logic and let $\mathsf{L} \nvdash \varphi$. Then there is a subdirectly irreducible Heyting algebra B such that $B \models \mathsf{L}$ and $B \not\models \varphi$. By the Filtration Lemma, there is a finite Heyting algebra A such that A is a bounded sublattice of B and $A \not\models \varphi$. Since B is subdirectly irreducible, so is A. As L is stable and $B \models \mathsf{L}$, it follows from Theorem 4.105 that $A \models \mathsf{L}$. Because A is finite and $A \not\models \varphi$, we conclude that L has the fmp.

Definition 4.111 Let Λ_{Stab} be the set of all stable logics.

Theorem 4.112

(1) Λ_{Stab} *is a complete sublattice of* Λ.
(2) *The cardinality of* Λ_{Stab} *is that of the continuum.*

Proof For (1) see (Bezhanishvili et al. (2019, Theorem 3.7)) and for (2) see (Bezhanishvili and Bezhanishvili (2017, Theorem 6.13)).

We conclude with some examples of stable logics. Recall that $\mathfrak{F}_n$, $\mathfrak{D}_n$, and $\mathfrak{C}_n$ denote the n-fork, n-diamond, and n-chain (see Fig. 4.1). For a rooted poset P we abbreviate $\gamma(P^*)$ with $\gamma(P)$.

Theorem 4.113 (Bezhanishvili and Bezhanishvili 2017, Theorem 7.5)

(1) $\mathsf{CPC} = \mathsf{LC} + \gamma(\mathfrak{C}_2)$.
(2) $\mathsf{KC} = \mathsf{IPC} + \gamma(\mathfrak{F}_2)$.
(3) $\mathsf{LC} = \mathsf{IPC} + \gamma(\mathfrak{F}_2) + \gamma(\mathfrak{D}_2)$.
(4) $\mathsf{LC}_n = \mathsf{LC} + \gamma(\mathfrak{C}_{n+1})$.
(5) $\mathsf{BW}_n = \mathsf{IPC} + \gamma(\mathfrak{F}_{n+1}) + \gamma(\mathfrak{D}_{n+1})$.
(6) $\mathsf{BTW}_n = \mathsf{IPC} + \gamma(\mathfrak{F}_{n+1})$.

On the other hand, there are intermediate logics that are not stable; e.g., BD_n for $n \geq 2$ (see Bezhanishvili and Bezhanishvili 2017, Theorem 7.4).

4.7.2 Cofinal Stable Rules and Formulas

As we have seen, cofinal subframe logics are axiomatizable by formulas of the form $\alpha(A)$ while subframe logics by formulas of the form $\beta(A)$. In analogy, stable logics are axiomatizable by formulas of the form $\gamma(A)$. These can be thought to be parallel to $\beta(A)$ because both $\beta(A)$ and $\gamma(A)$ do not encode the behavior of negation. On the other hand, $\alpha(A)$ does encode it. It is natural to seek a stable analogue of $\alpha(A)$.

For this we need to work with the pseudocomplemented lattice reduct of Heyting algebras, instead of just the bounded lattice reduct like in the case of stable logics. Fortunately, the corresponding variety of pseudocomplemented distributive lattices remains locally finite (see, e.g., Bezhanishvili et al. 2016b, Theorem 6.1), and hence the algebraic approach is applicable. This allows us to develop the theory of cofinal stable logics, which generalizes the theory of stable logics. However, there is a key difference, which is due to the fact that an analogue of Lemma 4.101 fails for the pseudocomplemented lattice reduct (see Bezhanishvili et al. 2016b, Example 7.5). This, in particular, forces us to work with cofinal stable rules, rather than formulas.

Definition 4.114 Let A be a finite Heyting algebra, and for $a \in A$ let p_a be a new variable.

1. The *cofinal stable rule* of A is the multiple-conclusion rule $\sigma(A) = \Gamma/\Delta$, where

$$\begin{aligned}\Gamma = &\{p_0 \leftrightarrow \bot\} \cup \\ &\{p_{a \vee b} \leftrightarrow p_a \vee p_b \mid a, b \in A\} \cup \\ &\{p_{a \wedge b} \leftrightarrow p_a \wedge p_b \mid a, b \in A\} \cup \\ &\{p_{\neg a} \leftrightarrow \neg p_a \mid a \in A\}\end{aligned}$$

   and

$$\Delta = \{p_a \leftrightarrow p_b \mid a, b \in A \text{ with } a \neq b\}.$$

2. If in addition A is subdirectly irreducible, then the *cofinal stable formula* of A is

$$\delta(A) = \bigwedge \Gamma \to \bigvee \Delta.$$

Remark 4.108 There is no need to add $p_1 \leftrightarrow \top$ to Γ because it is derivable from Γ.

We then have the following generalization of the Stable Jankov Lemma.

Theorem 4.115 (Cofinal Stable Jankov Lemma) *Let A, B be Heyting algebras with A finite.*

(1) *$B \not\models \sigma(A)$ iff A is isomorphic to a pseudocomplemented sublattice of B.*
(2) *If A is subdirectly irreducible, then $B \not\models \delta(A)$ iff there is a subdirectly irreducible homomorphic image C of B such that A is isomorphic to a pseudocomplemented sublattice of C.*

Proof See (Bezhanishvili et al. 2016b, Propositions 6.4 and 7.1).

Similarly, we have the following generalization of the Filtration Lemma.

Lemma 4.116 (Cofinal Filtration Lemma) *Let B be a Heyting algebra such that $B \not\models \varphi$. Then there is a finite Heyting algebra A such that A is a pseudocomplemented sublattice of B and $A \not\models \varphi$. In addition, if B is well-connected, then A is subdirectly irreducible.*

However, we no longer have an analogue of Theorem 4.105. To see why, we need the following definition.

Definition 4.117 We say that a class $\mathcal{K}$ of subdirectly irreducible Heyting algebras is *closed under pseudocomplemented sublattices* if for any subdirectly irreducible Heyting algebras A, B, from $B \in \mathcal{K}$ and A being isomorphic to a pseudocomplemented sublattice of B it follows that $A \in \mathcal{K}$.

As follows from Bezhanishvili et al. (2016b, Example 7.9) it is no longer the case that an intermediate logic L is axiomatizable by cofinal stable formulas iff the class $\mathcal{V}(\mathsf{L})_{\mathsf{si}}$ of subdirectly irreducible L-algebras is closed under pseudocomplemented sublattices. Because of this we define cofinal subframe logics as those intermediate logics that are axiomatizable by stable canonical rules. We recall that an intermediate logic L is *axiomatizable* by a set $\mathcal{R}$ of multiple-conclusion rules if $\mathsf{L} \vdash \varphi$ iff the rule $/\varphi$ is derivable from $\mathcal{R}$. For more details about multiple-conclusion consequence relations we refer to Jeřábek (2009) and Iemhoff (2016).

Definition 4.118 An intermediate logic L is a *cofinal stable logic* if it is axiomatizable by cofinal stable rules.

We can utilize the Cofinal Stable Jankov Lemma and the Cofinal Filtration Lemma to prove the following analogue of Theorem 4.105.

Theorem 4.119 *For an intermediate logic L the following are equivalent.*

(1) *L is a cofinal stable logic.*
(2) *$\mathcal{V}(\mathsf{L})_{\mathsf{si}}$ is closed under pseudocomplemented sublattices.*
(3) *There is a class $\mathcal{K}$ of subdirectly irreducible Heyting algebras that is closed under pseudocomplemented sublattices and generates $\mathcal{V}(\mathsf{L})$.*

Further characterizations of cofinal stable logics can be found in Bezhanishvili et al. (2016b, Theorem 7.11).

Definition 4.120 Let Λ_{CStab} be the set of cofinal stable logics.

Clearly $\Lambda_{\mathsf{Stab}} \subsetneq \Lambda_{\mathsf{CStab}} \subsetneq \Lambda$.

Theorem 4.121

(1) *Each cofinal stable logic has the fmp.*
(2) *The cardinality of $\Lambda_{\mathsf{CStab}} \setminus \Lambda_{\mathsf{Stab}}$ is that of the continuum.*

Proof For (1) see (Bezhanishvili et al. (2016b, Remark 7.8)) and for (2) see (Bezhanishvili et al. (2016b, Proposition 8.3)).

In particular, the Maksimova logics ND_n, for $n \geq 2$, are examples of cofinal stable logics that are not stable logics (see Bezhanishvili et al. 2016b, Lemmas 9.4 and 9.5).

As follows from Bezhanishvili et al. (2016b, Proposition 7.7), each cofinal stable logic is axiomatizable by cofinal stable formulas. However, as we have already pointed out, the converse is not true in general. As far as we know, the problem mentioned in Bezhanishvili et al. (2016b, Remark 7.10)—whether all intermediate logics that are axiomatizable by cofinal stable formulas have the fmp—remains open, as does the problem of a convenient characterization of this class of intermediate logics.

Dual spaces of pseudocomplemented distributive lattices were described by Priestley (1975). Since pseudocomplemented distributive lattices are situated between bounded distributive lattices and Heyting algebras, their dual spaces are situated between Esakia spaces and Priestley spaces. Of interest to our considerations is the dual description of pseudocomplemented lattice homomorphisms between Esakia spaces. These are special stable maps $f : X \to Y$ that in addition satisfy the following *cofinality condition*:

$$\max \uparrow f(x) = f\,(\max \uparrow x)\,.$$

for each $x \in X$. Utilizing this, cofinal stable logics can be characterized as those intermediate logics for which the class of strongly rooted Esakia spaces is closed under cofinal stable images (meaning that, if X, Y are strongly rooted Esakia spaces such that $X \models \mathsf{L}$ and Y is a cofinal stable image of Y, then $Y \models \mathsf{L}$). We skip the details and refer the interested reader to Bezhanishvili et al. (2016b).

4.8 Subframization and Stabilization

As we pointed out in Theorems 4.87(1) and 4.112(1), the lattices of subframe and stable logics form complete sublattices of the lattice of all intermediate logics. Therefore, for each intermediate logic L, there is a greatest subframe logic contained in L and a least subframe logic containing L, called the downward and upward subframizations of L. Similarly, there is a greatest stable logic contained in L and a least stable logic containing L, called the downward and upward stabilizations of L. These are closest subframe and stable "neighbors" of L and were studied in Bezhanishvili et al. (2019), where connections with the Lax Logic and intuitionistic $\mathsf{S4}$ were also explored. The operation of subframization in modal logic was first studied by Wolter (1993).

4.8.1 Subframization

Definition 4.122 For an intermediate logic L, define the *downward subframization of* L as

$$\mathsf{Subf}_{\downarrow}(\mathsf{L}) = \bigvee\{\mathsf{M} \in \Lambda_{\mathsf{Subf}} \mid \mathsf{M} \subseteq \mathsf{L}\}$$

and the *upward subframization of* L as

$$\mathsf{Subf}_{\uparrow}(\mathsf{L}) = \bigwedge\{\mathsf{M} \in \Lambda_{\mathsf{Subf}} \mid \mathsf{L} \subseteq \mathsf{M}\}.$$

Lemma 4.123 (Bezhanishvili et al. 2019, Lemma 4.2) $\mathsf{Subf}_{\downarrow}$ *is an interior operator and* $\mathsf{Subf}_{\uparrow}$ *a closure operator on* Λ.

A semantic characterization of the downward and upward subframizations was given in Bezhanishvili et al. (2019, Proposition 4.3). We next use subframe canonical formulas to give a syntactic characterization. For this we require the following lemma.

Lemma 4.124 *Let A, B be finite Heyting algebras with A subdirectly irreducible and $D \subseteq A^2$. If $B \models \beta(A)$, then $B \models \alpha(A, D)$.*

Proof Suppose $B \not\models \alpha(A, D)$. By the Subframe Jankov Lemma, there is a homomorphic image C of B and a bounded implicative semilattice embedding $h : A \to C$ satisfying $\mathrm{CDC}_{\vee}$ for D. Since B is finite, Lemma 4.83(1) yields an implicative semilattice embedding of A into B. Therefore, $B \not\models \beta(A)$ by Theorem 4.84(1).

Let L be an intermediate logic. By Theorem 4.43, there are subframe canonical formulas $\alpha(A_i, D_i)$, $i \in I$, such that $\mathsf{L} = \mathsf{IPC} + \{\alpha(A_i, D_i) \mid i \in I\}$.

Theorem 4.125 (Bezhanishvili et al. 2019, Theorem 4.4) *For an intermediate logic* $\mathsf{L} = \mathsf{IPC} + \{\alpha(A_i, D_i) \mid i \in I\}$ *we have*

(1) $\mathsf{Subf}_{\downarrow}(\mathsf{L}) = \mathsf{IPC} + \{\beta(A) \mid \mathsf{L} \vdash \beta(A)\}$.
(2) $\mathsf{Subf}_{\uparrow}(\mathsf{L}) = \mathsf{IPC} + \{\beta(A_i) \mid i \in I)\}$.

Proof (1). By Definition 4.86(1), every subframe logic is axiomatizable by subframe formulas. Therefore, every subframe logic contained in L is axiomatizable by a set of subframe formulas that are provable in L. Thus, $\mathsf{IPC} + \{\beta(A) \mid \mathsf{L} \vdash \beta(A)\}$ is the largest subframe logic contained in L.

(2). Let $\mathsf{M} = \mathsf{IPC} + \{\beta(A_i) \mid i \in I\}$. Then M is a subframe logic by definition. Let B be a finite Heyting algebra. If $B \models \mathsf{M}$, then $B \models \beta(A_i)$ for all $i \in I$. By Lemma 4.124, $B \models \beta(A_i, D_i)$ for all $i \in I$. Thus, $B \models \mathsf{L}$, and so $\mathsf{L} \subseteq \mathsf{M}$ because M has the fmp (see Theorem 4.92). It remains to show that M is the least subframe logic containing L. If not, then there is a subframe logic $\mathsf{N} \supseteq \mathsf{L}$ and a Heyting algebra B such that $B \models \mathsf{N}$ and $B \not\models \mathsf{M}$. Therefore, $B \not\models \beta(A_i)$ for some $i \in I$. By Theorem 4.84(1), A_i is isomorphic to a $(\wedge, \to)$-subalgebra of B. Since N is a subframe logic, $A_i \models \mathsf{N}$ by Theorem 4.90. But $A_i \not\models \beta(A_i, D_i)$. Consequently, $A_i \not\models \mathsf{L}$, which is a contradiction since $\mathsf{N} \supseteq \mathsf{L}$.

Remark 4.126

1. The above proof of Theorem 4.125(2) is different from the one given in Bezhanishvili et al. (2019).

2. As was pointed out in Bezhanishvili et al. (2019, Remark 4.6), Theorem 4.125 can also be derived from the theory of describable operations of Wolter (1993).

In the next theorem we axiomatize the downward and upward subframizations for many well-known intermediate logics.

Theorem 4.127 (Bezhanishvili et al. 2019, Proposition 4.7)

(1) $\mathsf{Subf}_{\downarrow}(\mathsf{KC}) = \mathsf{IPC}$ *and* $\mathsf{Subf}_{\uparrow}(\mathsf{KC}) = \mathsf{LC}$.
(2) $\mathsf{Subf}_{\downarrow}(\mathsf{KP}) = \mathsf{IPC}$ *and* $\mathsf{Subf}_{\uparrow}(\mathsf{KP}) = \mathsf{BW}_2$.
(3) $\mathsf{Subf}_{\downarrow}(\mathsf{T}_n) = \mathsf{IPC}$ *and* $\mathsf{Subf}_{\uparrow}(\mathsf{T}_n) = \mathsf{BW}_n$ *for every* $n \geq 2$.
(4) $\mathsf{Subf}_{\downarrow}(\mathsf{BTW}_n) = \mathsf{IPC}$ *and* $\mathsf{Subf}_{\uparrow}(\mathsf{BTW}_n) = \mathsf{BW}_n$ *for every* $n \geq 2$.
(5) $\mathsf{Subf}_{\downarrow}(\mathsf{ND}_n) = \mathsf{IPC}$ *and* $\mathsf{Subf}_{\uparrow}(\mathsf{ND}_n) = \mathsf{BW}_2$ *for every* $n \geq 2$.

Remark 4.128 Since subframes are closely related to nuclei (Bezhanishvili and Ghilardi 2007), they are also connected to the propositional lax logic PLL (Fairtlough and Mendler 1997; Goldblatt 1981), which is the intuitionistic modal logic whose algebraic models are nuclear Heyting algebras. There is a nucleic Gödel–Gentzen translation of IPC into PLL which is extended to an embedding of the lattice of intermediate logics into the lattice of extensions of PLL. This yields a new characterization of subframe logics in terms of the propositional lax logic. We refer to Bezhanishvili et al. (2019, Sect. 6) for details.

4.8.2 *Stabilization*

In this section we obtain similar results for the downward and upward stabilizations.

Definition 4.129 For an intermediate logic L, define the *downward stabilization of* L as

$$\mathsf{Stab}_{\downarrow}(\mathsf{L}) = \bigvee \{\mathsf{M} \in \Lambda_{\mathsf{Stab}} \mid \mathsf{M} \subseteq \mathsf{L}\}$$

and the *upward stabilization of* L as

$$\mathsf{Stab}_{\uparrow}(\mathsf{L}) = \bigwedge \{\mathsf{M} \in \Lambda_{\mathsf{Stab}} \mid \mathsf{L} \subseteq \mathsf{M}\}.$$

Lemma 4.130 (Bezhanishvili et al. 2019, Lemma 7.2) $\mathsf{Stab}_{\downarrow}$ *is an interior operator and* $\mathrm{Stab}_{\uparrow}$ *a closure operator on* Λ.

A semantic characterization of the downward and upward stabilizations was given in Bezhanishvili et al. (2019, Proposition 7.3). We use stable canonical formulas to give a syntactic characterization, which requires the following lemma.

Lemma 4.131 *Let* A, B *be finite subdirectly irreducible Heyting algebras and* $D \subseteq A^2$. *If* $B \models \gamma(A)$, *then* $B \models \gamma(A, D)$.

Proof Suppose $B \not\models \gamma(A, D)$. By the Stable Jankov Lemma, there is a subdirectly irreducible homomorphic image C of B and a bounded lattice embedding $h : A \to C$ satisfying $\mathsf{CDC}_{\to}$ for D. Since B is finite, Lemma 4.101 yields a bounded lattice embedding of A into B. Therefore, $B \not\models \gamma(A)$ by Theorem 4.102.

Let L be an intermediate logic. By Theorem 4.61, there are stable canonical formulas $\gamma(A_i, D_i)$, $i \in I$, such that $\mathsf{L} = \mathsf{IPC} + \{\gamma(A_i, D_i) \mid i \in I\}$.

Theorem 4.132 (Bezhanishvili et al. 2019, Theorem 7.4) *For an intermediate logic* $\mathsf{L} = \mathsf{IPC} + \{\gamma(A_i, D_i) \mid i \in I\}$ *we have*

(1) $\mathsf{Stab}_{\downarrow}(\mathsf{L}) = \mathsf{IPC} + \{\gamma(A) \mid \mathsf{L} \vdash \gamma(A)\}$.
(2) $\mathsf{Stab}_{\uparrow}(\mathsf{L}) = \mathsf{IPC} + \{\gamma(A_i) \mid i \in I\}$.

Proof (1). By Definition 4.104, $\mathsf{IPC} + \{\gamma(A) \mid \mathsf{L} \vdash \gamma(A)\}$ is a stable logic, and clearly it is the largest stable logic contained in L. Therefore, $\mathsf{Stab}_{\downarrow}(\mathsf{L}) = \mathsf{IPC} + \{\gamma(A) \mid \mathsf{L} \vdash \gamma(A)\}$.

(2). Let $\mathsf{M} = \mathsf{IPC} + \{\gamma(A_i) \mid i \in I\}$. Then M is a stable logic by definition. Let B be a finite subdirectly irreducible Heyting algebra such that $B \models \mathsf{M}$. Then $B \models \gamma(A_i)$ for all $i \in I$. By Lemma 4.131, $B \models \gamma(A_i, D_i)$ for all $i \in I$. Thus, $B \models \mathsf{L}$, and so $\mathsf{L} \subseteq \mathsf{M}$ since M has the fmp (see Theorem 4.110). Suppose N is a stable extension of L, and B is a subdirectly irreducible Heyting algebra such that $B \models \mathsf{N}$. If $B \not\models \gamma(A_i)$ for some $i \in I$, then A_i is isomorphic to a bounded sublattice of B by Theorem 4.102. Therefore, $A_i \models \mathsf{N}$ by Theorem 4.105. But $A_i \not\models \gamma(A_i, D_i)$. So $A_i \not\models \mathsf{L}$, which contradicts N being an extension of L. Thus, $B \models \gamma(A_i)$ for all $i \in I$, and so $\mathsf{M} \subseteq \mathsf{N}$. Consequently, M is the least stable extension of L, and hence $\mathsf{Stab}_{\uparrow}(\mathsf{L}) = \mathsf{M}$.

Remark 4.133

1. The above proof of Theorem 4.132(2) is different from the one given in Bezhanishvili et al. (2019).
2. As was pointed out in Bezhanishvili et al. (2019, Remark 7.6), an alternative proof of Theorem 4.132 can be obtained using Wolter's describable operations.

The next theorem axiomatizes the downward and upward stabilizations of several intermediate logics.

Theorem 4.134 (Bezhanishvili et al. 2019, Proposition 7.7)

(1) $\mathsf{Stab}_{\downarrow}(\mathsf{BD}_n) = \mathsf{IPC}$ *and* $\mathsf{Stab}_{\uparrow}(\mathsf{BD}_n) = \mathsf{BC}_n$ *for all* $n \geq 2$.
(2) $\mathsf{Stab}_{\downarrow}(\mathsf{T}_n) = \mathsf{IPC}$ *and* $\mathsf{Stab}_{\uparrow}(\mathsf{T}_n) = \mathsf{BW}_n$ *for all* $n \geq 2$.
(3) *If* L *has the disjunction property, then* $\mathsf{Stab}_{\downarrow}(\mathsf{L}) = \mathsf{IPC}$.

Remark 4.135 In Bezhanishvili et al. (2019, Sect. 8) the Gödel translation was utilized to embed the lattice of intermediate logics into the lattice of multiple-conclusion consequence relations over the intuitionistic $\mathsf{S4}$, and it was shown that this provides a new characterization of stable logics.

References

Balbes, R., & Dwinger, P. (1974). *Distributive lattices*. Columbia: University of Missouri Press.

Berdzenishvili, L. (2019). *Sacred darkness* (Europa ed.).

Bezhanishvili, G., & Bezhanishvili, N. (2008). Profinite Heyting algebras. *Order, 25*(3), 211–227.

Bezhanishvili, G., & Bezhanishvili, N. (2009). An algebraic approach to canonical formulas: Intuitionistic case. *Review of Symbolic Logic, 2*(3), 517–549.

Bezhanishvili, G., & Bezhanishvili, N. (2011). An algebraic approach to canonical formulas: Modal case. *Studia Logica, 99*(1–3), 93–125.

Bezhanishvili, G., & Bezhanishvili, N. (2012). Canonical formulas for wK4. *Review of Symbolic Logic, 5*(4), 731–762.

Bezhanishvili, G., & Bezhanishvili, N. (2016). An algebraic approach to filtrations for superintuitionistic logics. In J. van Eijk, R. Iemhoff, & J. Joosten (Eds.), *Liber Amicorum Albert Visser*. Tribute series (Vol. 30, pp. 47–56). College Publications.

Bezhanishvili, G., & Bezhanishvili, N. (2017). Locally finite reducts of Heyting algebras and canonical formulas. *Notre Dame Journal of Formal Logic, 58*(1), 21–45.

Bezhanishvili, G., Bezhanishvili, N., & de Jongh, D. (2008). The Kuznetsov-Gerčiu and Rieger-Nishimura logics. *Logic and Logical Philosophy, 17*(1–2), 73–110.

Bezhanishvili, G., Bezhanishvili, N., Gabelaia, D., & Kurz, A. (2010). Bitopological duality for distributive lattices and Heyting algebras. *Mathematical Structures in Computer Science, 20*(3), 359–393.

Bezhanishvili, G., Bezhanishvili, N., & Iemhoff, R. (2016a). Stable canonical rules. *Journal of Symbolic Logic, 81*(1), 284–315.

Bezhanishvili, G., Bezhanishvili, N., & Ilin, J. (2016b). Cofinal stable logics. *Studia Logica, 104*(6), 1287–1317.

Bezhanishvili, G., Bezhanishvili, N., & Ilin, J. (2018a). Stable modal logics. *Review of Symbolic Logic, 11*(3), 436–469.

Bezhanishvili, G., Bezhanishvili, N., & Ilin, J. (2019). Subframization and stabilization for superintuitionistic logics. *Journal of Logic and Computation, 29*(1), 1–35.

Bezhanishvili, G., Bezhanishvili, N., Moraschini, T., Stronkowski, M. (2021). Profiniteness in varieties of Heyting algebras. *Advances in Mathematics, 391*, paper 107959, 47 pp.

Bezhanishvili, N. (2006). Lattices of intermediate and cylindric modal logics. Ph.D. thesis, ILLC, University of Amsterdam.

Bezhanishvili, N., & de Jongh, D. (2018). Stable formulas in intuitionistic logic. *Notre Dame Journal of Formal Logic, 59*(3), 307–324.

Bezhanishvili, N., Gabelaia, D., Ghilardi, S., & Jibladze, M. (2016). Admissible bases via stable canonical rules. *Studia Logica, 104*(2), 317–341.

Bezhanishvili, N., Galatos, N., & Spada, L. (2017). Canonical formulas for k-potent commutative, integral, residuated lattices. *Algebra Universalis, 77*(3), 321–343.

Bezhanishvili, G., & Ghilardi, S. (2007). An algebraic approach to subframe logics. Intuitionistic case. *Annals of Pure and Applied Logic, 147*(1–2), 84–100.

Bezhanishvili, G., Ghilardi, S., Jibladze, M. (2011). An algebraic approach to subframe logics. Modal case. *Notre Dame Journal of Formal Logic, 52*(2), 187–202.

Bezhanishvili, G., Harding, J., Ilin, J., & Lauridsen, F. (2018b). MacNeille transferability and stable classes of Heyting algebras. *Algebra Universalis, 79*(3), Paper No. 55, 21 pp.

Bezhanishvili, N., & Moraschini, T. (2019). Hereditarily structurally complete intermediate logics: Citkin's theorem via Esakia duality. Submitted. https://staff.fnwi.uva.nl/n.bezhanishvili/Papers/IPC-HSC.pdf.

Blackburn, P., de Rijke, M., & Venema, Y. (2001). *Modal logic*. Cambridge: Cambridge University Press.

Blok, W. (1978). On the degree of incompleteness of modal logics. *Bulletin of the Section of Logic, 7*(4), 167–175.

Blok, W. J., & Pigozzi, D. (1982). On the structure of varieties with equationally definable principal congruences. I. *Algebra Universalis, 15*(2), 195–227.
Burris, S., & Sankappanavar, H. (1981). *A course in universal algebra*. Berlin: Springer.
Chagrov, A., & Zakharyaschev, M. (1997). *Modal logic*. New York: The Clarendon Press.
Chang, C. C., & Keisler, H. J. (1990). *Model theory* (3rd ed.). Amsterdam: North-Holland Publishing Co.
Ciabattoni, A., Galatos, N., & Terui, K. (2012). Algebraic proof theory for substructural logics: Cut-elimination and completions. *Annals of Pure and Applied Logic, 163*(3), 266–290.
Citkin, A. (1978). Structurally complete superintuitionistic logics. *Soviet Mathematics - Doklady, 19*(4), 816–819.
Citkin, A. (1986). Finite axiomatizability of locally tabular superintuitionistic logics. *Mathematical Notes, 40*(3–4), 739–742.
Citkin, A. (1987). Structurally complete superintuitionistic logics and primitive varieties of pseudo-Boolean algebras. *Matematicheskie Issledovaniya, 98,* 134–151 (In Russian).
Citkin, A. (2013a). Characteristic formulas of partial Heyting algebras. *Logica Universalis, 7*(2), 167–193.
Citkin, A. (2013b). Jankov-style formulas and refutation systems. *Reports on Mathematical Logic, 48,* 67–80.
Citkin, A. (2015). Characteristic inference rules. *Logica Universalis, 9*(1), 27–46.
de Jongh, D. (1968). Investigations on the intuitionistic propositional calculus. Ph.D. thesis, University of Wisconsin.
de Jongh, D. H. J., & Troelstra, A. S. (1966). On the connection of partially ordered sets with some pseudo-Boolean algebras. *Indagationes Mathematicae, 28*, 317–329.
Diego, A. (1966). Sur les Algèbres de Hilbert. Translated from the Spanish by Luisa Iturrioz. Collection de Logique Mathématique, Sér. A, Fasc. XXI. Gauthier-Villars, Paris.
Esakia, L. (1974). Topological Kripke models. *Soviet Mathematics - Doklady, 15*, 147–151.
Esakia, L. (1979). On the theory of modal and superintuitionistic systems. *Logical inference (Moscow, 1974)* (pp. 147–172). Moscow: Nauka (In Russian).
Esakia, L. (2019). *Heyting algebras: Duality theory*. In G. Bezhanishvili & W. H. Holliday (Eds.), Trends in logic—Studia Logica library (Vol. 50). Berlin: Springer. Translated from the Russian edition by A. Evseev.
Fairtlough, M., & Mendler, M. (1997). Propositional lax logic. *Information and Computation, 137*(1), 1–33.
Fine, F. (1974). An ascending chain of S4 logics. *Theoria, 40*(2), 110–116.
Fine, K. (1985). Logics containing K4. II. *Journal of Symbolic Logic, 50*(3), 619–651.
Goldblatt, R. I. (1981). Grothendieck topology as geometric modality. *Z. Math. Logik Grundlag. Math., 27*(6), 495–529.
Grätzer, G. (1978). *General lattice theory*. New York: Academic.
Hosoi, T. (1967). On intermediate logics. I. *Journal of the Faculty of Science, University of Tokyo, Section I, 14*, 293–312.
Iemhoff, R. (2016). Consequence relations and admissible rules. *Journal of Philosophical Logic, 45*(3), 327–348.
Ilin, J. (2018). Filtration revisited: Lattices of stable non-classical logics. Ph.D. thesis, ILLC, University of Amsterdam.
Jankov, V. A. (1963). On the relation between deducibility in intuitionistic propositional calculus and finite implicative structures. *Doklady Akademii Nauk SSSR, 151,* 1293–1294 (In Russian).
Jankov, V. A. (1968). The construction of a sequence of strongly independent superintuitionistic propositional calculi. *Soviet Mathematics - Doklady, 9*, 806–807.
Jankov, V. A. (1969). Conjunctively irresolvable formulae in propositional calculi. *Izvestiya Akademii Nauk SSSR Series Mathematica, 33,* 18–38 (In Russian).
Jeřábek, E. (2009). Canonical rules. *Journal of Symbolic Logic, 74*(4), 1171–1205.
Jónsson, B. (1968). Algebras whose congruence lattices are distributive. *Mathematica Scandinavica, 21*, 110–121.

Kracht, M. (1990). An almost general splitting theorem for modal logic. *Studia Logica, 49*(4), 455–470.
Kuznetsov, A. V. (1974). On superintuitionistic logics. In *Proceedings of the International Congress of Mathematicians* (Vol. 1, pp. 243–249). Vancouver.
Kuznetsov, A. V., & Gerčiu, V. J. A. (1970). The superintuitionistic logics and finitary approximability. *Soviet Mathematics - Doklady, 11*, 1614–1619.
Lauridsen, F. (2019). Intermediate logics admitting a structural hypersequent calculus. *Studia Logica, 107*(2), 247–282.
Litak, T. (2002). A continuum of incomplete intermediate logics. *Reports on Mathematical Logic, 36*, 131–141.
McKay, C. (1968). The decidability of certain intermediate propositional logics. *Journal of Symbolic Logic, 33*, 258–264.
McKenzie, R. (1972). Equational bases and nonmodular lattice varieties. *Transactions of the American Mathematical Society, 174*, 1–43.
McKinsey, J. C. C. (1941). A solution of the decision problem for the Lewis systems S2 and S4, with an application to topology. *Journal of Symbolic Logic, 6*, 117–134.
McKinsey, J. C. C., & Tarski, A. (1944). The algebra of topology. *Annals of Mathematics, 45*, 141–191.
McKinsey, J. C. C., & Tarski, A. (1946). On closed elements in closure algebras. *Annals of Mathematics, 47*, 122–162.
Priestley, H. A. (1970). Representation of distributive lattices by means of ordered Stone spaces. *Bulletin of the London Mathematical Society, 2*, 186–190.
Priestley, H. A. (1972). Ordered topological spaces and the representation of distributive lattices. *Proceedings of the London Mathematical Society, 24*, 507–530.
Priestley, H. A. (1975). The construction of spaces dual to pseudocomplemented distributive lattices. *Quarterly Journal of Mathematics Oxford Series, 26*(102), 215–228.
Rasiowa, H., & Sikorski, R. (1963). *The mathematics of metamathematics*. Monografie Matematyczne (Vol. 41). Warsaw: Państwowe Wydawnictwo Naukowe.
Rautenberg, W. (1980). Splitting lattices of logics. *Archiv für Mathematische Logik und Grundlagenforschung, 20*(3–4), 155–159.
Rybakov, V. (1985). Bases of admissible rules of the logics S4 and Int. *Algebra and Logic, 24*(1), 55–68.
Rybakov, V. (1995). Hereditarily structurally complete modal logics. *Journal of Symbolic Logic, 60*(1), 266–288.
Shehtman, V. (1977). Incomplete propositional logics. *Doklady Akademii Nauk SSSR, 235*(3), 542–545 (In Russian).
Stone, M. H. (1936). The theory of representation for Boolean algebras. *Transactions of the American Mathematical Society, 40*(1), 37–111.
Tomaszewski, E. (2003). On sufficiently rich sets of formulas. Ph.D. thesis, Institute of Philosophy, Jagiellonian University, Kraków.
Troelstra, A. (1965). On intermediate propositional logics. *Indagationes Mathematicae, 27*, 141–152.
Umezawa, T. (1959). On intermediate propositional logics. *Journal of Symbolic Logic, 24*, 20–36.
Visser, A., van Benthem, J., de Jongh, D., Renardel de Lavalette, G. R. (1995). NNIL, a study in intuitionistic propositional logic. *Modal logic and process algebra (Amsterdam, 1994)* (pp. 289–326). Stanford: CSLI Publications.
Wolter, F. (1993). Lattices of modal logics. Ph.D. thesis, Free University of Berlin.
Wronski, A. (1973). Intermediate logics and the disjunction property. *Reports on Mathematical Logic, 1*, 39–51.
Zakharyaschev, M. (1989). Syntax and semantics of superintuitionistic logics. *Algebra and Logic, 28*(4), 262–282.

Zakharyaschev, M. (1992). Canonical formulas for K4. I. Basic results. *Journal of Symbolic Logic, 57*(4), 1377–1402.
Zakharyaschev, M. (1996). Canonical formulas for K4. II. Cofinal subframe logics. *Journal of Symbolic Logic, 61*(2), 421–449.

Chapter 5
Yankov Characteristic Formulas (An Algebraic Account)

Alex Citkin

Abstract The Yankov (characteristic) formulas were introduced by V. Yankov in 1963. Nowadays, the Yankov (or frame) formulas are used in virtually every branch of propositional logic: intermediate, modal, fuzzy, relevant, many-valued, etc. All these different logics have one thing in common: in one form or the other, they admit the deduction theorem. From a standpoint of algebraic logic, this means that their corresponding varieties have a ternary deductive (TD) term. It is natural to extend the notion of a characteristic formula to such varieties and, thus, to apply this notion to an even broader class of logics, namely, the logics in which the algebraic semantic is a variety with a TD term.

Keywords Yankov's formula · Characteristic formula · Intermediate logic · Algebraic semantic · Ternary deductive term · Independent axiomatizability · Finitely presentable algebra

5.1 Introduction

In this paper we study the Yankov characteristic formulas from algebraic standpoint. The characteristic formulas were introduced in a short, two-page-long note (Jankov 1963b). In Jankov (1969), the proofs of the announced results were given. Yankov observed the following: with any finite Heyting algebra **A** having a pretop element, one can associate a formula $Y(\mathbf{A})$—the Yankov formula—in such a way that for any formula B,

$$\mathsf{IPC} + B \vdash Y(\mathbf{A}) \text{ if and only if } \mathbf{A} \not\models B, \qquad \text{(Y1)}$$

where $\vdash$ is the derivation in IPC (intuitionistic propositional logic) with the postulated rules Modus Ponens and Uniform Substitution. In other words, the lattice ExtIPC is divided into two parts: the logics in which $Y(\mathbf{A})$ holds, and the logics for which **A** is

A. Citkin (✉)
Metropolitan Telecommunications, New York, USA
e-mail: acitkin@gmail.com

A. Citkin and I. M. Vandoulakis (eds.), *V. A. Yankov on Non-Classical Logics, History and Philosophy of Mathematics*, Outstanding Contributions to Logic 24,
https://doi.org/10.1007/978-3-031-06843-0_5

a model. Thus, if $\boldsymbol{L}(\mathbf{A})$ is a logic of all formulas valid in $\mathbf{A}$ and $\boldsymbol{L}(Y(A))$ is the logic of $\mathsf{IPC} + Y(\mathbf{A})$, for each $\boldsymbol{L} \in \mathsf{ExtIPC}$,

$$\boldsymbol{L} \subseteq \boldsymbol{L}(\mathbf{A}) \text{ or } \boldsymbol{L}(Y(\mathbf{A})) \subseteq \boldsymbol{L}. \tag{Y2}$$

For some time the technique developed in Jankov (1963b) went unnoticed. And even a more comprehensive paper (Jankov 1969), in which the technique was described in detail, was not immediately appreciated. But this was about to change (Fig. 5.1).

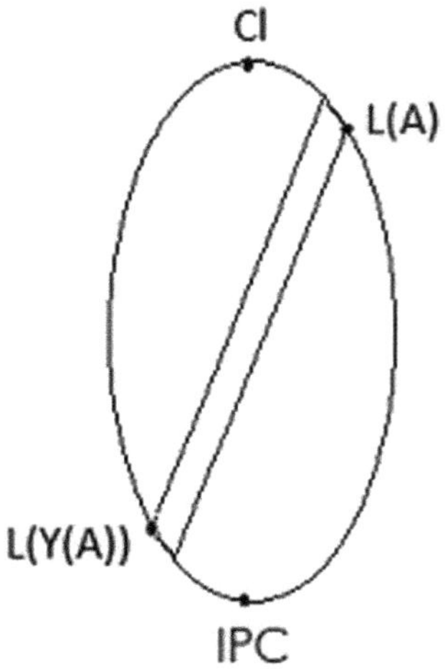

Fig. 5.1 Splitting in ExtIPC

Independently, in his Ph.D. thesis (de Jongh 1968), for intuitionistic frames, de Jongh introduced a notion of a frame formula that possesses the same properties as the Yankov formula. This inspired a separate line of research into different flavors of the frame and subframe formulas (cf., e.g., de Jongh and Yang 2010; Bezhanishvili 2006).

Independently, in Fine (1974), Fine introduced a notion of a frame formula for **S4**-frames. Like Yankov, he constructed an infinite independent set of **S4**-formulas and proved that there is a continuum of normal extensions of **S4** and, hence, that there exist normal extensions of **S4** that are not finitely axiomatizable. In addition, Fine proved that there exist normal extensions of **S4** that lack the finite model property. The definition of the Yankov formula was also extended to modal algebras in Rautenberg (1979, 1980), Blok (1976).

Different authors use different terms for these formulas: the Yankov formula, the Yankov-de Jongh formula, the Yankov–Fine formula, etc. To avoid confusion, we will be using the term "Yankov formula" only for formulas defined by a positive diagram of a finite subdirectly irreducible algebra, and for all the different flavors of this notion we will use the original term suggested in Jankov (1963b): a characteristic formula.

Clearly, for any class of logics, any given model divides the class of all extensions of these logics into two subclasses: the logics admitting this model and the logics, not admitting it. But only when the latter class contains the smallest logic there is a formula possessing properties similar to (Y2).

Let us note that in (Y2) the derivation is not important if we view the logic as a set of formulas closed under substitution and inference rules. Often, the classes of models for the logics form varieties. In algebraic terms, (Y2) can be restated: given a variety $\mathbb{V}$, a pair $(\mathbb{V}_0, \mathbb{V}_1)$ of subvarieties of $\mathbb{V}$ is called a *splitting pair* (cf., e.g., Galatos et al. 2007, Chap. 10) if $\mathbb{V}_0 \not\subseteq \mathbb{V}_1$ and for every subvariety $\mathbb{V}' \subseteq \mathbb{V}$,

$$\text{either } \mathbb{V}_0 \subseteq \mathbb{V}', \text{ or } \mathbb{V}' \subseteq \mathbb{V}_1. \tag{Spl}$$

A systematic study of splittings in the lattices of subvarieties of a given variety was started in McKenzie (1972). It was observed in McKenzie (1972) that for each splitting pair, $\mathbb{V}_0$ is generated by a single finitely generated subdirectly irreducible algebra—a $\mathbb{V}$-*splitting algebra*, while $\mathbb{V}_1$ is defined relative to $\mathbb{V}$ by a single identity—a $\mathbb{V}$-*splitting identity*, and every $\mathbb{V}$-splitting identity is $\wedge$-*prime relative to* $\mathbb{V}$; that is, if $\mathfrak{i}$ is a $\mathbb{V}$-splitting identity, then for any set of identities $\mathfrak{I}$,

$$\mathfrak{I} \vdash_{\mathbb{V}} \mathfrak{i} \text{ entails } \mathfrak{i}' \vdash_{\mathbb{V}} \mathfrak{i} \text{ for some } \mathfrak{i}' \in \mathfrak{I}.$$

Moreover, any $\wedge$-prime relative to the $\mathbb{V}$ identity is a splitting identity.

Let us note that for every variety $\mathbb{V}$, subvariety $\mathbb{V}' \subseteq \mathbb{V}$, and algebra $\mathbf{A} \in \mathbb{V}'$, if $\mathbf{A}$ is $\mathbb{V}$-splitting then it is also $\mathbb{V}'$-splitting (the converse does not need to be true), and every $\mathbb{V}'$-splitting identity is $\mathbb{V}$-splitting (the converse does not to be true).

Let us recall that Heyting algebras are algebraic models for **IPC** and that a class $\mathbb{HA}$ of all Heyting algebras forms a variety. In algebraic terms, the main result of Jankov (1969) is that a propositional formula A (or more precisely, an identity $A \approx \mathbf{1}$) is $\wedge$-prime in **IPC** (in $\mathbb{HA}$) if and only if A is interderivable with the Yankov formula of a finite subdirectly irreducible Heyting algebra—an $\mathbb{HA}$-splitting algebra. The finiteness of a splitting algebra in $\mathbb{HA}$ follows from the observation (cf. Blok and Pigozzi 1982) that in congruence-distributive varieties generated by finite members every splitting algebra is finite. And it is well known that variety $\mathbb{HA}$ is indeed congruence-distributive and that it is generated by finite Heyting algebras.

Nowadays, the Yankov (or frame) formulas are used in virtually every branch of propositional logic: intermediate, modal, fuzzy, relevant, many-valued, etc. All these different logics have one thing in common: in one form or the other, they admit the Deduction Theorem. From a standpoint of algebraic logic, this means that their equivalent algebraic semantics have a ternary deductive (TD) term (cf. Blok and Pigozzi 1994). It is natural to extend the notion of a characteristic formula to such varieties and, thus, to apply this notion to an even broader class of logics, namely, the logics for which the equivalent algebraic semantics are varieties with a TD term. In Sect. 5.5, we give such a generalization (cf., Agliano 2019).

Because Yankov's formulas are $\wedge$-prime relative to IPC, it is natural to ask whether any formula non-derivable in IPC is interderivable with a conjunction of some Yankov formulas (similarly to the decomposition of a given natural number into a product of primes). Or more generally, whether any logic from ExtIPC (any subvariety of $\mathbb{HA}$) can be axiomatized by the Yankov formulas (splitting identities). The answer to this question is negative, and in Sect. 5.4 we study the cases in which such an axiomatization is possible.

The paper is organized in following way. In Sect. 5.2, we recall the algebraic notions that are used in the subsequent sections. In Sect. 5.3, we study independent sets of splitting identities; in particular, we establish a link between independent sets of identities and antichains of splitting algebras (initially observed in Jankov 1969). Then, in Sect. 5.4, we use the results of Sect. 5.3 to construct independent bases of subvarieties. In Sect. 5.5, we define the characteristic identities for the varieties with a TD term. These identities can be explicitly described. They are a direct generalization of Yankov's formulas, and they form a subclass of splitting identities.

In the examples throughout the paper, we primarily use intermediate logics and varieties of Heyting algebras. The examples of the use of splittings in the lattices of modal logics the can be found in Rautenberg (1979), Wolter (1996), Kracht (1999). For the broader classes of logics, the reader is referred to Galatos et al. (2007), Agliano (2019).

5.2 Background

In this section, we recall the definitions and facts used in the following sections.

5.2.1 Basic Definitions

We consider algebras in an arbitrary but fixed signature Σ. The terms are built in a regular way from a set of variables X and operation signs from Σ. A map $\nu : X \longrightarrow \mathbf{A}$ into an algebra $\mathbf{A}$ is called a *valuation*. A valuation ν *satisfies identity* $t(x_1, \dots, x_n) \approx r(x_1, \dots, x_n)$ if $t(\nu(x_1), \dots, \nu(x_n)) = r(\nu(x_1), \dots, \nu(x_n))$ (in symbols, $\mathbf{A} \models_\nu t \approx r$); otherwise, ν *refutes identity* $t \approx r$ (in symbols, $\mathbf{A} \not\models_\nu t \approx r$). An identity is said to be *valid* in $\mathbf{A}$ if every valuation in $\mathbf{A}$ satisfies this identity; otherwise, the identity is *refuted* in $\mathbf{A}$. Accordingly, a valuation ν *satisfies a set of identities* $\mathfrak{I}$, if ν satisfies every $\mathrm{i} \in \mathfrak{I}$ (in symbols, $\mathbf{A} \models_\nu \mathfrak{I}$); otherwise, ν *refutes* $\mathfrak{I}$ (in symbols, $\mathbf{A} \not\models \mathfrak{I}$).

For an algebra $\mathbf{A}$, by $\mathrm{Con}(\mathbf{A})$ we denote a set of all congruences of $\mathbf{A}$, and by $\mathrm{Con}^\circ(\mathbf{A})$ we denote a set of all congruences from $\mathrm{Con}(\mathbf{A})$ that are distinct from identity.

If $\theta \in \mathrm{Con}(\mathbf{A})$ and $\mathsf{a} \in \mathbf{A}$, then a/θ is a θ-congruence class (coset) containing element a. For any two elements $\mathsf{a}, \mathsf{b} \in \mathbf{A}$, there is a smallest congruence $\theta(\mathsf{a}, \mathsf{b})$

such that $\mathsf{a} \equiv \mathsf{b} \pmod{\theta}$. Congruence $\theta(\mathsf{a}, \mathsf{b})$ is called a *principal congruence generated by the pair of elements* a, b (cf. Grätzer 2008). Likewise, for any finite lists of elements $\underline{a} \overset{\text{def}}{\iff} \mathsf{a}_1, \dots, \mathsf{a}_n$ and $\underline{b} \overset{\text{def}}{\iff} \mathsf{b}_1, \dots, \mathsf{b}_n$ of the same length, there is a smallest congruence $\theta(\underline{a}, \underline{b})$ such that $\mathsf{a}_i \equiv \mathsf{b}_i \pmod{\theta}$ for all $i = 1, \dots, n$. A congruence $\theta(\underline{a}, \underline{b})$ is called a *compact (or finitely generated) congruence generated by* $\underline{a}, \underline{b}$ (cf. Grätzer 2008).

Let us recall that an algebra $\mathbf{A}$ is called *subdirectly irreducible (s.i.)* if $\mathrm{Con}^{\circ}(\mathbf{A})$ contains a smallest congruence, which is called a *monolith* and is denoted by $\mu(\mathbf{A})$. It is not hard to see that an algebra $\mathbf{A}$ is s.i. if and only if it has two elements that are not distinguishable by any proper homomorphism. For instance, a Heyting algebra is s.i. if and only if it has a pretop element: any proper homomorphism always sends a pretop element to $\mathbf{1}$.

In addition, for any algebra $\mathbf{A}$, $\mathrm{Con}(\mathbf{A})$ forms a lattice, and $\mathbf{A}$ is said to be *congruence-distributive* if $\mathrm{Con}(\mathbf{A})$ is distributive. Accordingly, if in a class of algebras $\mathbb{K}$ all members are congruence-distributive, $\mathbb{K}$ is said to be *congruence-distributive* as well.

Let $\mathbf{A}$ be an algebra, $\mathsf{a}, \mathsf{b} \in \mathbf{A}$ and $\mathsf{a} \neq \mathsf{b}$. We call elements a and b *h-indistinguishable* if for any congruence $\theta \in \mathrm{Con}^{\circ}(\mathbf{A})$,

$$\mathsf{a} \equiv \mathsf{b} \pmod{\theta}.$$

Clearly, an algebra is s.i. if and only if it contains a pair of h-indistinguishable elements. It is easy to see that the following holds.

Proposition 5.1 *Let* $\mathbf{A}$ *be an algebra,* $\mathsf{a}, \mathsf{b} \in \mathsf{A}$ *and* $\mathsf{a} \neq \mathsf{b}$. *Then the following assertions are equivalent:*

(a) elements a, b *are h-indistinguishable;*
(b) $\mathsf{a} \not\equiv \mathsf{b} \pmod{\mu(\mathbf{A})}$*;*
(c) $\mu(\mathbf{A}) = \theta(\mathsf{a}, \mathsf{b})$*;*
(d) $\varphi(\mathsf{a}) = \varphi(\mathsf{b})$ *for any homomorphism* $\varphi : \mathbf{A} \longrightarrow \mathbf{B}$ *that is not an embedding.*

Let us also note the following corollary.

Corollary 5.1 *Let* a, b *be h-indistinguishable elements of algebra* $\mathbf{A}$*, and let* $\varphi : \mathbf{A} \longrightarrow \mathbf{B}$ *be a homomorphism of* $\mathbf{A}$ *into* $\mathbf{B}$*. Then,* $\varphi(\mathsf{a}) \neq \varphi(\mathsf{b})$ *yields that* φ *is an embedding.*

Suppose that $\mathbf{A} = (\mathsf{A}, \Sigma)$ is an algebra and $\mathrm{i} \overset{\text{def}}{\iff} t(x_1, \dots, x_m) \approx t'(x_1, \dots, x_m)$ is an identity. If $\nu : x_i \mapsto \mathsf{a}, i = 1, \dots, n$ and $t(\nu(x_1), \dots, \nu(x_n)) = t'(\nu(x_1), \dots, \nu(x_n))$, we express this fact by $\mathbf{A} \models_{\nu} \mathrm{i}$ and call ν a *valuation satisfying* i; and by $\mathbf{A} \models \mathrm{i}$ we express the fact that i *is valid* in $\mathbf{A}$; that is, for any $\mathsf{a}_1, \dots, \mathsf{a}_m \in \mathsf{A}$, $t(\mathsf{a}_1, \dots, \mathsf{a}_m) = t'(\mathsf{a}_1, \dots, \mathsf{a}_m)$. If $\mathbb{K}$ is a class of algebras and $\mathbf{A} \models \mathrm{i}$ for every $\mathbf{A} \in \mathbb{K}$, we write $\mathbb{K} \models \mathrm{i}$.

Suppose that $\mathbb{K}$ is a class of algebras. By $\mathbf{S}\mathbb{K}$, $\mathbf{H}\mathbb{K}$, $\mathbf{P}\mathbb{K}$, and $\mathbf{P}_u\mathbb{K}$ we denote, respectively, classes of all isomorphic copies of subalgebras, homomorphic images, direct products, and ultraproducts of algebras from $\mathbb{K}$. A *variety generated by* $\mathbb{K}$ is denoted by $\mathbf{V}(\mathbb{K})$; that is, $\mathbf{V}(\mathbb{K}) = \mathbf{HSP}\mathbb{K}$.

An algebra $\mathbf{A}$ has the *congruence extension property* (CEP) if for every subalgebra $\mathbf{B}$ and $\theta \in \mathrm{Con}(\mathbf{B})$, there is a congruence $\theta' \in \mathrm{Con}(\mathbf{A})$ such that $\theta = \theta' \cap \mathsf{B}^2$. A class $\mathbb{K}$ of algebras has the CEP if every algebra in the class has the CEP.

Example 5.1 The vast majority of varieties associated with propositional logics has the CEP. In particular, the variety $\mathbb{HA}$ of all Heyting algebras has the CEP: if $\mathbf{B}$ is a subalgebra of a Heyting algebra $\mathbf{A}$ and F is a filter of $\mathbf{B}$, then F generates in $\mathbf{A}$ a filter F' such that $\mathsf{F} = \mathsf{F}' \cap \mathsf{B}$.

Note 5.1 In all examples throughout the paper we consider Heyting algebras in the signature $\wedge, \vee, \to, \neg, \mathbf{1}$, and we use $x \leftrightarrow y$ as an abbreviation for $(x \to y) \wedge (y \to x)$; the variety of all Heyting algebras is denoted by $\mathbb{HA}$.

Let $\mathbb{K}$ be a class of algebras and $\mathfrak{I}$ be a set of identities. An identity $\mathfrak{i}$ is called a $\mathbb{K}$-*consequence* of $\mathfrak{I}$ (in symbols, $\mathfrak{I} \vdash_{\mathbb{K}} \mathfrak{i}$) if for every $\mathbf{A} \in \mathbb{K}$ and every valuation ν in $\mathbf{A}$ satisfying $\mathfrak{I}$, ν satisfies $\mathfrak{i}$. If $\mathfrak{I} = \{\mathfrak{i}_1, \dots, \mathfrak{i}_n\}$, then $\mathfrak{i}_1, \dots, \mathfrak{i}_n \vdash_{\mathbb{K}} \mathfrak{i}$ if and only if a quasi-identity $\mathfrak{i}_1, \dots, \mathfrak{i}_n \Longrightarrow \mathfrak{i}$ holds in $\mathbb{K}$ (cf. Mal'cev 1973). Identities $\mathfrak{i}_1$ and $\mathfrak{i}_2$ are $\mathbb{K}$-*equivalent* if $\mathfrak{i}_1 \vdash_{\mathbb{K}} \mathfrak{i}_2$ and $\mathfrak{i}_2 \vdash_{\mathbb{K}} \mathfrak{i}_1$ (in symbols, $\mathfrak{i}_1 \sim_{\mathbb{K}} \mathfrak{i}_2$).

If $\mathfrak{I}_1, \mathfrak{I}_2$ are sets of identities, we say that $\mathfrak{I}_1$ and $\mathfrak{I}_2$ are $\mathbb{K}$-*equivalent* (in symbols, $\mathfrak{I}_1 \sim_{\mathbb{K}} \mathfrak{I}_2$) if $\mathfrak{I}_1 \vdash_{\mathbb{K}} \mathfrak{i}$ for every $\mathfrak{i} \in \mathfrak{I}_2$ and $\mathfrak{I}_2 \vdash_{\mathbb{K}} \mathfrak{i}$ for every $\mathfrak{i} \in \mathfrak{I}_1$. If one of the sets, say $\mathfrak{I}_2$, consists of a single identity $\mathfrak{i}$, we omit braces and we write $\Gamma_1 \sim \mathfrak{i}$.

We also say that $\mathfrak{i}$ $\mathbb{K}$-*follows* from a set of identities $\mathfrak{I}$ (in symbols, $\mathfrak{I} \models_{\mathbb{K}} \mathfrak{i}_2$) if $\mathbb{K} \models \mathfrak{i}$ as long as $\mathbb{K} \models \mathfrak{I}$, that is, $\mathbb{K} \models \mathfrak{i}'$ for each $\mathfrak{i}' \in \mathfrak{I}$. If $\mathfrak{i}_1 \models_{\mathbb{K}} \mathfrak{i}_2$ and $\mathfrak{i}_2 \models_{\mathbb{K}} \mathfrak{i}_1$, we say that identities $\mathfrak{i}_1$ and $\mathfrak{i}_2$ are $\mathbb{K}$-*equipotent* (in symbols, $\mathfrak{i}_1 \approx_{\mathbb{K}} \mathfrak{i}_2$). $\mathfrak{I}_1 \approx_{\mathbb{K}} \mathfrak{I}_2$ is understood similarly to $\mathfrak{I}_1 \sim_{\mathbb{K}} \mathfrak{I}_2$.

Clearly, if $\mathbb{V}$ is a variety, then two identities are $\mathbb{V}$-equipotent if and only if these identities define the same subvariety of $\mathbb{V}$. Also, it is easily seen that

$$\mathfrak{i}_1 \sim_{\mathbb{K}} \mathfrak{i}_2 \Longrightarrow \mathfrak{i}_1 \approx_{\mathbb{K}} \mathfrak{i}_2. \tag{5.1}$$

Example 5.2 Let us consider identities $\neg\neg x \approx \mathbf{1}x$ and $x \vee \neg x \approx \mathbf{1}$.

$$\neg\neg x \approx x \ \approx \mathbb{HA}x \ \vee \neg x, \approx \ \text{while} \ \neg\neg x \approx x \ \not\sim \mathbb{HA} \ x \vee \neg x \approx \mathbf{1}.$$

Indeed, $\mathbb{HA} + \neg\neg x \approx x = \mathbb{HA} + x \vee \neg x \approx \mathbf{1} = \mathbb{BA}$, where $\mathbb{BA}$ is a variety of all Boolean algebras, while $\neg\neg x \to x \approx \mathbf{1} \nvdash \mathbb{HA}x \vee \neg x \approx \mathbf{1}$.

5.2.2 Finitely Presentable Algebras

There are different definitions of finitely presentable algebras. For our purpose, the most convenient one is a definition from Mal'cev (1973, Chap. 5 §11).

Let X be a set of variables and $\mathfrak{D}$ be a set of identities in variables from X. Then a pair $(X, \mathfrak{D})$ is called a *defining pair* and $\mathfrak{D}$ is a set of *defining relations*. An algebra $\mathbf{A}$ from a class of algebras $\mathbb{K}$ is said to be *presentable by a pair* $\Delta = (X, \mathfrak{D})$ and a *defining map* $\delta : X \longrightarrow \mathbf{A}$ *relative to* $\mathbb{K}$ if the following hold

(a) set $\{\delta(x) \mid x \in X\}$ generates algebra $\mathbf{A}$;
(b) for any identity $\mathfrak{i}$, if $\delta(\mathfrak{i})$ is satisfied in $\mathbf{A}$, then $\mathfrak{D} \vdash_{\mathbb{K}} \mathfrak{i}$.

Example 5.3 Let $\mathbf{A}$ be a Heyting algebra generated by elements $\mathsf{a}_1, \dots, \mathsf{a}_n$, and let $A(x_1, \dots, x_n)$ be a propositional formula. Then, $\mathbf{A}$ is presentable by a defining pair $\Delta = (\{x_1, \dots, x_n\}, A(x_1, \dots, x_n) \approx \mathbf{1})$ and valuation $\delta : x_i \mapsto \mathsf{a}_i$, provided that for each formula $B(x_1, \dots, x_n)$ such that $B(\mathsf{a}_1, \dots, \mathsf{a}_n) = \mathbf{1}, \vdash A \to B$ holds in IPC.

Let us observe that by the definition of $\vdash_{\mathbb{K}}$, for any set of identities $\mathfrak{I}$ and any identity $\mathfrak{i}$, $\mathfrak{I} \vdash_{\mathbb{K}} \mathfrak{i}$ entails $\mathfrak{I} \vdash_{\mathbb{K}'} \mathfrak{i}$ for any $\mathbb{K}' \subseteq \mathbb{K}$. Hence, immediately from the definition of presentable algebra, we can derive the following statement.

Proposition 5.2 *If algebra* $\mathbf{A}$ *is presentable relative to a class of algebras* $\mathbb{K}$ *by a defining pair* $(X, \mathfrak{D})$ *and a defining valuation* δ*, then* $\mathbf{A}$ *is presentable relative to any subclass* $\mathbb{K}' \subseteq \mathbb{K}$ *containing* $\mathbf{A}$ *by the same defining pair and valuation.*

The following property of finitely presentable algebra is very useful.

Proposition 5.3 (Mal'cev 1973, §11 Theorem 1) *Suppose that* $(X, \mathfrak{D})$ *is a defining pair and* $\delta : X \longrightarrow \mathbf{A}$ *is a map to an algebra* $\mathbf{A}$ *from a class* $\mathbb{K}$*. Then* $\mathbf{A}$ *is presentable by* $(X, \mathfrak{D})$ *and* δ *if and only if*

(a) set $\{\delta(x) \mid x \in X\}$ *generates algebra* $\mathbf{A}$*;*
(b) valuation δ *satisfies all identities from* $\mathfrak{D}$*;*
(c) for every $\mathbf{B} \in \mathbb{K}$ *and any valuation* $\varphi : X \longrightarrow \mathbf{B}$ *satisfying each identity* $\mathfrak{i} \in \mathfrak{D}$*, there is a homomorphism* $\psi : \mathbf{A} \longrightarrow \mathbf{B}$ *such that* $\psi(\delta(x)) = \varphi(x)$ *for each* $x \in X$*:*

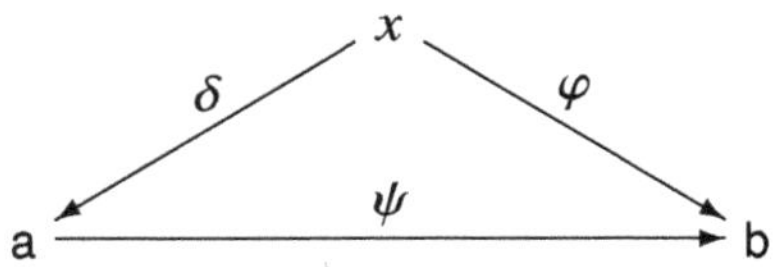

Immediately from the definition it follows that all algebras defined by the same pair relative to the same class $\mathbb{K}$, are mutually isomorphic. By $\mathbf{F}_{\mathbb{K}}(\Delta, \nu)$ we denote an algebra defined relative to $\mathbb{K}$ by the pair Δ and valuation ν; in the cases in which a particular ν is not important, we simply write $\mathbf{F}_{\mathbb{K}}(\Delta)$. It is clear that when the set of defining identities is empty, then $\mathbf{F}_{\mathbb{K}}(\{X, \varnothing\}, \delta)$ is a free algebra with free generators $\delta(X)$.

Example 5.4 Let $\mathsf{Z}_3 = (\{\mathbf{0}, \omega, \mathbf{1}\}; \wedge, \vee, \to, \neg, \mathbf{1})$ be a 3-element Heyting algebra, the Hasse diagram of which is depicted in Fig. 5.2. Element ω generates Z_3. If we take $\Delta = (\{x\}, \{\neg\neg x \approx \mathbf{1}\})$ as a defining pair and $\delta : x \mapsto \omega$ as a defining valuation, then formula $\neg\neg x$ defines algebra Z_3 relative to $\mathbb{HA}$. Hence, algebra Z_3 is finitely presentable.

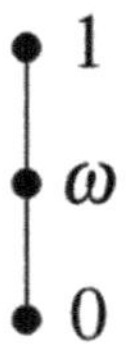

Fig. 5.2 Algebra Z_3

The following simple corollaries from Proposition 5.3 will be useful in what follows.

Proposition 5.4 *Suppose that* $\mathbb{V}$ *is a variety and that an algebra* $\mathbf{A}$ *is defined relative to* $\mathbb{V}$ *by a pair* $(X, \mathfrak{D})$ *and valuation* δ. *If* $\mathbf{B} \in \mathbb{V}$ *and valuation* $\varphi : X \longrightarrow \mathbf{B}$ *satisfies all identities from* $\mathfrak{D}$, *then the map* $\psi : \delta(x) \mapsto \varphi(x)$ *can be extended to a homomorphism* $\psi' : \mathbf{A} \longrightarrow \mathbf{B}$.

Recall that the defining valuation satisfies all identities from the defining set. Then, by properties of homomorphisms, we obtain the following corollary:

Corollary 5.2 *Algebra* $\mathbf{A}$ *defined by a pair* $(X, \mathfrak{D})$ *relative to variety* $\mathbb{V}$ *admits a homomorphism into algebra* $\mathbf{B} \in \mathbb{V}$ *if and only if there is a valuation in* $\mathbf{B}$ *that satisfies identities* $\mathfrak{D}$.

A defining pair (X, Δ) is *finite*, if X and $\mathfrak{D}$ are finite. An algebra $\mathbf{A}$ from a class $\mathbb{K}$ is said to be *finitely presentable relative to* $\mathbb{K}$ if $\mathbf{A}$ is presentable relative to $\mathbb{K}$ by a finite defining pair.

If $\mathbb{K}$ is a class of algebras of finite similarity type, then every finite algebra from $\mathbb{K}$ is finitely presentable relative to $\mathbb{K}$. Indeed, if $\mathbb{K}$ is a class of algebras of signature $\Sigma = \{f_i, i = 1, \dots, k\}$, $\mathbf{A} = (\mathsf{A}, \Sigma) \in \mathbb{K}$ and $\mathsf{A} = \{\mathsf{a}_j, j = 1, \dots, n\}$, one can take $X(\mathbf{A}) = \{x_{\mathsf{a}_j}, j = 1, \dots, n\}$ and let

$$\Delta(\mathbf{A}) \overset{\mathsf{def}}{\Longleftrightarrow} \{f(x_{a_{j1}}, \dots, x_{a_{jm}}) \approx x_{f(\mathsf{a}_{j1}, \dots, \mathsf{a}_{jm})} \mid f \in \Sigma, \mathsf{a}_{j1}, \dots, \mathsf{a}_{jm} \in \mathsf{A}\}.$$

It is not hard to verify that $(X(\mathbf{A}) = \{x_\mathsf{a} \mid \mathsf{a} \in \mathbf{A}\}, \Delta(\mathbf{A}))$ and the map $\delta : x_\mathsf{a} \mapsto \mathsf{a}$ are indeed a defining pair and a defining map.

The following observation is very important for what follows: if a congruence-distributive variety $\mathbb{V}$ is generated by its finite members, then every s.i. algebra that is finitely presentable relative to $\mathbb{V}$ is finite (cf. Blok and Pigozzi 1982).

Example 5.5 It is known that variety $\mathbb{HA}$ of all Heyting algebras is generated by finite members. Let $\mathbf{A}$ be a finitely presentable algebra from Example 5.3 and b, c be two distinct elements from $\mathbf{A}$. Because elements $\mathsf{a}_1, \dots, \mathsf{a}_n$ generate algebra $\mathbf{A}$, there are formulas B and C such that $\mathsf{b} = B(\mathsf{a}_1, \dots, \mathsf{a}_n)$ and $\mathsf{c} = C(\mathsf{a}_1, \dots, \mathsf{a}_n)$. It is clear that $A \to (B \leftrightarrow C)$ does not hold in $\mathbb{HA}$. Hence, because $\mathbb{HA}$ is generated by finite members, there is a valuation φ in a finite algebra $\mathbf{B}$ such that $\varphi(A) = \mathbf{1}$

and $\varphi(B \leftrightarrow C) \neq \mathbf{1}$. By Proposition 5.3, there is a homomorphism $\psi : \mathbf{A} \longrightarrow \mathbf{B}$ and $\psi(\mathsf{b}) \neq \psi(\mathsf{c})$. Thus, $\mathbf{A}$ is residually finite and hence, every s.i. finitely presented algebra is finite.

Proposition 5.5 (e.g., Mal'cev 1973) *Suppose that* $\mathbf{A}$ *is a finitely presentable algebra relative to a class* $\mathbb{K}$. *Then for any finite system of generators* $\mathsf{G} \subseteq \mathbf{A}$ *there is a defining pair* $(X, \mathfrak{D})$ *and a defining valuation* $\delta : X \longrightarrow \mathsf{G}$ *presenting* $\mathbf{A}$ *relative to* $\mathbb{K}$.

5.2.3 Splitting

Following Blok and Pigozzi (1982), a pair $(\mathbb{V}_1, \mathbb{V}_2)$ of subvarieties of a variety $\mathbb{V}$ is called a *splitting pair* if

(a) $\mathbb{V}_1 \not\subseteq \mathbb{V}_2$;
(b) for every variety $\mathbb{V}' \subseteq \mathbb{V}$, either $\mathbb{V}_1 \subseteq \mathbb{V}'$ or $\mathbb{V}' \subseteq \mathbb{V}_2$.

Thus $(\mathbb{V}_1, \mathbb{V}_2)$ is a splitting pair if and only if $\mathbb{V}_2$ is the largest subvariety of $\mathbb{V}$ not containing $\mathbb{V}_1$. If $(\mathbb{V}_1, \mathbb{V}_2)$ is a splitting pair, then $\mathbb{V}_1$ is generated by a single finitely generated s.i. algebra—a *splitting algebra*, while $\mathbb{V}_2$ is defined relative to $\mathbb{V}$ by a single identity—a *splitting identity* (cf., e.g., Galatos et al. 2007, Chap. 10). It was observed in McKenzie (1972) that every splitting identity $\mathfrak{i}$ is $\wedge$-prime; that is, if $\mathfrak{i}$ $\mathbb{V}$-follows from a set of identities $\mathfrak{I}$, then $\mathfrak{i}$ $\mathbb{V}$-follows from some identity from $\mathfrak{I}$.

Let us note that every splitting algebra $\mathbf{A}$ defines a splitting pair $(\mathbb{V}_1, \mathbb{V}_2)$, where $\mathbb{V}_1 = \mathbf{V}(\mathbf{A})$ and $\mathbb{V}_2 = \overline{\mathbf{V}}(\mathbf{A})$, where $\overline{\mathbf{V}}(\mathbf{A})$ is the largest subvariety of $\mathbb{V}$ not containing $\mathbf{A}$. All splitting identities are $\mathbb{V}$-equipotent because they define the same subvariety of $\mathbb{V}$. For a splitting algebra $\mathbf{A}$, by $\mathfrak{s}(\mathbf{A})$ we denote a corresponding splitting identity, and by $\mathbb{V}_{spl}$ we denote a class of all splitting algebras from $\mathbb{V}$.

Example 5.6 Without an explicit use of Yankov's formulas, Hosoi introduced the notion of slices of intermediate logics (cf. Hosoi 1967). Namely, if $\mathbf{C}_n$ is a linearly ordered Heyting algebra of cardinality n (a *chain algebra*), then an intermediate logic $\boldsymbol{L}$ belongs to the nth slice if $\mathbf{C}_{n+1}$ is a model of $\boldsymbol{L}$, while $\mathbf{C}_{n+2}$ is not a model of $\boldsymbol{L}$. In algebraic terms, $\mathbf{C}_{n+2}$ defines in $\mathbb{HA}$ a splitting pair $(\mathbf{V}(\mathbf{C}_{n+2}), \overline{\mathbf{V}}(\mathbf{C}_{n+2}))$; thus $\overline{\mathbf{V}}(\mathbf{C}_{n+2})$ is an algebraic counterpart of the smallest logic of nth slice. For instance, the classical logic Cl belongs to the 1st slice, because the 2-element chain algebra (which is the 2-element Boolean algebra) is a model of Cl, while the 3-element chain $\mathbf{C}_3$ is not a model of Cl.

Example 5.7 Suppose that $\mathbb{V}_1$ is a variety having the largest proper subvariety $\mathbb{V}_2$. Then, the pair $(\mathbb{V}_1, \mathbb{V}_2)$ is a splitting pair. Indeed, $\mathbb{V}_1 \not\subseteq \mathbb{V}_2$ and if $\mathbb{V}' \subseteq \mathbb{V}$ is a subvariety of $\mathbb{V}$, then either $\mathbb{V}'$ is a proper subvariety (and hence, $\mathbb{V}' \subseteq \mathbb{V}_2$), or $\mathbb{V}' = \mathbb{V}_1$. For instance, suppose that $\mathbb{V}_1$ is a variety not generated by finite algebras but all its proper subvarieties are generated by finite algebras (there are continuum many varieties of Heyting algebras with this property (Mardaev 1987, Theorem 5). Then,

$\mathbb{V}_1$ contains the largest proper subvariety $\mathbb{V}_2$, namely, the subvariety generated by all finite algebras of $\mathbb{V}_1$, and $(\mathbb{V}_1, \mathbb{V}_2)$ is a splitting pair.

Immediately from the definition it follows that for any splitting algebra $\mathbf{A}$ any corresponding splitting identity is refuted in $\mathbf{A}$; that is,

$$\mathbf{A} \not\models \mathfrak{s}(\mathbf{A}). \tag{SelfRft}$$

The following proposition shows that there is a link between a number of generators in a splitting algebra and number of distinct variables in a splitting identity.

Proposition 5.6 *Suppose that* $(\mathbb{V}_1, \mathbb{V}_2)$ *is a splitting pair of variety* $\mathbb{V}$, $\mathbb{S}$ *is a class of all splitting algebras generating* $\mathbb{V}_1$, *and* $\mathfrak{S}$ *is a set of all splitting identities defining* $\mathbb{V}_2$ *relative to* $\mathbb{V}$. *Then,*

(a) If $\mathbb{S}$ *contains an n-generated algebra, then* $\mathfrak{S}$ *contains an identity having at most n distinct variables;*
(b) If $\mathfrak{S}$ *contains an identity having n distinct variables, then* $\mathbb{S}$ *contains a k-generated algebra such that* $k \leq n$.

Proof (a) Suppose $\mathbf{A} \in \mathbb{S}$ is an n-generated algebra, $\mathfrak{s}(x_1, \dots, x_k) \in \mathfrak{S}$ and $k > n$. Let $\mathfrak{I}$ be a set of all identities in variables $y_1, \dots, y_n$ valid in $\mathbb{V}_2$, and let $\mathbb{V}'$ be the subvariety defined relative to $\mathbb{V}$ by $\mathfrak{I}$. It is clear that $\mathbb{V}_2 \subseteq \mathbb{V}'$. Let us verify that $\mathbb{V}_2 = \mathbb{V}'$.

For contradiction, suppose that $\mathbb{V}' \not\subseteq \mathbb{V}_2$. Then because $(\mathbb{V}_1, \mathbb{V}_2)$ is a splitting pair,

$$\mathbb{V}_1 \subseteq \mathbb{V}' \text{ and consequently, } \mathbf{A} \in \mathbb{V}'. \tag{5.2}$$

By (SelfRft), $\mathbf{A} \not\models \mathfrak{s}(x_1, \dots, x_k)$. Hence, there are elements $\mathsf{a}_1, \dots, \mathsf{a}_k$ such that $\mathfrak{s}(\mathsf{a}_1, \dots, \mathsf{a}_k)$ is not true. Each element a_j, $j \in [1, k]$, can be expressed via generators $\mathsf{g}_1, \dots, \mathsf{g}_n$; that is, for some terms $t_i(y_1, \dots, y_n)$,

$$\mathsf{a}_j = t_j(\mathsf{g}_1, \dots, \mathsf{g}_n),\ j \in [1, k].$$

Let $\mathfrak{i}(y_1, \dots, y_n) = \mathfrak{s}(t_1(y_1, \dots, y_n), \dots, t_k((y_1, \dots, y_n)))$. It is clear that $\mathfrak{s} \vdash_{\mathbb{V}} \mathfrak{i}$, for $\mathfrak{i}$ was obtained from $\mathfrak{s}$ by substitution. Recall that $\mathbb{V}_2$ is defined by $\mathfrak{s}$; hence, $\mathbb{V}_2 \models \mathfrak{s}$ and consequently, $\mathbb{V}_2 \models \mathfrak{i}$. Thus, $\mathfrak{i} \in \mathfrak{I}$, and it is clear that $\mathbf{A} \not\models \mathfrak{i}$, because $\mathfrak{i}(\mathsf{g}_1, \dots, \mathsf{g}_n)$ is not true. Hence, by the definition of $\mathbb{V}'$, $\mathbf{A} \notin \mathbb{V}'$ and we have arrived at a contradiction with (5.2).

Next, let us note that $\mathbb{V}' = \mathbb{V}_2$ yields $\mathfrak{s} \vdash_{\mathbb{V}} \mathfrak{i}'$ for all $\mathfrak{i}' \in \mathfrak{I}$.

On the other hand, $\mathbb{V}' = \mathbb{V}_2$ entails that identities $\mathfrak{I}$ define $\mathbb{V}_2$ relative to $\mathbb{V}$ and hence, $\mathfrak{I} \vdash_{\mathbb{V}} \mathfrak{s}$. Recall that $\mathfrak{s}$ is a splitting identity and, therefore, $\wedge$-prime. Hence, there is an $\mathfrak{i}' \in \mathfrak{I}$ such that $\mathfrak{i}' \vdash_{\mathbb{V}} \mathfrak{s}$. Thus, $\mathfrak{i}'$ is $\mathbb{V}$-interderivable with $\mathfrak{s}$, and it is a desired splitting identity containing at most n distinct variables.

(b) Suppose that $\mathfrak{s}(x_1, \dots, x_n) \in \mathfrak{S}$ and $\mathbf{A} \in \mathbb{S}$. By (SelfRft), $\mathbf{A} \not\models \mathfrak{s}$. Hence, there is an n-generated subalgebra $\mathbf{A}' \leq \mathbf{A}$ in which $\mathfrak{s}$ is refuted, and subsequently, there is an s.i. factor $\mathbf{B}$ of $\mathbf{A}'$ in which $\mathfrak{s}$ is refuted. Let us show that $\mathbf{B}$ generates $\mathbb{V}_1$ and thus, $\mathbf{B} \in \mathbb{S}$.

Consider a variety $\mathbb{V}' = \mathbf{V}(\mathbf{B})$ generated by $\mathbf{B}$. Because $\mathbf{B} \in \mathbf{HSA}$, $\mathbf{B} \in \mathbb{V}_1$ and hence, $\mathbb{V}' \subseteq \mathbb{V}_1$. On the other hand, if $\mathbb{V}_1 \nsubseteq \mathbb{V}'$, by the definition of splitting, $\mathbb{V}' \subseteq \mathbb{V}_2$ and therefore, $\mathbb{V}' \models \mathfrak{s}$, which is not true for $\mathbf{B} \not\models \mathfrak{s}$. Thus, $\mathbb{V}' = \mathbb{V}_1$; that is, $\mathbf{B}$ generates $\mathbb{V}_1$ and it is a desired splitting algebra.

Moreover, the following holds.

Proposition 5.7 *Suppose that* $\mathbf{A} \in \mathbb{V}$ *is a splitting algebra from a variety* $\mathbb{V}$. *Then for any identity* $\mathfrak{i}$,

$$\mathbf{A} \not\models \mathfrak{i} \textit{ if and only if } \mathfrak{i} \vdash_{\mathbb{V}} \mathfrak{s}(\mathbf{A}). \tag{Spl1}$$

Moreover, if $\mathbb{V}$ *is congruence-distributive, then for any algebra* $\mathbf{B} \in \mathbb{V}$,

$$\mathbf{B} \not\models \mathfrak{s}(\mathbf{A}) \textit{ if and only if } \mathbf{A} \in \mathbf{HSP}_u\mathbf{B}. \tag{Spl2}$$

If in addition $\mathbf{B}$ *is finite, then*

$$\mathbf{B} \not\models \mathfrak{s}(\mathbf{A}) \textit{ if and only if } \mathbf{A} \in \mathbf{HSB}. \tag{Spl3}$$

And if $\mathbb{V}$ *has the CEP, then* $\mathbf{HS} = \mathbf{SH}$ *(cf., e.g., Burris and Sankappanavar 1981) and*

$$\mathbf{B} \not\models \mathfrak{s}(\mathbf{A}) \textit{ if and only if } \mathbf{A} \in \mathbf{SHB}. \tag{Spl4}$$

Proof Suppose that $\mathbf{A}$ is a splitting algebra defining a splitting pair $(\mathbb{V}_1, \mathbb{V}_2)$ and $\mathbf{A} \not\models \mathfrak{i}$. Let $\mathbb{V}'$ be a variety defined relative to $\mathbb{V}$ by $\mathfrak{i}$. By the definition of a splitting pair, $\mathbb{V}' \subseteq \mathbb{V}_2$ and hence, $\mathfrak{i} \vdash_{\mathbb{V}} \mathfrak{s}(\mathbf{A})$, because $\mathbb{V}' \models \mathfrak{s}(\mathbf{A})$ and $\mathfrak{i}$ defines $\mathbb{V}'$ relative to $\mathbb{V}$.

The converse immediately follows from (SelfRft).

Suppose that $\mathbb{V}$ is a congruence-distributive variety, $\mathbf{B} \in \mathbb{V}$ and $\mathbf{B} \not\models \mathfrak{s}(\mathbf{A})$. Then, by the definition of a splitting pair, $\mathbf{A} \in \mathbf{V}(\mathbf{A}) \subseteq \mathbf{V}(\mathbf{B})$. Because $\mathbf{A}$ is s.i., by Corollary 3.2 from Jónsson (1967), $\mathbf{A} \in \mathbf{HSP}_u\mathbf{B}$ and if $\mathbf{B}$ is finite, then $\mathbf{A} \in \mathbf{HSB}$ (cf. Jónsson 1967, Corollary 3.4).

If $\mathbf{A} \in \mathbf{HSPB}$ (or $\mathbf{A} \in \mathbf{HSB}$), then $\mathbf{A} \in \mathbf{V}(\mathbf{B})$ and therefore, $\mathbf{B} \not\models \mathfrak{s}(\mathbf{A})$ because $\mathbf{A} \not\models \mathfrak{s}(\mathbf{A})$.

In general, the following holds.

Proposition 5.8 *Suppose that* $\mathbb{V}$ *is a variety and* $\mathbf{A}, \mathbf{B} \in \mathbb{V}_{spl}$. *Then,*

$$\mathfrak{s}(\mathbf{A}) \vdash_{\mathbb{V}} \mathfrak{s}(\mathbf{B}) \textit{ if and only if } \mathbf{V}(\mathbf{B}) \subseteq \mathbf{V}(\mathbf{A}). \tag{Spl5}$$

Hence, if $\mathbb{V}$ *is congruence-distributive,*

$$\mathfrak{s}(\mathbf{A}) \vdash_{\mathbb{V}} \mathfrak{s}(\mathbf{B}) \textit{ if and only if } \mathbf{B} \in \mathbf{HSP}_u\mathbf{A}. \tag{Spl6}$$

Proof Indeed, by (SelfRft), $\mathbf{B} \not\models \mathfrak{s}(\mathbf{B})$ and hence, $\mathfrak{s}(\mathbf{A}) \vdash_{\mathbb{V}} \mathfrak{s}(\mathbf{B})$ entails $\mathbf{B} \not\models \mathfrak{s}(\mathbf{A})$; that is, $\mathbf{V}(\mathbf{B}) \not\models \mathfrak{s}(\mathbf{A})$. Then, by the definition of splitting, $\mathbf{V}(\mathbf{A}) \subseteq \mathbf{V}(\mathbf{B})$.

Conversely, if $\mathbf{V}(\mathbf{A}) \subseteq \mathbf{V}(\mathbf{B})$, then $\mathbf{A} \in \mathbf{V}(\mathbf{B})$ and $\mathbf{A} \not\models \mathfrak{s}(\mathbf{A})$ entails $\mathbf{B} \not\models \mathfrak{s}(\mathbf{A})$. Hence by (Spl1), $\mathfrak{s}(\mathbf{A}) \vdash_{\mathbb{V}} \mathfrak{s}(\mathbf{B})$.

Equation (Spl6) follows from (Spl5) by Jonsson's Lemma.

If $\mathbb{V}$ is a congruence-distributive variety generated by finite algebras, then from (SelfRft) and (Spl3) it follows that every splitting algebra is finite (cf., e.g., Blok and Pigozzi 1982).

Example 5.8 Because variety $\mathbb{HA}$ is generated by finite algebras, every splitting algebra from $\mathbb{HA}$ is finite. It was observed in Jankov (1968a) that there are subvarieties of $\mathbb{HA}$ not generated by finite algebras. In such subvarieties, the splitting algebras may be infinite, and in fact, there are some infinite splitting algebras (cf. Citkin 2012). For modal algebras such an example was found by Kracht (cf. Kracht 1999).

Let us note that relative to a variety $\mathbb{V}$ a splitting identity $\mathfrak{s}(\mathbf{A})$ of any splitting algebra $\mathbf{A} \in \mathbb{V}_{spl}$ defines the largest subvariety not containing algebra $\mathbf{A}$. More generally, the following holds.

Proposition 5.9 *Suppose that* $\mathbb{V}$ *is a variety. Then for any class* $\mathbb{S} \subseteq \mathbb{V}_{spl}$ *of splitting algebras, identities* $\mathfrak{I} = \{\mathfrak{s}(\mathbf{A}) \mid \mathbf{A} \in \mathbb{S}\}$ *define relative to* $\mathbb{V}$ *the largest subvariety not containing any algebra from* $\mathbb{A}$.

Proof Suppose that $\mathbb{V}'$ is a subvariety defined relative to $\mathbb{V}$ by identities $\mathfrak{I}$. Clearly, because $\mathbf{A} \not\models \mathfrak{s}(\mathbf{A})$, neither algebra from $\mathbb{S}$ belongs to $\mathbb{V}'$.

On the other hand, if $\mathbf{B} \notin \mathbb{V}'$, for some $\mathbf{A} \in \mathbb{S}$, $\mathbf{B} \not\models \mathfrak{s}(\mathbf{A})$ and hence, by (Spl2), $\mathbf{A} \in \mathbf{HSP}_u\mathbf{B}$. Therefore, $\mathbf{A}$ belongs to every variety containing algebra $\mathbf{B}$; that is, $\mathbb{V}'$ is the largest subvariety of $\mathbb{V}$ not containing any algebras from $\mathbb{S}$.

Example 5.9 Two algebras, the Hasse diagrams of which are depicted in Fig. 5.3, are splitting algebras in $\mathbb{HA}$. Their splitting identities define the largest variety of Heyting algebras not containing these algebras, namely, the variety of algebras satisfying identity $(x \to y) \vee (y \to x) \approx \mathbf{1}$ (cf. Idziak and Idziak 1988).

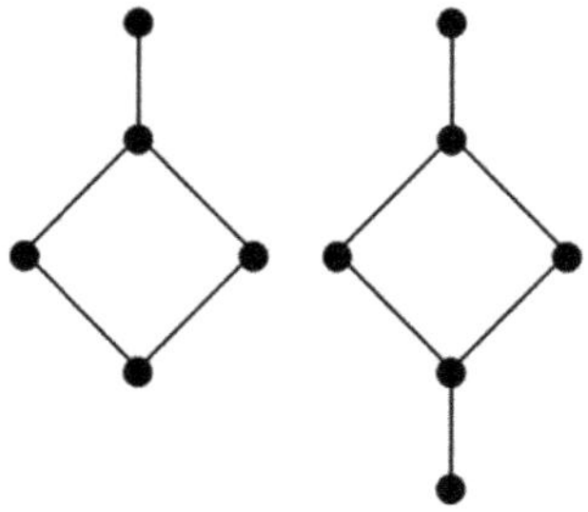

Fig. 5.3 Two splitting algebras

5.3 Independent Sets of Splitting Identities

One of the important applications of the characteristic formulas is the ability to construct independent sets of identities. For a variety $\mathbb{V}$, a set of identities $\mathfrak{I}$ is $\mathbb{V}$-*independent* if not any identity $\mathfrak{i} \in \mathfrak{I}$ $\mathbb{V}$-follows from the rest of the identities from $\mathfrak{I}$. The independent sets are important because they provide independent bases for subvarieties of $\mathbb{V}$ (cf. McKenzie 1972).

In Jankov (1968a) it was observed that there is a continuum of intermediate logics. In order to prove this, Yankov introduced a notion of strongly independent logics[1]: a set of logics $\mathcal{L} = \{\boldsymbol{L}_i \mid i \in I\}$ is called *strongly independent* if not any logic $\boldsymbol{L}_i$ is not included in the logic generated by the rest of the logics from $\mathcal{L}$; that is $\boldsymbol{L}_i \notin \cup'\{\boldsymbol{L}_j \mid j \neq i, j \in J\}$ (where $\cup'$ is a closed union). A set of formulas F is called *independent* if neither formula $A \in \mathsf{F}$ is derivable from the rest of the formulas from F; that is $\mathsf{F} \setminus \{A\} \nvdash A$. Given an independent set of formulas F, one can easily construct a strongly independent set of logics: indeed, logics $\boldsymbol{L}(A)$, $A \in \mathsf{F}$ defined by A as an axiom, form an independent set. Moreover, it is not hard to see that the logics defined by distinct subsets of an independent set of formulas F (as axioms) are distinct. The latter observation is often employed for constructing continuum logics possessing certain properties. For instance, Yankov used this approach to show that there are continuum many intermediate logics (cf. Jankov 1968a), and Blok used the same idea to prove the continuality of the class of extension of the normal modal logic **S4** (cf. Blok 1977).

Because splitting identities enjoy the same properties as Yankov's formulas, we can use the regular technique of constructing irredundant bases consisting of splitting formulas (cf., e.g., Jankov 1969; Citkin 1986; Skvortsov 1999; Tomaszewski 2002; Bezhanishvili 2006).

Let us note the following property of independent sets of identities.

Proposition 5.10 *Suppose* $\mathbb{V}$ *is a variety and* $\mathfrak{I}$ *is a* $\mathbb{V}$*-independent set of identities. Then, for any two distinct subsets* $\mathfrak{I}_1, \mathfrak{I}_2 \subseteq \mathfrak{I}$*, the subvarieties* $\mathbb{V}_i = \mathbb{V}(\mathfrak{I}_i)$ *defined relative to* $\mathbb{V}$ *by identities* $\mathfrak{I}_i$*, are distinct.*

Corollary 5.3 *Suppose that* $\mathfrak{I}$ *is an infinite* $\mathbb{V}$*-independent set of identities. Then* $\mathbb{V}$ *contains continuum many distinct subvarieties.*

Indeed, by Proposition 5.10, any subset of $\mathfrak{I}$ defines relative to $\mathbb{V}$ a distinct subvariety.

One of the important properties of a variety is whether it has an independent base. Often the splitting identities are instrumental in constructing such bases. We start with the following observation.

---

[1] In Troelstra (1965) the logics $\boldsymbol{L}_1$ and $\boldsymbol{L}_2$ are called independent just in case if they are incomparable, i.e. $\boldsymbol{L}_1 \nsubseteq \boldsymbol{L}_2$ and $\boldsymbol{L}_2 \nsubseteq \boldsymbol{L}_1$.

Proposition 5.11 *Suppose $\mathbb{V}$ is a congruence-distributive variety and $\mathbb{S} \subseteq \mathbb{V}_{spl}$. Let $\mathfrak{I} = \{\mathfrak{s}(\mathbf{A}) \mid \mathbf{A} \in \mathbb{S}\}$. Then the following assertions are equivalent:*

(a) $\mathfrak{I}$ is independent;
(b) for any distinct $\mathbf{A}, \mathbf{B} \in \mathbb{S}$, $\mathfrak{s}(\mathbf{B}) \nvdash_{\mathbb{V}} \mathfrak{s}(\mathbf{A})$;
(c) for any distinct $\mathbf{A}, \mathbf{B} \in \mathbb{S}$, $\mathbf{B} \notin \mathsf{HSP}_u\mathbf{A}$.

Proof (a) $\Rightarrow$ (b) is trivial.

(b) $\Leftrightarrow$ (c) follows immediately from Corollary 5.8.

(c) $\Rightarrow$ (a). For contradiction, assume that for some $\mathbf{A} \in \mathbb{S}$, $\mathfrak{I}^- \vdash_{\mathbb{V}} \mathfrak{s}(\mathbf{A})$, where $\mathfrak{I}^- = \mathfrak{I} \setminus \{\mathfrak{s}(\mathbf{A})\}$. Recall that by (SelfRft), $\mathbf{A} \not\models \mathfrak{s}(\mathbf{A})$. Hence, there is an identity $\mathfrak{i} \in \mathfrak{I}^-$ such that $\mathbf{A} \not\models \mathfrak{i}$. That is, for some distinct from $\mathbf{A}$ algebra $\mathbf{B} \in \mathbb{S}$, $\mathbf{A} \not\models \mathfrak{s}(\mathbf{B})$ and hence, by (Spl2), $\mathbf{B} \in \mathsf{HSP}_u\mathbf{A}$, which contradicts (c).

5.3.1 Quasi-order

In Jankov (1963b), Yankov introduced a quasi-order on the set of splitting algebras (more details can be found in Jankov 1969). In this section, we study the properties of this quasi-order in a general setting.

Suppose that $\mathbb{V}$ is a variety. Let us introduce on $\mathbb{V}_{spl}$ a quasi-order: for any $\mathbf{A}, \mathbf{B} \in \mathbb{V}_{spl}$,

$$\mathbf{A} \preceq \mathbf{B} \leftrightharpoons \mathfrak{s}(\mathbf{A}) \vdash_{\mathbb{V}} \mathfrak{s}(\mathbf{B}), \tag{5.3}$$

and by Proposition 5.8,

$$\mathbf{A} \preceq \mathbf{B} \text{ if and only if } \mathbf{V}(\mathbf{A}) \subseteq \mathbf{V}(\mathbf{B}). \tag{5.4}$$

Let us observe that if $\mathbb{V}$ is congruence-distributive, by Corollary 5.8,

$$\mathbf{A} \preceq \mathbf{B} \text{ if and only if } \mathbf{A} \in \mathsf{HSP}_u\mathbf{B}. \tag{5.5}$$

If, in addition, $\mathbb{V}$ is generated by finite algebras, all members of $\mathbb{V}_{spl}$ are finite and

$$\mathbf{A} \preceq \mathbf{B} \text{ if and only if } \mathbf{A} \in \mathsf{HS}\mathbf{B}. \tag{5.6}$$

In the varieties enjoying the CEP, $\mathsf{HS} = \mathsf{SH}$ and hence, in such varieties,

$$\mathbf{A} \preceq \mathbf{B} \text{ if and only if } \mathbf{A} \in \mathsf{SH}\mathbf{B}. \tag{5.7}$$

Proposition 5.12 *Suppose that the congruence-distributive variety $\mathbb{V}$ is generated by finite algebras. Then the quasi-order defined on $\mathbb{V}_{spl}$ by (5.3) is a partial order (assuming, we consider algebras modulo isomorphism).*

Proof Indeed, if $\mathbb{V}$ is generated by finite algebras, then all members of $\mathbb{V}_{spl}$ are finite. Hence, (5.6) yields that $\mathbf{A} \preceq \mathbf{B}$ entails $|\mathbf{A}| \leq |\mathbf{B}|$, where $|\mathbf{A}|$, $|\mathbf{B}|$ are cardinalities of $\mathbf{A}$ and $\mathbf{B}$. Thus, $\mathbf{A} \preceq \mathbf{B}$ and $\mathbf{B} \preceq \mathbf{A}$ yield $|\mathbf{A}| = |\mathbf{B}|$ and therefore by (5.6), $\mathbf{A} \cong \mathbf{B}$.

Already in Jankov (1963b, Theorem 6), Yankov observed that for Heyting algebras the above quasi-order is, in fact, a partial order. For modal algebras the reader can find additional information in Kracht (1999).

In a natural way, the quasi-order $\preceq$ induces the equivalence relation $\mathbf{A} \simeq \mathbf{B} \leftrightharpoons \mathbf{A} \preceq \mathbf{B}$ and $\mathbf{B} \preceq \mathbf{A}$. If $\mathbb{S}$ is a class of splitting algebras, by $\mathbb{S}_{min}$ we denote a class of minimal, relative to $\preceq$, members of $\mathbb{S}$ containing a single representative from each $\simeq$-equivalence class. We say that $\mathbb{S}$ is *m-complete* if for each $\mathbf{A} \in \mathbb{S}$ there is a $\mathbf{B} \in \mathbb{S}_{min}$ such that $\mathbf{B} \preceq \mathbf{A}$. Let us note that $\mathbf{B} \preceq \mathbf{A}$ entails $|\mathbf{B}| \leq |\mathbf{A}|$ and hence, every class of finite splitting algebra is always m-complete.

5.3.2 Antichains

Algebras $\mathbf{A}, \mathbf{B} \in \mathbb{V}_{spl}$ are said to be *comparable*, if $\mathbf{A} \preceq \mathbf{B}$ or $\mathbf{B} \preceq \mathbf{A}$, otherwise these algebras are *incomparable*. A subset $\mathbb{A} \subseteq \mathbb{V}_{spl}$ is an *antichain* if any two distinct members of $\mathbb{A}$ are incomparable. Also, for a class of splitting algebras $\mathbb{S}$, by $\mathfrak{s}(\mathbb{A})$ we denote $\{\mathfrak{s}(\mathbf{A}) \mid \mathbf{A} \in \mathbb{S}\}$.

The following proposition follows from the definition of quasi-order and Proposition 5.11.

Proposition 5.13 *Suppose that $\mathbb{V}$ is a variety and $\mathbb{A} \subseteq \mathbb{V}_{spl}$ is an antichain. Then $\mathfrak{s}(\mathbb{A})$ is a $\mathbb{V}$-independent set of identities .*

Thus, to construct a $\mathbb{V}$-independent set of identities it suffices to construct an antichain in $\mathbb{V}_{spl}$.

Corollary 5.4 *Suppose that $\mathbb{V}$ is a variety and $\mathbb{V}_{spl}$ contains an infinite antichain. Then $\mathbb{V}$ contains at least continuum many subvarieties.*

Example 5.10 The Heyting algebras depicted in Fig. 5.4. by the Hasse diagrams and frames, form an infinite antichain.[2] Hence, the characteristic identities of these algebras form an independent set. Therefore, there are continuum many varieties of Heyting algebras (there are continuum many intermediate logics (cf. Jankov 1968a, Corollary 1) and consequently, there are continuum many not finitely axiomatizable intermediate logics (cf. Jankov 1968a, Corollary 2). Let us note that the existence of not finitely axiomatizable intermediate logic, follows from Umezawa (1959): Umezawa exhibited a chain of intermediate logics of type ω^{ω}, and the limit of any strongly ascending chain of logics cannot be finitely axiomatized (due to the finitarity of inference).

---

[2] These algebras are due to Wroński (1974) and Blok (1977); in Jankov (1968a) the different algebras were employed.

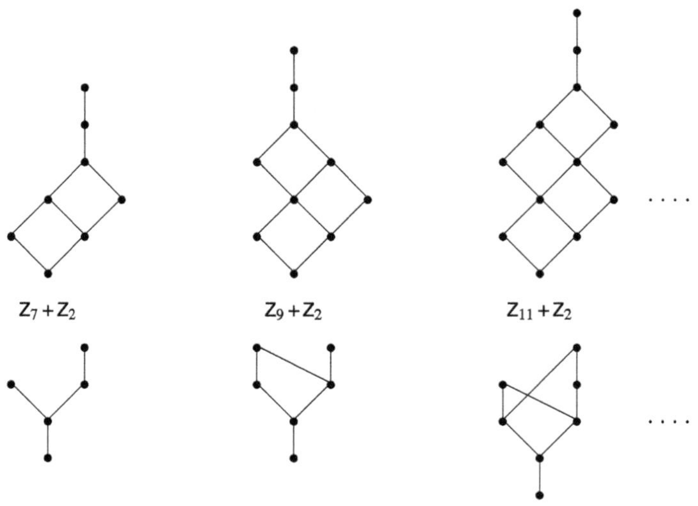

Fig. 5.4 An antichain in $\mathbb{HA}_{spl}$

5.4 Independent Bases

In this section, we consider the following problem: given a variety $\mathbb{V}$ and its subvariety $\mathbb{V}'$, how to construct an *independent* (or *irredundant*) base relative to $\mathbb{V}$—a $\mathbb{V}$-*base* of $\mathbb{V}'$. To that end, when it is possible, we will employ the splitting identities and use Theorem 5.1.

The varieties in which every finitely generated algebra is finite are called *locally finite*. Clearly, every locally finite variety is generated by its finite members.

We will consider following three classes of varieties:

variety $\mathbb{V}$ enjoys the *finite splitting (Fsi-Spl)* property if $\mathbb{V}_{fsi} \subseteq \mathbb{V}_{spl}$;
variety $\mathbb{V}$ enjoys the *splitting finite (Spl-Fsi)* property if $\mathbb{V}_{spl} \subseteq \mathbb{V}_{fsi}$;
variety $\mathbb{V}$ enjoys the *(Spl=Fsi)* property if $\mathbb{V}_{fsi} = \mathbb{V}_{spl}$.

The examples of classes of varieties possessing the above properties are:

(1) Spl-Fsi: locally finite varieties;
(2) Spl-Fsi: congruence-distributive varieties generated by finite members;
(3) Fsi-Spl: varieties with equationally definable principal congruences (EDPC) of a finite type generated by finite members;
(4) Spl=Fsi: congruence-distributive locally finite varieties;
(5) Spl=Fsi: varieties of a finite type with EDPC and generated by finite members.

Reasons:

(1) all splitting algebras are finitely generated and, therefore, finite;
(2) cf. McKenzie (1972);

(3) cf. Blok and Pigozzi (1982, Corollary 3.2);
(4) cf. Day (1973);
(5) immediately from (2) and (3).

Proposition 5.14 *Let $\mathbb{V}$ be a congruence-distributive variety enjoying the Fsi-Spl property and generated by finite algebras. Then, $\mathbb{V}$ enjoys the Spl=Fsi property.*

The proof is straightforward.

Given a variety $\mathbb{V}$ and a set of identities $\mathfrak{I}$, by $\mathbb{V}(\mathfrak{I})$ we denote the subvariety defined by $\mathfrak{I}$ relative to $\mathbb{V}$; that is, $\mathbb{V}(\mathfrak{I}) = \{\mathbf{A} \in \mathbb{V} \mid \mathbf{A} \models \mathfrak{I}\}$.

Theorem 5.1 *Suppose that $\mathbb{V}$ enjoys the Spl-Fsi property and $\mathfrak{I}$ is a set of $\mathbb{V}$-splitting identities. Then, $\mathfrak{I}$ contains an independent subset $\mathfrak{I}'$ such that $\mathbb{V}(\mathfrak{I}) = \mathbb{V}(\mathfrak{I}')$.*

Proof Suppose that $\mathfrak{I} = \{\mathfrak{s}(\mathbf{A}), \mathbf{A} \in \mathbb{S}\}$, where $\mathbb{S} \subseteq \mathbb{V}_{spl}$. Because $\mathbb{V}$ enjoys the Spl-Fsi property, all algebras from $\mathbb{S}$ are finite and hence, $\mathbb{S}_{min}$ is m-complete. Let $\mathfrak{I}' = \mathfrak{s}(\mathbb{S}_{min})$.

First, we observe that set $\mathbb{S}_{min}$ forms an antichain, for $\mathbb{S}_{min}$ consists of minimal elements of $\mathbb{S}$ that are clearly mutually incomparable. Hence, by Proposition 5.13, set $\mathfrak{I}'$ is independent.

Next, we note that $\mathfrak{I}' \subseteq \mathfrak{I}$ and hence, $\mathbb{V}(\mathfrak{I}) \subseteq \mathbb{V}(\mathfrak{I}')$. On the other hand, for every $\mathbf{A} \in \mathbb{S}$, there is an algebra $\mathbf{B} \in \mathbb{S}_{min}$ such that $\mathbf{B} \preceq \mathbf{A}$. Hence, by (5.3), $\mathfrak{s}(\mathbf{B}) \vdash_{\mathbb{V}} \mathfrak{s}(\mathbf{A})$ and therefore, for any algebra $\mathbf{C}$, $\mathbf{C} \models \mathfrak{s}(\mathbf{B})$ entails $\mathbf{C} \models \mathfrak{s}(\mathbf{A})$; that is, $\mathbb{V}(\mathfrak{I}') \subseteq \mathbb{V}(\mathfrak{I})$. Thus, $\mathbb{V}(\mathfrak{I}') = \mathbb{V}(\mathfrak{I})$.

Theorem 5.1 yields that in the varieties enjoying the Spl-Fsi property every subvariety defined by a set of splitting identities has an independent base.

Example 5.11 Variety $\mathbb{HA}$ has EDPC and is generated by finite algebras. Yet it contains a subvariety that has no independent $\mathbb{HA}$-base (cf. Chagrov and Zakharyaschev 1995). Hence, not every variety of Heyting algebras can be defined by splitting identities.

5.4.1 Subvarieties Defined by Splitting Identities

In this section, we study the subvarieties of variety $\mathbb{V}$ that have $\mathbb{V}$-base consisting of splitting identities.

Suppose that $\mathbb{V}$ is a variety and $\mathbb{V}' \subseteq \mathbb{V}$ is a subvariety. Then, for every algebra $\mathbf{A} \in \mathbb{V}_{spl} \setminus \mathbb{V}'_{spl}$, $\mathbb{V}' \models \mathfrak{s}(\mathbf{A})$. Indeed, consider a splitting pair $(\mathbb{V}_1, \mathbb{V}_2)$ where $\mathbb{V}_1$ is generated by $\mathbf{A}$. Because $\mathbf{A} \notin \mathbb{V}'$, we have $\mathbb{V}' \subseteq \mathbb{V}_2$. By the definition of splitting, $\mathbb{V}_2 \models \mathfrak{s}(\mathbf{A})$ and therefore, $\mathbb{V}' \models \mathfrak{s}(\mathbf{A})$. Hence, for $\mathfrak{I} = \{\mathfrak{s}(\mathbf{A}) \mid \mathbf{A} \in \mathbb{V}_{spl} \setminus \mathbb{V}'_{spl}\}$, we always have $\mathbb{V}' \subseteq \overline{\mathbb{V}'}$, where $\overline{\mathbb{V}'} = \mathbb{V}(\mathfrak{I})$. Thus, the following holds.

Proposition 5.15 *A subvariety $\mathbb{V}'$ of variety $\mathbb{V}$ has a $\mathbb{V}$-base consisting of splitting identities if and only if $\overline{\mathbb{V}'} = \mathbb{V}'$.*

Proposition 5.15 entails the following important corollary.

Corollary 5.5 (cf. Tomaszewski 2003, Corollary 3 p. 57) *Suppose that* $\mathbb{V}$ *is a variety and* $\{\mathbb{V}_i \subseteq \mathbb{V}, i \in I\}$ *is a class of subvarieties having the same splitting algebras. Then, at most one subvariety* $\mathbb{V}_i$ *has a* $\mathbb{V}$*-base consisting of splitting identities.*

Example 5.12 The first example of an intermediate logic without the finite model property (and hence, an example of a variety of Heyting algebras not generated by finite algebras) was given in Jankov (1968a). Hence, there is an intermediate logic (a variety of Heyting algebras) which cannot be axiomatized by Yankov's formulas (Yankov's identities).

Example 5.13 In Tomaszewski (2003), Tomaszewski constructed a class of continuum many varieties of Heyting algebras having the same splitting algebras. Thus, he proved that there are continuum many varieties of Heyting algebras that do not have an $\mathbb{HA}$-base consisting of splitting (Yankov's) identities.

Corollary 5.6 (cf. Tomaszewski 2003, Corollary 4 p. 57) *Suppose that* $\mathbb{V}$ *is a variety enjoying the Spl=Fsi property generated by finite members. Then the following are equivalent:*

(a) *every subvariety of* $\mathbb{V}$ *is generated by finite members;*
(b) *every subvariety of* $\mathbb{V}$ *can be defined by splitting identities.*

Let us turn to the varieties whose subvarieties have bases consisting of splitting identities.

5.4.2 *Independent Bases in the Varieties Enjoying the Fsi-Spl Property*

Let us take a closer look at the varieties enjoying the Spl=Fsi property. First, we observe the following simple, and yet useful, property of such varieties.

Proposition 5.16 *Suppose* $\mathbb{V}$ *is a variety enjoying the Fsi-Spl property and* $\mathbb{V}' \subset \mathbb{V}$ *is a proper subvariety. Then* $\overline{\mathbb{V}'} \setminus \mathbb{V}'$ *does not contain finite algebras.*

Proof For contradiction, suppose that $\mathbf{B} \in \overline{\mathbb{V}'} \setminus \mathbb{V}'$ is a finite algebra. Without loss of generality we can assume that $\mathbf{B}$ is an s.i. algebra. Then, by Fsi-Spl, $\mathbf{B} \in \mathbb{V}_{spl} \setminus \mathbb{V}'_{spl}$ and by the definition of $\overline{\mathbb{V}'}$, we have $\overline{\mathbb{V}'} \models \mathfrak{s}(\mathbf{B})$. Hence, $\mathbf{B} \models \mathfrak{s}(\mathbf{B})$ and we have arrived at a contradiction with (SelfRft).

Theorem 5.2 *Suppose that* $\mathbb{V}$ *is a congruence-distributive variety such that*

(i) $\mathbb{V}$ *enjoys the Fsi-Spl property;*
(ii) $\mathbb{V}$ *and each of its subvarieties are generated by their finite members.*

Then, every proper subvariety $\mathbb{V}' \subset \mathbb{V}$ admits an independent $\mathbb{V}$-base consisting of splitting identities.

Proof To prove the theorem, we will demonstrate that $\mathbb{V}'$ can be axiomatized relative to $\mathbb{V}$ by splitting identities, and then we can apply Theorem 5.1.

To this end, it is sufficient to verify that $\overline{\mathbb{V}'} = \mathbb{V}'$. We already know that $\mathbb{V}' \subseteq \overline{\mathbb{V}'}$. Thus, we only need to demonstrate that $\overline{\mathbb{V}'} \setminus \mathbb{V}' = \varnothing$.

Indeed, assume the contrary: $\overline{\mathbb{V}'} \setminus \mathbb{V}' \neq \varnothing$. By (ii), varieties $\overline{\mathbb{V}'}$ and $\mathbb{V}'$ are generated by their finite algebras; hence, there is a finite algebra $\mathbf{B} \in \overline{\mathbb{V}'} \setminus \mathbb{V}'$. But (i) and Proposition 5.16 entail that class $\overline{\mathbb{V}'} \setminus \mathbb{V}'$ does not contain finite algebras. Thus, we have arrived at a contradiction.

Corollary 5.7 *Suppose that $\mathbb{V}$ is a locally finite congruence-distributive variety. Then, every proper subvariety of $\mathbb{V}$ admits an independent $\mathbb{V}$-base.*

Example 5.14 Any subvariety of a locally finite variety of Heyting algebras admits an independent $\mathbb{V}$-base.

Under certain conditions, all locally finite subvarieties of a congruence-distributive variety enjoying the Fsi-Spl property admit an independent base. More precisely, the following holds.

Theorem 5.3 *Suppose that $\mathbb{V}$ is a congruence-distributive variety such that*

(i) $\mathbb{V}$ enjoys the Fsi-Spl property;
(ii) for every infinite finitely generated algebra $\mathbf{A} \in \mathbb{V}$ and any given natural number n, $\mathbf{A}$ has a quotient of cardinality exceeding n.

Then, every proper locally finite subvariety of $\mathbb{V}$ admits an independent $\mathbb{V}$-base consisting of splitting identities.

Proof Suppose that $\mathbb{V}' \subset \mathbb{V}$ is a proper locally finite subvariety. Our goal is to show that $\overline{\mathbb{V}'} = \mathbb{V}'$, and then we can apply Theorem 5.1 and complete the proof.

We already know that $\mathbb{V}' \subseteq \overline{\mathbb{V}'}$, and we only need to show that $\overline{\mathbb{V}'} \setminus \mathbb{V}' = \varnothing$.

For contradiction, suppose that $\overline{\mathbb{V}'} \setminus \mathbb{V}' \neq \varnothing$ and $\mathbf{B} \in \overline{\mathbb{V}'} \setminus \mathbb{V}'$. Without loss of generality we can assume that $\mathbf{B}$ is a finitely generated algebra; let us say, $\mathbf{B}$ is a k-generated algebra. By (i) and Proposition 5.16, class $\overline{\mathbb{V}'} \setminus \mathbb{V}'$ does not contain any finite algebras. Hence, $\mathbf{B}$ is infinite.

Let us recall that $\mathbb{V}'$ is a locally finite variety. Hence, the cardinalities of all its k-generated algebras are bounded by a natural number n (one can take $n = |\mathbf{F}_{\mathbb{V}'}(k)|$—the cardinality of a free k-generated algebra of $\mathbb{V}'$). By assumption (ii), there is a finite quotient algebra $\mathbf{B}/\theta$ such that $|\mathbf{B}/\theta| > n$. Hence, $\mathbf{B}/\theta \in \overline{\mathbb{V}}$, because $\mathbf{B} \in \overline{\mathbb{V}'}$, and on the other hand, $\mathbf{B}/\theta \notin \mathbb{V}'$, because $|\mathbf{B}/\theta| > n$. Thus, $\mathbf{B}/\theta$ is a finite algebra that is in $\overline{\mathbb{V}'} \setminus \mathbb{V}'$ and thus, we have arrived at a contradiction with Proposition 5.16.

Example 5.15 (Tomaszewski 2003, Theorem 4.5; Bezhanishvili 2006, Theorem 3.4.24) For Heyting algebras condition (i) of Theorem 5.3 was established in Jankov (1963b), while condition (ii) follows from Kuznetsov's Theorem (cf. Kuznetsov 1973). Hence, any locally finite variety of Heyting algebras has an independent $\mathbb{HA}$-base consisting of splitting identities.

Let us note the role played by minimal, relative to $\preceq$, algebras from $\mathbb{V}_{spl} \setminus \mathbb{V}'_{spl}$ in constructing independent bases. Namely, we will show that in the varieties enjoying the Spl-Fsi property, each independent $\mathbb{V}$-base of $\mathbb{V}'$ consisting of splitting identities is precisely the set of splitting identities of all minimal algebras from $\mathbb{V}_{spl} \setminus \mathbb{V}'_{spl}$.

Proposition 5.17 *Suppose that $\mathbb{V}$ is a variety enjoying the Spl-Fsi property and $\mathbb{V}' \subset \mathbb{V}$ is a proper subvariety. If $\mathfrak{I}$ is an independent $\mathbb{V}$-base of $\mathbb{V}'$ consisting of splitting identities, then for each algebra $\mathbf{B} \in \left(\mathbb{V}_{spl} \setminus \mathbb{V}'_{spl}\right)_{min}$, $\mathfrak{I}$ contains an identity $\mathbb{V}$-equipotent with $\mathfrak{s}(\mathbf{B})$.*

Proof Suppose that $\mathbb{V}$ is a variety and $\mathbb{V}' \subset \mathbb{V}$ is a proper subvariety defined relative to $\mathbb{V}$ by a set $\mathfrak{I}$ of splitting identities, and suppose that $\mathfrak{i} \in \mathfrak{I}$. Then, $\mathfrak{i}$ is $\mathbb{V}$-equipotent with $\mathfrak{s}(\mathbf{A})$ of some algebra $\mathbf{A} \in \mathbb{V}_{spl} \setminus \mathbb{V}'_{spl}$. The Spl-Fsi property yields that all members of $\mathbb{V}_{spl}$ are finite and hence, class $\mathbb{V}_{spl} \setminus \mathbb{V}'_{spl}$ is m-complete. Therefore, $\mathbf{B} \preceq \mathbf{A}$ for some $\mathbf{B} \in \left(\mathbb{V}_{spl} \setminus \mathbb{V}'_{spl}\right)_{min}$, which means that $\mathfrak{s}(\mathbf{B}) \vdash_{\mathbb{V}} \mathfrak{s}(\mathbf{A})$; that is $\mathfrak{s}(\mathbf{B}) \vdash_{\mathbb{V}} \mathfrak{i}$.

On the other hand, because $\mathbf{B} \notin \mathbb{V}'$, there is an $\mathfrak{i}' \in \mathfrak{I}$ such that $\mathbf{B} \not\models \mathfrak{i}'$ and therefore, by (Spl1), $\mathfrak{i}' \vdash_{\mathbb{V}} \mathfrak{s}(\mathbf{B})$. Recall that $\mathfrak{I}$ is independent; hence, $\mathfrak{s}(\mathbf{B}) \vdash \mathfrak{i}$ and $\mathfrak{i}' \vdash \mathfrak{s}(\mathbf{B})$ yields $\mathfrak{i} = \mathfrak{i}'$ so $\mathfrak{i}$ is $\mathbb{V}$-equipotent with $\mathfrak{s}(\mathbf{B})$.

Proposition 5.17 means that the only way to construct a dependent $\mathbb{V}$-base of a subvariety $\mathbb{V}'$ consisting of $\wedge$-prime identities is to take $\mathfrak{s}(\left(\mathbb{V}_{spl} \setminus \mathbb{V}'_{spl}\right)_{min})$.

5.4.3 Finite Bases in the Varieties Enjoying the Fsi-Spl Property

From Theorem 5.2 we know that in any congruence-distributive variety $\mathbb{V}$ having the Fsi-Spl property, if all subvarieties are generated by finite algebras, then, every subvariety has an independent $\mathbb{V}$-base. The following theorem answers the question of when every subvariety of $\mathbb{V}$ has a finite $\mathbb{V}$-base.

Theorem 5.4 (cf. Wolter 1993, Theorem 2.4.8) *Suppose that $\mathbb{V}$ is a congruence-distributive variety such that*

(i) $\mathbb{V}$ enjoys the Fsi-Spl property;
(ii) $\mathbb{V}$ and each of its subvarieties are generated by their finite members.

Let $\mathcal{L}(\mathbb{V})$ be a lattice of all subvarieties of $\mathbb{V}$. Then, the following are equivalent:

(a) every subvariety of $\mathbb{V}$ is finitely based relative to $\mathbb{V}$;
(b) $\mathcal{L}(\mathbb{V})$ is countable;
(c) $\mathbb{V}_{spl}$ has no infinite antichains;
(d) $\mathcal{L}(\mathbb{V})$ enjoys the descending chain condition.

Proof First, we observe that Proposition 5.14 yields that all varieties from $\mathcal{L}(\mathbb{V})$ enjoy the Spl=Fsi property.

(a) $\Rightarrow$ (b) is trivial.

(b) $\Rightarrow$ (c) (cf. Grätzer and Quackenbush 2010, Theorem 5.1) From Corollary 5.4 it follows that if $\mathbb{V}_{spl}$ contains an infinite antichain, then $\mathcal{L}(\mathbb{V})$ is not countable.

(c) $\Rightarrow$ (d) For contradiction, assume that $\mathbb{V}_{spl}$ has no infinite antichains and $\mathbb{V}_0 \supset \mathbb{V}_1 \supset \cdots$ is a strongly descending chain of subvarieties of $\mathbb{V}$. Then, for each $i \geq 0$, there is a finite s.i. algebra $\mathbf{A}_i \in \mathbb{V}_i \setminus \mathbb{V}_{i+1}$ and hence, $\mathbf{A}_i \in \mathbb{V}_{spl}$. As we pointed out earlier, the set $\mathbf{A}_i, i \geq 0$, contains an m-complete subset of minimal, relative to $\preceq$, members. Let $\{\mathbf{A}_i, i \in I\}$ be a set of all minimal elements. Then,

$$\text{for any } n \geq 0, \text{ there is, } j \in I \text{ such that } \mathbf{A}_j \preceq \mathbf{A}_n;$$

that is,

$$\text{for any } n \geq 0, \text{ there is a } j \in I \text{ such that } \mathbf{A}_j \in \mathbf{HSA}_n \subseteq \mathbf{V}(A_n),$$

or, equivalently,

$$\text{for any } n \geq 0, \text{ there is } j \in I \text{ such that } \mathbf{V}(\mathbf{A}_j) \subseteq \mathbf{V}(\mathbf{A}_n). \tag{5.8}$$

Next, we observe that, because of the minimality of its members, the set $\{\mathbf{A}_i, i \in I\}$ forms an antichain, and hence, by our assumption, it is finite. Suppose that I does not contain numbers exceeding k. Then, by our selection of algebras $\mathbf{A}_i$, we have

$$\mathbf{A}_j \notin \mathbb{V}_{k+1} \text{ for all } j \leq k. \tag{5.9}$$

On the other hand, consider $\mathbf{A}_{k+1}$. By selection, $\mathbf{A}_{k+1} \in \mathbb{V}_{k+1} \setminus \mathbb{V}_{k+2}$ and hence, $\mathbf{A}_{k+1} \in \mathbb{V}_{k+1}$ and

$$\mathbf{V}(\mathbf{A}_{k+1}) \subseteq \mathbb{V}_{k+1}. \tag{5.10}$$

By (5.8), for some $j \leq k$, there is a minimal algebra $\mathbf{A}_j$ such that

$$\mathbf{V}(\mathbf{A}_j) \subseteq \mathbf{V}(\mathbf{A}_{k+1}). \tag{5.11}$$

Thus, from (5.11) and (5.10), we have

$$\mathbf{A}_j \in \mathbf{V}(\mathbf{A}_j) \subseteq \mathbf{V}(\mathbf{A}_{k+1}) \subseteq \mathbb{V}_{k+1}$$

and this contradicts (5.9).

(d) $\Rightarrow$ (a) Suppose that $\mathcal{L}(\mathbb{V})$ enjoys the descending chain condition. We need to demonstrate that every proper subvariety $\mathbb{V}' \subset \mathbb{V}$ is finitely based relative to $\mathbb{V}$.

Indeed, let $\{\mathrm{i}_\mu, \mu < \sigma\}$ be a set of all identities such that $\mathbb{V}' \models \mathrm{i}_\mu$ and $\mathbb{V} \not\models \mathrm{i}_\mu$. For each $\kappa < \sigma$, consider a variety $\mathbb{V}_\kappa$ defined relative to $\mathbb{V}$ by identities $\mathrm{i}_\mu, \mu \leq \kappa$. It is clear that $\kappa \leq \kappa'$ yields $\mathbb{V}_\kappa \supseteq \mathbb{V}_{\kappa'}$; hence,

$$\{\mathbb{V}_\kappa \mid \kappa < \sigma\} \text{ is a descending chain such that } \bigcap_{m \geq 0} \mathbb{V}_m = \mathbb{V}'. \tag{5.12}$$

Because $\mathcal{L}(\mathbb{V})$ enjoys the descending chain condition, $\{\mathbb{V}_\kappa \mid \kappa < \sigma\}$ is finite. Hence, by (5.12), for some $n < \omega$, $\mathbb{V}'$ coincides with $\mathbb{V}_n$, and this means that $\mathbb{V}'$ is defined relative to $\mathbb{V}$ by a finite set of identities, namely by identities $\mathfrak{i}_k$, for all $k < n$.

Corollary 5.8 *Let $\mathbb{V}$ be a locally finite congruence-distributive variety and $\mathcal{L}(\mathbb{V})$ be a lattice of subvarieties of $\mathbb{V}$. Then, the following are equivalent:*

(a) *every subvariety of $\mathbb{V}$ is finitely based relative to $\mathbb{V}$;*
(b) *$\mathcal{L}(\mathbb{V})$ is countable;*
(c) *$\mathbb{V}_{spl}$ has no infinite antichains;*
(d) *$\mathcal{L}(\mathbb{V})$ enjoys the descending chain condition.*

5.4.4 Reduced Bases

Every subvariety can be defined relative to a variety that contains it by a set of identities. However, as we saw, not every variety admits an independent base. In addition, as we saw, not every variety has a finite axiomatic rank. But as we have established in the previous sections, in many cases the subvarieties of locally finite varieties have independent bases consisting of splitting formulas. This gives us a way to find the reduced bases: the independent bases consisting of $\wedge$-prime formulas and containing the least possible number of variables.

5.4.4.1 Edge Algebras

We recall from Mal'cev (1973) that the *rank* of an identity is the number of distinct variables occurring in it. Suppose that $\mathbb{V}'$ is a subvariety of a variety $\mathbb{V}$. An *axiomatic rank* of subvariety $\mathbb{V}'$ relative to $\mathbb{V}$ is the smallest natural number r (which we denote by $r_a(\mathbb{V}'/\mathbb{V})$) such that $\mathbb{V}'$ can be defined (axiomatized) relative to $\mathbb{V}$ by identities of a rank not exceeding r. In this section, we show that in locally finite varieties any subvariety $\mathbb{V}'$ can be defined by an independent set of characteristic identities of a rank not exceeding $r_a(\mathbb{V}'/\mathbb{V})$. To that end, we introduce the notion of *edge algebra*, which is very similar to the notion of a critical algebra form (Tomaszewski 2003, Definition 3.9); we prefer the term "edge algebra" because the term "critical algebra" is already used with different meanings.

Suppose that $\mathbb{V}$ is a variety and $\mathbf{A}$ is an algebra. We say that $\mathbf{A}$ is an *edge algebra of* $\mathbb{V}$ (or $\mathbb{V}$-*edge*) if $\mathbf{A} \notin \mathbb{V}$, while all proper subalgebras and homomorphic images of $\mathbf{A}$ are in $\mathbb{V}$.

Similarly to splitting algebras, the following holds.

Proposition 5.18 *Suppose that $\mathbb{V}$ is a variety and $\mathbf{A}$ is a $\mathbb{V}$-edge algebra. Then, $\mathbf{A}$ is a finitely generated s.i. algebra.*

The proof is straightforward, and it is left for the reader.

Recall from Mal'cev (1973) that the *basis rank* of an algebra **A** is the minimal cardinality of the set of elements generating **A**. The basis rank of algebra **A** we denote by $r_b(\mathbf{A})$, and any set of generators of cardinality $r_b(\mathbf{A})$ we will call a *basis* (of **A**).

The splitting identities of edge algebras are instrumental in constructing the bases of subvarieties.

Proposition 5.19 *Suppose that* $\mathbb{V}$ *is a variety,* $\mathbb{V}' \subseteq \mathbb{V}$ *is a subvariety, and* $\mathbf{A} \in \mathbb{V}$ *is a* $\mathbb{V}'$*-edge algebra. Then*

$$r_b(\mathbf{A}) \leq r_a(\mathbb{V}'/\mathbb{V}).$$

Proof Suppose that $\mathbf{A} \in \mathbb{V}$ is a $\mathbb{V}'$-edge algebra. Then, it is finitely generated and subsequently, it has a finite basis rank, say $r = r_b(\mathbf{A})$.

For contradiction, assume that $r_a(\mathbb{V}) < r_b(\mathbf{A}) = r$. Then, there is a set of identities of a rank strongly lower than r and defining the subvariety $\mathbb{V}'$ relative to $\mathbb{V}$. Recall that $\mathbf{A} \in \mathbb{V} \setminus \mathbb{V}'$. Therefore, there is an identity $\mathrm{i}(x_1, \ldots, x_m)$, where $m < r$, such that $\mathbb{V}' \models \mathrm{i}$ but $\mathbf{A} \not\models \mathrm{i}$. Assume that $\mathsf{a}_1, \ldots, \mathsf{a}_m \in \mathbf{A}$ and $\mathrm{i}(\mathsf{a}_1, \ldots, \mathsf{a}_m)$ does not hold. Let $\mathbf{A}'$ be a subalgebra of $\mathbf{A}$ generated by the elements $\mathsf{a}_1, \ldots, \mathsf{a}_m$. Because $m < r_b(\mathbf{A})$, by virtue of the definition of the basis rank, $\mathbf{A}'$ is a proper subalgebra of $\mathbf{A}$. By the definition of an edge algebra, $\mathbf{A}' \in \mathbb{V}'$. Hence, $\mathbf{A}' \not\models \mathrm{i}$ and this contradicts that i is valid in $\mathbb{V}'$.

Example 5.16 Suppose that **A** is a Heyting algebra of the nth slice; that is, **A** contains a subalgebra isomorphic to the $n+1$-element chain algebra $\mathbf{C}_{n+1}$, while $\mathbf{C}_{n+2}$ is not embedded in **A** (cf. Example 5.6). Because all proper subalgebras and homomorphic images of a chain algebra are chain algebras of a strongly lower cardinality, $\mathbf{C}_{n+2}$ is a $\mathbf{V}(\mathbf{A})$-edge algebra. Thus, $r_b(\mathbf{C}_{n+1}) \leq r_a(\mathbf{V}(\mathbf{A}))$. Observe that $r_b(\mathbf{C}_n) = 1$ when $n = 2$ and $r_b(\mathbf{C}_n) = n - 2$ for all $n > 2$. Hence, $r_a(\mathbf{V}(\mathbf{A})) \geq n$ (cf. Bellissima 1988, Theorem 2.2).

Corollary 5.9 *If* $\mathbb{V}'$ *is a subvariety of a variety* $\mathbb{V}$ *and if for any natural number* m *there is a* $\mathbb{V}'$*-edge algebra* $\mathbf{A} \in \mathbb{V}$ *of basis rank exceeding* m*, then* $\mathbb{V}'$ *has an infinite axiomatic rank relative to* $\mathbb{V}$ *and, hence,* $\mathbb{V}'$ *does not have a finite* $\mathbb{V}$*-base.*

For example (cf. Maksimova et al. 1979), the variety of Heyting algebras corresponding to Medvedev's Logic has the edge algebras of a basis rank exceeding any given natural number and, thus, this variety is not finitely axiomatizable.

Proposition 5.20 *Suppose that* $\mathbb{V}'$ *is a proper subvariety of a variety* $\mathbb{V}$ *enjoying the Spl=Fsi property. Then,* $\left(\mathbb{V}_{spl} \setminus \mathbb{V}'_{spl}\right)_{min}$ *is precisely the class of all* $\mathbb{V}'$*-edge algebras from* $\mathbb{V}$*.*

Proof By Proposition 5.18, every $\mathbb{V}'$-edge algebra **A** is finitely generated and s.i. Hence, **A** is a finite s.i. algebra and therefore, by the Spl=Sfi property, it is a splitting algebra. It is not hard to see that **A** is a minimal splitting algebra.

Conversely, suppose that $\mathbf{A} \in \left(\mathbb{V}_{spl} \setminus \mathbb{V}'_{spl}\right)_{min}$ and thus, **A** is a finite s.i. algebra. For contradiction, suppose that **B** is a proper homomorphic image of **A** or a proper subalgebra of $\mathbf{A}$, and $\mathbf{B} \notin \mathbb{V}'$. Then $\mathbf{B}$ has an s.i. factor $\mathbf{C}$ such that $\mathbf{C} \notin \mathbb{V}'$. Thus, $\mathbf{C} \in \mathsf{HSA}$ and by (5.6), $\mathbf{C} \preceq \mathbf{A}$. Let us note that $|\mathbf{C}| < |\mathbf{A}|$, that is $\mathbf{A} \npreceq \mathbf{C}$, and we have arrived at a contradiction with the minimality of $\mathbf{A}$.

5.4.4.2 Uniqueness of Reduced Base

Suppose that $\mathfrak{I}$ is a base of subvariety $\mathbb{V}'$ relative to variety $\mathbb{V}$ that contains it. Then $\mathfrak{I}$ is *reduced* if

(a) it is independent;
(b) every identity in $\mathfrak{I}$ is $\wedge$-prime;
(c) identities from $\mathfrak{I}$ contain at most $r_a(\mathbb{V})$ distinct variables.

Let us recall that by Theorem 5.1, in the varieties enjoying the Spl-Fsi property, from every base consisting of splitting identities one can extract an independent subbase defining this variety. Thus, we obtain a base satisfying properties (a) and (b) of the definition of a reduced base. To construct a reduced base, it suffices to find an interdependent base consisting of splitting identities and having the least possible number of distinct variables.

First, let us observe that an independent base consisting of $\wedge$-prime identities is unique in the following sense.

Proposition 5.21 *Let $\mathfrak{I}_1$ and $\mathfrak{I}_2$ be two independent bases of a subvariety $\mathbb{V}'$ of a variety $\mathbb{V}$ consisting of $\wedge$-prime identities. Then, $|\mathfrak{I}_1| = |\mathfrak{I}_2|$, and there is a 1-1-correspondence φ between $\mathfrak{I}_1$ and $\mathfrak{I}_2$ such that $\mathfrak{i}$ and $\varphi(\mathfrak{i})$ are $\mathbb{V}$-equipotent for all $\mathfrak{i} \in \mathfrak{I}_1$.*

Proof Suppose that $\mathfrak{i} \in \mathfrak{I}_1$ and hence, $\mathbb{V}' \models \mathfrak{i}$. Then, because $\mathfrak{I}_2$ is a base of $\mathbb{V}'$,

$$\mathfrak{I}_2 \vdash_{\mathbb{V}} \mathfrak{i}.$$

Recall that $\mathfrak{i}$ is $\wedge$-prime. Hence, there is an identity $\mathfrak{i}' \in \mathfrak{I}_2$ such that

$$\mathfrak{i}' \vdash_{\mathbb{V}} \mathfrak{i}.$$

On the other hand, $\mathfrak{I}_1 \vdash_{\mathbb{V}} \mathfrak{i}'$ and hence, for some $\mathfrak{i}'' \in \mathfrak{I}_1$, we have

$$\mathfrak{i}'' \vdash_{\mathbb{V}} \mathfrak{i}'.$$

Because set $\mathfrak{I}_1$ is independent, we can conclude that

$$\mathfrak{i} = \mathfrak{i}''.$$

5.4.4.3 Reduced Bases in Locally Finite Varieties

In this section, we will show that any subvariety $\mathbb{V}'$ of a given congruence-distributive locally finite variety $\mathbb{V}$ admits a reduced base relative to $\mathbb{V}$, and we use the splitting identities to construct such a base (cf. Bezhanishvili 2004, Remark 2.8).

Proposition 5.22 *Suppose that $\mathbb{V}$ is a variety enjoying the Fsi-Spl property, $\mathbb{V}' \subseteq \mathbb{V}$ is a subvariety, and $\mathbf{A}$ is a finite algebra from $\mathbb{V}_{spl} \setminus \mathbb{V}'_{spl}$. Then, $\mathbf{A}$ is minimal relative to $\preceq$ if and only if it is a $\mathbb{V}'$-edge algebra.*

Proof Suppose that $\mathbf{A}$ is a finite minimal algebra in $\mathbb{V}_{spl} \setminus \mathbb{V}'_{spl}$. For contradiction, assume that $\mathbf{A}$ is not a $\mathbb{V}'$-edge algebra. Then, $\mathbb{V}_{spl} \setminus \mathbb{V}'_{spl}$ contains a proper subalgebra $\mathbf{B}$ or a proper homomorphic image $\mathbf{C}$ of $\mathbf{A}$. In any case, there is an s.i. homomorphic image $\mathbf{D}$ of $\mathbf{B}$ or of $\mathbf{C}$ such that $\mathbf{D} \in \mathbb{V} \setminus \mathbb{V}'$. Because $\mathbb{V}$ enjoys the Fsi-Spl property, $\mathbf{D} \in \mathbb{V}_{spl} \setminus \mathbb{V}'_{spl}$. But $\mathbf{D} \in \mathsf{HSA}$ and hence by (5.5), $\mathbf{D} \preceq \mathbf{A}$, and this contradicts the minimality of $\mathbf{A}$: because $|\mathbf{D}| < |\mathbf{A}|$, we have $\mathbf{D} \prec \mathbf{A}$.

If $\mathbf{A}$ is a finite $\mathbb{V}'$-edge algebra, then $\mathbf{A}$ is s.i. and finitely generated, and hence by Fsi-Spl, it is a splitting algebra. It is clear that $\mathbf{A}$ is minimal relative to $\preceq$.

Theorem 5.5 *Suppose that $\mathbb{V}$ is a congruence-distributive variety enjoying the Spl=Fsi property, and that $\mathbb{V}$ and each of its subvarieties are generated by finite algebras. If $\mathbb{V}' \subset \mathbb{V}$ is a proper subvariety, then $\mathbb{V}'$ has a reduced $\mathbb{V}$-base consisting of splitting formulas.*

Proof Suppose that $\mathbb{V}$ is a congruence-distributive variety enjoying the Spl=Fsi property. Let us consider $(\mathbb{V}_{spl} \setminus \mathbb{V}'_{spl})_{min}$. From Proposition 5.17 it follows that $\mathfrak{s}((\mathbb{V}_{spl} \setminus \mathbb{V}'_{spl})_{min})$ forms an independent $\mathbb{V}$-base. By Proposition 5.20, each algebra from $(\mathbb{V}_{spl} \setminus \mathbb{V}'_{spl})_{min}$ is a $\mathbb{V}'$-edge algebra. Suppose that $n = r_a(\mathbb{V}'/\mathbb{V})$. Then by Proposition 5.19, each algebra $\mathbf{A} \in (\mathbb{V}_{spl} \setminus \mathbb{V}'_{spl})_{min}$ can be generated by a set containing at most n elements. Hence by Proposition 5.6(a), there is a splitting identity $\mathfrak{s}(\mathbf{A})$ containing at most n variables, and this observation completes the proof.

Let us note that in general, even if we are given a splitting algebra, we do not know how to write down a splitting identity. In the following section, we consider the varieties with a TD term, that is, varieties in which some sort of deduction theorem holds. In such varieties, it is possible to obtain a splitting identity by a given splitting algebra.

5.5 Varieties with a TD Term

In this section we generalize the notion of a characteristic formula to varieties with a ternary deductive term (TD term for short) introduced in Blok and Pigozzi (1994). As we will see in Sect. 5.5.1.1, most logics have as an algebraic semantic a variety with a TD term.

5.5.1 Definition of the TD Term

The following definitions are due to Blok and Pigozzi (1994).

Let $\mathbb{V}$ be a variety of algebras. A ternary term $td(x, y, z)$ is called a *ternary deductive term of variety* $\mathbb{V}$ (a TD term) if for any algebra $\mathbf{A} \in \mathbb{V}$ and any elements $\mathsf{a}, \mathsf{b}, \mathsf{c}, \mathsf{d}$

$$\begin{aligned} &td(\mathsf{a}, \mathsf{a}, \mathsf{b}) = \mathsf{b}, \\ &td(\mathsf{a}, \mathsf{b}, \mathsf{c}) = td(\mathsf{a}, \mathsf{b}, \mathsf{d}) \text{ if } \mathsf{c} \equiv \mathsf{d} \pmod{\theta(\mathsf{a}, \mathsf{b})}. \end{aligned} \tag{5.13}$$

Suppose that $\mathbf{A}$ is an algebra and $td(x, y, z)$ is a TD term. Then for any $\mathsf{a}, \mathsf{b} \in \mathbf{A}$, in each $\theta(\mathsf{a}, \mathsf{b})$-congruence class, the TD term selects a representative in a uniform way. For instance, for $\mathbb{HA}$ there are two TD terms: $td_{\to}(x, y, z) = (x \leftrightarrow y) \to z$ and $td_{\wedge}(x, y, z) = (x \leftrightarrow y) \wedge z$. If $\mathbf{A}$ is a Heyting algebra and θ is a principal congruence of $\mathbf{A}$, each θ-congruence class contains the largest and the smallest elements. $td_{\to}$ selects in each θ-congruence class the largest element, while $td_{\wedge}$ selects the smallest element. By $td_{\to}$, two elements c, d belong to the same congruence class, if the largest elements in the congruence classes containing a and b coincide.

It is important to remember that the varieties with a TD term are congruence-distributive, have EDPC, and enjoy the CEP (cf. Blok and Pigozzi 1994). Hence, if $\mathbb{V}$ is a variety with a TD term generated by finite algebras, then all its splitting algebras are finite; that is $\mathbb{V}$ enjoys the Spl-Fsi property. If in addition $\mathbb{V}$ is of a finite similarity type, each finite s.i. algebra is a splitting algebra; that is, $\mathbb{V}$ enjoys the Spl=Fsi property.

5.5.1.1 Examples of Varieties with a TD Term

In the following table we give some examples of varieties and their td-terms. For the details we refer the reader to Blok and Pigozzi (1994), Blok and Raftery (1997), Agliano (1998), Berman and Blok (2004), Odintsov (2005), Spinks and Veroff (2007).

We start with noting that all discriminator varieties (cf. Blok and Pigozzi 1994) have a TD term.

Consider the following ternary terms (used in Table 5.1):

(a) $td \overset{\text{def}}{\iff} (p \to q) \to ((q \to p) \to r);$

(b) $td \overset{\text{def}}{\iff} (p \to q) \wedge (q \to p) \wedge r$, or $p \leftrightarrow q \wedge r;$

(c) $td \overset{\text{def}}{\iff} (p \to q) \overset{n-1}{\longrightarrow} ((q \to p) \overset{n-1}{\longrightarrow} r);$

(d) $td \overset{\text{def}}{\iff} (p \leftrightarrow q) \wedge \Box(p \leftrightarrow q) \wedge \cdots \wedge \Box^{n-1}(p \leftrightarrow q) \to r;$

(e) $td \overset{\text{def}}{\iff} (p \leftrightarrow q) \wedge \Box(p \leftrightarrow q) \wedge \cdots \wedge \Box^{n-1}(p \leftrightarrow q) \wedge r;$

(f) $td \overset{\text{def}}{\iff} \Box(p \leftrightarrow q) \to r;$

(g) $td \overset{\text{def}}{\iff} \Box(p \leftrightarrow q) \wedge r;$

Table 5.1 Examples of TD terms

Variety	td-term
Hilbert algebras	(a)
Brouwerian semilattices	(a), (b)
Heyting algebras	(a), (b)
KM algebras	(a), (b)
Johansson's algebras	(a), (b)
n-potent hoops	(c)
n-transitive modal algebras	(d), (e)
Interior algebras	(f), (g)
BCI monoids	(h), (i)

(h) $td \overset{\text{def}}{\Longleftrightarrow} (p \leftrightarrow q)^n \cdot r;$

(i) $td \overset{\text{def}}{\Longleftrightarrow} (p \leftrightarrow q) \overset{n}{\rightarrow} r.$

5.5.1.2 Iterated TD Terms

First, we recall from Blok and Pigozzi (1994) that every variety $\mathbb{V}$ with a TD term is congruence-distributive and hence, every s.i. algebra is a splitting algebra as long as it is finitely presented relative to $\mathbb{V}$.

As we see, a TD term gives us a uniform way to define the principal congruences. By iterating the TD term, a (5.13)-like characterization of principal congruences can be extended to compact congruences (cf. Blok and Pigozzi 1994, Theorem 2.6): if $\underline{a}, \underline{b}$ are lists of elements of an algebra $\mathbf{A} \in \mathbb{V}$ of the same length, then

$$\mathsf{c} \equiv \mathsf{d} \pmod{\theta(\underline{a}, \underline{b})} \text{ if and only if } td(\underline{a}, \underline{b}, \mathsf{c}) = td(\underline{a}, \underline{b}, \mathsf{d}), \tag{5.14}$$

where

$$td(\underline{a}, \underline{b}, c) = td(\mathsf{a}_1, \mathsf{b}_1, td(\mathsf{a}_2, \mathsf{b}_2, \ldots, td(\mathsf{a}_m, \mathsf{b}_m, \mathsf{c}))\ldots). \tag{5.15}$$

Using a simple induction, it is not hard to prove that for any $\underline{a} \subseteq \mathbf{A}$ and any $\mathsf{c} \in \mathbf{A}$

$$td(\underline{a}, \underline{a}, \mathsf{c}) = \mathsf{c}. \tag{5.16}$$

Clearly, (5.14) defines compact congruences similarly to how (5.13) defines principal congruences.

Let us note that even if $td(\mathsf{a}_2, \mathsf{b}_2, td(\mathsf{a}_1, \mathsf{b}_1, \mathsf{c})) \neq td(\mathsf{a}_1, \mathsf{b}_1, td(\mathsf{a}_2, p_2, \mathsf{c}))$, the TD terms $td(\mathsf{a}_2, \mathsf{b}_2, td(\mathsf{a}_1, \mathsf{b}_1, z))$ and $td(\mathsf{a}_1, \mathsf{b}_1, td(\mathsf{a}_2, \mathsf{b}_2, z))$ define the same congruence (these terms may select different representatives of congruence classes).

5.5.2 Definition and Properties of Characteristic Identities

Suppose that $\mathbb{V}$ is a variety with a TD term td and containing an s.i. algebra $\mathbf{A}$ that is presentable relative to $\mathbb{V}$ by a pair $\Delta = (X, \mathfrak{D})$. Suppose also that $\mathfrak{D} = \{t_1 \approx t'_1, \ldots, t_n \approx t'_n\}$ and $\underline{t} = t_1, \ldots, t_n$ and $\underline{t}' = t'_1, \ldots, t'_n$. Because $\mathbf{A}$ is s.i., the monolith $\mu(\mathbf{A})$ is not trivial. Let $\mathsf{a}_1, \mathsf{a}_2 \in \mu(\mathbf{A})$ be distinct members of the monolith. As elements $\delta(X)$ generate $\mathbf{A}$, there are two terms r_1 and r_2 in the variables from X such that $\delta(r_1) = \mathsf{a}_1$ and $\delta(r_2) = \mathsf{a}_2$. We define a $\mathbb{V}$ *-characteristic identity of algebra* $\mathbf{A}$ (and we omit the reference to $\mathbb{V}$ when no confusion can arise) by letting

$$\mathfrak{ch}_{\mathbb{V}}(\mathbf{A}, td) \overset{\text{def}}{\iff} td(\underline{t}, \underline{t}', r_1) \approx td(\underline{t}, \underline{t}', r_2), \tag{5.17}$$

and we will omit the reference to the TD term and the variety when no confusion can arise.

Later (cf. Theorem 5.6), we show that every characteristic identity is a splitting identity. Hence, it is unique up to $\mathbb{V}$-equipotence, so it does not depend on the selection of elements of the monolith or on terms representing them.

Let us note that Proposition 5.2 entails that if $\mathfrak{ch}_{\mathbb{V}}(\mathbf{A})$ is a characteristic identity of $\mathbf{A}$, then for any subvariety $\mathbb{V}' \subseteq \mathbb{V}$ containing $\mathbf{A}$, $\mathfrak{ch}_{\mathbb{V}}(\mathbf{A})$ is a characteristic identity of $\mathbf{A}$ relative to $\mathbb{V}'$; that is, $\mathfrak{ch}_{\mathbb{V}}(\mathbf{A}) = \mathfrak{ch}_{\mathbb{V}'}(\mathbf{A})$.

Example 5.17 Let us consider algebra Z_3 from Example 5.4. We have $X = \{x\}$, $\mathfrak{D} = \{\neg\neg x \approx \mathbf{1}\}$, and $\omega, \mathbf{1} \in \mu(\mathsf{Z}_3)$. Then, for the TD term $td_{\to} = (x \leftrightarrow y) \to z$, we have

$$\mathfrak{ch}_{\mathbb{HA}}(\mathsf{Z}_3, td_{\to}) = (\neg\neg x \leftrightarrow \mathbf{1}) \to x \approx (\neg\neg x \leftrightarrow \mathbf{1}) \to \mathbf{1}.$$

It is not hard to see that the above identity is equivalent to the following identity:

$$\neg\neg x \to x \approx \mathbf{1}.$$

For the TD term $td_{\wedge}$, we have

$$\mathfrak{ch}_{\mathbb{HA}}(\mathsf{Z}_3, td_{\wedge}) = (\neg\neg x \leftrightarrow \mathbf{1}) \wedge x \approx (\neg\neg x \leftrightarrow \mathbf{1}) \wedge \mathbf{1},$$

and the above identity is equivalent to the following identity:

$$x \approx \neg\neg x.$$

Proposition 5.23 *Suppose that* $\mathbb{V}$ *is a variety with a TD term td and* $\mathbf{A} \in \mathbb{V}$ *is an s.i. algebra finitely presented by a pair* $(X, \mathfrak{D})$ *and a valuation* δ. *Then*

$$\mathbf{A} \not\models \mathfrak{ch}_{\mathbb{V}}(\mathbf{A}, td).$$

Proof Suppose that $\mathfrak{D} = \{t_j \approx t'_j, j \in [1, n]\}$. Then, by the definition,

$$\mathfrak{ch}_{\mathbb{V}}(\mathbf{A}, td) = td(\underline{t}, \underline{t}', r_1) \approx td(\underline{t}, \underline{t}', r_2), \tag{5.18}$$

where $\underline{t} = t_1, \ldots, t_n$ and $\underline{t}' = t'_1, \ldots, t'_n$.

Let us confirm that δ is a refuting valuation. Indeed,

$$\delta(td(\underline{t}, \underline{t}', r_1)) = td(\delta(\underline{t}), \delta(\underline{t}'), \delta(r_1)).$$

By the definition of finitely presentable algebras, δ satisfies all the defining identities, that is;

$$\delta(\underline{t}) = \delta(\underline{t}').$$

Hence, from the above equality and by (5.16),

$$td(\delta(\underline{t}), \delta(\underline{t}'), \delta(r_1)) = td(\delta(\underline{t}), \delta(\underline{t}), \delta(r_1)) = \delta(r_1).$$

By the definition of a characteristic identity, $\delta(r_1) \neq \delta(r_2)$, and this observation completes the proof.

Let us observe that by Proposition 5.5, given a finitely presentable s.i. algebra, one can write down a characteristic identity in any given set of generators. This observation is a generalization of Theorem 4 from Jankov (1963b).

Example 5.18 If $\mathbf{A}$ is a finite s.i. Heyting algebra, one can take as a set of generators the set G of all $\vee$-irreducible elements that are distinct from $\mathbf{0}$: any element of $\mathbf{A}$ can be expressed as a disjunction of $\vee$-irreducible elements (and $\mathbf{0} = \neg\mathbf{1}$). If we define a characteristic formula of $\mathbf{A}$ in variables $\{x_{\mathsf{g}}, \mathsf{g} \in \mathsf{G}\}$ and defining valuation $\delta : x_{\mathsf{g}} \mapsto \mathsf{g}, \mathsf{g} \in \mathsf{G}$, we obtain the de Jongh formula of $\mathbf{A}$ (cf. de Jongh 1968; Bezhanishvili 2006). Similarly, for closure algebras we will obtain the Fine formula (cf. Fine 1974).

5.5.2.1 The Main Theorem About Characteristic Identities

The theorem and the corollaries of this section extend the results by Yankov (cf. Jankov 1963b, 1969 to varieties with a TD term.

Theorem 5.6 *Suppose that $\mathbb{V}$ is a variety with a TD term $td(x, y, z)$ and that $\mathbf{A}$ is an s.i. algebra finitely presented relative to $\mathbb{V}$ by a defining pair $(X, \mathfrak{D})$ and defining valuation δ. Then, for every $\mathbf{B} \in \mathbb{V}$,*

$$\mathbf{B} \not\models \mathfrak{ch}_{\mathbb{V}}(\mathbf{A}, td) \text{ if and only if } \mathbf{A} \in \mathbf{SHB}.$$

Proof Suppose that $X = \{x_i, i \in [1, m]\}$ and $\mathfrak{D} = \{t_j \approx t'_j, j \in [1, n]\}$. First, we prove that $\mathbf{B} \not\models \mathfrak{ch}_{\mathbb{V}}(\mathbf{A}, td)$ entails $\mathbf{A} \in \mathbf{SHB}$. To this end, we will construct a congruence $\theta \in \mathrm{Con}(\mathbf{B})$ and a homomorphism $\psi : \mathbf{A} \longrightarrow \mathbf{B}/\theta$, such that $\psi(\mathsf{a}) \neq \psi(\mathsf{b})$,

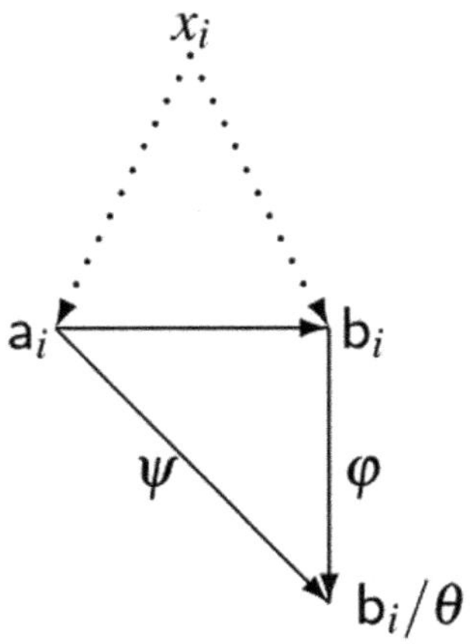

Fig. 5.5 Embedding

where $\mathsf{a}, \mathsf{b} \in \mu(\mathbf{A})$ are distinct elements from the monolith that are expressed via generators by terms r_1 and r_2. Thus, by Corollary 5.1, ψ will be an embedding.

Suppose that $\mathbf{B} \in \mathbb{V}$ and $\underline{b} \overset{\text{def}}{\Longleftrightarrow} \mathsf{b}_1, \ldots, \mathsf{b}_m$ are elements of $\mathbf{B}$ refuting $\mathfrak{ch}_{\mathbb{V}}(\mathbf{A}, td)$; that is,

$$td(\underline{t}(\underline{b}), \underline{t}'(\underline{b}), r_1(\underline{b})) \neq td(\underline{t}(\underline{b}), \underline{t}'(\underline{b}), r_2(\underline{b})). \tag{5.19}$$

Let $\underline{c} \overset{\text{def}}{\Longleftrightarrow} t_1(\underline{b}), \ldots, t_n(\underline{b})$ and $\underline{c}' \overset{\text{def}}{\Longleftrightarrow} t_1'(\underline{b}), \ldots, t_n'(\underline{b})$. Then, from (5.19), by virtue of (5.14), we can conclude that

$$r_1(\underline{b}) \not\equiv r_2(\underline{b}) \pmod{\theta(\underline{c}, \underline{c}')}. \tag{5.20}$$

Let $\varphi : \mathbf{B} \to \mathbf{B}/\theta(\underline{c}, \underline{c}')$ be a natural homomorphism of $\mathbf{B}$ onto quotient algebra $\mathbf{B}/\theta(\underline{c}, \underline{c}')$. Recall that congruence $\theta(\underline{c}, \underline{c}')$ is generated by pairs $(t_j(\underline{b}), t_j'(\underline{b}))$, $j \in [1, n]$ and hence, $t_j(\underline{b}) \equiv t_j'(\underline{b}) \pmod{\theta(\underline{c}, \underline{c}')}$ for every $j \in [1, n]$. Therefore, for every $j \in [1, \ldots, n]$,

$$\varphi(t_j(\underline{b})) = \varphi(t_j'(\underline{b}),$$

and because φ is a homomorphism,

$$t_j(\varphi(\underline{b})) = t_j'(\varphi(\underline{b})).$$

Let $\mathsf{b}_i' = \varphi(\mathsf{b}_i)$, $i = 1, \ldots, m$. Then, in $\mathbf{B}/\theta(\underline{c}, \underline{c}')$,

$$t_j(\underline{b}') = t_j'(\underline{b}') \text{ for all } j \in [1, n], \tag{5.21}$$

while from (5.20)

$$r_1(\underline{b}') \neq r_2(\underline{b}'). \tag{5.22}$$

By the definition of finitely presentable algebras, elements $\mathsf{a}_i = \delta(x_i), i \in [1, m]$ are the generators of algebra $\mathbf{A}$. Let us consider the mapping

$$\psi : \mathsf{a}_i \mapsto \mathsf{b}_i'; i \in [1, , m]. \tag{5.23}$$

By (5.21) and Proposition 5.3, there is a homomorphism $\psi : \mathbf{A} \longrightarrow \mathbf{B}/\theta(\underline{c}, \underline{c}')$ (cf. Fig. 5.5). Let us observe that from (5.22),

$$\psi(r_1(\underline{b}) \neq \psi(r_2(\underline{b})).$$

Recall that $r_1(\underline{b})$ and $r_1(\underline{b})$ are distinct elements from $\mu(\mathbf{A})$. Thus, we can apply Corollary 5.1 and conclude that ψ is an embedding.

Conversely, assume that $\mathbf{A} \in \mathbf{SHB}$. By Proposition 5.23, $\mathbf{A} \not\models_{\mathbb{V}} \mathfrak{ch}_{\mathbb{V}}(\mathbf{A}, td)$ and hence, $\mathbf{B} \not\models \mathfrak{ch}_{\mathbb{V}}(\mathbf{A}, td)$.

Example 5.19 As we saw in Example 5.17,

$$\mathfrak{ch}_{\mathbb{HA}}(\mathsf{Z}_3, td_{\rightarrow}) = \neg\neg x \rightarrow x \approx \mathbf{1},$$

and the defining valuation $\delta : x \mapsto \omega$ (see Fig. 5.2) refutes the characteristic identity.

5.5.2.2 Properties of Characteristic Identities

First, let us observe that Theorem 5.6 entails that a characteristic identity is a splitting identity.

Theorem 5.7 *Suppose that $\mathbb{V}$ is a variety with a TD term. Then, every characteristic identity is a splitting identity.*

Proof Indeed, suppose that an algebra $\mathbf{A} \in \mathbb{V}$ is finitely presentable relative to a variety $\mathbb{V}$ and $\mathbb{V}' \subseteq \mathbb{V}$ is a subvariety. Let $\mathbb{V}_1 = \mathbf{V}(\mathbf{A})$ and $\mathbb{V}_2$ be a variety defined relative to $\mathbb{V}$ by $\mathfrak{ch}_{\mathbb{V}}(\mathbf{A})$ Then, either $\mathbb{V}' \models \mathfrak{ch}_{\mathbb{V}}(\mathbf{A})$ and therefore, $\mathbb{V}' \subseteq \mathbb{V}_2$, or $\mathbb{V}' \not\models \mathfrak{ch}_{\mathbb{V}}(\mathbf{A})$ and hence, there is an algebra $\mathbf{B} \in \mathbb{V}'$ such that $\mathbf{B} \not\models \mathfrak{ch}_{\mathbb{V}}(\mathbf{A})$. In the latter case, by Theorem 5.6, $\mathbf{A} \in \mathbf{SHB}$; that is, $\mathbf{A} \in \mathbf{V}(\mathbf{B}) \subseteq \mathbb{V}'$. Thus, $\mathbb{V}_1 \subseteq \mathbb{V}'$.

Theorem 5.7 yields that characteristic identities possess all the properties of splitting identities. In particular, the following holds.

Let $\mathbb{V}$ be a variety with a TD term. By $\mathbb{V}_{fpsi}$ we denote a class of all s.i. algebras that are finitely presentable relative to $\mathbb{V}$. Then, any algebra from $\mathbb{V}_{fpsi}$ is a splitting algebra; that is, $\mathbb{V}_{fpsi} \subseteq \mathbb{V}_{spl}$. The converse does not need to be true (the reader can find examples in Kracht 1999, Theorem 7.5.16 or in Citkin 2012). But if $\mathbb{V}$ is generated by finite algebras, then $\mathbb{V}_{fpsi} = \mathbb{V}_{fsi} = \mathbb{V}_{spl}$.

Theorem 5.8 *Suppose that $\mathbb{V}$ is a variety with a TD term, $\mathbf{A}$ is an s.i. algebra that is finitely presentable relative to $\mathbb{V}$, and $\mathfrak{ch}_{\mathbb{V}}(\mathbf{A})$ is its characteristic identity. Then the following holds:*

(a) *for any identity* $\mathfrak{i}$,

$$\mathbf{A} \not\models \mathfrak{i} \text{ if and only if } \mathfrak{i} \vdash_{\mathbb{V}} \mathfrak{ch}_{\mathrm{v}}(\mathbf{A});$$

(b) *all characteristic identities of* $\mathbf{A}$ *are* $\mathbb{V}$*-equipotent;*

(c) *every characteristic identity is* $\wedge$*-prime relative to* $\mathbb{V}$*; moreover, if* $\mathbb{V}$ *is generated by finite members and it is of a finite type, then an identity is* $\wedge$*-prime if and only if it is* $\mathbb{V}$*-equipotent with a characteristic identity of some finite s.i. algebra from* $\mathbb{V}$.

Proof (a) If $\mathfrak{i}$ is an identity such that $\mathbf{A} \not\models \mathfrak{i}$, then $\mathfrak{i}$ defines a subvariety $\mathbf{V}(\mathfrak{i}) \subseteq \mathbb{V}$. Because $\mathfrak{ch}_{\mathrm{v}}(\mathbf{A})$ is a splitting identity and $\mathbf{A} \not\models \mathfrak{i}$, $\mathbf{V}(\mathfrak{i}) \subseteq \mathbf{V}(\mathfrak{ch}_{\mathrm{v}}(\mathbf{A}))$. Hence, $\mathfrak{i} \vdash_{\mathbb{V}} \mathfrak{ch}_{\mathrm{v}}(\mathbf{A})$.

The converse is trivial.

(b) follows immediately from (a) and Proposition 5.23 stating that all characteristic identities of $\mathbf{A}$ are refutable in $\mathbf{A}$.

(c) It was observed in McKenzie (1972) that all splitting identities are $\wedge$-prime.

If $\mathbb{V}$ is a variety of a finite similarity type with a TD term and generated by finite algebras, then all its splitting algebras are finite (cf. Blok and Pigozzi 1982) and hence, each splitting identity is equipotent with a characteristic formula.

Example 5.20 It is known that $\mathbb{HA}$ is generated by finite algebras. In addition, any identity $t \approx p$ is $\mathbb{HA}$-equivalent to identity $t \leftrightarrow p \approx \mathbf{1}$, and the latter identity is $\wedge$-prime relative to $\mathbb{HA}$ if and only if $t \leftrightarrow p$ is $\wedge$-prime as a formula in the sense of Jankov (1969). Hence, the class of $\wedge$-prime formulas relative to $\mathbb{HA}$ coincides with a class of formulas that are IPC-interderivable with the characteristic formulas of finite s.i. Heyting algebras, and this is one of the main results of Jankov (1969).

Example 5.21 As we saw in Example 5.19, $\neg\neg x \to x \approx \mathbf{1}$ is a characteristic identity of Heyting algebra Z_3. Hence, this identity defines in $\mathbb{HA}$ a splitting: for each subvariety $\mathbb{V} \subseteq \mathbb{HA}$, either $\mathbb{V} \models \neg\neg x \to x \approx \mathbf{1}$ (and thus, it is a variety of Boolean algebras or a trivial variety) or $\mathsf{Z}_3 \in \mathbb{V}$. Hence, for any classical tautology A that is refuted in Z_3, $\mathsf{IPC} + A$ defines the classical logic. This was stated without a proof in Jankov (1963a, Theorem 3(a)) and proved in Troelstra (1965, Theorem 5.3) and Jankov (1968b, Theorem 1). A similar statement for the extensions of **S4** is proven in Rautenberg (1980, Criterion 1).

Proposition 5.24 *Suppose that* $\mathbb{V}$ *is a variety with a TD term and* $\mathbf{A}$ *is an s.i. algebra finitely presented relative to* $\mathbb{V}$ *by a pair* $(X, \mathfrak{D})$ *and defining valuation* δ*. Let* $\mathfrak{ch}_{\mathrm{v}}(\mathbf{A})$ *be a characteristic identity of* $\mathbf{A}$*. Then, if* $\mathfrak{i}$ *is an identity refutable in* $\mathbf{A}$*, then there is a substitution* σ *such that* $\sigma(\mathfrak{i}) \vdash_{\mathbb{V}} \mathfrak{ch}_{\mathrm{v}}(\mathbf{A})$.

Proof Suppose that $\mathfrak{i}(y_1, \ldots, y_k)$ is an identity refuted in $\mathbf{A}$. Because $\mathbf{A} \not\models \mathfrak{i}$, there are elements c_i, $i \in [1, k]$, such that $\mathfrak{i}(\underline{\mathsf{c}})$ does not hold. Then elements $\mathsf{a}_j = \delta(x_j)$, $j \in [1, n]$, where δ is a defining valuation, are the generators of algebra $\mathbf{A}$. Hence, we can express every element c_i via generators; that is, $\mathsf{c}_i = t_i(\underline{\mathsf{a}})$, $i \in [1, k]$ for some terms t_i. Let us consider the substitution

$$\sigma : y_i \mapsto t_i(x_1, \ldots, x_n), i \in [1, t].$$

It is clear that

$$\sigma(\mathfrak{i}(y_1, \ldots, y_k)) = \mathfrak{i}(\sigma(y_1), \ldots, \sigma(y_k)) = \mathfrak{i}(t_1(\underline{x}), \ldots, t_k(\underline{x})),$$

and therefore, $\sigma(\mathfrak{i})(\mathsf{a}_1, \ldots, \mathsf{a}_n)$ does not hold.

Next, we want to demonstrate that given an algebra $\mathbf{B} \in \mathbb{V}$ and a valuation $\nu : x_i \mapsto \mathsf{b}_i, \in \mathbf{B}, i \in [1, n]$ such that $\mathfrak{ch}_{\mathbb{V}}(\mathbf{A})(\mathsf{b}_1, \ldots, \mathsf{b}_n)$ does not hold, identity $\sigma(\mathfrak{i})(\mathsf{b}_1, \ldots, \mathsf{b}_n)$ does not hold too.

Indeed, if $\mathfrak{ch}_{\mathbb{V}}(\mathbf{A})(\mathsf{b}_1, \ldots, \mathsf{b}_n)$ does not hold, one can use a reasoning similar to the proof of Theorem 5.6 and obtain the embedding $\psi : \mathbf{A} \longrightarrow \mathbf{B}/\theta$ such that $\psi(\mathsf{a}_i) = \varphi(\mathsf{b}_i) = \mathsf{b}/\theta, i \in [1, n]$ (cf. Fig. 5.5). Hence, $\sigma(\mathfrak{i})(\varphi(\mathsf{b}_1), \ldots, \varphi(\mathsf{b}_n))$ does not hold, because $\sigma(\mathfrak{i})(\mathsf{a}_1, \ldots, \mathsf{a}_n)$ does not hold. Thus, $\sigma(\mathfrak{i})(\mathsf{b}_1, \ldots, \mathsf{b}_n)$ cannot hold, because φ is a homomorphism.

Because in the varieties with a TD term all s.i. finitely presentable algebras are splitting algebras, the following holds (cf. Jankov 1969, Theorem About Ordering).

Theorem 5.9 *Suppose that $\mathbb{V}$ is a variety with a TD term and $\mathbf{A}, \mathbf{B} \in \mathbb{V}_{fpsi}$. Then the following assertions are equivalent:*

(a) $\mathbf{A} \preceq \mathbf{B}$;
(b) $\mathbf{B} \not\models \mathfrak{ch}_{\mathbb{V}}(\mathbf{A})$;
(c) every formula refutable in $\mathbf{A}$ *is refutable in* $\mathbf{B}$;
(d) $\mathbf{A} \in \mathsf{SHB}$.

Theorem 5.9 is an immediate consequence of Theorem 5.8.

The following is the main property of antichains that is employed in most applications of characteristic identities or formulas.

Theorem 5.10 *Suppose that $\mathbb{V}$ is a variety with a TD term and $\mathbb{S} \subseteq \mathbb{V}_{fpsi}$ is an antichain. Then, the set of identities $\mathfrak{s}(\mathbb{S})$ is $\mathbb{V}$-independent.*

Proof Let $\mathbf{A} \in \mathbb{S}$ and let $\mathbf{V}(\mathbf{A})$ be a variety generated by $\mathbf{A}$. By Proposition 5.23, $\mathbf{A} \not\models \mathfrak{ch}_{\mathbb{V}}(\mathbf{A})$ and hence, $\mathbf{V}(\mathbf{A}) \not\models \mathfrak{ch}_{\mathbb{V}}(\mathbf{A})$. On the other hand, $\mathbb{C}$ is an antichain and hence, every algebra $\mathbf{A}' \in \mathbb{C}$ distinct from $\mathbf{A}$ is incomparable with $\mathbf{A}$. Therefore, $\mathbf{A} \models \mathfrak{ch}_{\mathbb{V}}(\mathbf{A}')$.

The following corollary is an immediate consequence of the above theorem and Corollary 5.3.

Corollary 5.10 *If a variety $\mathbb{V}$ with a TD term contains an infinite antichain of s.i. algebras that are finitely presentable relative to $\mathbb{V}$, then $\mathbb{V}$ contains at least continuum many subvarieties.*

5.5.3 Independent Bases in Subvarieties Generated by Finite Algebras

It follows from Blok and Pigozzi (1982, 1994) that in varieties of a finite type with a TD term generated by finite algebras, every finite s.i. algebra is a splitting algebra. Thus, using the argument employed in the proof of Theorem 5.5, we can prove the following theorem.

Theorem 5.11 *Suppose that* $\mathbb{V}$ *is a variety of a finite type with a TD term generated by finite algebras. If all subvarieties of* $\mathbb{V}$ *are generated by finite algebras, then all subvarieties of* $\mathbb{V}$ *have reduced* $\mathbb{V}$*-bases consisting of characteristic identities.*

Let us observe that if $\mathbb{V}$ is a variety of a finite type with a TD term generated by finite algebras and $\mathbb{V}' \subseteq \mathbb{V}$ is a subvariety generated by finite algebras, then each s.i. algebra that is finitely presentable relative to $\mathbb{V}'$ is a finite s.i. algebra and, thus, finitely presentable relative to $\mathbb{V}$. Hence, if a subvariety $\mathbb{V}^*$ of $\mathbb{V}'$ has a $\mathbb{V}'$-base consisting of $\mathbb{V}'$-characteristic formulas, then $\mathbb{V}^*$ has a $\mathbb{V}'$-base consisting of $\mathbb{V}$-characteristic formulas. Therefore, if $\mathbb{V}'$ has a $\mathbb{V}$-base consisting of $\mathbb{V}$-characteristic formulas, then $\mathbb{V}^*$ has a $\mathbb{V}$-base consisting of $\mathbb{V}$-characteristic formulas. Thus, the following holds.

Corollary 5.11 *Suppose that* $\mathbb{V}$ *is a variety of a finite type with a TD term generated by finite algebras. If* $\mathbb{V}' \subseteq \mathbb{V}$ *is a subvariety such that* $\mathbb{V}'$ *and all its subvarieties are generated by finite algebras, then every subvariety of* $\mathbb{V}'$ *has a reduced* $\mathbb{V}'$*-base consisting of* $\mathbb{V}$*-characteristic identities.*

Example 5.22 (cf. Skvortsov 1999; Tomaszewski 2002) It is well known that any finite Heyting algebra $\mathbf{A}$ belongs to a finite slice; that is, for some n, $\mathbf{A} \in \mathbb{V}_n$ where $\mathbb{V}_n$ is defined by a characteristic identity of the $(n+2)$-element chain algebra. It is also well known that varieties from $\mathbb{V}_n$ are locally finite. Therefore, all the varieties generated by a finite Heyting algebra admit optimal bases constructed of $\mathbb{HA}$-characteristic identities. To construct such a base, one can take a characteristic identity the $\mathfrak{ch}_{\mathbb{HA}}(\mathbf{C}_{n+2})$ of $n+2$-element chain algebra and the characteristic identities of all finite s.i. algebras from $\mathbb{V}_n \setminus \mathbf{V}(\mathbf{A})$. The local finiteness of $\mathbb{V}_n$ guarantees that there are only finitely many non-isomorphic s.i. algebras in $\mathbb{V}_n \setminus \mathbf{V}(\mathbf{A})$. When we construct a characteristic identity, we use the smallest (by cardinality) generating set of the algebra.

5.5.4 A Note on Iterated Splitting

Iterated splitting can be defined as follows (cf. Wolter 1993, Sect. 2.4 or Kracht 1999, Chap. 7). Suppose that $\mathbb{V}_2^{(0)}$ is a variety and $(\mathbb{V}_1^{(1)}, \mathbb{V}_2^{(1)})$ is a splitting pair. $\mathbb{V}_2^{(1)}$ is a subvariety of $\mathbb{V}_2^{(0)}$ defined relative to $\mathbb{V}_2^{(0)}$ by a $\mathbb{V}_2^{(0)}$-splitting identity. Next, we consider a splitting pair $(\mathbb{V}_1^{(2)}, \mathbb{V}_2^{(2)})$ of variety $\mathbb{V}_2^{(1)}$. Observe that $\mathbb{V}_2^{(2)}$ can be defined relative to $\mathbb{V}_2^{(0)}$ by $\mathbb{V}_2^{(1)}$-splitting and $\mathbb{V}_2^{(1)}$-splitting identities. We can continue this process: if we have a splitting pair $(\mathbb{V}_1^{(i)}, \mathbb{V}_2^{(i)})$ of the variety $\mathbb{V}_2^{(i-1)}$, we consider a splitting pair $(\mathbb{V}_1^{(i+1)}, \mathbb{V}_2^{(i+1)})$ of the variety $\mathbb{V}_2^{(i)}$. In such a way we obtain a descending chain $\mathbb{V}_2^{(0)} \supseteq \mathbb{V}_2^1 \supseteq \mathbb{V}_2^2 \supseteq \cdots$. We say that variety $\mathbb{V}' = \bigcap_{i\geq 0} \mathbb{V}_2^{(i)}$ is defined by *iterated splitting*; it is clear that $\mathbb{V}'$ has a $\mathbb{V}_2^{(0)}$-base consisting of $\mathbb{V}_2^{(i)}$-splitting identities $(i \geq 0)$.

As we know, not all subvarieties of a given variety $\mathbb{V}$ can be defined by a splitting or characteristic identities. In varieties generated by finite algebras, a characteristic identity defines a subvariety that is the largest subvariety from a class having the same finite algebras. Suppose that $\mathbb{V}$ is a variety of a finite type generated by finite algebras and $\mathbb{V}' \subset \mathbb{V}^* \subset \mathbb{V}$, where $\mathbb{V}^*$ is a subvariety of $\mathbb{V}$ defined by some characteristic identities and $\mathbb{V}'$ is a subvariety of $\mathbb{V}^*$ generated by all finite algebras from $\mathbb{V}^*$. We know that $\mathbb{V}'$ cannot be defined by $\mathbb{V}$-characteristic identities, but on the other hand, $\mathbb{V}^*$ is not generated by finite algebras and can contain some $\mathbb{V}^*$-finitely presentable s.i. algebras that are not $\mathbb{V}$-finitely presentable. In other words, there may be some $\mathbb{V}^*$-characteristic identities that we can use to define $\mathbb{V}'$ relative to $\mathbb{V}^*$. In Citkin (2018) it was proven that (in algebraic terms) there are continuum many varieties of Heyting algebras that can be defined by iterated splittings, while they have no $\mathbb{HA}$-bases consisting of $\mathbb{HA}$-characteristic identities. On the other hand, there are varieties of Heyting algebras that cannot be defined by iterated splitting.

5.6 Final Remarks

Soon after the characteristic formulas were introduced, it became clear that they give us a very convenient and powerful tool for studying intermediate logics, or varieties of Heyting algebras. So, it was natural to generalize this notion.

5.6.1 From Characteristic Identities to Characteristic Rules

The ability to construct a characteristic identity rests on the properties of TD term. If we move from varieties (that is, form logics understood as closed sets of formulas) to quasivarieties (that is, to logics understood as single conclusion consequence relations), the situation becomes simpler, because there is no need to use a TD term. Indeed, if $\mathbb{Q}$ is a quasivariety and $\mathbf{A} \in \mathbb{Q}$ is an algebra finitely presented relative to $\mathbb{Q}$ by a pair $(X, \mathfrak{D})$ and defining valuation δ, one can take a *characteristic* quasi-identity

$$\mathsf{q}_{\mathbb{Q}}(\mathbf{A}) = \mathfrak{D} \Rightarrow x_1 \approx x_2,$$

where $\delta(x_1), \delta(x_2)$ are two distinct elements from the monolith $\mu(\mathbf{A})$. It is not hard to see that for any algebra $\mathbf{B} \in \mathbb{Q}$,

$$\mathbf{B} \not\models \mathsf{q}_{\mathbb{Q}}(\mathbf{A}) \text{ if and only if } \mathbf{B} \in \mathbf{SA}.$$

Logical counterparts of characteristic quasi-identities—the quasi-characteristic rules—were introduced for intermediate logics in Citkin (1977). Using the technique developed by Yankov, in Citkin (1977) it was proven that there are continuum many quasivarieties the equational closure of which is $\mathbb{HA}$. Later, in Rybakov (1997a, b), the notion of the quasi-characteristic rule was extended to modal logics. The algebraic properties of corresponding quasi-identities are studied in Budkin and Gorbunov (1975). The reader can find more information in Gorbunov (1998).

It is clear that every quasi-identity that is characteristic relative to a quasivariety $\mathbb{Q}$ defines a splitting in the lattice $\mathcal{L}(\mathbb{Q})$ of all subquasivarieties of $\mathbb{Q}$. If $\mathbb{Q}$ is a locally finite quasivariety of a finite type, all splitting algebras from $\mathbb{Q}$ are precisely finite s.i. relative to the $\mathbb{Q}$ algebras. Denote by $\mathbb{Q}_{spl}$ the class of all splitting algebras of $\mathbb{Q}$, and define on $\mathbb{Q}_{spl}$ a partial order

$$\mathbf{A} \preceq_q \mathbf{B} \text{ if and only if } \mathbf{B} \in \mathbf{SPA}.$$

Using the same arguments that were used in the proof of Theorem 5.4, one can prove the following theorem.

Theorem 5.12 *Let $\mathbb{Q}$ be a locally finite quasivariety of finite type and $\mathcal{L}(\mathbb{Q})$ be a lattice of subquasivarieties of $\mathbb{V}$. Then, the following are equivalent:*

(a) every subquasivariety of $\mathbb{Q}$ is finitely based relative to $\mathbb{Q}$;
(b) $\mathcal{L}(\mathbb{Q})$ is countable;
(c) $\mathbb{Q}_{spl}$ has no infinite antichains;
(d) $\mathcal{L}(\mathbb{Q})$ enjoys the descending chain condition.

5.6.2 From Characteristic Quasi-identities to Characteristic Implications

Another way to generalize the notion of a characteristic formula was suggested in Wroński (1974). More recently, the same approach was used for Johansson's algebras in Odintsov (2005, Sect. 7) and Odintsov (2006, Sect. 6.3): instead of a formula, one can use a consequence relation defined by an algebra. More precisely, each algebra $\mathbf{A}$ defines a consequence relation in the following way: a formula A (an identity $\mathfrak{i}$) is a *consequence of a set of formulas (identities)* Γ if every valuation that refutes A (respectively $\mathfrak{i}$) refutes at least one formula (identity) from Γ, in symbols $\Gamma \models_{\mathbf{A}} A$ (or $\Gamma \models_{\mathbf{A}} \mathfrak{i}$). If we take any countable s.i. algebra $\mathbf{A}$ and take its diagram

$$\delta^+(\mathbf{A}) = \{f(x_{\mathsf{a}_1}, \ldots, x_{\mathsf{a}_n}) \approx x_{f(\mathsf{a}_1,\ldots,\mathsf{a}_n)} \mid \mathsf{a}_1, \ldots, \mathsf{a}_n \in \mathsf{A} \text{ and } f \in \Sigma\}$$

one can prove the following Proposition.

Proposition 5.25 *(cf. Wroński 1974, Lemma 3) Let* **A** *be a countable s.i. algebra and* **B** *be an algebra from a variety* $\mathbb{V}$. *Then the following conditions are equivalent:*

(a) **A** *is embedded in* **B**;
(b) $\delta^+(\mathbf{A}) \not\models_{\mathbf{B}} x_{\mathsf{b}_1} \approx x_{\mathsf{b}_2}$, *where* $\mathsf{b}_1, \mathsf{b}_2$ *are any two distinct elements from the monolith* $\mu(\mathbf{A})$.

5.6.3 From Algebras to Complete Algebras

Let us note that a transition from a formula to a consequence relation in logical terms means a transition from formulas to (structural) rules (with, perhaps, countably many premises). In algebraic terms, it signifies a transition from the varieties to implicative classes (cf. Budkin and Gorbunov 1973). Note that if we restrict Wroński's definition to finite algebras, we will not arrive at Yankov's formulas. Instead, we will obtain quasi-characteristic rules.

In Tanaka (2007, Definition 5.1) the notion of a Yankov formula is extended to complete Heyting algebras, that is, the Heyting algebras admitting infinite joins and meets. If **A** is a complete Heyting algebra, then a subset $\mathsf{C} \in 2^{\mathbf{A}}$ is called the basis of **A** if

1. for any $\mathsf{A}' \subseteq \mathbf{A}$ there exists a $\mathsf{C}' \in \mathsf{C}$ such that $\vee\mathsf{A}' = \vee\mathsf{C}'$ and for any $\mathsf{c} \in \mathsf{C}'$ there exists an $\mathsf{a} \in \mathsf{A}'$ such that $\mathsf{a} \geq \mathsf{c}$;
2. for any $\mathsf{A}' \subset \mathsf{A}$ there exists a $\mathsf{C} \in \mathbb{C}$ such that $\wedge\mathsf{A}' = \wedge\mathsf{C}$ and for any $\mathsf{c} \in \mathsf{C}$ there exists an $\mathsf{a} \in \mathsf{A}'$ such that $\mathsf{a} \leq \mathsf{c}$.

A characteristic formula for a complete s.i. Heyting algebra **A** and its basis $\mathbb{C}$ is defined as

$$\chi(\mathbf{A}, \mathbb{C}) \overset{\text{def}}{\Longleftrightarrow} \bigwedge_{\mathsf{C}\in\mathbb{C}} (p_{\vee\mathsf{C}} \leftrightarrow \bigvee_{\mathsf{c}\in\mathsf{C}} p_{\mathsf{c}}) \wedge \bigwedge_{\mathsf{C}\in\mathbb{C}} (p_{\wedge\mathsf{C}} \leftrightarrow \bigwedge_{\mathsf{c}\in\mathsf{C}} p_{\mathsf{c}})$$
$$\wedge \bigwedge_{\mathsf{a},\mathsf{b}\in\mathsf{A}} (p_{\mathsf{a}\to\mathsf{b}} \leftrightarrow (p_{\mathsf{a}} \to p_{\mathsf{b}})) \wedge \bigwedge_{\mathsf{a}\in\mathsf{A}} (p_{\neg\mathsf{a}} \to \neg p_{\mathsf{a}}) \to p_{\omega}.$$

Using the argument similar to the one that was used in the proof of Yankov's Theorem, one can prove (cf. Tanaka 2007, Proposition 5.1) that an s.i. complete Heyting algebra **A** is embedded into a homomorphic image of a complete Heyting algebra **B** if and only if $\mathbf{B} \not\models \chi(\mathbf{A}, \mathbb{C})$ and the homomorphism and the embedding are continuous.

5.6.4 From Finite Algebras to Infinite Algebras

As we mentioned before, all splitting algebras in the varieties generated by their finite members are finite. One of the ways to extend the notion of a characteristic identity to infinite s.i. algebras is to use the finite partial subalgebras (cf. Tomaszewski 2003; Citkin 2013).

References

Agliano, P. (1998). Ternary deduction terms in residuated structures. *Acta Scientiarum Mathematicarum (Szeged), 64*(3–4), 397–429.
Agliano, P. (2019). Varieties of BL-algebras III: Splitting algebras. *Studia Logica.*
Bellissima, F. (1988). Finite and finitely separable intermediate propositional logics. *Journal of Symbolic Logic, 53*(2), 403–420. https://doi.org/10.2307/2274513.
Berman, J., & Blok, W. J. (2004). Free łukasiewicz and hoop residuation algebras. *Studia Logica, 77*(2), 153–180. https://doi.org/10.1023/B:STUD.0000037125.49866.50.
Bezhanishvili, N. (2004). De Jongh's characterization of intuitionistic propositional calculus. *Liber Amicorum Dick de Jongh* (pp. 1–10). ILLC University of Amsterdam.
Bezhanishvili, N. (2006). Lattices of intermediate and cylindric modal logics. Ph.D. thesis, Institute for Logic, Language and Computation University of Amsterdam.
Blok, W. (1976). Varieties of interior algebras. Ph.D. thesis, University of Amsterdam.
Blok, W. J. (1977). $2^{\aleph_0}$ varieties of Heyting algebras not generated by their finite members. *Algebra Universalis, 7*(1), 115–117.
Blok, W. J., & Pigozzi, D. (1982). On the structure of varieties with equationally definable principal congruences. I. *Algebra Universalis, 15*(2), 195–227. https://doi.org/10.1007/BF02483723.
Blok, W. J., & Pigozzi, D. (1994). On the structure of varieties with equationally definable principal congruences. III. *Algebra Universalis, 32*(4), 545–608. https://doi.org/10.1007/BF01195727.
Blok, W. J., & Raftery, J. G. (1997). Varieties of commutative residuated integral pomonoids and their residuation subreducts. *Journal of Algebra, 190*(2), 280–328. https://doi.org/10.1006/jabr.1996.6834.
Budkin, A. I., & Gorbunov, V. A. (1973). Implicative classes of algebras. *Algebra i Logika, 12,* 249–268, 363.
Budkin, A. I., & Gorbunov, V. A. (1975). On the theory of quasivarieties of algebraic systems. *Algebra i Logika, 14*(2), 123–142, 240.
Burris, S., & Sankappanavar, H. P. (1981). *A course in universal algebra.* Graduate Texts in Mathematics (Vol. 78). New York: Springer.
Chagrov, A., & Zakharyaschev, M. (1995). On the independent axiomatizability of modal and intermediate logics. *Journal of Logic and Computation, 5*(3), 287–302. https://doi.org/10.1093/logcom/5.3.287.
Citkin, A. (1977). On admissible rules of intuitionistic propositional logic. *Mathematics of the USSR-Sbornik, 31,* 279–288 (A. Tsitkin).
Citkin, A. (2012). Not every splitting Heyting or interior algebra is finitely presentable. *Studia Logica, 100*(1–2), 115–135. https://doi.org/10.1007/s11225-012-9391-1.
Citkin, A. (2013). Characteristic formulas of partial Heyting algebras. *Logica Universalis, 7*(2), 167–193. https://doi.org/10.1007/s11787-012-0048-7.
Citkin, A. (2018). Characteristic formulas over intermediate logics. *Larisa Maksimova on implication, interpolation, and definability.* Outstanding contributions to logic (Vol. 15, pp. 71–98). Cham: Springer.

Citkin, A. I. (1986). Finite axiomatizability of locally tabular superintuitionistic logics. *Matematicheskie Zametki, 40*(3), 407–413, 430.

Day, A. (1973). Splitting algebras and a weak notion of projectivity. In *Proceedings of the University of Houston Lattice Theory Conference (Houston, Tex., 1973)* (pp. 466–485), Department of Mathematics, University of Houston, Houston, Tex.

Fine, K. (1974). An ascending chain of S4 logics. *Theoria, 40*(2), 110–116.

Galatos, N., Jipsen, P., Kowalski, T., & Ono, H. (2007). *Residuated lattices: An algebraic glimpse at substructural logics*. Studies in logic and the foundations of mathematics (Vol. 151). Amsterdam: Elsevier B. V.

Gorbunov, V. A. (1998). Algebraic theory of quasivarieties. Siberian School of Algebra and Logic, Consultants Bureau, New York, translated from the Russian.

Grätzer, G. (2008). *Universal algebra* (2nd ed.). New York: Springer. https://doi.org/10.1007/978-0-387-77487-9, with appendices by Grätzer, Bjarni Jónsson, Walter Taylor, Robert W. Quackenbush, Günter H. Wenzel, and Grätzer and W. A. Lampe.

Grätzer, G., & Quackenbush, R. W. (2010). Positive universal classes in locally finite varieties. *Algebra Universalis, 64*(1–2), 1–13. https://doi.org/10.1007/s00012-010-0089-9010-0089-9.

Hosoi, T. (1967). On intermediate logics. I. *Journal of the Faculty of Science, University of Tokyo, Section I, 14*, 293–312.

Idziak, K., & Idziak, P. (1988). Decidability problem for finite Heyting algebras. *Journal of Symbolic Logic, 53*(3), 729–735. https://doi.org/10.2307/2274568.

Jankov, V. A. (1963a). On certain superconstructive propositional calculi. *Doklady Akademii Nauk SSSR, 151*, 796–798, English translation in *Soviet Mathematics - Doklady, 4*, 1103–1105.

Jankov, V. A. (1963b). On the relation between deducibility in intuitionistic propositional calculus and finite implicative structures. *Doklady Akademii Nauk SSSR, 151*, 1293–1294, English translation in *Soviet Mathematics - Doklady, 4*, 1203–1204.

Jankov, V. A. (1968a). The construction of a sequence of strongly independent superintuitionistic propositional calculi. *Doklady Akademii Nauk SSSR, 181*, 33–34, English translation in *Soviet Mathematics - Doklady, 9*, 806–807.

Jankov, V. A. (1968b). On an extension of the intuitionistic propositional calculus to the classical one and of the minimal one to the intuitionistic one. *Izvestiya Akademii Nauk SSSR Series Mathematica, 32*, 208–211, English translation in *Mathematics of the USSR-Izvestiya, 2*(1), 205–208.

Jankov, V. A. (1969). Conjunctively irresolvable formulae in propositional calculi. *Izvestiya Akademii Nauk SSSR Series Mathematica, 33*, 18–38, English translation in *Mathematics of the USSR-Izvestiya, 3*(1), 17–35.

de Jongh, D. (1968). Investigations on intuitionistic propositional calculus. Ph.D. thesis, University of Wisconsin.

Jongh, D. D., & Yang, F. (2010). *Jankov's theorems for intermediate logics in the settings of universal models*. ILLC publication series (Vol. PP-2010-07). Amsterdam: Institute for Logic, Language and Computation, University of Amsterdam.

Jónsson, B. (1967). Algebras whose congruence lattices are distributive. *Mathematica Scandinavica, 21*, 110–121.

Kracht, M. (1999). *Tools and techniques in modal logic*. Studies in logic and the foundations of mathematics (Vol. 142). Amsterdam: North-Holland Publishing Co.

Kuznetsov, A. V. (1973). On finitely generated pseudo-Boolean algebras and finitely approximable varieties. In *Proceedings of the 12th USSR Algebraic Colloquium, Sverdlovsk* (p. 281) (in Russian).

Maksimova, L. L., Skvorcov, D. P., & Šehtman, V. B. (1979). Impossibility of finite axiomatization of Medvedev's logic of finite problems. *Doklady Akademii Nauk SSSR, 245*(5), 1051–1054.

Mal'cev, A. (1973). *Algebraic systems*. Die Grundlehren der mathematischen Wissenschaften (Vol. 192). Berlin: Springer, Akademie-Verlag. XII, 317 p.

Mardaev, S. I. (1987). Embedding of implicative lattices and superintuitionistic logics. *Algebra i Logika, 26*(3), 318–357, 399.

McKenzie, R. (1972). Equational bases and nonmodular lattice varieties. *Transactions of the American Mathematical Society, 174*, 1–43.
Odintsov, S. P. (2005). On the structure of paraconsistent extensions of Johansson's logic. *Journal of Applied Logic, 3*(1), 43–65. https://doi.org/10.1016/j.jal.2004.07.011.
Odintsov, S. P. (2006). The lattice of extensions of minimal logic. *Matematicheskije Trudy, 9*(2), 60–108.
Rautenberg, W. (1979). *Klassische und nichtklassische Aussagenlogik*. Logik und Grundlagen der Mathematik [Logic and foundations of mathematics] (Vol. 22). Braunschweig: Friedr. Vieweg & Sohn.
Rautenberg, W. (1980). Splitting lattices of logics. *Archiv für Mathematische Logik und Grundlagenforschung, 20*(3–4), 155–159. https://doi.org/10.1007/BF02021134.
Rybakov, V. V. (1997a). *Admissibility of logical inference rules*. Studies in logic and the foundations of mathematics (Vol. 136). Amsterdam: North-Holland Publishing Co.
Rybakov, V. V. (1997b). Quasi-characteristic inference rules for modal logics. In *Logical foundations of computer science (Yaroslavl, 1997)*. Lecture notes in computer science (Vol. 1234, pp. 333–341). Berlin: Springer.
Skvortsov, D. (1999). Remark on a finite axiomatization of finite intermediate propositional logics. *Journal of Applied Non-classical Logics, 9*(2–3), 381–386. https://doi.org/10.1080/11663081.1999.10510973, issue in memory of George Gargov.
Spinks, M., & Veroff, R. (2007). Characterisations of Nelson algebras. *Revista de la Unión Matemática Argentina, 48*(1), 27–39.
Tanaka, Y. (2007). An infinitary extension of Jankov's theorem. *Studia Logica, 86*(1), 111–131. https://doi.org/10.1007/s11225-007-9048-7.
Tomaszewski, E. (2002). An algorithm for finding finite axiomatizations of finite intermediate logics by means of Jankov formulas. *Bulletin of the Section of Logic University of Łódź, 31*(1), 1–6.
Tomaszewski, E. (2003). On sufficiently rich sets of formulas. Ph.D. thesis, Institute of Philosophy, Jagiellonian University, Krakov.
Troelstra, A. S. (1965). On intermediate propositional logics. *Nederlandse Akademie van Wetenschappen Proceedings Series A* 68=*Indagationes Mathematicae, 27*, 141–152.
Umezawa, T. (1959). On intermediate many-valued logics. *Journal of the Mathematical Society of Japan, 11*, 116–128.
Wolter, F. (1993). Lattices of modal logics. Ph.D. thesis, Freien Universität Berlin.
Wolter, F. (1996). Tense logic without tense operators. *Mathematical Logic Quarterly, 42*(2), 145–171.
Wroński, A. (1974). On cardinality of matrices strongly adequate for the intuitionistic propositional logic. *Polish Academy of Sciences, Institute of Philosophy and Sociology, Bulletin of the Section of Logic, 3*(1), 34–40.

Chapter 6
The Invariance Modality

Silvio Ghilardi

Abstract In Gerla (1987), G. Gerla introduced the so-called transformational semantics for predicate modal logic and considered in particular semantic frameworks given by a classical model endowed with a group of automorphisms, where a boxed formula is true iff it holds invariantly (i.e. it remains true whenever an automorphism is applied to the individuals it is talking about). With this interpretation, *de dicto* modalities collapse, but *de re* modalities remain quite informative. We handle the axiomatization problem of such modal structures, by employing classic model-theoretic tools (iterated ultrapowers and double chains).

Keywords Quantified modal logic · Presheaf semantics · Invariance modality · Ultrapowers

6.1 Introduction

The distinction between *de re* and *de dicto* modalities is a classical topic in the philosophical investigations concerning modal predicate logic. We recall that *de dicto* modalities are represented by formulae of the kind $\Box\phi$ where ϕ is a *sentence*; in case ϕ has free variables, the modality $\Box\phi$ is called a *de re* modality. Whereas *de dicto* modalities can be interpreted without referring to some kind of "essential" properties of individuals, the same does not apply to *de re* modalities: that's why *de re* modalities appear to be compromised with some sort of essentialist metaphysics.

However, looking at mathematical contexts, as pointed out in Gerla (1987), *de re* modalities assume a rather natural interpretation: $\Box\phi(x)$ means that x enjoys the property ϕ in a way that is *invariant* with respect to transformations that might be applied to x. In particular, when transformations are applied inside a given domain, only *de re* modalities survive and *de dicto* modalities collapse (i.e. it happens exactly the contrary of what anti-essentialist philosophy would consider to be desirable).

S. Ghilardi (✉)
Dipartimento di Matematica, Università degli Studi di Milano, Milano, Italy
e-mail: silvio.ghilardi@unimi.it

A. Citkin and I. M. Vandoulakis (eds.), *V. A. Yankov on Non-Classical Logics, History and Philosophy of Mathematics*, Outstanding Contributions to Logic 24,
https://doi.org/10.1007/978-3-031-06843-0_6

To this aim, in Gerla (1987) a suitable *invariance logic* is introduced and shown to be recursively axiomatizable (via a reduction to two-sorted predicate logic). In this paper, we show that the axiomatization implicitly suggested in Gerla (1987) is complete: we consider both the case where transformations are assumed to be functions and the case where transformations are assumed to be bijections (we use the name of 'transformation semantics' for the former case and of 'invariance semantics' for the latter).[1]

The main results of this paper were already contained in Ghilardi (1990); they were listed in Ghilardi and Meloni (1991) (together with other results in modal predicate logic), but never published. The proofs presented here are a deep revisitation and clarification of the original proofs in Ghilardi (1990). The completeness results in Sect. 6.4 can be obtained also via the saturated/special models technique of Ghilardi (1992); however the technique presented here does not depend on cardinal arithmetics and has the merit of reinterpreting classical model theoretic methods in the specific context of modal logic.

6.2 Preliminaries

We consider a first order modal language $\mathcal{L}$ with identity: terms are restricted to be only variables, whereas formulae are built up from atoms using the connectives $\neg, \bot, \vee, \Box$ and the quantifier $\exists$ (further connectives $\wedge, \rightarrow, \Diamond$ and the quantifier $\forall$ are defined in the standard way).

We shall use languages expanded with constants (called *parameters*) to introduce our semantic notions; parameters are taken from a set specified in the context. If we are given a function μ acting on the set of parameters, we use the notation $\phi[\mu]$ to mean the formula obtained from ϕ by replacing every parameter a occurring in it by $\mu(a)$. When we speak of $\mathcal{L}$-sentences, $\mathcal{L}$-formulae, etc. we mean sentences, formulae, etc. *without parameters*; on the contrary, when we speak of sentences, formulae, etc. parameters are allowed to occur.

We fix for the whole paper an $\mathcal{L} - theory\ T$, i.e. a set of $\mathcal{L}$-sentences. Our calculus has modus ponens (from ϕ and $\phi \rightarrow \psi$ infer ψ) and necessitation (from ϕ infer $\Box\phi$) as rules; as axioms we take an axiomatic base for first order classical logic with identity, the $\mathcal{L}$-sentences from T, the $S4$-axiom schemata

$$\Box(\phi \rightarrow \psi) \rightarrow (\Box\phi \rightarrow \Box\psi), \qquad \Box\phi \rightarrow \phi, \qquad \Box\phi \rightarrow \Box\Box\phi \tag{6.1}$$

and the further schema

$$\phi \rightarrow \Box\phi \tag{dDC}$$

[1] Thus our use of the name of 'transformation semantics' does not fully agree with Gerla (1987), whereas our 'invariance semantics' is the same as the semantics for 'invariance logic' of Gerla (1987). In this paper, for simplicity, we completely leave apart the question of the interpretation of constants and function symbols (if they are not rigid, they need to be handled via the approach of Ghilardi and Meloni 1988, see also Braüner and Ghilardi 2007).

restricted to $\mathcal{L}$-sentences. Such a schema is called the *de dicto collapse schema.* We write $\vdash_T \phi$ or $T \vdash \phi$ to mean that ϕ has a derivation in this calculus. We assume that our T is *consistent*, i.e. that $T \nvdash \bot$.

Notice that our calculus does not give rise to a logic, according to the standard definition of a modal predicate logic Gabbay et al. (2009), Braüner and Ghilardi (2007), because *it is not closed under uniform substitution*: in fact, the schema (dDC) is not assumed to hold in case ϕ contains free variables. In addition, notice also that when we expand the language with parameters, the schema (dDC) *does not apply to sentences containing parameters*, but only to sentences in the original language $\mathcal{L}$.[2]

An immediate consequence of the de dicto collapse schema is that the standard form of the deduction theorem holds:

Proposition 6.1 *Let ψ be an $\mathcal{L}$-sentence and ϕ be an arbitrary formula; we have that $T \cup \{\psi\} \vdash \phi$ holds iff $T \vdash \psi \to \phi$ holds.*

As a corollary, we can prove a Lindenbaum Lemma (T is said to be *maximal* iff either $T \vdash \psi$ or $T \vdash \neg\psi$ holds for every $\mathcal{L}$-sentence ψ):

Lemma 6.1 *Our consistent $\mathcal{L}$-theory T can be extended to a maximal consistent $\mathcal{L}$-theory in the same language $\mathcal{L}$.*

6.2.1 *Transformational and Invariance Models*

A *modal transformational model* (or just a *transformational model*) for $\mathcal{L}$ is a triple $\mathfrak{M} = (M, \mathcal{I}, E)$, where $(M, \mathcal{I})$ is a Tarski structure for the first-order language $\mathcal{L}$ and E is a set of functions from M into M closed under composition and containing the identity function; a *modal invariance model* (or just an *invariance model*) is a modal transformation model $\mathfrak{M} = (M, \mathcal{I}, E)$ such that all functions in E are bijective.

Sentences in the expanded language $\mathcal{L} \cup M$ (containing a parameter name for each element of M) are evaluated inductively in a modal transformation model $\mathfrak{M} = (M, \mathcal{I}, E)$ as follows:

(i) $\mathfrak{M} \models P(a_1, \dots, a_n)$ iff $(a_1, \dots, a_n) \in \mathcal{I}(P)$ (for every n-ary predicate symbol P from $\mathcal{L}$);
(ii) $\mathfrak{M} \models a_1 = a_2$ iff a_1 is equal to a_2;
(iii) $\mathfrak{M} \not\models \bot$;
(iv) $\mathfrak{M} \models \neg\phi$ iff $\mathfrak{M} \not\models \phi$;
(v) $\mathfrak{M} \models \phi_1 \vee \phi_2$ iff ($\mathfrak{M} \models \phi_1$ or $\mathfrak{M} \models \phi_2$);
(vi) $\mathfrak{M} \models \exists x\phi$ iff there is some $a \in M$ s.t. $\mathfrak{M} \models \phi(a/x)$.
(vii) $\mathfrak{M} \models \Box\phi$ iff for all $\mu \in E$, $\mathfrak{M} \models \phi[\mu]$.

---

[2] Obviously, the addition of parameters is conservative: if an $\mathcal{L}$-formula has a proof in a language with parameters, it also has a proof in $\mathcal{L}$ (just replace parameters with free variables in such a proof).

$\mathfrak{M}$ is said to be a modal transformation model *of the $\mathcal{L}$-theory T* iff we have

(o) $\mathfrak{M} \models \phi$ for every $\mathcal{L}$-sentence $\phi \in T$.

The following soundness property is easily established by induction on the proof witnessing $\vdash_T \phi$:

Proposition 6.2 *Take an $\mathcal{L}$-formula $\phi(x_1, \ldots, x_n)$ whose free variables are among $x_1, \ldots, x_n$ and suppose that $\vdash_T \phi$; then for every modal transformational model $\mathfrak{M} = (M, I, E)$ of T, for every $a_1, \ldots, a_n \in M$ we have that $\mathfrak{M} \models \phi(a_1 \ldots, a_n)$.*

Notice that we use the notation $\phi(x_1, \ldots, x_n)$ to express that ϕ contains free variables among $x_1, \ldots, x_n$ and the notation $\phi(a_1, \ldots, a_n)$ for the formula obtained from ϕ by replacing x_i by a_i (for $i = 1, \ldots, n$). When we use such notation, the $x_1, \ldots, x_n$ are assumed to be distinct, whereas the $a_1, \ldots, a_n$ may not be distinct. To have a more compact notation, we may use underlined letters for tuples of unspecified length: then $\phi(\underline{a})$ means the sentence obtained from $\phi(\underline{x})$ by replacing componentwise the tuple of variables $\underline{x}$ by the tuple of parameters $\underline{a}$ (the latter is supposed to be of equal length as $\underline{x}$).

In the definition of a modal transformation model, we took E to be just a set of endofunctions. One may wonder whether we can ask more for them: the answer is 'yes' in case we suitably enrich out theory T. In fact, if we take as further axioms for T the universal closure of the $\mathcal{L}$-formulae

$$A \to \Box A \tag{6.2}$$

where A is atomic, then because of (o) and (vii), all functions in E are forced to be *homomorphisms*; to force them to be *embeddings* (in the standard model-theoretic sense Chang and Keisler 1990), it is sufficient to put in T the universal closure of all the $\mathcal{L}$-formulae of the kind

$$L \to \Box L \tag{6.3}$$

where L is a literal (i.e. an atom or the negation of an atom).

We did not consider function symbols in our language $\mathcal{L}$ for simplicity; however, an n-ary function symbol h can be represented via an $n + 1$-ary predicate symbol P_h via existence and uniqueness axioms; assuming (6.2) for atoms rooted at P_h, restricts all $f \in E$ to be *h-homomorphisms* in the algebraic sense.

In general, a modal collapse axiom

$$\theta \to \Box\theta \tag{6.4}$$

expresses the fact that θ *is an invariant* with respect to the class of endomorphisms considered in E. More information on connections between the collapse of modalities and model-theoretic notions is available in Gerla and Vaccaro (1984).

6.3 Classical Models and Ultrapowers

A *classical model* is a pair $\mathcal{M} = (M, \models_{\mathcal{M}})$, where M is a set and $\models_{\mathcal{M}}$ is a set of sentences in the expanded language $\mathcal{L} \cup M$ satisfying the condition (o) and the conditions (ii)-(vi) above (we usually write $\models_{\mathcal{M}} \phi$ as $\mathcal{M} \models \phi$).

Thus a classical model is in fact nothing but a model (in the classical sense) for our calculus rewritten in (non modal) first-order logic as follows. We let $\mathcal{L}_{ext}$ be the first order language obtained by expanding $\mathcal{L}$ by an extra predicate symbol $P_{\Box\phi}$ for every boxed $\mathcal{L}$-formula $\Box\phi$ (the arity of $P_{\Box\phi}$ is the number of free variables occurring in ϕ); similarly, we let T_{ext} be the first-order theory having as axioms all the $\mathcal{L}_{ext}$-formulae θ such that, replacing in θ the subformulae of the kind $P_{\Box\phi}(\underline{t})$ by $\Box\phi(\underline{t})$, one obtains an $\mathcal{L}$-formula provable in T (here $\underline{t}$ is a tuple of variables matching the length of the tuple of free variables of ϕ). Then a classical model in the above sense is just a first-order $\mathcal{L}_{ext}$-structure which is a model of T_{ext} according to Tarski semantics. Once we view classical models in this sense, *it is clear that we can apply to them standard model-theoretic constructions* (we shall be interested in particular into ultrapowers and chain limits).

Given two classical models $\mathcal{M} = (M, \models_{\mathcal{M}})$ and $\mathcal{N} = (N, \models_{\mathcal{N}})$, an *elementary morphism* among them is a map $\mu : M \longrightarrow N$ among the support sets satisfying the condition

$$\mathcal{M} \models \phi \quad \Rightarrow \quad \mathcal{N} \models \phi[\mu] \tag{6.5}$$

for all $\mathcal{L} \cup M$-sentences ϕ.

A *modal morphism* among $\mathcal{M} = (M, \models_{\mathcal{M}})$ and $\mathcal{N} = (M, \models_{\mathcal{M}})$ is a map $\mu : M \longrightarrow N$ among the support sets satisfying the condition

$$\mathcal{M} \models \Box\phi \quad \Rightarrow \quad \mathcal{N} \models \phi[\mu] \tag{6.6}$$

for all $\mathcal{L} \cup M$-sentences ϕ. Notice that modal morphisms need not be injective, unlike elementary morphisms.[3]

It is useful to extend the notion of a modal morphisms to subsets of classical models. Suppose that we are given two classical models $\mathcal{M} = (M, \models_{\mathcal{M}})$, $\mathcal{N} = (N, \models_{\mathcal{N}})$ and a subset $A \subseteq M$; a *partial modal morphism* of domain A is a map $\mu : A \longrightarrow N$ satisfying (6.6) for all $\mathcal{L} \cup A$-sentences ϕ (a partial modal morphism is usually indicated as $\mu : A \longrightarrow \mathcal{N}$, leaving $\mathcal{M}$ as understood). Modal (partial) morphisms are coloured in *red* in the diagrams below in the digital version of the book.

Given an ultrafilter D (on any set of indices) and a classical model $\mathcal{M}$, we can form the *ultrapower* $\Pi_D \mathcal{M}$ of $\mathcal{M}$ (as an $\mathcal{L}_{ext}$-structure) in the standard way Chang and Keisler (1990); we recall also that we have a canonical elementary morphism $\iota_D : \mathcal{M} \longrightarrow \Pi_D \mathcal{M}$.

We shall make extensive use of $\Box$-ultrafilters, which are defined as follows. Take a classical model $\mathcal{M}$ and consider the set $I_{\mathcal{M}}$ formed by the $\mathcal{L} \cup M$-sentences ϕ

---

[3] However, a modal morphism must be an h-homomorphism for every operation h which is definable in T via a predicate P_h, in case (6.2) is assumed for all atoms rooted at P_h, as above explained.

such that $\mathcal{M} \models \Box\phi$. A $\Box$-*ultrafilter* over $\mathcal{M}$ is any ultrafilter extending the finite intersections closed family given by the subsets of $I_{\mathcal{M}}$ of the kind $\downarrow \phi = \{\psi \in I_{\mathcal{M}} \mid T \vdash \psi \to \phi\}$, varying ϕ in $I_{\mathcal{M}}$.

The following Lemma explains a typical use of $\Box$-ultrafilters:

Lemma 6.2 *Let $\mathcal{M} = (M, \models_{\mathcal{M}})$, $\mathcal{N} = (N, \models_{\mathcal{N}})$ be classical models and let $A \subseteq M$ be a subset of M. Suppose we are given a partial modal morphism $\nu : A \longrightarrow \mathcal{N}$ and a $\Box$-ultrafilter D over $\mathcal{M}$. Then ν can be extended to a full modal morphism into $\Pi_D\mathcal{N}$, in the sense that there exists a modal morphism $\bar{\nu} : \mathcal{M} \longrightarrow \Pi_D\mathcal{N}$ such that $\bar{\nu}(a) = \iota_D(\nu(a))$ holds for all $a \in A$.*

$$\begin{array}{ccc} A & \xrightarrow{\ \nu\ } & \mathcal{N} \\ \downarrow & & \downarrow{\scriptstyle \iota_D} \\ \mathcal{M} & \xrightarrow{\ \bar{\nu}\ } & \Pi_D\mathcal{N} \end{array}$$

Proof For every $\phi \in I_{\mathcal{M}}$, we define a map $h_\phi : M \longrightarrow N$ in the following way. Let $\underline{b}$ be the parameters from M occurring in ϕ: suppose that we have $\underline{b} = \underline{a}, \underline{b}'$, where the $\underline{b}'$ are the (distinct) elements from $\underline{b}$ not belonging to A. We have $\mathcal{M} \models \Box\phi(\underline{a}, \underline{b}')$ by definition of $I_{\mathcal{M}}$; it follows that $\mathcal{M} \models \exists \underline{y} \Box\phi(\underline{a}, \underline{y})$. Since $T \vdash \exists \underline{y}\Box\phi(\underline{a}, \underline{y}) \to \Box\exists\underline{y}\phi(\underline{a}, \underline{y})$,[4] we get that $\mathcal{M} \models \Box\exists\underline{y}\phi(\underline{a}, \underline{y})$ and also that there exist $\underline{c}$ in N such that $\mathcal{N} \models \phi(\nu(\underline{a}), \underline{c})$, because ν is modal. We take $h_\phi : M \longrightarrow N$ to be any extension of the partial map sending d to $\nu(d)$ (for all $d \in A$) and the $\underline{b}'$ to the $\underline{c}$. As a consequence of this definition we have

$$\mathcal{N} \models \phi[h_\phi], \qquad \text{for all } \phi \in I_{\mathcal{M}}\,. \tag{6.7}$$

Let us now define $\bar{\nu} : \mathcal{M} \longrightarrow \Pi_D\mathcal{N}$ as the map sending $d \in M$ to the equivalence class of the $I_{\mathcal{M}}$-indexed tuple $\langle h_\phi(d)\rangle_\phi$. We need to prove that for every ψ with parameters in $\mathcal{M}$, we have

$$\mathcal{M} \models \Box\psi \quad \Rightarrow \quad \Pi_D\mathcal{N} \models \psi[\bar{\nu}]\ .$$

If $\mathcal{M} \models \Box\psi$, then $\psi \in I_{\mathcal{M}}$ and so, since $\downarrow \psi \in D$, it is sufficient to check that for every $\phi \in \downarrow \psi$ we have $\mathcal{N} \models \psi[h_\phi]$. Now $\phi \in \downarrow \psi$ means that we have $\mathcal{M} \models \Box\phi$ (i.e. $\phi \in I_{\mathcal{M}}$) and $T \vdash \phi \to \psi$. Since $\mathcal{N}$ is a classical model of T and (6.7) holds, we get $\mathcal{N} \models \psi[h_\phi]$, as wanted. $\dashv$

Corollary 6.1 *Let $\mathcal{M}_0, \mathcal{M}, \mathcal{N}$ be classical models; suppose we are given an elementary morphism $\mu : \mathcal{M}_0 \longrightarrow \mathcal{M}$, a modal morphism $\nu : \mathcal{M}_0 \longrightarrow \mathcal{N}$ and a $\Box$-ultrafilter D over $\mathcal{M}$. Then there exists a modal morphism $\bar{\nu} : \mathcal{M} \longrightarrow \Pi_D\mathcal{N}$ such that $\bar{\nu} \circ \mu = \iota_D \circ \nu$.*

---

[4] This is provable in all quantified normal modal systems (Hughes and Cresswell 1968).

$$\begin{array}{ccc} \mathcal{M}_0 & \xrightarrow{\nu} & \mathcal{N} \\ \big\downarrow{\scriptstyle\mu} & & \big\downarrow{\scriptstyle\iota_D} \\ \mathcal{M} & \xrightarrow{\bar{\nu}} & \Pi_D \mathcal{N} \end{array}$$

Proof This is easily reduced to the previous lemma, considering that, up to an isomorphism, μ is an inclusion (because it is injective). $\dashv$

Recall the symmetry axiom schemata

$$\Diamond\Box\phi \rightarrow \phi \tag{6.8}$$

which axiomatizes the modal logic $S5$, once added to the $S4$ axiom schemata (6.1). If our T is an extension of $S5$ (i.e. if it contains all the examples of the above schema (6.8)), a variant of Lemma 6.2 holds:

Lemma 6.3 *Suppose that T is an extension of $S5$. Let $\mathcal{M}$, $\mathcal{N}$ be classical models and $A \subseteq N$ be a subset of N. Suppose we are given a partial modal morphism $\mu : A \longrightarrow \mathcal{M}$ and a $\Box$-ultrafilter D over $\mathcal{M}$. Then there exists a modal morphism $\theta : \mathcal{M} \longrightarrow \Pi_D \mathcal{N}$ such that $\theta(\mu(a)) = \iota_D(a)$ for all $a \in A$.*

$$\begin{array}{ccc} A & \hookrightarrow & \mathcal{N} \\ \big\downarrow{\scriptstyle\mu} & & \big\downarrow{\scriptstyle\iota_D} \\ \mathcal{M} & \xrightarrow{\theta} & \Pi_D \mathcal{N} \end{array}$$

Proof Let us preliminarily check that the implication

$$\mathcal{M} \models \exists \underline{y} \Box\phi(\mu(\underline{a}), \underline{y}) \quad \Rightarrow \quad \mathcal{N} \models \exists \underline{y} \phi(\underline{a}, \underline{y}) \tag{6.9}$$

holds for all $\phi(\underline{a}, \underline{y})$ with parameters $\underline{a}$ from A. To show this, assume that $\mathcal{M} \models \exists \underline{y} \Box\phi(\mu(\underline{a}), \underline{y})$; then (since μ is modal—by contraposition of the partial modal morphism definition) we have $\mathcal{N} \models \Diamond\exists \underline{y} \Box\phi(\underline{a}, \underline{y})$ and also $\mathcal{N} \models \Diamond\Box\exists \underline{y} \phi(\underline{a}, \underline{y})$. By the symmetry axiom (6.8), $\mathcal{N} \models \exists \underline{y} \phi(\underline{a}, \underline{y})$ follows.

As a second observation, we notice that μ *is injective*: this is because T is an extension of $S5$ and the necessity of the difference is a theorem in quantified $S5$ (see Hughes and Cresswell 1968).

Now we can proceed similarly as in the proof of Lemma 6.2. For every $\phi \in I_{\mathcal{M}}$, we define a map $h_\phi : M \longrightarrow N$ in the following way. Let $\underline{b}$ be the parameters from M occurring in ϕ; we can decompose $\underline{b}$ as $\underline{b} = \mu(\underline{a}), \underline{b}'$, where the $\underline{b}'$ are the (distinct) elements from $\underline{b}$ not belonging to $\mu(A)$. We have $\mathcal{M} \models \Box\phi(\mu(\underline{a}), \underline{b}')$ by definition of $I_{\mathcal{M}}$; it follows that $\mathcal{M} \models \exists \underline{y} \Box\phi(\mu(\underline{a}), \underline{y})$ and consequently $\mathcal{N} \models \exists \underline{y} \phi(\underline{a}, \underline{y})$ by (6.9), so there are $\underline{c} \in N$ such that $\mathcal{N} \models \phi(\underline{a}, \underline{c})$. We take $h_\phi : M \longrightarrow N$ to be any extension of the partial map sending $\mu(d)$ to d (for all $d \in A$) and the $\underline{b}'$ to the $\underline{c}$: this is possible because, as noticed above, μ is injective. As a consequence of this definition we have

$$\mathcal{N} \models \phi[h_\phi], \qquad \text{for all } \phi \in I_{\mathcal{M}}. \tag{6.10}$$

We finally define $\bar{\nu} : \mathcal{M} \longrightarrow \Pi_D \mathcal{N}$ as the map sending $d \in M$ to the equivalence class of the $I_{\mathcal{M}}$-indexed tuple $\langle h_\phi(d)\rangle_\phi$: the proof now continues as in Lemma 6.2. ⊣

Corollary 6.2 *Suppose that T is an extension of S5. Let $\mathcal{M}_0$, $\mathcal{M}$, $\mathcal{N}$ be classical models. Suppose we are given a modal morphism $\mu : \mathcal{M}_0 \longrightarrow \mathcal{M}$, an elementary morphism $\nu : \mathcal{M}_0 \longrightarrow \mathcal{N}$ and a $\Box$-ultrafilter D over $\mathcal{M}$. Then there exists a modal morphism $\theta : \mathcal{M} \longrightarrow \Pi_D \mathcal{N}$ such that $\theta \circ \mu = \iota_D \circ \nu$.*

$$\begin{array}{ccc} \mathcal{M}_0 & \overset{\nu}{\hookrightarrow} & \mathcal{N} \\ \downarrow \mu & & \downarrow \iota_D \\ \mathcal{M} & \overset{\theta}{\longrightarrow} & \Pi_D \mathcal{N} \end{array}$$

6.4 Strong Completeness Theorems

In this section, we prove our main results, namely that our transformational semantics is axiomatized by the de dicto collapse schema (together with S4 axiom schemata (6.1)) and that the addition of the symmetry axiom schema (6.8) axiomatizes modal invariance models.

Theorem 6.1 *For a given sentence ϕ, we have that if $T \nvdash \phi$ then there is a transformational model $\mathfrak{M}$ of T such that $\mathfrak{M} \not\models \phi$.*

Proof We prove the theorem in the following equivalent form (the equivalence is guaranteed by Proposition 6.1 and by the Lindenbaum Lemma 6.1): if T is maximal consistent, then there is a transformational model for T.

Notice that to show the claim it is sufficient to produce a classical model $\mathcal{M} = (M, \models_{\mathcal{M}})$ for T satisfying the following additional condition for every $\mathcal{L} \cup M$-sentence ϕ:

(*) if $\mathcal{M} \not\models \Box\phi$, then there exists a modal endomorphism $\nu : \mathcal{M} \longrightarrow \mathcal{M}$ such that $\mathcal{M} \not\models \phi[\nu]$.

In fact, once such a classical model is found, we can turn it into the transformational model $\mathfrak{M} = (M, \mathcal{I}, E)$, where E is the set of modal endomorphisms of $\mathcal{M}$ and $\mathcal{I}$ is the interpretation function mapping an n-ary predicate symbol P into the set of n-tuples $(a_1, \ldots, a_n) \in M^n$ such that $\mathcal{M} \models P(a_1, \ldots, a_n)$: an easy induction then proves a standard 'truth lemma', namely that $\mathcal{M} \models \phi$ holds iff $\mathfrak{M} \models \phi$ holds for all $\mathcal{L} \cup M$-sentences ϕ.

Thus we are left to the task of finding a classical model satisfying $(*)$ above for our maximal consistent T. We shall build $\mathcal{M}$ as a chain limit of ultrapowers.

We start with a classical model $\mathcal{M}_0$ having some saturation properties. In fact, we need a weaker variant of ω-saturation, which we are going to explain. An *n -ary type* for T is a set of formulae $\tau(\underline{x})$ having at most the $\underline{x}$ as free variables (here $\underline{x} = x_1, \ldots, x_n$) such that for every finite subset $\tau_0 \subseteq \tau$, we have that $T \cup \{\exists \underline{x} \bigwedge \tau_0(\underline{x})\}$ is consistent. An n-ary type $\tau(\underline{x})$ is *realized* in a classical model $\mathcal{M} = (M, \models_{\mathcal{M}})$ iff there is a tuple $\underline{a} \in M^n$ such that we have $\mathcal{M} \models \theta(\underline{a})$ for every $\theta(\underline{x}) \in \tau(\underline{x})$. Since T is maximal consistent, by a simple compactness argument, it is possible to show that there is a classical model $\mathcal{M}_0$ for T realizing all n-ary types for T (for all n). This $\mathcal{M}_0$ is the starting model of our chain.

Having already defined the classical model $\mathcal{M}_i$, we let $\mathcal{M}_{i+1}$ be $\Pi_{D_i}\mathcal{M}_i$, where D_i is a $\Box$-ultrafilter of $\mathcal{M}_i$. Now let us take the limit $\mathcal{M}$ of the chain given by the $\mathcal{M}_i$ and the elementary embeddings ι_{D_i}

$$\mathcal{M}_0 \xrightarrow{\iota_{D_0}} \cdots \xrightarrow{\iota_{D_{i-1}}} \mathcal{M}_i \xrightarrow{\iota_{D_i}} \cdots \tag{6.11}$$

We prove that $\mathcal{M}$ satisfies condition $(*)$. Let ϕ be a sentence with parameters from $\mathcal{M}$ such that $\mathcal{M} \not\models \Box\phi$. Let the parameters occurring in ϕ be $\underline{a}$ and let all of them be from a certain $\mathcal{M}_i$. We claim that the set of formulae

$$\{\neg\phi(\underline{x})\} \cup \{\theta(\underline{x}) \mid \mathcal{M} \models \Box\theta(\underline{a})\} \tag{6.12}$$

is a type. Otherwise there are formulae $\theta_1(\underline{x}), \ldots, \theta_m(\underline{x})$ such that we have both $\mathcal{M} \models \bigwedge_{k=1}^{m} \Box\theta_k(\underline{a})$ and $T \cup \{\exists \underline{x}(\neg\phi(\underline{x}) \wedge \bigwedge_{k=1}^{m} \theta_k(\underline{x})\} \vdash \bot$. By the deduction theorem (Proposition 6.1), classical validities, necessitation rule, the converse of the Barcan formula (available in quantified normal systems Hughes and Cresswell 1968) and the distribution axiom (6.1), we get $T \vdash \forall \underline{x}(\bigwedge_{k=1}^{m} \Box\theta_k(\underline{x}) \to \Box\phi(\underline{x}))$, contradicting $\mathcal{M} \models \bigwedge_{k=1}^{m} \Box\theta_k(\underline{a})$ and $\mathcal{M} \not\models \Box\phi(\underline{a})$.

Let the type (6.12) be realized by some tuple $\underline{b}$ from $\mathcal{M}_i$ (actually, there is such a tuple already in $\mathcal{M}_0$ by the above weak saturation property of $\mathcal{M}_0$). We let $A = \{\underline{a}\}$ and ν be the partial modal morphism $\nu : A \longrightarrow \mathcal{M}_i$ mapping the $\underline{a}$ to the $\underline{b}$. By Lemma 6.2, there is a modal morphism $\nu_i : \mathcal{M}_i \longrightarrow \mathcal{M}_{i+1}$ such that $\nu_i(\underline{a}) = \iota_{D_i}(\nu(\underline{a})) = \iota_{D_i}(\underline{b})$.

$$\begin{array}{ccc} A & \xrightarrow{\nu} & \mathcal{M}_i \\ \big\downarrow & & \big\downarrow{\scriptstyle \iota_{D_i}} \\ \mathcal{M}_i & \xrightarrow{\nu_i} & \mathcal{M}_{i+1} \end{array}$$

If we now apply repeatedly Corollary 6.1, for all $j \geq i$, we can find modal morphisms ν_{j+1} such that $\nu_{j+1} \circ \iota_{D_j} = \iota_{D_{j+1}} \circ \nu_j$.

$$\begin{array}{ccc} \mathcal{M}_j & \xrightarrow{\nu_j} & \mathcal{M}_{j+1} \\ \big\downarrow{\scriptstyle \iota_{D_j}} & & \big\downarrow{\scriptstyle \iota_{D_{j+1}}} \\ \mathcal{M}_{j+1} & \xrightarrow{\nu_{j+1}} & \mathcal{M}_{j+2} \end{array}$$

Putting all these ν_j together in the chain limit, we get a modal morphism $\nu : \mathcal{M} \longrightarrow \mathcal{M}$ which maps (the colimit equivalence class of) $\underline{a}$ into (the colimit equivalence class of) $\underline{b}$, so that we have $\mathcal{M} \not\models \phi[\nu]$, as required. ⊣

6.4.1 Invariance Models

We now consider strong completeness for invariance models:

Theorem 6.2 *Suppose that T is an extension of $S5$. For a given sentence ϕ, we have that if $T \nvdash \phi$ then there is an invariance model $\mathfrak{M}$ of T such that $\mathfrak{M} \not\models \phi$.*

Proof Again, we can freely suppose that T is maximal consistent and the theorem is proved if we find a classical model $\mathcal{M} = (M, \models_{\mathcal{M}})$ for T satisfying the following condition for every $\mathcal{L} \cup M$-sentence ϕ:

(**) if $\mathcal{M} \not\models \Box\phi$, then there exists a *bijective* modal endomorphism $\nu : \mathcal{M} \longrightarrow \mathcal{M}$ such that $\mathcal{M} \not\models \phi[\nu]$.

Notice in fact that, if T is an extension of $S5$, the inverse of a bijective modal morphism is also a modal morphism. Thus, if (**) holds, then we can turn the classical model $\mathcal{M}$ into the invariance model $\mathfrak{M} = (M, \mathcal{I}, E)$ by taking as E the set of bijective modal endomorphisms of $\mathcal{M}$ and by defining $\mathcal{I}$ as in the proof of Theorem 6.1.

To find a classical model satisfying (**), we proceed as in the proof of Theorem 6.1: we first build the sufficiently saturated model $\mathcal{M}_0$, the chain of models (6.11) and its chain colimit $\mathcal{M}$. Also, given ϕ with parameters in $\mathcal{M}_i$ such that $\mathcal{M}_i \not\models \Box\phi$, we build a modal morphism $\nu_i : \mathcal{M}_i \longrightarrow \mathcal{M}_{i+1}$ such that $\mathcal{M}_{i+1} \not\models \phi[\nu_i]$. The question is now how to extend this modal morphism to a bijective modal morphism $\mathcal{M} \longrightarrow \mathcal{M}$. To this aim we shall use Corollary 6.2 and a double chain argument. Because of Corollary 6.2 we can in fact inductively define for every $j \geq i$ a modal morphism $\nu_{j+1} : \mathcal{M}_{j+1} \to \mathcal{M}_{j+2}$ so that we have $\nu_{j+1} \circ \nu_j = \iota_{D_{j+1}} \circ \iota_{D_j}$.

$$\begin{array}{ccc} \mathcal{M}_j & \overset{\iota_{D_j}}{\hookrightarrow} & \mathcal{M}_{j+1} \\ \downarrow{\scriptstyle \nu_j} & & \downarrow{\scriptstyle \iota_{D_{j+1}}} \\ \mathcal{M}_{j+1} & \overset{\nu_{j+1}}{\longrightarrow} & \mathcal{M}_{j+2} \end{array}$$

This equality holds for all $j \geq i$; thus, applying it to k and $k+1$, we get (for every $k \geq i$)

$$\iota_{D_{k+2}} \circ \iota_{D_{k+1}} \circ \nu_k = \nu_{k+2} \circ \nu_{k+1} \circ \nu_k = \nu_{k+2} \circ \iota_{D_{k+1}} \circ \iota_{D_k}$$

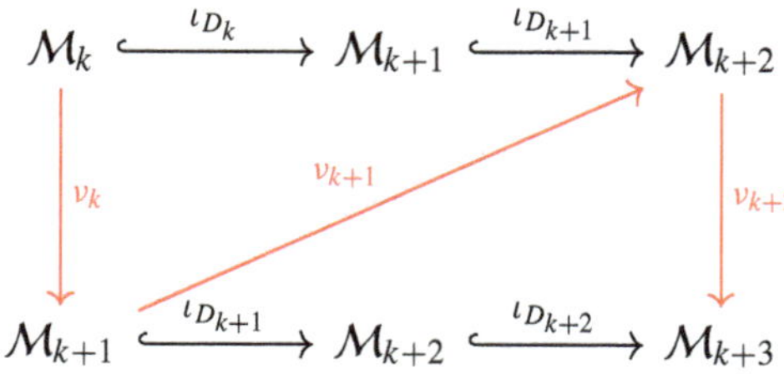

This means that the family of modal morphisms $\{\nu_{i+2s}\}_{s\geq 0}$ determines the required modal morphism $\mathcal{M} \longrightarrow \mathcal{M}$ extending ν_i: this morphism is bijective because its inverse is the modal morphism determined by the family of modal morphisms $\{\nu_{i+2s+1}\}_{s\geq 0}$. ⊣

6.5 Conclusions

We proved strong completeness theorems for transformational and invariance models. Such models are special cases of presheaf models, where the domain category of presheaves is a monoid (resp. a group). Presheaf models have been shown to be quite effective in proving the weakness of Kripke semantics in quantified modal logic (Ghilardi 1989, 1991), however a systematic semantic investigation on them (covering e.g. crucial topics like correspondence theory) still waits for substantial development.

Another potentially interesting (although difficult) research direction would be that of identifying classes of monoids and groups whose associated transformational and invariance models are sensible to a transparent modal axiomatization. Invariance theory is at the heart of mathematics in various areas and it would be nice if modal logic could contribute to it in some respect.

References

Braüner, T., Ghilardi, S. (2007). First-order modal logic. In *Handbook of modal logic*, volume 3 of Stud. Log. Pract. Reason. (pp. 549–620). Amsterdam: Elsevier B. V.

Chang, C.-C., & Keisler, J. H. (1990). *Model theory* (3rd ed.). Amsterdam: North-Holland.

Gabbay, D. M., Shehtman, V. B., & Skvortsov D. P. (2009). *Quantification in nonclassical logic* (Vol. 1), volume 153 of Studies in logic and the foundations of mathematics. Amsterdam: Elsevier B. V.

Gerla, G. (1987). Transformational semantics for first order logic. *Logique et Analyse (N.S.)*, *30*(117-118), 69–79.

Gerla, G., & Vaccaro, V. (1984). Modal logic and model theory. *Studia Logica, 43*(3), 203–216.

Ghilardi, S. (1989). Presheaf semantics and independence results for some non-classical first-order logics. *Archive for Mathematical Logic, 29*, 125–136.

Ghilardi, S. (1990). *Modalità e categorie*. Ph.D thesis, Università degli Studi di Milano.

Ghilardi, S. (1991). Incompleteness results in Kripke semantics. *Journal of Symbolic Logic, 56*(2), 517–538.

Ghilardi, S. (1992). Quantified extensions of canonical propositional intermediate logics. *Studia Logica, 51*(2), 195–214.

Ghilardi, S., & Meloni, G. (1988). Modal and tense predicate logic: Models in presheaves and categorical conceptualization. In *Categorical algebra and its applications*, volume 1348 of Lecture notes in mathematics (pp. 130–142). Springer.

Ghilardi, S., & Meloni, G. (1991). Relational and topological semantics for modal and temporal first order predicative logic. In *Nuovi problemi della logica e della filosofia della scienza* (pp. 59–77). CLUED - Bologna.

Hughes, G. E., & Cresswell, M. J. (1968). *An introduction to modal logic*. London: Methuen.

Chapter 7
The Lattice NExtS41 as Composed of Replicas of NExtInt, and Beyond

Alexei Muravitsky

Dedicated to Vadim A. Yankov (Jankov)

Abstract We study the lattice of all normal consistent extensions of the modal logic **S4**, NExt**S4**, from a structural point of view. We show that a pattern isomorphic to the lattice of intermediate logics is present as a sublattice in NExt**S4** in many, in fact, in infinitely many places, and this pattern itself is isomorphic to a quotient lattice of NExt**S4**. We also designate three "dark spots" of NExt**S4**, three sublattices of it, where, although we can characterize the logics belonging to each of these sublattices, their structural picture is invisible at all.

Keywords Intuitionistic propositional calculus · Modal propositional calculus S4 · The lattice of all normal extensions of a calculus · Distributive lattice · Heyting algebra

2020 Mathematics Subject Classification Primary 03B45 · Secondary 03B55 · 03G1

7.1 Introduction

The present study is devoted to the structural analysis of the collection of consistent "normal" extensions of the modal logic **S4**. Our terminology and notation are slightly different from the usual. Namely, we apply the term *normal extension* only in relation to calculus. Thus, given a calculus **C** in a formal language L, a collection

A. Muravitsky (✉)
Louisiana Scholars' College, Northwestern State University, Natchitoches, LA 71497, USA
e-mail: alexeim@nsula.edu

A. Citkin and I. M. Vandoulakis (eds.), *V. A. Yankov on Non-Classical Logics, History and Philosophy of Mathematics*, Outstanding Contributions to Logic 24,
https://doi.org/10.1007/978-3-031-06843-0_7

L of L-formulas is a *normal extension of* **C** if, and only if, L contains all axioms of **C** and closed under all rules of inference postulated in **C**. According to this definition, given a calculus **C**, we denote by NExt**C** the set of all normal extensions of **C**. Thus we employ NExt**S4**, NExt**Grz**, and NExt**Int** referring to the set of the normal extensions of modal logic **S4** (defined as a calculus), that of the normal extensions of modal logic **Grz** (also given as a calculus), and that of the normal extensions of intuitionistic propositional calculus **Int** defined with two postulated rules of inference, (uniform) substitution and modus ponens. We take into account only *inconsistent extensions*, that is those which differ from the set of all formulas. With this agreement, the elements of NExt**Int** are commonly referred to as *intermediate logics* (to distinguish them from *superintuitionistic logics* which include the *inconsistent logic*).

This paper is dedicated to Vadim Anatol'evich Yankov, whose last name, originally in Russian, was spelled in the English translations of his earlier works as *Jankov*. In what follows, it may seem unclear to what extent the present study is based on Yankov's research. Therefore, it would be correct to immediately indicate such an influence.

In the proof of the claim that each nth S-slice is an interval (see (7.13)), we use Yankov's characteristic formulas defined in Jankov (1963) (see also Jankov 1969, § 3).[1] It is well known that characteristic formulas can identify one of the two logics of a *splitting pair* (see Definition 7.1). We will deal not only with splitting pairs of NExt**S4**, but also with the splitting pairs of some of its sublattices. Certain splitting pairs will allow us to divide NExt**S4**, as well as some of its sublattices, into smaller pieces, which are easier to analyze from a structural point of view.

The structural complexity of NExt**S4** was noted in comparison with other partially ordered structures. For instance, we know that:

- any finite distributive lattice is embedded into NExt**S4**;
- any countable partially ordered set is embedded into NExt**S4**;
- the free distributive lattice of countable rank is embedded into NExt**S4**;

cf. Maksimova and Rybakov (1974), corollary of Theorem 3. The above properties hold simply because $\mathcal{L}$ (see definition in the next section, Pr 4), which is isomorphic to NExt**Int**, is a sublattice of NExt**S4** and, on the other hand, the analogous properties are true for NExt**Int**; about the latter see Gerĉiu (1970), corollary, and Gerĉiu and Kuznecov (1970), corollaries of Theorems 5 and 6.

The purpose of this study is to consider the structure of NExt**S4** modulo NExt**Int**. Since the mid-1970s, it has been known two sublattices of NExt**S4**, namely $\mathcal{L}$ and NExt**Grz**, which are isomorphic to NExt**Int**. Although the former is not an interval in NExt**S4**, the latter is. The first fact eventually leads to the existence of a quotient

[1] Yankov introduced these formulas relating them to finite subdirectly irreducible Heyting algebras. Independently, similar characteristic formulas associated with finite rooted posets were defined in de Jongh (1968). This approach has been extended further for propositional monomodal language and **S4**-frames in Fine (1974); see more in this section below in Pr 2.

of NExt**S4** isomorphic to NExt**Int**, the second is known as the "Blok–Esakia theorem;"[2] see about the latter, e.g., in Wolter and Zakharyaschev (2014) and Muravitsky (2006). In Sect. 7.3, we show that a quotient of the filter generated by $\mathbf{Grz} \cap \mathbf{S5}$ is also isomorphic to NExt**Int**. In addition, $\mathcal{L}$ allows one to spot at least two more isomorphic copies of NExt**Int**, which are sublattices of the interval [**S4**, **Grz**]. However, the most unexpected feature of NExt**S4** comes from the discovery of an infinite sequence of intervals, each of which is isomorphic to NExt**Int**, where NExt**Grz** is one of them; this will be discussed in Sect. 7.3. This makes the Blok–Esakia theorem an episode, pointing at a deeper relationship between the structures of NExt**Int** and NExt**S4**.

In conclusion, we note that the idea of studying the lattice $\mathfrak{L}_\sigma$ of all equational theories with a fixed signature σ from a structural point of view is not new. Back in 1968, A. Tarski wrote in Tarski (1968):

> It would be interesting to provide a full intrinsic characterization of the lattice $\mathfrak{L}_\sigma$, using exclusively lattice-theoretical terms.

7.2 Preliminaries

Although the main characters of our discussion are lattices of logics, we start with main calculi which generate these lattices.

The calculi involved in our discussion are formulated in of the following propositional languages which are based on an infinite set ***Var*** of propositional variables. Metavariables for propositional variables are letters $p, q, r \dots$. When using metavariables, different metavariables represent different variables. The calculi **Int** and **Cl** (below) are formulated in an assertoric (modality-free) propositional language, L_a, with logical constants $\wedge, \vee, \rightarrow, \neg$ and $\bot$. Metavariables for formulas of this language are $A, B, \dots$. The modal logics are formulated in a mono-modal extension L_m of the language L_a by the enrichment of the latter with modality $\Box$. The metavariables for the formulas of the modal language L_m are Greek letters $\alpha, \beta, \dots$ As usual,

$$\Diamond\alpha := \neg\Box\neg\alpha,$$

for any formula α.

We will employ the following main calculi:

- intuitionistic propositional calculus, **Int**, with two postulated rules of inference, (uniform) substitution and modus ponens; cf., e.g., Chagrov and Zakharyaschev (1997), Sect. 2.6;

[2] W. Blok established this isomorphism in terms of the corresponding varieties of algebras in Blok (1976), and L. Esakia stated the existence of such an isomorphism in Esakia (1976), but never published the proof.

- classical propositional calculus, **Cl**, is defined as a normal extension of **Int** by adding the axiom $\neg\neg p \rightarrow p$;
- modal propositional calculus **S4** is defined with postulated rules of inference substitution, modus ponens and necessitation ($\alpha/\Box\alpha$); cf., e.g., Chagrov and Zakharyaschev (1997), Sect. 4.3, Table 4.2;
- modal propositional calculus **S5** is defined as a normal extension of **S4** by adding the axiom $p \rightarrow \Box\Diamond p$; cf., e.g., Chagrov and Zakharyaschev (1997), Sect. 4.3, Table 4.2;
- modal propositional calculus **Grz** (*Grzegorczyk logic*) is defined as a normal extension of **S4** by adding the axiom $\Box(\Box(p \rightarrow \Box p) \rightarrow p) \rightarrow p$; cf., e.g., Chagrov and Zakharyaschev (1997), Sect. 4.3, Table 4.2;
- modal propositional calculus **S4.1** (*McKinsey logic*) is defined as a normal extension of **S4** by adding the axiom $\Box\Diamond p \rightarrow \Diamond\Box p$; cf., e.g., Chagrov and Zakharyaschev (1997), Sect. 4.3, Table 4.2.

Other calculi will appear as needed.

To obtain a normal extension of a calculus **C**, we add a set of formulas Γ to the axioms of **C**. The resulting logic, not necessarily finitely axiomatizable, is denoted by $\mathbf{C} + \Gamma$. We will be considering only *consistent* normal extensions, that is those which are not coincident with the set of all formulas of the language in which **C** is formulated.

Pr 1: Although in this paper we consider the lattice NExt**Int** as a whole, but on one occasion, in Sect. 7.4, we refer to its structure.
Namely, we remind that if we denote by $\mathbf{G}_n$, for each integer $n \geq 2$, the intermediate logic of an n-element linear Heyting algebra and by **LC** the logic of a denumerable linear Heyting algebra, then the following compound inclusion holds:

$$\mathbf{LC} \subset \cdots \subset \mathbf{G}_3 \subset \mathbf{G}_2;$$

moreover, the logics $\mathbf{G}_n$ are only normal extensions of **LC**; cf. Dummett (1959).

Pr 2: V. Yankov introduced the characteristic formula of any finite subdirectly irreducible Heyting algebra in Jankov (1963); see a more comprehensive account in Jankov (1969), § 3. Later on, Fine (1974) has extended the notion of characteristic formula for any finite subdirectly irreducible **S4**-*algebra* (aka *topological Boolean algebra* or *topo-Boolean algebra*). Namely, the following properties (a)–(b) are originated with Jankov (1969). Let $\mathfrak{A}$ be any finite subdirectly irreducible **S4**-algebra and $\chi_{\mathfrak{A}}$ be its characteristic formula. Then the following equivalences hold.

(a) *For any formula* α, $\mathfrak{A}$ *refutes* α *if, and only if,* $\mathbf{S4} + \alpha \vdash \chi_{\mathfrak{A}}$.
(b) *For any* **S4**-*algebra* $\mathfrak{B}$, $\mathfrak{B}$ *refutes* $\chi_{\mathfrak{A}}$ *if, and only if,* $\mathfrak{A} \in \mathbf{HS}(\mathfrak{B})$, *where* **H** *and* **S** *are class operators of formation of homomorphic images and subalgebras, respectively.*

See more in Rautenberg (1979) and Citkin (2013, 2014) for a more general setting.

Pr 3: Throughout this paper, we use the fact that NExt**S4** is a distributive lattice with respect to set-intersection $\cap$ as meet and union closure with respect to postulated inference rules, $\oplus$, as join.[3]

The lattice operations of NExt**S4** mentioned above can be specified as follows.

- $(\mathbf{S4} + \Gamma) \cap (\mathbf{S4} + \Delta) = \mathbf{S4} + \mathbf{Set}\Box\alpha \vee' \Box\beta\alpha \in \Gamma$ and $\beta \in \Delta$, where $\vee'$ is *non-repetitional disjunction*; cf. Maksimova and Rybakov (1974), § 1, or Gerĉiu and Kuznecov (1970).[4]
- $(\mathbf{S4} + \Gamma) \oplus (\mathbf{S4} + \Delta) = \mathbf{S4} + (\Gamma \cup \Delta)$.

Pr 4: The following maps were defined in Maksimova and Rybakov (1974).

$$\rho : \text{NExt}\mathbf{S4} \longrightarrow \text{NExt}\mathbf{Int} : \ M \mapsto \mathbf{Set}AM \vdash \mathbf{t}(A),$$
$$\tau : \text{NExt}\mathbf{Int} \longrightarrow \text{NExt}\mathbf{S4} : \ L \mapsto \mathbf{S4} + L^{\mathbf{t}},$$

where $\mathbf{t}$ is the Gödel–McKinsey–Tarski translation and, given a set Γ of assertoric formulas,

$$\Gamma^{\mathbf{t}} := \mathbf{Set}\mathbf{t}(A)A \in \Gamma.$$

An impotent property of the map τ is that for any logic $L \in \text{NExt}\mathbf{Int}$ and any assertoric formula A,

$$A \in L \Longleftrightarrow \mathbf{S4} + L^{\mathbf{t}} \vdash \mathbf{t}(A); \tag{7.1}$$

cf. Dummett and Lemmon (1959), Theorem 1.

It was proved there that ρ is a lattice epimorphism and τ is a lattice embedding, as well as that

$$\rho(\tau(L)) = L, \ \textit{for any } L \in \text{NExt}\mathbf{Int}. \tag{7.2}$$

We denote the isomorphic image of NExt**Int** in NExt**S4** with respect to the map τ by $\mathcal{L}$.

We will also be using a well-known map

$$\sigma(L) := \mathbf{Grz} \oplus \tau(L).$$

[3] At first, the distributivity of NExt**Int** was noted in Hosoi (1969), Theorem 1.6. Then, the distributivity of NExt**S4** was established in Maksimova and Rybakov (1974), Theorem 2.

[4] It is believed that the idea of this equality was borrowed from Miura (1966).

We remind that

$$\text{(a) } \tau(\mathbf{Cl}) = \mathbf{S5} \text{ and (b) } \sigma(\mathbf{Cl}) = \mathbf{S4} + p \to \Box p. \tag{7.3}$$

It is well known that σ is a lattice isomorphism of NExt**Int** onto [**Grz**, S_0]; cf. Chagrov and Zakharyaschev (1997), Theorem 9.66. Also, it is well known that $\rho^{-1}(\mathbf{Int}) =$ [**S4**, **Grz**]; cf. Esakia (1979). This implies that for any **S4**-logics M and N,

$$M \cap N \in [\mathbf{S4}, \mathbf{Grz}] \Longleftrightarrow either\ M \in [\mathbf{S4}, \mathbf{Grz}]\ or N \in [\mathbf{S4}, \mathbf{Grz}]. \tag{7.4}$$

The $\Leftarrow$-part is obvious. To prove the $\Rightarrow$-part, we assume that $M \cap N \in [\mathbf{S4}, \mathbf{Grz}]$. Then, since $\rho(M \cap N) = \rho(M) \cap \rho(N)$, $\rho(M) \cap \rho(N) = \mathbf{Int}$. This implies that either $\rho(M) = \mathbf{Int}$ or $\rho(N) = \mathbf{Int}$.[5] Then, we apply the above observation from Esakia (1979).

We also need the following observation which is absolutely obvious:

For any **S4**-logics M and N,

$$M \oplus N \in [\mathbf{S4}, \mathbf{Grz}] \Longleftrightarrow M \in [\mathbf{S4}, \mathbf{Grz}]\ and N \in [\mathbf{S4}, \mathbf{Grz}]. \tag{7.5}$$

Pr 5: Let $\{S_n\}_{n \geq 0}$ be the collection of Scroggs' logics; see Scroggs (1951). We call it *S-series* and depict it as follows:

$$\mathbf{S5} = S_0 \subset \cdots \subset S_2 \subset S_1 = \mathbf{S4} + p \to \Box p, \qquad (S\text{-}series)$$

We remind that each S_n, where $n \geq 1$, is the logic of an 2^n-element **S4**-algebra with only two open elements, **0** and **1**. We denote these algebras by $\mathfrak{B}_n$.

Scroggs showed that the logics S_n form the interval [**S5**, S_1], that is to say, there is no **S4**-logic between each pair S_n and S_{n+1}, for any $n \geq 1$, and among proper extensions of **S5** there are only logics S_n, as well as

$$\bigcap_{n \geq 1} S_n = S_0.$$

Pr 6: Given $n \geq 1$, we define the nth S-slice as a set

$$\mathscr{S}_n := \mathbf{Set} M \in \text{NExt}\mathbf{S4} M \subseteq S_n \text{ and } M \not\subseteq S_{n+1};$$

further, the 0th S-slice is defined as the interval

[5] The last argument is based on Miura's theorem on the intersection of two intermediate logics (see Miura 1966) and the fact that **Int** possesses the disjunction property.

$$\mathscr{S}_0 := [\mathbf{S4}, \mathbf{S5}].$$

Then, we observe that $\{\mathscr{S}_n\}_{n\geq 0}$ is a partition of NExt**S4**; moreover,

$$M \in \mathscr{S}_n \Longleftrightarrow M \oplus S_0 = S_n; \tag{7.6}$$

cf. Muravitsky (2018), Proposition 4.

Pr 7: The algebras we are going to deal with have the signatures which are restrictions of the languages L_a and L_m. However, those restrictions allow one to interpret all the formulas of L_a and L_m in the corresponding algebras.

An algebra $\mathfrak{A} = \langle \mathcal{A}; \wedge, \vee, \rightarrow, \mathbf{0}\rangle$ is a ***Heyting algebra***[6] if it is a distributive lattice, relative to $\wedge$ (meet) and $\vee$ (join), with a relative pseudocomplementation $\rightarrow$ and a constant $\mathbf{0}$ representing a least element with respect to the following relation:

$$x \leq y \overset{\text{df}}{\Longleftrightarrow} x \wedge y = x.$$

It is convenient to remember that

$$x \leq y \Longleftrightarrow x \vee y = y.$$

It is also convenient to remember that the operation

$$\neg x := x \rightarrow \mathbf{0},$$

necessary to interpret the formulas of the form $\neg A$, is a pseudocomplementation of x with respect to $\leq$.

Since each element of the form $x \rightarrow x$ is a greatest element with respect to $\leq$, we denote:

$$\mathbf{1} := x \rightarrow x. \tag{7.7}$$

We remind that a Heyting algebra is ***subdirectly irreducible*** if, and only if, it has a pre-top element.

An algebra $\mathfrak{B} = \langle \mathcal{B}; \wedge, \vee, \sim, \Box, \mathbf{0}\rangle$, where its restriction $\mathfrak{B} = \langle \mathcal{B}; \wedge, \vee, \sim, \mathbf{0}\rangle$ is a Boolean algebra, is called an **S4-*algebra***[7] if the operation $\Box$ satisfies the following identities:

---

[6] Some authors use the term pseudo-Boolean algebra instead; see, e.g., Rasiowa and Sikorski (1970).

[7] Some authors prefer the term *topological Boolean algebra* or *topoboolean algebra*; see, e.g., Rasiowa and Sikorski (1970) and Chagrov and Zakharyaschev (1997), for the former, and Gabbay and Maksimova (2005), for the latter.

(a) $\Box(x \wedge y) = \Box x \wedge \Box y$;
(b) $\Box x \leq x$;
(c) $\Box x \leq \Box\Box x$;
(d) $\Box \mathbf{1} = \mathbf{1}$.

(We note that in any Boolean algebra the identity $x \to y =\sim x \vee y$ holds.)

An element x of an **S4**-algebra is called ***open*** if $\Box x = x$. The set $\mathcal{B}^\circ$ of all open elements of an **S4**-algebra $\mathfrak{B}$ is a sublattice of $\langle \mathcal{B}; \wedge, \vee \rangle$. More than that, defining on $\mathcal{B}^\circ$ the operation

$$x \to y := \Box(\sim x \vee y)$$

we obtain a Heyting algebra $\mathfrak{B}^\circ = \langle \mathcal{B}^\circ, \wedge, \vee, \to, \mathbf{0} \rangle$. An **S4**-algebra $\mathfrak{B}$ is ***subdirectly irreducible*** if, and only if, the Heyting algebra $\mathfrak{B}^\circ$ is subdirectly irreducible.[8]

Interpreting in any Heyting algebra the logical constant $\neg$ as

$$\neg x := x \to \mathbf{0}$$

and in any **S4**-algebra as $\sim x$ (complementation), we can count any assertoric formula A a "Heyting term" and any L_m-formula α as a term of **S4**-algebra. Taking this into account, we, given a Heyting algebra $\mathfrak{A}$ and an L_a-formula A or given an **S4**-algebra $\mathfrak{B}$ and an L_m-formula α, say that A is ***valid*** in $\mathfrak{A}$ and α is ***valid*** in $\mathfrak{B}$ if the identity $A = \mathbf{1}$ is true in $\mathfrak{A}$ or, respectively, the identity $\alpha = \mathbf{1}$ is true in $\mathfrak{B}$. Otherwise, we say that $\mathfrak{A}$ ***refutes*** A or, respectively, $\mathfrak{B}$ ***refutes*** α.

Give $L \in \mathrm{NExt}\mathbf{Int}$, a formula A and a Heyting algebra $\mathfrak{A}$, we write $L \not\models_{\mathfrak{A}} A$ to say that the algebra $\mathfrak{A}$ validates all formulas of L and refutes A. A similar meaning has the notation $M \not\models_{\mathfrak{B}} \alpha$, $M \in \mathrm{NExt}\mathbf{S4}$ and $\mathfrak{B}$ is an **S4**-algebra.

In connection with the notion of validity, we recall the following important property which will be used in Sects. 7.5, 7.4 and 7.9: *Given an* **S4**-*algebra* $\mathfrak{B}$ *and an* L_a-*formula* A, *the formula* $\mathbf{t}(A)$ *is valid in* $\mathfrak{B}$ *if, and only if,* A *is valid in* $\mathfrak{B}^\circ$; cf. Rasiowa and Sikorski (1970), Chap. xi, § 8, or Rasiowa (1974), Chap. xiii, Sect. 5.3.

Let $\langle W, R \rangle$ be a quasi-ordered set. A set $X \subseteq W$ is called ***upward closed*** (with respect to R) if for any $x \in X$ and $y \in W$, xRy implies $y \in X$. The set of all upward closed sets is denoted by $\mathcal{U}(W)$. It is easy to check that the following operations are closed on $\mathcal{U}(W)$:

[8] All these facts are well-known and can be found, e.g. in Rasiowa and Sikorski (1970), Chagrov and Zakharyaschev (1997) or Gabbay and Maksimova (2005).

- $X \cap Y$,
- $X \cup Y$,
- $X \rightarrow Y := \mathbf{Set} x \in W \forall y \in W.\ xRy$ and $y \in X \Longrightarrow y \in Y$.

(We note that the empty set $\emptyset$ is upward closed with respect to any quasi-ordering.)

Now we define on the power set $\mathcal{P}(W)$ a unary operation $\Box$ as follows:

$$\Box X := \mathbf{Set} x \in W \forall y.\ xRy \Longrightarrow y \in X.$$

(We note that for any X, $\Box X$ is upward closed with respect to R.)

Using these operations, we associate with any quasi-ordered set $\langle W, R\rangle$ two types of algebras—$\mathrm{H}(W) = \langle \mathcal{U}(W), \cap, \cup, \rightarrow, \mathbf{0}\rangle$ and $\mathrm{B}(W) = \langle \mathcal{P}(W), \cap, \cup, \sim, \mathbf{0}\rangle$, where '$\sim$' denotes the unary operation of complementation over the subsets of W.

It is well-known facts that $\mathrm{H}(W)$ is a Heyting algebra and $\mathrm{B}(W)$ is an **S4**-algebra; moreover, $\mathrm{B}(W)^{\circ} = \mathrm{H}(W)$.

Other semantic notions and techniques will be employed as needed. Also, we will introduce other logics later.

7.3 The Interval $[M_0, S_1]$

Beginning with this section, we employ the notions of a splitting pair and that of a cosplitting pair (defined below in this section) as a regulator of our analysis of the structure of NExt**S4**.

We denote:

$$M_0 := \mathbf{Grz} \cap \mathbf{S5} \text{ and } S_1 := \mathbf{S4} + p \rightarrow \Box p.$$

(We remind that S_1 is the greatest logic in the S-series.)

Now we define:

$$\sigma_0(L) := M_0 \oplus \tau(L).$$

Using the definition of σ from Sect. 7.2, we obtain the following.

Proposition 7.1 (cf. Muravitsky 2006, Theorem 6.1) *For any $L \in$ NExt**Int**, the intervals $[M_0, \mathbf{Grz}]$ and $[\sigma_0(L), \sigma(L)]$ are isomorphic.*

Sketch of proof. The maps

$$g : [\sigma_0(L), \sigma(L)] \longrightarrow [M_0, \mathbf{Grz}] :\ M \mapsto M \cap \mathbf{Grz},$$
$$h : [M_0, \mathbf{Grz}] \longrightarrow [\sigma_0(L), \sigma(L)] :\ M \mapsto M \oplus \tau(L)$$

are lattice homomorphisms and inverses of one another.

Corollary 7.1 *The interval $[M_0, \mathbf{Grz}]$ is ordered by $\subseteq$ as order type $1 + \omega^*$.*

Proof Applying the last proposition for $L = \mathbf{Cl}$, we conclude that $[M_0, \mathbf{Grz}]$ is isomorphic to the interval $[\mathbf{S5}, S_1]$ of Scroggs' logics.

We denote:

$$M_n := g(S_n) = \mathbf{Grz} \cap S_n, \text{ for any } n \geq 0.$$

According to Corollary 7.1, the interval $[M_0, \mathbf{Grz}]$ can be depicted as the following chain:

$$\mathbf{Grz} \cap \mathbf{S5} = M_0 \subset \cdots \subset M_2 \subset M_1 = \mathbf{Grz}. \qquad (M\text{-}series)$$

The following properties hold:

$$\bigcap_{n \geq 1} M_n = M_0; \tag{7.8}$$

$$M_n = S_n \cap M_1; \tag{7.9}$$

$$S_n = M_n \oplus S_0; \tag{7.10}$$

$$(M_l \subseteq M_n \text{ or } S_l \subseteq S_n) \Longrightarrow M_n \cap S_l = M_l; \tag{7.11}$$

cf. Muravitsky (2018), the Properties (10.10)–(10.13), respectively.
Next, we define:

$$\sigma_n(L) := M_n \oplus \tau(L), \text{ for any } L \in \text{NExt}\mathbf{Int}.$$

We note that $\sigma_1 = \sigma$.

Proposition 7.2 (cf. Muravitsky 2006, Proposition 7.3) *For any $n \geq 0$, the map σ_n is a lattice isomorphism of* NExt**Int** *onto* $[M_n, S_n]$.

Sketch of proof. This is a routine task to check that σ_n is a lattice homomorphism.
Then, since $\mathbf{Grz} \oplus \sigma_n(L) = \sigma(L)$, the map σ_n is injective.
Finally, if $M \in [M_n, S_n]$, then $M = M_n \oplus \tau(\rho(M))$. Hence $M = \sigma_n(\rho(M))$.

For each $n \geq 0$, we denote:

$$\mathscr{M}_n := [M_n, S_n]. \tag{7.12}$$

Thus $\mathscr{M}_0 = [\mathbf{Grz} \cap \mathbf{S5}, \mathbf{S5}] = [M_0, S_0]$ and $\mathscr{M}_1 = [\mathbf{Grz}, \mathbf{S4} + p \rightarrow \Box p] = [M_1, S_1]$.

To complete the description of $[M_0, S_1]$, we observe the following.

Proposition 7.3 (cf. Muravitsky 2018, Proposition 14)[9] *For any number $n \geq 0$ and logic $M \in [M_0, S_1]$, the following conditions are equivalent:*

[9] In Muravitsky (2018), we use a different, but equivalent, definition of $\mathscr{M}_n$.

(a) $M \in \mathscr{M}_n$;
(b) $M \subseteq S_n$ *and* $M \not\subseteq S_{n+1}$.

Sketch of proof. Let $M \in \mathscr{M}_n$. For contradiction, assume that $M \subseteq S_{n+1}$. Then $M_n \subseteq S_{n+1}$ and hence, in virtue of (7.9), $M_n \subseteq M_{n+1}$. A contradiction.

Conversely, let $M \subseteq S_n$ and $M \not\subseteq S_{n+1}$. If it were that for any $k \geq 1$, $M \subseteq M_k$, then, according to (7.8), we would have that $M = M_0$ and hence $M \subseteq S_{n+1}$. This implies that there is $k \geq 1$ such that $M_k \subseteq M$ and $M \not\subseteq M_{k+1}$. This implies that $M_k \subseteq M \oplus S_0 = S_n$ and hence $M_k \subseteq S_n$. From the last inclusion we in turn derive that $M_k = M_k \cap \mathbf{Grz} = g(S_n) = M_n$.

Corollary 7.2 *The collection* $\{\mathscr{M}_n\}_{n\geq 0}$ *is a partition of the interval* $[M_0, S_1]$.

Definition 7.1 (splitting pair; cf. McKenzie 1972) Let $\mathscr{K}$ be a sublattice of NExt**S4**. Given two **S4**-logics M and N in $\mathscr{K}$ we call $\langle M, N\rangle$ a splitting pair in $\mathscr{K}$ if for any M' in $\mathscr{K}$ either $M \subseteq M'$ or $M' \subseteq N$.

Corollary 7.3 *For any* $n \geq 1$, *the pair* $\langle M_n, S_{n+1}\rangle$ *is a splitting pair in* $[M_0, S_1]$.

Proof Straightforwardly follows from Corollary 7.2 and (7.12).

Definition 7.2 (*cosplitting pair*) Let $\mathscr{K}$ be a sublattice of NExt**S4**. And let **S4**-logics M, N, M^* and N^* belong in $\mathscr{K}$. We call $\langle M, N\rangle$ a cosplitting pair in $\mathscr{K}$ with a switch $\langle M^*, N^*\rangle$ if $\langle M \cap M^*, N \oplus N^*\rangle$ is a splitting pair in $\mathscr{K}$.

Corollary 7.4 *For any* $n \geq 1$, *the pair* $\langle S_n, M_{n+1}\rangle$ *is a cosplitting pair in* $[M_0, S_1]$ *with a switch* $\langle$**Grz**, **S5**$\rangle$.

Proof Follows from (7.9), (7.10) and Corollary 7.3.

Next, we show that NExt**Int** is "present within" $[M_0, S_1]$ not only as a sublattice, but also as latter's homomorphic image. To emphasize that now we deal with $[M_0, S_1]$ as an independent lattice, we use in this task its other notation, NExtM_0.[10]

Let $L \in$ NExt**Int**. We define:

$$\mathscr{L}_L := \mathbf{Set}\sigma_n(L)n \geq 0,$$

and call the last set an *L-layer*.

We note that, since $\sigma_n(L) \in \mathscr{M}_n$, each L-layer is a countable set which is ordered by $\subseteq$ as order type $1 + \omega^*$; namely:

$$\sigma_0(L) \subset \cdots \subset \sigma_2(L) \subset \sigma_1(L).$$

[10] We note that, in view of Maksimova and Rybakov (1974), Theorem 1, M_0 is a calculus.

Now, for any logics $M, N \in \text{NExt}M_0$, we define:

$$(M, N) \in \Theta \overset{\text{df}}{\iff} M \text{ and } N \text{ belong to one and the same } L\text{-layer}.$$

We aim to prove that Θ is a lattice congruence on $\text{NExt}M_0$ such that the quotient $\text{NExt}M_0/\Theta$ is isomorphic to NExt**Int**.

We start with the following lemma.

Lemma 7.1 *Given a logic* $L \in \text{NExt}\mathbf{Int}$, $\mathscr{L}_L = \rho^{-1}(L) \cap \text{NExt}M_0$.

Proof Indeed,

$$\begin{aligned} M \in \mathscr{L}_L &\iff M = \tau_n(L), \text{ for some } n \geq 0; \\ &\iff M = M_n \oplus \tau(L), \text{ for some } n \geq 0; \\ &\iff \rho(M) = L \text{ and } M \in \text{NExt}M_0; \text{ [by (7.2)]} \\ &\iff M \in \rho^{-1}(L) \cap \text{NExt}M_0. \end{aligned}$$

Lemma 7.1 implies that each L-layer is a congruence class with respect to the homomorphism ρ restricted to $\text{NExt}M_0$, and Θ is the congruence induced by the last homomorphism. Hence, we obtain the following.

Proposition 7.4 $\text{NExt}M_0/\Theta$ *is isomorphic to* NExt**Int**.

7.4 The Interval [S4, S5]

At first, the presence of NExt**Int** within NExt**S4** has been discovered in Maksimova and Rybakov (1974), Theorem 3, namely, as the image of the former with respect to the mapping τ. We denote this image by $\boldsymbol{\mathcal{L}}$ and unspecified elements from the last lattice by $\boldsymbol{\mathcal{T}}$ (perhaps with subscripts). We note that $\boldsymbol{\mathcal{L}}$ is a sublattice of the interval [**S4**, **S5**]. From Sect. 7.3, we know that another sublattice of the last interval, which is also isomorphic to NExt**Int**, is the interval [M_0, **S5**]. In this section, we will show another one replica of NExt**Int**, this time a sublattice of the interval [**S4**, M_0].

Proposition 7.5 (cf. Muravitsky 2018, Proposition 3) *Lat* $\boldsymbol{\mathcal{T}} \in \boldsymbol{\mathcal{L}}$. *Then* (A) *the following conditions are equivalent:*

(a) $\boldsymbol{\mathcal{T}} \subseteq \mathbf{Grz}$;
(b) $\boldsymbol{\mathcal{T}} \subseteq M_0$;
(c) $\boldsymbol{\mathcal{T}} = \mathbf{S4}$;

(B) *if* $M_0 \subseteq \boldsymbol{\mathcal{T}}$, *then* $\boldsymbol{\mathcal{T}} = \mathbf{S5}$.

We note that $[M_0, \mathbf{S5}] \cup [\mathbf{S4}, M_0] \cup \boldsymbol{\mathcal{L}} \subseteq [\mathbf{S4}, \mathbf{S5}]$. According to Proposition 7.5–A, $[M_0, \mathbf{S5}] \cap \boldsymbol{\mathcal{L}} = \{\mathbf{S5}\}$ and $[\mathbf{S4}, M_0] \cap \boldsymbol{\mathcal{L}} = \{\mathbf{S4}\}$. There might seem that

$[M_0, \mathbf{S5}] \cup [\mathbf{S4}, M_0] \cup \mathcal{L} = [\mathbf{S4}, \mathbf{S5}]$. However, the lattice $[\mathbf{S4}, \mathbf{S5}]$ is more complex that it might be expected. As we show below, the above inclusion is proper.

We define:

$$M^* := (M_0 \cap \tau(\mathbf{G}_3)) \oplus \tau(\mathbf{LC}).$$

(See about **LC** and $\mathbf{G}_n$, $n \geq 2$, in Pr 1.)

Lemma 7.2 *The following logics are equal to each other:*

$$\begin{array}{l}(\mathrm{m}_1)\ M^*,\\ (\mathrm{m}_2)\ (M_0 \cap \tau(\mathbf{G}_3)) \oplus \tau(\mathbf{LC}),\\ (\mathrm{m}_3)\ (\mathbf{Grz} \cap \tau(\mathbf{G}_3)) \oplus \tau(\mathbf{LC}),\\ (\mathrm{m}_4)\ (M_0 \oplus \tau(\mathbf{LC})) \cap \tau(\mathbf{G}_3),\\ (\mathrm{m}_5)\ (\mathbf{Grz} \oplus \tau(\mathbf{LC})) \cap \tau(\mathbf{G}_3).\end{array}$$

Proof The equality $(\mathrm{m}_1) = (\mathrm{m}_2)$ holds by definition. The equality $(\mathrm{m}_2) = (\mathrm{m}_3)$ is true, because $M_0 \cap \tau(\mathbf{G}_3) = \mathbf{Grz} \cap \mathbf{S5} \cap \tau(\mathbf{G}_3) = \mathbf{Grz} \cap \tau(\mathbf{G}_3)$. Finally, the equalities $(\mathrm{m}_2) = (\mathrm{m}_4)$ and $(\mathrm{m}_3) = (\mathrm{m}_5)$ are true, since the lattice NExt**S4** is distributive and $\tau(\mathbf{LC}) \subseteq \tau(\mathbf{G}_3)$.

Proposition 7.6 $M^* \in [\mathbf{S4}, \mathbf{S5}] \setminus [M_0, \mathbf{S5}] \cup [\mathbf{S4}, M_0] \cup \mathcal{L}$

Proof We notice that, by definition, $M^* \in [\mathbf{S4}, \mathbf{S5}]$.

If it were that $M_0 \subseteq M^*$, then we would have $M_0 \subseteq \tau(\mathbf{G}_3)$ which is a contradiction with Proposition 7.5–B.

Next, if it were that $M^* \subseteq M_0$, then, using the equality $(\mathrm{m}_1) = (\mathrm{m}_2)$, we would derive that $\tau(\mathbf{G}_3) \subseteq M_0$ which would yield a contradiction with Proposition 7.5–A.

Now we aim to show that $M^* \notin \mathcal{L}$.

From the equalities $(\mathrm{m}_1) = (\mathrm{m}_2)$ and $(\mathrm{m}_1) = (\mathrm{m}_5)$, we conclude that

$$\tau(\mathbf{LC}) \subseteq M^* \subseteq \tau(\mathbf{G}_3). \qquad \text{(P 7.6–}*\text{)}$$

Then, from the equality $(\mathrm{m}_1) = (\mathrm{m}_3)$, we would have

$$\mathbf{Grz} \cap \tau(\mathbf{G}_3) \subseteq \tau(\mathbf{LC}). \qquad \text{(P 7.6–}**\text{)}$$

Let formulas A and B axiomatize $\mathbf{G}_3$ and **LC**, respectively; that is, $\mathbf{G}_3 = \mathbf{Int} + A$ and $\mathbf{LC} = \mathbf{Int} + B$. According to Pr 3, the inclusion (P 7.6–**) means that

$$\mathbf{S4} + \mathbf{t}(B) \vdash \mathbf{t}(A) \vee' \Box(\Box(\Box(p \to \Box p) \to p) \to p). \qquad \text{(P 7.6–}***\text{)}$$

We refute (P 7.6–* * *) as follows. Let $\langle W, R \rangle$ be a quasi-ordered set corresponding to the following diagram.

It should be clear that the algebra H(W) validates the formula B and, hence, the algebra B(W) validates the formula $\mathbf{t}(B)$, for $\mathrm{B}(W)^\circ = \mathrm{H}(W)$; cf. Pr 7. On the other hand, since H(W) is a 4-element chain, the formula A will be refuted in H(W) and, hence, the formula $\mathbf{t}(A)$ will be refuted in B(W). Grzegorczyk formula is also refuted in B(W), because $\langle W, R\rangle$ contains a nontrivial cluster. We note that the algebra B(W) is subdirectly irreducible and, therefore, the formula $\mathbf{t}(A) \vee' \Box(\Box(\Box(p \to \Box p) \to p) \to p)$ is invalid in B(W). This completes the refutation of (P 7.6–$***$) and, therefore, of (P 7.6–$**$).

Thus, it were that $M^* = \tau(L)$, for some some $L \in \mathrm{NExt}\mathbf{Int}$, then, by (P 7.6–$*$), would have that

$$\tau(\mathbf{LC}) \subset M^* = \tau(L) \subseteq \tau(\mathbf{G}_3). \qquad \text{(P 7.6–}*****\text{)}$$

This, in virtue of Pr 1, implies that $M^* = \tau(\mathbf{G}_n)$, for some $n \geq 3$. Using the last equality, we obtain the following sequence of consecutive conclusions.

$$\tau(\mathbf{G}_n) = (\mathbf{Grz} \oplus \tau(\mathbf{LC})) \cap \tau(\mathbf{G}_3) \text{ [replacement of } M^* \text{ with } \tau(\mathbf{G}_n) \text{ in}(\mathrm{m}_1) = (\mathrm{m}_5)]$$
$$\Downarrow$$
$$\tau(\mathbf{G}_n) = (\mathbf{Grz} \oplus \tau(\mathbf{LC})) \cap \tau(\mathbf{G}_n) \text{ [for } \tau(\mathbf{G}_n) \subseteq \tau(\mathbf{G}_3)\text{; see Pr 1]}$$
$$\Downarrow$$
$$\tau(\mathbf{G}_n) \subseteq \mathbf{Grz} \oplus \tau(\mathbf{LC})$$
$$\Downarrow$$
$$\mathbf{Grz} \oplus \tau(\mathbf{G}_n) \subseteq \mathbf{Grz} \oplus \tau(\mathbf{LC})$$
$$\Downarrow$$
$$\mathbf{G}_n \subseteq \mathbf{LC} \qquad \text{[cf. Pr 4].}$$

A contradiction.

As we have seen, the lattice [**S4**, **S5**] contains two distinct replicas ot NExt**Int**. In the next section we will show that it contains two more, and possibly infinitely many, distinct replicas. However, the presence of NExt**Int** in [**S4**, **S5**] finds itself, though implicitly, also in the following form.

Let use denote by ρ^* the homomorphism ρ restricted to [**S4**, **S5**]; see about ρ in Pr 4. Since $\mathcal{L} \subseteq$ [**S4**, **S5**], ρ^* is an epimorphism of [**S4**, **S5**] onto NExt**Int**. Hence the quotient of [**S4**, **S5**] with respect to the kernel of ρ^* is isomorphic to NExt**Int**.

We conclude this section with the following remark related to the map ρ^*. Since the lattice [**S4**, **S5**] is a principal ideal in NExt**S4**, according Rasiowa and Sikorski (1970), Theorem iv.8.2, [**S4**, **S5**] is an image of NExt**Int** with respect to a Heyting homomorphism and, hence, is itself a Heyting algebra. The question is **open**, if ρ^* is a Heyting homomorphism.

7.5 The Interval [S4, Grz]

For each $n \geq 1$, we define

$$K_n := \mathbf{S4} + \Box\chi_{n+1},$$

where χ_{n+1} is the characteristic formula of the algebra $\mathfrak{B}_{n+1}$; also, we define

$$K_0 := \mathbf{S4}.$$

From Muravitsky (2006), Proposition 9, we know that

$$\mathbf{S4} \subset \cdots \subset K_2 \subset K_1$$

and from ibid, Proposition 10, that

$$\bigcap_{n\geq 1} K_n = \mathbf{S4}.$$

It turns out that for each $n \geq 0$,

$$\mathscr{S}_n = [K_n, S_n]; \tag{7.13}$$

cf. Muravitsky (2018), Proposition 5. We note that $K_1 = \mathbf{S4.1}$; cf. Muravitsky (2018), Proposition 6. Thus $\mathscr{S}_1 = [\mathbf{S4.1}, \mathbf{S4} + p \to \Box p]$.

The following properties hold:

$$S_l \subseteq S_n \Longrightarrow K_n \oplus S_l = S_n; \tag{7.14}$$

$$K_n \subseteq M_m; \tag{7.15}$$

$$M_l \subseteq M_n \Longrightarrow K_n \oplus M_l = M_n; \tag{7.16}$$

cf. Muravitsky (2018), the Properties (10.14), Proposition 8, and (10.15), respectively.

We observe the following.

Proposition 7.7 (cf. Muravitsky 2018, Corollary 7) *For any* $n \geq 1$, $\langle K_n, M_{n+1}\rangle$ *is a splitting pair in* [**S4**, **Grz**].

For convenience, we denote

$$\mathscr{E}_n := [K_n, M_n] \text{ where } n \geq 1, \text{ and } \mathscr{E}_0 := [\mathbf{S4}, M_0].$$

From Proposition 7.7, we obtain that $\{\mathscr{E}_n\}_{n\geq 0}$ is a partition of the lattice [**S4**, **Grz**]. This is implies the following.

Corollary 7.5 *For any $n \geq 1$, $\langle M_n, K_{n+1}\rangle$ is a cosplitting pair in* [**S4**, **Grz**] *with a switch* $\langle \mathbf{S4.1}, M_0\rangle$.

Proof Along with Proposition 7.7, employs (7.15) and (7.16).

Now we show that the structural complexity of each $\mathscr{E}_n$ with $n \geq 1$ increases as n increases.

For any $n \geq 1$, we denote

$$T_n := K_n \cap S_0$$

We observe in Muravitsky (2018), Corollary 4, that

$$\bigcap_{n\geq 1} T_n = \mathbf{S4} \tag{7.17}$$

Definition 7.3 (*maps h_n^* and s_n^**, for $n \geq 0$) We successively define the following maps: for any $n \geq 0$,

$$\begin{aligned} h^n &: M \mapsto M \cap S_n, \\ h_n^* &:= h^0 \upharpoonright \mathscr{E}_n, \\ s^n &: M \mapsto M \oplus K_n, \\ s_n^* &:= s^n \upharpoonright \mathscr{E}_0. \end{aligned}$$

(See also Muravitsky 2018, Sect. 10.7, where we give a slightly different, but equivalent, definition of these maps.)

Using (7.14), we prove the following.

Proposition 7.8 (cf. Muravitsky 2018, Proposition 31) *For any $n \geq 1$, the lattice $[T_n, M_0]$ is an isomorphic image of $\mathscr{E}_n$ with respect to h_n^*. Moreover, the map s_n^* restricted to $[T_n, M_0]$ is the inverse of h_n^*.*

Now we turn to $\mathscr{E}_0$. We show that the latter lattice contains two distinct replicas of NExt**Int**. It will allow us to propose a conjecture.

Let $M = \mathbf{S4} + \alpha$ and $\boldsymbol{\mathcal{T}} \in \boldsymbol{\mathcal{L}}$ so that $\boldsymbol{\mathcal{T}} = \tau(L)$, where $L \in$ NExt**Int**. According to [Pr 3], we have:

$$\boldsymbol{\mathcal{T}} \cap M = \mathbf{S4} + \mathbf{Sett}(A) \vee' \Box\alpha A \in L. \tag{7.18}$$

Lemma 7.3 *Let $M = M_0$ or $M = T_1$. Then for any $\boldsymbol{\mathcal{T}}_1, \boldsymbol{\mathcal{T}}_2 \in \boldsymbol{\mathcal{L}}$, $\boldsymbol{\mathcal{T}}_1 \neq \boldsymbol{\mathcal{T}}_2$ implies $\boldsymbol{\mathcal{T}}_1 \cap M \neq \boldsymbol{\mathcal{T}}_2 \cap M$.*

Proof [11]We give one proof for both cases, when

$$\alpha := \Box(\Box(p \to \Box p) \to p) \to p \text{ or } \alpha := \Box\Diamond p \to \Diamond\Box p.$$

[11] We note that for any $\boldsymbol{\mathcal{T}} \in \boldsymbol{\mathcal{L}}$, $M_0 \cap \boldsymbol{\mathcal{T}} = \mathbf{Grz} \cap \boldsymbol{\mathcal{T}}$. A sketch of proof of the implication ($\boldsymbol{\mathcal{T}}_1 \neq \boldsymbol{\mathcal{T}}_2 \Longrightarrow \mathbf{Grz} \cap \boldsymbol{\mathcal{T}}_1 \neq \mathbf{Grz} \cap \boldsymbol{\mathcal{T}}_2$) was concisely outlined in Muravitsky (2018), Lemma 1. Here we present a detailed proof, some details of which, however, are given in appendix (Sect. 7.9).

Let $M := (\mathbf{S4} + \alpha) \cap \mathbf{S5}$. Then, when the former is the case, $M = M_0$; when the latter, then $M = T_1$.

Suppose that $\boldsymbol{\mathcal{T}}_1 \neq \boldsymbol{\mathcal{T}}_2$, where $\boldsymbol{\mathcal{T}}_1 = \tau(L_1)$ and $\boldsymbol{\mathcal{T}}_2 = \tau(L_2)$. This implies that $L_1 \neq L_2$. Without loss of generality, we assume that there is an assertoric formula $A \in L_1 \setminus L_2$. Then there is a finitely generated subdirectly irreducible Heyting algebra $\mathfrak{A}$ such that $L_2 \not\models_{\mathfrak{A}} A$. According to Propositions 7.14 and 7.15, there is an **S4**-algebra $\mathfrak{B}$ such that $\mathfrak{B}^\circ = \mathfrak{A}$ and $\mathfrak{B}$ refutes $\Box\alpha$. Because of the former the algebra $\mathfrak{B}$ is subdirectly irreducible. In virtue of (7.1), the former also implies that $\boldsymbol{\mathcal{T}}_2 \not\models_{\mathfrak{B}} \mathbf{t}(A)$. Since $\mathfrak{B}$ is subdirectly irreducible, $\boldsymbol{\mathcal{T}}_2 \not\models_{\mathfrak{B}} \mathbf{t}(A) \vee' \Box\alpha$. However, in view of (7.1), $\mathbf{t}(A) \in \boldsymbol{\mathcal{T}}_1$ and hence $\mathbf{t}(A) \vee' \Box\alpha \in \boldsymbol{\mathcal{T}}_1$.

Lemma 7.3 straightforwardly implies the following.

Proposition 7.9 *Let $M = M_0$ or $M = T_1$. Then the map $\varphi : \boldsymbol{\mathcal{T}} \mapsto \boldsymbol{\mathcal{T}} \cap M$ is a lattice embedding of $\boldsymbol{\mathcal{L}}$ into* $[\mathbf{S4}, M_0]$.

Proof Lemma 7.3 implies that φ is injective. On the other hand, φ is a lattice homomorphism. Indeed, for any $\boldsymbol{\mathcal{T}}_1, \boldsymbol{\mathcal{T}}_2 \in \boldsymbol{\mathcal{L}}$, we have:

$$(\boldsymbol{\mathcal{T}}_1 \cap \boldsymbol{\mathcal{T}}_2) \cap M = (\boldsymbol{\mathcal{T}}_1 \cap M) \cap (\boldsymbol{\mathcal{T}}_2 \cap M),$$

$$(\boldsymbol{\mathcal{T}}_1 \oplus \boldsymbol{\mathcal{T}}_2) \cap M = (\boldsymbol{\mathcal{T}}_1 \cap M) \oplus (\boldsymbol{\mathcal{T}}_2 \cap M).$$

Thus, the lattices $[\mathbf{S4}, M_0]$ and $[\mathbf{S4}, T_1]$ contain, respectively, two replicas of $\boldsymbol{\mathcal{L}}$.

We propose a **conjecture** that the lattice $\boldsymbol{\mathcal{L}}$ can be embedded in each segment $[\mathbf{S4}, T_n]$, for any $n > 0$. If the conjecture were true, in view of (7.17), we would observe a countable descending sequence of replicas of $\boldsymbol{\mathcal{L}}$ in the interval $[\mathbf{S4}, M_0]$, the least element of each of which is **S4** and the greatest are T_n, respectively. Along with Proposition 7.8, it would give us the following structural picture of $[\mathbf{S4}, M_0]$: for any $n \geq 1$, the ordinal sum of *a replica of* $\boldsymbol{\mathcal{L}}$ and *a replica of* $\mathscr{E}_n$ is a sublattice of $[\mathbf{S4}, M_0]$ with the correction that the greatest element of the former is identified with the least element of the latter.

7.6 Sublattices $\mathcal{S}$, $\mathcal{R}$, and $\mathcal{T}$

In this section we discuss three fragments of NExt**S4**. Our goal is to find characterizations of these fragments.

We begin with the following two lemmas.

Lemma 7.4 *For any $n, p \geq 1$ and any* **S4**-*logic M, $(M_n \oplus M) \cap S_{n+p} = (M_{n+p} \oplus M) \cap S_{n+p}$.*

Proof Indeed, we have:

$$\begin{aligned}(M_n \oplus M) \cap S_{n+p} &= (M_n \cap S_{n+p}) \oplus (M \cap S_{n+p}) \\ &= M_{n+p} \oplus (M \cap S_{n+p}) \quad \text{[in virtue of (7.11)]} \\ &= (M_{n+p} \oplus M) \cap (M_{n+p} \oplus S_{n+p}) \\ &= (M_{n+p} \oplus M) \cap S_{n+p}.\end{aligned}$$

Lemma 7.5 *For any $n \geq 1$, if $M \in \mathscr{S}_n$, then for any $p \geq 0$, $M_n \oplus M = M_{n+p} \oplus M$.*

Proof If $p = 0$, the equality in question is obvious. So we assume that $p \geq 1$.

Let us take $M \in [K_n, S_n]$, where $n \geq 1$. Since $M_{n+p} \subseteq M_n$ (see [Pr 3]), $M_{n+p} \oplus M \subseteq M_n \oplus M$. For contradiction, assume that $M_{n+p} \oplus M \subset M_n \oplus M$.

Now we observe that $M \oplus M_n \subseteq S_n$ (see [P2 2]–[Pr 4]). For contradiction, we assume that $M_n \oplus M = S_n$. This implies the following:

$$\begin{aligned}S_{n+p} &= S_n \cap S_{n+p} = (M_n \oplus M) \cap S_{n+p} \\ &= (M_n \cap S_{n+p}) \oplus (M \cap S_{n+p}) \\ &= M_{n+p} \oplus (M \cap S_{n+p}) \quad \text{[in virtue of (7.11)]} \\ &= (M_{n+p} \oplus M) \cap (M_{n+p} \oplus S_{n+p}) \\ &\subseteq M_{n+p} \oplus M.\end{aligned}$$

The equality $S_{n+p} = M_{n+p} \oplus M$ is impossible, for $M \not\subseteq S_{n+1}$ (see [Pr 2]). Thus $S_{n+p} \subset M_{n+p} \oplus M$. Since $S_{n+1} \subset S_n$ and there is no logic in between, we have: $S_n \subseteq M_{n+p} \oplus M \subseteq S_n$. This implies that $M_n \oplus M = S_n = M_{n+p} \oplus M$. A contradiction. Hence $M \oplus M_n \subset S_n$.

Further, we observe that $M_{n+p} \oplus M \not\subseteq S_{n+p}$ (for $M \not\subseteq S_{n+1}$). Thus we have the following two chains:

$$\left.\begin{aligned}&(M_{n+p} \oplus M) \cap S_{n+p} \subset M_{n+p} \oplus M \subset M_n \oplus M \subset S_n, \\ &(M_{n+p} \oplus M) \cap S_{n+p} \subset S_{n+p} \subset S_n.\end{aligned}\right\} \qquad \text{(L 7.5–}*\text{)}$$

We aim to show that (L 7.5–∗) is a pentagon. For this, it suffices to prove the following two equalities:

$$(M_n \oplus M) \cap S_{n+p} = (M_{n+p} \oplus M) \cap S_{n+p} \qquad \text{(L 7.5–}{*}{*}\text{)}$$

and

$$(M_{n+p} \oplus M) \oplus S_{n+p} = S_n. \qquad \text{(L 7.5–}{*}{*}{*}\text{)}$$

To prove (7.5–∗∗), we use Lemma 7.4.

To prove (7.5–∗ ∗ ∗), we first observe that $(M_{n+p} \oplus M) \oplus S_{n+p} = M \oplus S_{n+p}$. Since $M \not\subseteq S_{n+p}$, $S_{n+p} \subset M \oplus S_{n+p}$. Therefore, we have: $S_n \subseteq M \oplus S_{n+p} \subseteq S_n$. Thus we have proved that (L 7.5–∗) is a pentagon. A contradiction.

We denote:

$$\mathcal{S} := \bigcup_{n \geq 1} \mathscr{S}_n.$$

It is obvious that $\mathcal{S}$ is a sublattice of NExt**S4**. Using Lemma 7.5, we obtain a characterization of the logics from $\mathcal{S}$.

Proposition 7.10 *For any* **S4**-*logic* M, *if* $M \in \mathcal{S}$, *then there is a least number* $n \geq 1$ *such that for any* $p \geq 0$, $M_n \oplus M = M_{n+p} \oplus M$. *Conversely, if for some* $n, p \geq 1$, $M_n \oplus M = M_{n+p} \oplus M$, *then* $M \in \mathcal{S}$.

Proof Let $M \in \mathcal{S}$. Then there is a unique number $n \geq 1$ such that $M \in \mathscr{S}_n$, that is $M \in [K_n, S_n]$. According to Lemma 7.5, for any $p \geq 0$, $M_n \oplus M = M_{n+p} \oplus M$. We aim to show that n is the least number among the numbers greater than or equal to 1, for which the last property holds.

For contradiction, assume that n is not the least number with the property in question. Then, firstly, $n > 1$ and hence $n - 1 \geq 1$. Secondly, we would have that $M_{n-1} \oplus M = M_n \oplus M$. Since, by premise, $M_n \oplus M \subseteq S_n$, we would also have that $M_{n-1} \oplus M \subseteq S_n$, which would imply that $M_{n-1} \subseteq S_n$. A contradiction.

Now we suppose that the equality

$$M_n \oplus M = M_{n+p} \oplus M \tag{P 7.10–$*$}$$

holds, for some $n, p \geq 1$. For contradiction, assume that $M \subseteq$ **S5**. This implies that $M_{n+p} \oplus M \subseteq S_{n+p}$ and hence, according to (P 7.10–$*$), $M_n \oplus M \subseteq S_{n+p}$. This implies that $M_n \subseteq S_{n+p}$ and hence $M_n \subseteq S_{n+1}$. A contradiction.

Corollary 7.6 *For any* **S4**-*logic* M, $M \in \mathcal{S}$ *if, and only if, there are numbers* $n, p \geq 1$ *such that* $M_n \oplus M = M_{n+p} \oplus M$.

Next we define:

$$\mathcal{R} := \mathcal{S} \setminus [\mathbf{S4}, \mathbf{Grz}].$$

Thus, for any **S4**-logic M,

$$M \in \mathcal{R} \iff M \in \mathcal{S} \textit{ and } M \not\subseteq M_1. \tag{7.19}$$

Now we have the following.

Proposition 7.11 $\mathcal{R}$ *is a sublattice of* NExt**S4**.

Proof Let us take logics M and N from $\mathcal{R}$. Since neither M nor N is in $\mathscr{S}_0$, there are unique numbers $k, n \geq 1$ such that $M \in \mathscr{S}_k$ and $N \in \mathscr{S}_n$. Without loss of generality, we assume that $k \leq n$.

It is obvious that $M \cap N \in \mathscr{S}_n$. Also, according to (7.4), $M \cap N \notin$ [**S4**, **Grz**]. Thus $M \cap N \in \mathcal{R}$.

Next, we observe that, in virtue of (7.5), $M \oplus N \notin [\mathbf{S4}, \mathbf{Grz}]$. Also, we notice that $M \oplus N \in \mathscr{S}_k$. Hence $M \oplus N \in \mathcal{R}$.

Now we define:

$$\mathcal{T} := \mathcal{R} \setminus [M_0, S_1].$$

Thus, for any **S4**-logic M,

$$M \in \mathcal{T} \iff M \in \mathcal{S},\ M \nsubseteq M_1 \text{ and } M_0 \nsubseteq M. \tag{7.20}$$

Proposition 7.12 *$\mathcal{T}$ is a sublattice of* NExt**S4**.

Proof Let logics M and N be in $\mathcal{T}$. Then, according to Proposition 7.11, both $M \cap N$ and $M \oplus N$ are in $\mathcal{R}$. Also, it should be obvious that $M_0 \nsubseteq M \cap N$. So, it remains to show that $M_0 \nsubseteq M \oplus N$.

Since $\mathcal{T} \subseteq \mathcal{S}$, there are numbers $k, n \geq 1$ such that $M \in \mathscr{S}_k$ and $N \in \mathscr{S}_n$. Without loss of generality, assume that $k \leq n$. Then it should be obvious that $M \oplus N \in \mathscr{S}_k$. For contradiction, suppose that $M_0 \subseteq M \oplus N$. This implies that $M_k \subseteq M \oplus N \subseteq S_k$. In virtue of Lemma 7.5, we derive that $M_k \oplus (M \oplus N) = M_{k+1} \oplus (M \oplus N)$. This in turn implies that $M_{k+1} \oplus (M \oplus N) = M \oplus N \subseteq S_k$ and hence $M_{k+1} \subseteq S_k$. A contradiction.

Although (7.19) and (7.20), in conjunction with Corollary 7.6, give us characterizations of the lattices $\mathcal{R}$ and $\mathcal{T}$, respectively, we have nothing more to say about these lattices. From a viewpoint of lattice theory, they are "dark spots" of NExt**S4** and we will not discuss them here any longer. Instead, in previous sections, we have focused on the parts of NExt**S4** that were taken off to define the aforementioned lattices, namely on the intervals $[M_0, S_1]$, [**S4**, **S5**], and [**S4**, **Grz**]. As to the lattice $\mathcal{S}$ as a whole, its structure can be described in terms of splitting pairs of $[M_0, S_1]$ and of [**S4**, **Grz**].

7.7 Mathematical Remarks

In Sect. 7.3 we showed that the maps σ_n, $n \geq 0$, establish lattice isomorphisms from NExt**Int** onto $\mathscr{M}_n$, respectively. Since each $\mathscr{M}_n$ is an interval of NExt**S4** and taking into account that NExt**Int** is a complete lattice, these isomorphisms also possess the following properties.

- *The isomorphism σ_1 preserves the* inf *of all sets and the* sup *of all nonempty sets* (for the units of NExt**S4** and $\mathscr{M}_1$ coincide).
- *The isomorphisms σ_n, where $n > 1$, and σ_0 preserve the* inf *and* sup *of all nonempty sets.*

Another replica of NExt**Int** is $\mathcal{L}$, but it is not an interval of NExt**S4**, for $M_0 \notin \mathcal{L}$ (Proposition 7.5). However, since the map τ preserves the sup of all sets (see Maksimova and Rybakov (1974), Theorem 3), $\mathcal{L}$ is a complete join semilattice of NExt**S4** and also a meet semilattice for nonempty sets. Further, it would be interesting to know whether

$$[\mathbf{S4}, \mathbf{S5}] = [\mathbf{S4}, M_0] \cup [M_0, \mathbf{S5}] \cup \mathcal{L}.$$

In Sect. 7.5 two more replicas of NExt**Int** have been obtained via the maps $\varphi : \boldsymbol{T} \mapsto \boldsymbol{T} \cap M$, where either $M = M_0$ or $M = T_1$. Since the property $a \cap \sup \mathbf{Set} b_t t \in T = \sup \mathbf{Set} a \cap b_t t \in T$ holds in any complete Heyting algebra (see Rasiowa and Sikorski (1970), Theorem iv.7.1), the obtained replicas are complete join subsemilattices of NExt**S4**.

7.8 Philosophical Remarks

The purpose of these remarks, if we paraphrase William Kneale, is the desire of a mathematical mind to make the truth seem attractive.[12]

Contemporary philosophers call the Context Principle the following viewpoint on the meanings of terms, which was originated with Frege (in his *Foundations of Arithmetic*, § 60, § 62, and § 106) and promoted by Wittgenstein both in his *Tractatus* (3.3, 3.314) and *Philosophical Investigations* (part I, § 49) that the term has meaning only in the context of a proposition.

But can the same be said about the logical system? Then the legitimate question arises: What can be called the context of a logical system? Our answer: one of the contexts of a logical system is a set of its extensions. We even consider these extensions as a whole, so to speak, as a unitary object, not excluding, however, the possibility for other contexts.

The study of the entire set of extensions, normal or otherwise, of a logical calculus raises some additional questions: What can we learn from this? And if research results do not entail immediate applications, can motivation be sought in philosophy (a justificatory philosophy in this case) and epistemology?

We remind that one of the contributions of the *Port Royal Logic* was the distinction between the *comprehension* and the *extension* (*étendue*) of a general term. According to this tradition, the comprehension of a general term is the set of attributes, or modes, or qualities, or characteristics, "which could not be removed without destruction of the idea;" the extension of the term, on the other hand, "is the set of things to which it is applicable." Also, it was stressed that "the comprehension and the extension of a term are not properties of it, but rather sets of entities to which it is related in certain ways;" in addition, the extension of a general term was considered as "the set of its inferiors" so that the term with respect to them is the superior. Later tradition

[12] Kneale writes: "... one at least of the marks of a philosophical mind is a desire to make the truth seem plausible;" cf. Kneale (1948), p. 160.

"followed Sir William Hamilton in replacing 'comprehension' by 'intension', which has no use in ordinary language."[13]

Further discussion of the distinction between intension and extension and further clarifications of these concepts can be found, e.g., in Leibniz (see Swoyer 1995), Mill and Jevons. Mill states: "The various objects denoted by the class name are what is meant by the Extension of the concept, while the attributes connoted are its Comprehension."[14] Jevons, for his part, provides a narrower, but clearer definition: "The meaning of a term in extension consists of the objects to which the term may be applied; its meaning in intension consists of the qualities which are necessarily possessed by objects bearing that name."[15]

According to modern tradition, an extension is thought of "as being either an individual, a class, or a truth-value" and an intension is thought of "as being either an individual concept (if this rather queer phrase may be allowed), a propositional function, or a proposition."[16] Also, modern tradition has extended the scope of applicability of the dualism *extension-intension* from general terms to predicates and, in general, to sentences. We will not follow Rescher (1959) in our discussion below of this dualism for predicates; however, we use some notions of his exposition.[17]

Instead of defining the notions *extension* and *intension* as such, modern logic concerns with the distinction between "predicates in extension" and "predicates in intension" (as it has been noted in Jevons; see the quote above). Rescher states: "Two predicates are said to have the same extension when the class of objects to which one can be ascribed is identical with the class of objects to which the other applies." Thus, according to this point of view, predicates $P(x)$ and $Q(x)$ are equal in extension if

$$\forall x.\ P(x) \Longleftrightarrow Q(x). \tag{7.21}$$

The point of view we defend is that the comparison of two predicates in extension can be subtler than the definition (7.21), but, in any event, comparison in extension can be reduced to equality relation.[18] In contrast, comparison in intension is harder to define, since the concept of an attribute of a predicate is rather vague and can take various specific forms. As a starting point of comparison in intension, we say that predicates $P(x)$ and $Q(x)$ stand in relation $\mathcal{R}$ if

$$\mathcal{R}(F(P), F(Q)), \tag{7.22}$$

---

[13] The quotes of this paragraph were taken from Kneale and Kneale (1971), Sect. v.1.

[14] Cf. Mill (1979), pp. 332–333.

[15] Cf. Jevons (1888), lesson v.

[16] See, e.g., Kneale and Kneale (1971), Sect. x.3.

[17] It is not our purpose here to discuss the extension-intension dualism in comparison with the reference-sense dualism of Frege and Russell. Also, we find that the explication of extension and intension by Carnap in terms of equivalence and L-equivalence is too narrow for our discussion; see Carnap (1956), Chap. 1, § 5.

[18] We note that (7.21) is equivalent to the equality $\mathbf{Set}x\, P(x) = \mathbf{Set}x\, Q(x)$.

"for every admissible, logically feasible context F". (The wording in the quotation marks is taken from Rescher 1959, but the reader should not worry about the vagueness of this wording, for it is our goal to make it clear through various precise conditions.)[19]

As Leibniz considered "propositions themselves to be complex terms, i.e., complex concepts" (cf. Ishiguro 1990, Sect. II.1), it would be fertile, we believe, to consider calculi as complex concepts and, therefore, apply to calculi that relate to concepts in the previous traditions of the philosophy of logic. In particular, below we defend the viewpoint that the distinction between extension and intension makes sense if it applies also to calculus.

Let us fix a propositional language, L, and denote the set of L-formulas by $\mathfrak{F}_L$. Now, let **C** be a calculus over L. We can propose at least two *explicata* of the extension of a calculus **C** as *explicandum*.

- Predicate 'A formula A is deducible in **C**'.
- The consequence operator that is determined by **C**.

We note, firstly, that both explicata are realized as sets—the first explicatum as a subset of $\mathfrak{F}_L$, and the second as a function of $\mathcal{P}(\mathfrak{F}_L)$ to $\mathfrak{F}_L$. Thus the comparison in extension is reduced to equality relation.

Secondly, there are calculi which are equal in extension according to the former, but are different according to the latter.[20] Therefore, it would be more accurate to speak of aspects of the extension of **C**, leaving the last term undefined. However, we believe that the comparison in extension for calculi, within any of its aspects, can be reduced to equality relation.

In contrast, the intension of a calculus **C** or comparison in intension of calculi $\mathbf{C}_1$ and $\mathbf{C}_2$ is a more difficult problem for definition or identification. Yet, let us use (7.22) as a starting point, only replacing P and Q with $\mathbf{C}_1$ and $\mathbf{C}_2$, respectively:

$$\mathcal{R}(F(\mathbf{C}_1), F(\mathbf{C}_2)), \tag{7.23}$$

"for every admissible, logically feasible context F" (Rescher, *ibd.*).

The first question, of course, is about the nature of F. It should not be surprising that answer may vary. In accordance with old traditional view on intension, all "interiors" of the extension of **C** must satisfy to an attribute F of **C**. Thus if we choose as an expicatum of the extension of **C** the predicate of deducibility, then a normal extension of **C** can play a role of F. Yet, we believe that the intension of calculus **C** is not just the set of its attributes, but has a certain unity, which in our case manifests itself in a structural way, namely as the lattice NExt**C**. Specifically, NExt**Int** and NExt**S4** are not merely sets; they are distributive lattices, even complete Heyting algebras. Even if NExt**C** does not represent the entire intension of **C**, it represents some its numerous aspects.

[19] Rescher (1959) uses the equality relation for $\mathcal{R}$.

[20] Take, for instance, the calculi **GL** and **GL*** of Muravitsky (2014).

The intension of a calculus becomes especially interesting, when one wants to compare it with the intension of another calculus. Even more so, if we want to employ (7.23). Specifically, we want to examine

$$\mathcal{R}(\text{NExt}\mathbf{Int}, \text{NExt}\mathbf{S4}). \tag{7.24}$$

A natural question: How should $\mathcal{R}$ be understood in (7.24)? We note that, dealing with the intension of **Int** and that of **S4** as structures, it does not matter that **Int** and **S4** are defined in different propositional languages. The "communication" between NExt**Int** and NExt**S4**, to some extent, is forced by the equivalence (7.1) which is a basis for an isomorphism between NExt**Int** and $\mathcal{L}$ discussed in Sect. 7.4. Thus, the present paper, from a philosophical point of view, concerns the interaction of the intension of **Int** and that of **S4**.

7.9 Appendix

The terminology, notions and facts used in this section are standard. The main references here are Rasiowa and Sikorski (1970), Rasiowa (1974) and Chagrov and Zakharyaschev (1997). Although we avoid using the term *Kripke frame*, this concept underlies our exposition below.

Lat $\mathfrak{A}$ be a Heyting algebra and $\mathsf{S}_{\mathfrak{A}}$ be the set of all prime filters of $\mathfrak{A}$. The map $h : a \mapsto \mathbf{Set} x \in \mathsf{S}_{\mathfrak{A}} a \in x$, where $a \in |\mathfrak{A}|$, is called ***Stone embedding***, because it is an embedding of $\mathfrak{A}$ into $\mathrm{H}(\mathsf{S}_{\mathfrak{A}})$, where the last algebra is a Heyting algebra associated with the partially ordered set $\langle \mathsf{S}_{\mathfrak{A}}, \subseteq \rangle$.

Let us fix $x^* \in \mathsf{S}_{\mathfrak{A}}$ and let x_0 and x_1 be two new distinct elements that do not belong to $\mathsf{S}_{\mathfrak{A}}$. We define

$$\mathsf{S}^*_{\mathfrak{A}} := (\mathsf{S}_{\mathfrak{A}} \setminus \{x^*\}) \cup \{x_0, x_1\};$$

and, then, define the following relation R on $\mathsf{S}^*_{\mathfrak{A}}$ as follows:

$$x R^* y \overset{\text{df}}{\Longleftrightarrow} \begin{cases} x, y \in \mathsf{S}_{\mathfrak{A}} \setminus \{x^*\} \text{ and } x \subseteq y; \\ x \in \{x_0, x_1\} \text{ and } y \in \{x_0, x_1\}; \\ x \in \mathsf{S}_{\mathfrak{A}} \setminus \{x^*\}, y \in \{x_0, x_1\}, \text{ and } x \subseteq x^*; \\ y \in \mathsf{S}_{\mathfrak{A}} \setminus \{x^*\}, x \in \{x_0, x_1\}, \text{ and } x^* \subseteq y. \end{cases}$$

We note that the relation R is transitive and reflexive on $\mathsf{S}^*_{\mathfrak{A}}$.

Next we define: for any $X \subseteq \mathsf{S}_{\mathfrak{A}}$,

$$X^{\oplus} := \begin{cases} X, & \text{if } x^* \notin X, \\ (X \setminus \{x^*\}) \cup \{x_1, x_2\}, & \text{if } x^* \in X; \end{cases} \tag{7.25}$$

and for any $X \subseteq \mathsf{S}^*_{\mathfrak{A}}$,

$$X^{\ominus} := \begin{cases} X, & \text{if } X \cap \{x_1, x_2\} = \emptyset, \\ X \setminus \{x_1, x_2\}, & \text{if } X \cap \{x_1, x_2\} = \{x_i\}, \text{ where } i = 1 \text{ or } i = 2, \\ (X \setminus \{x_1, x_2\}) \cup \{x^*\}, & \text{if } X \cap \{x_1, x_2\} \neq \emptyset. \end{cases} \tag{7.26}$$

We note that for any $X \subseteq \mathsf{S}^*_{\mathfrak{A}}$,

$$\forall y \in \mathsf{S}_{\mathfrak{A}} \setminus \{x^*\}.\ y \in X^{\ominus} \Longleftrightarrow y \in X. \tag{7.27}$$

Also, for any $X \subseteq \mathsf{S}_{\mathfrak{A}}$,

$$X \setminus \{x^*\} = X^{\oplus} \setminus \{x_1, x_2\}; \tag{7.28}$$

and for any $X \subseteq \mathsf{S}^*_{\mathfrak{A}}$,

$$X \setminus \{x_1, x_2\} = X^{\ominus} \setminus \{x^*\}. \tag{7.29}$$

It is obvious that if $X \subseteq \mathsf{S}_{\mathfrak{A}}$, then $X^{\oplus} \subseteq \mathsf{S}^*_{\mathfrak{A}}$; and if $X \subseteq \mathsf{S}^*_{\mathfrak{A}}$, then $X^{\ominus} \subseteq \mathsf{S}_{\mathfrak{A}}$. The last observation can be refined as follows.

Lemma 7.6 *For any $X \in \mathcal{U}(S_{\mathfrak{A}})$, $X^{\oplus} \in \mathcal{U}(S^*_{\mathfrak{A}})$; and for any $X \in \mathcal{U}(S^*_{\mathfrak{A}})$, $X^{\ominus} \in \mathcal{U}(S_{\mathfrak{A}})$.*

Proof Let $X \in \mathcal{U}(\mathsf{S}_{\mathfrak{A}})$. If $x^* \notin X$, then $X = X^{\oplus}$. Let us take $x \in X$ and $y \in \mathsf{S}^*_{\mathfrak{A}}$ with xRy. For contradiction assume that $y = x_i$. Then $x \subseteq x^*$ and hence $x^* \in X$. This implies that $y \in \mathsf{S}_{\mathfrak{A}}$ and $x \subseteq y$. Hence $y \in X$.

Now we suppose that $x^* \in X$. Then $\{x_1, x_2\} \subseteq X^{\oplus}$. Let us take $x \in X^{\oplus}$ and $y \in \mathsf{S}^*_{\mathfrak{A}}$ with xRy. We have to consider the following cases.

Case: $y \in \{x_1, x_2\}$. Then, obviously, $y \in X^{\oplus}$.

Case: $x \in X \setminus \{x_1, x_2\}$ and $y \notin \{x_1, x_2\}$. Then $x \subseteq y$ and, by premise, $y \in X$. Since $y \neq x^*$, $y \in X^{\oplus}$.

Case: $x \in \{x_1, x_2\}$ and $y \notin \{x_1, x_2\}$. Then $x^* \subseteq y$ and, by supposition, $y \in X$. And since $y \neq x^*$, $y \in X^{\oplus}$.

The other part of the statement can be proven in a similar way.

The last observation can be extended as follows.

Proposition 7.13 *Restricted to the upward sets of $S_{\mathfrak{A}}$ and, respectively, to that of $S^*_{\mathfrak{A}}$, the maps $g : X \mapsto X^{\oplus}$ and $g^{-1} : X \mapsto X^{\ominus}$ are mutually inverse isomorphisms between $\langle \mathcal{U}(S_{\mathfrak{A}}), \subseteq \rangle$ and $\langle \mathcal{U}(S^*_{\mathfrak{A}}), \subseteq \rangle$, and hence between* $\mathrm{H}(S_{\mathfrak{A}})$ *and* $\mathrm{H}(S^*_{\mathfrak{A}})$.

Proof It should be obvious that for any X and Y of $\mathcal{U}(\mathsf{S}_{\mathfrak{A}})$,

$$X \subseteq Y \Longrightarrow X^{\oplus} \subseteq Y^{\oplus},$$

and for any X and Y of $\mathcal{U}(\mathsf{S}^*_{\mathfrak{A}})$,

$$X \subseteq Y \Longrightarrow X^{\ominus} \subseteq Y^{\ominus}.$$

Also, for any $X \in \mathcal{U}(\mathsf{S}_{\mathfrak{A}})$ and any $Y \in \mathcal{U}(\mathsf{S}^*_{\mathfrak{A}})$,

$$X^{\oplus\ominus} = X \text{ and } Y = Y^{\ominus\oplus}.$$

Lemma 7.7 *For any* $X \cup Y \subseteq \mathsf{S}^*_{\mathfrak{A}}$, *if* $X \cap \{x_1, x_2\} = \{x_i\} \neq \{x_j\} = Y \cap \{x_1, x_2\}$, *then*

$$(X \cup Y)^{\ominus} = X^{\ominus} \cup Y^{\ominus} \tag{7.30}$$

Proof We consider the following cases.

Case: $X \cap \{x_1, x_2\} = \{x_1, x_2\}$ or $Y \cap \{x_1, x_2\} = \{x_1, x_2\}$. Then $(X \cup Y) \cap \{x_1, x_2\} = \{x_1, x_2\}$. This yields:

- to obtain $X^{\ominus}$, x^* replaces $\{x_1, x_2\}$ in X;
- or to obtain $Y^{\ominus}$, x^* replaces $\{x_1, x_2\}$ in Y;
- and to obtain $(X \cup Y)^{\ominus}$, x^* replaces $\{x_1, x_2\}$ in $X \cup Y$.

Thus (7.30) holds.

Case: $X \cap \{x_1, x_2\} \subset \{x_1, x_2\}$ and $Y \cap \{x_1, x_2\} \subset \{x_1, x_2\}$. Taking into account the premise, this leads to the following three cases:

- $X \cap \{x_1, x_2\} = \{x_i\}$ and $Y \cap \{x_1, x_2\} = \emptyset$;
- $X \cap \{x_1, x_2\} = \emptyset$ and $Y \cap \{x_1, x_2\} = \{x_i\}$; and
- $X \cap \{x_1, x_2\} = \emptyset$ and $Y \cap \{x_1, x_2\} = \emptyset$.

In all these cases, (7.30) is true.

Lemma 7.8 *For any* $X \subseteq \mathsf{S}^*_{\mathfrak{A}}$, $(\boxplus\, X)^{\ominus} = \Box X^{\ominus}$.

Proof We consider two cases.

Case: $X \cap \{x_1, x_2\} = \emptyset$. Then $\boxplus\, X \cap \{x_1, x_2\} = \emptyset$. This also implies that, although $\boxplus\, X \subseteq X \subseteq \mathsf{S}_{\mathfrak{A}}$, $x^* \notin X$ and hence $x^* \notin \boxplus\, X$.

Next, we observe that if $x \neq x^*$, then

$$(\forall y \in \mathsf{S}^*_{\mathfrak{A}}.\, xRy \Longrightarrow y \in X) \Longleftrightarrow (\forall y \in \mathsf{S}_{\mathfrak{A}}.\, x \subseteq y \Longrightarrow y \in X). \tag{L 7.8–$*$}$$

Indeed, suppose the left-hand implication holds and $x \subseteq y$ with $y \in \mathsf{S}_{\mathfrak{A}}$. It is clear that $y \notin \{x_1, x_2\}$ and xRy. This implies that $y \in X$. Now we assume that the right-hand implication holds and xRy with $y \in \mathsf{S}^*_{\mathfrak{A}}$. If it were that $y \in \{x_1, x_2\}$, then, we would have that $x \subseteq x^*$ and, by the right-hand implication, $x^* \in X$. Thus, $y \notin \{x_1, x_2\}$. This means that $y \in \mathsf{S}_{\mathfrak{A}}$ and hence, by the right-hand implication, $y \in X$.

We continue considering the first case as follows.

$$\begin{aligned} x \in (\boxplus X)^{\ominus} &\Longleftrightarrow x \in \boxplus X \text{ [since } x^* \notin \boxplus X] \\ &\Longleftrightarrow \forall y \in \mathsf{S}^*_{\mathfrak{A}}.\, xRy \Longrightarrow y \in X \\ &\Longleftrightarrow \forall y \in \mathsf{S}_{\mathfrak{A}}.\, x \subseteq y \Longrightarrow y \in X \text{ [by (L 7.8} - *)] \\ &\Longleftrightarrow x \in \Box X \\ &\Longleftrightarrow x \in \Box X^{\ominus} \text{ [since } x^* \notin X]. \end{aligned}$$

Case: $X \cap \{x_1, x_2\} = \{x_i\}$. Without loss of generality, we count that $X \cap \{x_1, x_2\} = \{x_1\}$. Hence, $X^{\ominus} \subseteq \mathsf{S}_{\mathfrak{A}}$ and $x^* \notin X^{\ominus}$. This implies that $x^* \notin \Box X^{\ominus}$. On the other hand, $\boxplus X \cap \{x_1, x_2\} = \emptyset$. Hence $\boxplus X \subseteq \mathsf{S}_{\mathfrak{A}}$ and $x^* \notin \boxplus X$ and, therefore, $x^* \notin (\boxplus X)^{\ominus}$.

Now, assume that $x \neq x^*$. We obtain:

$$\begin{aligned} x \in (\boxplus X)^{\ominus} &\Longleftrightarrow x \in \boxplus X \text{ [since } x^* \notin \boxplus X] \\ &\Longleftrightarrow \forall y \in \mathsf{S}^*_{\mathfrak{A}}.\, xRy \Longrightarrow y \in X \\ &\Longleftrightarrow \forall y \in \mathsf{S}_{\mathfrak{A}}.\, x \subseteq y \Longrightarrow y \in X \text{ [by (L 7.8} - *)] \\ &\Longleftrightarrow \forall y \in \mathsf{S}_{\mathfrak{A}}.\, x \subseteq y \Longrightarrow y \in X^{\ominus} \text{ [by (7.27)]} \\ &\Longleftrightarrow x \in \Box X^{\ominus} \text{ [since } x^* \notin X]. \end{aligned}$$

Case: $X \cap \{x_1, x_2\} = \{x_1, x_2\}$. Then $x^* \in X^{\ominus}$. It should be clear that

$$x^* \in (\boxplus X)^{\ominus} \Longleftrightarrow x^* \in \Box X^{\ominus}.$$

Now assume that $x \neq x^*$. First, we show that

$$(\forall y \in \mathsf{S}^*_{\mathfrak{A}}.\, xRy \Longrightarrow y \in X) \Longleftrightarrow (\forall y \in \mathsf{S}_{\mathfrak{A}}.\, x \subseteq y \Longrightarrow y \in X^{\ominus}). \tag{L 7.8$-**$}$$

Indeed, suppose the left-hand side is valid and $x \subseteq y$ with $y \in \mathsf{S}_{\mathfrak{A}}$. Assume first that $y \neq x^*$. Then, in virtue of the premise, $y \in X$ and, applying (7.27), we conclude that $y \in X^{\ominus}$. If $y = x^*$, then, by definition (7.26), $x^* \in X^{\ominus}$.

Next, assume that the right-hand side of (L 7.8–$**$) is true. Let xRy, where $y \in \mathsf{S}^*_{\mathfrak{A}}$. If $y \in \{x_1, x_2\}$, then, by premise, $y \in X$. Now assume that $y \notin \{x_1, x_2\}$, that is $y \in \mathsf{S}_{\mathfrak{A}} \setminus \{x^*$. This implies that $y \in X^{\ominus}$ and, in virtue of (7.27), $y \in X$.

Now, we obtain for $x \neq x^*$:

$$\begin{aligned} x \in (\boxplus X)^{\ominus} &\Longleftrightarrow x \in \boxplus X \text{ [in virtue of (7.27)]} \\ &\Longleftrightarrow \forall y \in \mathsf{S}^*_{\mathfrak{A}}.\, xRy \Longrightarrow y \in X \\ &\Longleftrightarrow \forall y \in \mathsf{S}_{\mathfrak{A}}.\, x \subseteq y \Longrightarrow y \in X^{\ominus} \text{ [by (L 7.8} - **)] \\ &\Longleftrightarrow x \in \Box X^{\ominus}. \end{aligned}$$

Lemma 7.9 *For any* $a \in \mathfrak{A}$, $(\mathsf{S}^*_{\mathfrak{A}} \setminus h(a)^{\oplus})^{\ominus} = \mathsf{S}_{\mathfrak{A}} \setminus h(a)$.

Proof We consider the following two cases.

Case: $x^* \in h(a)$. Then $\mathsf{S}^*_{\mathfrak{A}} \setminus h(a)^{\oplus} = \mathsf{S}_{\mathfrak{A}} \setminus h(a)$. The conclusion is obvious.

Case: $x^* \notin h(a)$. Then $h(a)^{\oplus} = h(a)$ and $\{x_1, x_2\} \subseteq (\mathsf{S}^*_{\mathfrak{A}} \setminus h(a)^{\oplus})$. The conclusion is obvious again.

Proposition 7.14 *Let $X \subseteq S^*_{\mathfrak{A}}$ and the following conditions be satisfied:*

$$(a)\ (S^*_{\mathfrak{A}} \setminus X)^{\ominus} = S_{\mathfrak{A}} \setminus h(a^*),\ \textit{for some } a^* \in \mathfrak{A},$$
$$(b)\ X^{\ominus} = h(b^*),\ \textit{for some } b^* \in \mathfrak{A}.$$

Then, if $\mathfrak{B}$ is the subalgebra of $B(S^*_{\mathfrak{A}})$ generated by the set $\mathbf{Set}h(a)^{\oplus}a \in \mathfrak{A} \cup \{X\}$, *then $\mathfrak{B}^{\circ} = g(h(\mathfrak{A}))$.*

Proof Generating the algebra $\mathfrak{B}$, if we use only Boolean operations, according to Rasiowa and Sikorski (1970), Theorem II.2.1, we get meets of terms of the following three categories:

$$\left.\begin{array}{l} \bullet\ (\mathsf{S}_{\mathfrak{A}}{}^* \setminus h(a)^{\oplus}) \cup h(b)^{\oplus},\ \text{for some}\, a, b \in \mathfrak{A}, \\ \bullet\ (\mathsf{S}_{\mathfrak{A}}{}^* \setminus X) \cup h(b)^{\oplus},\ \text{for some}\, b \in \mathfrak{A}, \\ \bullet\ (\mathsf{S}_{\mathfrak{A}}{}^* \setminus h(a)^{\oplus}) \cup X,\ \text{for some}\, a \in \mathfrak{A}. \end{array}\right\} \tag{7.31}$$

If we apply the operation ⊞ to those meets and, then, distribute ⊞, applying it to each of the terms listed above (which is possible, for $\mathrm{B}(\mathsf{S}^*_{\mathfrak{A}})$ is an **S4**-algebra), we obtain meets of terms of the following three forms:

$$\bullet\ \boxplus\ ((\mathsf{S}^*_{\mathfrak{A}} \setminus h(a)^{\oplus}) \cup h(b)^{\oplus}),\ \text{for some } a, b \in \mathfrak{A},$$
$$\bullet\ \boxplus\ ((\mathsf{S}^*_{\mathfrak{A}} \setminus X) \cup h(b)^{\oplus}),\ \text{for some } b \in \mathfrak{A},$$
$$\bullet\ \boxplus\ ((\mathsf{S}^*_{\mathfrak{A}} \setminus h(a)^{\oplus}) \cup X),\ \text{for some } a \in \mathfrak{A}.$$

We aim to show that these terms belong to $g(h(\mathfrak{A}))$. This conclusion will be reached, if we show that the g^{-1}-images of these terms belong to $h(\mathfrak{A})$. Thus, according to Proposition 7.13, we have to consider the following terms:

$$\bullet\ (\boxplus\ ((\mathsf{S}^*_{\mathfrak{A}} \setminus h(a)^{\oplus}) \cup h(b)^{\oplus}))^{\ominus},\ \text{for some } a, b \in \mathfrak{A},$$
$$\bullet\ (\boxplus\ ((\mathsf{S}^*_{\mathfrak{A}} \setminus X) \cup h(b)^{\oplus}))^{\ominus},\ \text{for some } b \in \mathfrak{A},$$
$$\bullet\ (\boxplus\ ((\mathsf{S}^*_{\mathfrak{A}} \setminus h(a)^{\oplus}) \cup X))^{\ominus},\ \text{for some } a \in \mathfrak{A}.$$

According to Lemma 7.8, we will focus on the terms:

$$\bullet\ \Box((\mathsf{S}^*_{\mathfrak{A}} \setminus h(a)^{\oplus}) \cup h(b)^{\oplus})^{\ominus},\ \text{for some } a, b \in \mathfrak{A},$$
$$\bullet\ \Box((\mathsf{S}^*_{\mathfrak{A}} \setminus X) \cup h(b)^{\oplus})^{\ominus},\ \text{for some } b \in \mathfrak{A},$$
$$\bullet\ \Box((\mathsf{S}^*_{\mathfrak{A}} \setminus h(a)^{\oplus}) \cup X)^{\ominus},\ \text{for some } a \in \mathfrak{A}.$$

We note that the premise of Lemma 7.7 is satisfied, when we apply the operation $X \mapsto X^{\ominus}$ to the terms (7.31). Thus, applying successively Lemmas 7.7, 7.9 and Proposition 7.13, we obtain that the above terms are equal, respectively, to the following terms:

$$\bullet\ \Box((\mathsf{S}_{\mathfrak{A}} \setminus h(a)) \cup h(b)) = h(a) \rightarrow h(b),$$
$$\bullet\ \Box((\mathsf{S}^*_{\mathfrak{A}} \setminus X)^{\ominus} \cup h(b)) = h(a^*) \rightarrow h(b)\ \text{[by the first premise]},$$
$$\bullet\ \Box((\mathsf{S}_{\mathfrak{A}} \setminus h(a)) \cup X^{\ominus}) = h(a) \rightarrow h(b^*)\ \text{[by the second premise]}.$$

Table 7.1 Refutation of $\Box(\Box(p \to \Box p) \to p) \to p$

$x_1 \in v(A)$ or $x_1 \notin v(A)$	$x_2 \in v(A)$ or $x_2 \notin v(A)$
$x_1 \in v(p)$	$x_2 \notin v(p)$
$x_1 \notin v(\Box p)$	$x_2 \notin v(\Box p)$
$x_1 \notin v(p \to \Box p)$	$x_2 \in v(p \to \Box p)$
$x_1 \notin v(\Box(p \to \Box p))$	$x_2 \notin v(\Box(p \to \Box p))$
$x_1 \in v(\Box(p \to \Box p)\to p)$	$x_2 \in v(\Box(p \to \Box p) \to p)$
$x_1 \in v(\Box(\Box(p \to \Box p)) \to p)$	$x_2 \in v(\Box(\Box(p \to \Box p) \to p))$ $x_2 \notin v(\Box(\Box(p \to \Box p) \to p) \to p)$

This completes the proof.

Proposition 7.15 *Let $\mathfrak{A}$ be a finitely generated Heyting algebra. Then there is a prime filter $x^* \in \mathsf{S}_{\mathfrak{A}}$ and a set $X \subseteq \mathsf{S}^*_{\mathfrak{A}}$ such that the conditions (a) and (b) of Proposition 7.14 are satisfied; in addition, the valuation $v : p \mapsto X$ refutes the formulas $\Box(\Box(p \to \Box p) \to p) \to p$ and $\Box\Diamond p \to \Diamond\Box p$ in the algebra $\mathfrak{B}$ of Proposition 7.14.*

Proof According to Kuznetsov (1973), the algebra $\mathfrak{A}$ is atomic.[21] Let a^* be an atom of $\mathfrak{A}$. Then we define

$$x^* := [a^*).$$

We note that, since a^* is an atom, $h(a^*) = \{x^*\}$.

Remembering that new elements x_1, x_2 replace x^* in $\mathsf{S}_{\mathfrak{A}}$ to generate $\mathsf{S}^*_{\mathfrak{A}}$, we define:

$$X := \{x_1\}.$$

It should be clear that $\mathsf{S}^*_{\mathfrak{A}} \setminus X = (\mathsf{S}_{\mathfrak{A}} \setminus \{x^*\}) \cup \{x_2\}$. Therefore, $(\mathsf{S}^*_{\mathfrak{A}} \setminus X)^{\ominus} = \mathsf{S}_{\mathfrak{A}} \setminus h(a^*)$; that is the condition (a) of Proposition 7.14 is satisfied.

On the other hand, $X^{\ominus} = \emptyset = h(\mathbf{0})$; that is the condition (b) of Proposition 7.14 is also fulfilled.

In Table 7.1 we show a refutation of the formula $\Box(\Box(p \to \Box p) \to p) \to p$ in the algebra $\mathfrak{B}$, where A denotes a subformula of this formula. We aim to prove that $v(\Box(\Box(p \to \Box p) \to p) \to p) \neq \mathsf{S}^*_{\mathfrak{A}}$.

In the next table we show a refutation of the formula $\Box\Diamond p \to \Diamond\Box p$ in $\mathfrak{B}$. We aim to prove that $v(\Box\Diamond p \to \Diamond\Box p) \neq \mathsf{S}^*_{\mathfrak{A}}$ (Table 7.2).

[21] Proof of this fact can be extracted, e.g., from the Proofs of Lemmas 2 and 3 in Tsitkin (1986); see also Bezhanishvili and Grigolia (2005), Lemma 2.2.

Table 7.2 Refutation of $\Box\Diamond p \to \Diamond\Box p$

$x_1 \in v(A)$ or $x_1 \notin v(A)$	$x_2 \in v(A)$ or $x_2 \notin v(A)$
$x_1 \in v(p)$	$x_2 \notin v(p)$
$x_1 \in v(\Diamond p)$	$x_2 \in v(\Diamond p)$
$x_1 \in v(\Box\Diamond p)$	
$x_1 \notin v(\Box p)$	$x_2 \notin v(\Box p)$
$x_1 \notin v(\Diamond\Box p)$	
$x_1 \notin v(\Box\Diamond p \to \Diamond\Box p)$	

References

Bezhanishvil, G., & Grigolia, R. (2005). Locally finite varieties of Heyting algebras. *Algebra Universalis, 54*, 465–473.

Blok, W. (1976). Varieties of interior algebras. Ph.D. thesis, University of Amsterdam.

Carnap, R. (1956). *Meaning and necessity*. Chicago: The University of Chicago Press.

Chagrov, A., & Zakharyaschev, M. (1997). *Modal logic*. Oxford logic guides (Vol. 35). New York: The Clarendon Press, Oxford University Press.

Citkin, A. (2013). Jankov formula and ternary deductive term. In *Proceedings of the Sixth Conference on Topology, Algebra, and Category in Logic (TACL 2013)* (pp. 48–51). Nashville: Vanderbilt University.

Citkin, A. (2014). Characteristic formulas 50 years later (An algebraic account). arXiv:1407583v1.

de Jongh, D. (1968). Investigations on the intuitionistic propositional calculus. Ph.D. thesis, University of Wisconsin, Madison.

Dummett, M. (1959). A propositional calculus with denumerable matrix. *Journal of Symbolic Logic, 24*, 97–106.

Dummett, M., & Lemmon, E. (1959). Modal logics between S4 and S5. *Zeitschrift für mathematische Logik und Grundlagen der Mathematik, 5*(3/4), 250–264.

Esakia, L. (1976). On modal "counterparts" of superintuitionistic logics. In *Abstracts of the 7th All-Union Symposium on Logic and Methodology of Science, Kiev* (pp. 135–136) (In Russian).

Esakia, L. (1979). On the variety of Grzegorczyk algebras. In A. I. Mikhailov (Ed.), *Studies in non-classical logics and set theory, Moscow, Nauka* (pp. 257–287). (In Russian) [Translation: Esakia, L. (1983/1983). *Selecta Mathematica Sovietica, 3*(4), 343–366].

Fine, K. (1974). An ascending chain of S4 logics. *Theoria, 40*(2), 110–116.

Gabbay, D., & Maksimova, L. (2005). *Interpolation and definability*. Oxford logic guides (Vol. 46). Oxford: The Clarendon Press, Oxford University Press.

Gerĉiu, V. (1970). Sverkhnezavisimye superintuitsionistskie logiki. *Matematicheskie Issledovaniya, 5*(3), 24–37 (In Russian). [Superindependent superintuitionistic logics].

Gerĉiu, V., & Kuznecov, A. (1970). The finitely axiomatizable superintuitionistic logics. *Soviet Mathematics - Doklady, 11*, 1654–1658.

Hosoi, T. (1969). In intermediate logics II. *Journal of the Faculty of Science, University of Tokyo, Section I, 14*, 1–12.

Ishiguro, H. (1990). *Leibniz's philosophy of logic and language* (2nd ed.). Cambridge: Cambridge University Press.

Jankov, V. (1963). On the relation between deducibility in intuitionistic propositional calculus and finite implicative structures. *Doklady Akademii Nauk SSSR, 151*, 1293–1294 (In Russian) [Translation: Jankov, V. (1963). *Soviet Mathematics - Doklady, 4*(4), 1203–1204].

Jankov, V. (1969). Conjunctively indecomposable formulas in propositional calculi. *Izvestiya Akademii Nauk SSSR, Series Mathematica, 33*, 18–38 (In Russian) [Translation: Jankov, V. (1969). *Mathematics of the USSR–Izvestiya, 3*(1), 17–35].

Jevons, W. (1888). *Elementary lessons in logic*. London: Macmillan and Co.; New York: The Mises Institute (2010).

Kneale, W. (1948). What can logic do for philosophy? In *Proceedings of the Aristotelian Society*, Supplementary Volumes *22*, 155–166.

Kneale, W., & Kneale, M. (1971). *The development of logic*. London: The Clarendon Press, Oxford University Press.

Kuznetsov, A. (1973). O konechno-porozhdennykh psevdobulevykh algebrakh i konechno-approksimiruevykh mnogoobraziyakh. In *Abstracts of the 12th All-Soviet Algebraic Colloquium, Sverdlovsk* (p. 281) (Russian) [On finitely generated pseudo-Boolean algebras and finitely approximable varieties].

Maksimova, L., & Rybakov, V. (1974). The lattice of normal modal logics. *Algebra i Logika, 13*(2), 188–216 (In Russian) [Translation: Maksimova, L., & Rybakov, V. (1974). *Algebra and Logic, 13*(2), 105–122].

McKenzie, R. (1972). Equational bases and nonmodular lattice varieties. *Transactions of the American Mathematical Society, 174*, 1–43.

Mill, J. (1979). An examination of Sir William Hamilton's philosophy and of the principle philosophical questions discussed in his writings. In J. M. Robson (Ed.), With an introduction by A. Ryan, Vol. 9 of Collected works of John Stuart Mill, gen. J. M. Robson et al. (Eds.). London: Routledge and Kegan Paul.

Miura, S. (1966). A remark on the intersection of two logics. *Nagoya Mathematical Journal, 26*, 167–171.

Muravitsky, A. (2006). The embedding theorem: Its further development and consequences. Part 1. *Notre Dame Journal of Formal Logic, 47*(4), 525–540.

Muravitsky, A. (2014). Logic KM: A biography. In G. Bezhanishvili (Ed.), *Leo Esakia on duality in modal and intuitionistic logics, outstanding contributions to logic* (Vol. 4, pp. 155–185). Berlin: Springer.

Muravitsky, A. (2018). Lattice NExtS4 from the embedding theorem viewpoint. In S. Odintsov (Ed.), *Larisa Maksimova on implication, interpolation, and definability, outstanding contributions to logic* (Vol. 15, pp. 185–218). Berlin: Springer.

Rasiowa, H. (1974). *An algebraic approach to non-classical logics*. Studies in logic and the foundations of mathematics (Vol. 78). Amsterdam: North-Holland.

Rasiowa, H., & Sikorski, R. (1970). *The mathematics of metamathematics*. Monografie Matematyczne (3rd ed.) (Vol. 41). Warsaw: PAN–Polish Scientific Publishers.

Rautenberg, W. (1979). *Klassische und nichtclassische Aussagenlogik*. Logik und Grundlagen der Mathematik (Vol. 22). Braunschweig: Friedr. Vieweg & Sohn (In German).

Rescher, N. (1959). The distinction between predicate intension and extension. *Revue Philosophique de Louvain, Troisième série, 57*(56), 623–636.

Scroggs, S. (1951). Extensions of the Lewis system S5. *Journal of Symbolic Logic, 16*, 112–120.

Swoyer, C. (1995). Leibniz on intension and extension. *Naûs, 1*, 96–114.

Tarski, A. (1968). Equational logic and equational theories of algebras. In H. Schmidt, K. Schütte, & H.-J. Thiele (Eds.), *Contributions to mathematical logic: Proceedings of the logic colloquium, Hannover 1966* (pp. 275–288). Amsterdam: North-Holland.

Tsitkin, A. (1986). Finite axiomatizability of locally tabular superintuitionistic logics. *Mathematical Notes of the Academy of Sciences of the USSR, 40*(3), 739–742.

Wolter, F., & Zakharyaschev, M. (2014). On the Blok-Esakia theorem. In G. Bezhanishvili (Ed.), *Leo Esakia on duality in modal and intuitionistic logics, outstanding contributions to logic* (Vol. 4, pp. 99–118). Berlin: Springer.

Chapter 8
An Application of the Yankov Characteristic Formulas

Valery Plisko

Abstract A detailed exposition of one of the author's old results concerning the relationship between the propositional logic of realizability and the logic of Medvedev is given. The characteristic formulas introduced by Yankov play a decisive role in the proof. Along the way, a brief overview of Yankov's contribution to the study of propositional logic of realizability is given.

Keywords Intuitionistic logic · Recursive realizability · Finite problems · Medvedev Logic · Characteristic formulas

8.1 Introduction

It is a great honor and pleasure for me to publish an article in a volume dedicated to V. A. Yankov. I don't know him personally, but his work in the field of mathematical logic has had a strong influence on my research. As a student, I carefully studied his articles. We can say that Vadim Anatolyevich was my correspondence teacher.

V.A.Yankov's works are devoted to non-classical logics. In this article we consider his contribution to the study of the propositional logic of realizability and one application of his ideas to the study of the relationship between the propositional logic of realizability and Medvedev's logic of finite problems.

8.2 Intuitionistic Propositional Logic

From the intuitionistic point of view, a proposition is true if it is proved. Thus the truth of a proposition is connected with its proof. In order to avoid any confusion with formal proofs, we shall use the term 'a verification' instead of 'a proof'. This

V. Plisko (✉)
Faculty of Mechanics and Mathmetics, Moscow State University, GSP-1, 1 Leninskie Gory, Moscow, Russia
e-mail: veplisko@yandex.ru

A. Citkin and I. M. Vandoulakis (eds.), *V. A. Yankov on Non-Classical Logics, History and Philosophy of Mathematics*, Outstanding Contributions to Logic 24,
https://doi.org/10.1007/978-3-031-06843-0_8

understanding of the meaning of propositions leads to an original interpretation of logical connectives and quantifiers stated by L. E. J. Brouwer, A. N. Kolmogorov, and A. Heyting. Namely for every true proposition A we can consider its verification as a text justifying A. Now, if A and B are propositions, then

- a verification of a conjunction $A \& B$ is a text containing a verification of A and a verification of B;
- a verification of a disjunction $A \vee B$ is a text containing a verification of A or a verification of B and indicating which of them is verified;
- a verification of an implication $A \to B$ is a text describing a general effective operation for obtaining a verification of B from every verification of A;
- a verification of a negative proposition $\neg A$ is a verification of the proposition $A \to \perp$, where $\perp$ is a proposition having no verification.

If $A(x)$ is a predicate with a parameter x over a domain M given in an appropriate way, then

- a verification of an universal proposition $\forall x\, A(x)$ is a text describing a general effective operation which allows to obtain a verification of $A(m)$ for every given $m \in M$;
- a verification of an existential proposition $\exists x\, A(x)$ is a text indicating a concrete $m \in M$ and containing a verification of the proposition $A(m)$.

Of course, this semantics is very informal and is not precise from the mathematical point of view. Nevertheless it is enough for formulating intuitionistically valid logical principles. We shall consider the principles of intuitionistic propositional logic, i.e., those expressible by propositional formulas.

Propositional formulas are constructed in the usual way from a countable set of propositional variables $p, q, r, \ldots$ (possibly with subscripts) and connectives $\&$, $\vee, \to, \neg$. $A(p_1, \ldots, p_n)$ will denote a propositional formula containing no variables other than $p_1, \ldots, p_n$. The formula $(A \to B)\ \&\ (B \to A)$ will be denoted as $A \equiv B$.

A propositional formula $A(p_1, \ldots, p_n)$ is called intuitionistically valid if every proposition of the form $A(A_1, \ldots, A_n)$, where $A_1, \ldots, A_n$ are arbitrary propositions, is intuitionistically true. Thus intuitionistically valid formulas can be considered as principles of intuitionistic propositional logic. The first axiomatization of intuitionistic propositional logic was proposed by A. N. Kolmogorov (1925). Another more wide axiom system of *intuitionistic propositional calculus* IPC was later proposed by A. Heyting (1930). If $A_1, \ldots, A_n$ are propositional formulas, let $\mathsf{IPC} + A_1 + \cdots + A_n$ be the calculus obtained by adding the formulas $A_1, \ldots, A_n$ as axiom schemes to IPC. Formulas A and B are *deductively equivalent* if $\mathsf{IPC} + A \vdash B$ and $\mathsf{IPC} + B \vdash A$.

The problem of completeness of the calculus IPC can not be stated in a precise mathematical form because intuitionistic semantics is very informal. It can be made more precise if we define in a mathematical mode two key notions used in the above description of the informal semantics, namely the notions of a verification and a general effective operation. We consider some interpretations of intuitionistic propositional logic.

8.3 Heyting Algebras and Yankov's Characteristic Formulas

We start with an abstract algebraic interpretation. A *logical matrix* $\mathfrak{M}$ is a non-empty set M with a distinguished element 1 equipped with an unary operation $\neg$ and binary operations &, $\vee$, and $\rightarrow$ such that for any elements $x, y \in M$ the following conditions hold:

- if $1 \rightarrow x = 1$, then $x = 1$;
- if $x \rightarrow y = y \rightarrow x = 1$, then $x = y$.

A valuation on $\mathfrak{M}$ is a function f assigning to each propositional variable p some element $f(p) \in M$. Any valuation is extended to all propositional formulas in a natural way. We say that a formula A is *valid* in $\mathfrak{M}$ and write $\mathfrak{M} \models A$ if $f(A) = 1$ for each valuation f on $\mathfrak{M}$. If $f(A) \neq 1$, we say that f *refutes* A on $\mathfrak{M}$. A formula A is *refutable* on a logical matrix $\mathfrak{M}$ iff there is a valuation refuting A on $\mathfrak{M}$.

A logical matrix $\mathfrak{M}$ is called a *model* of a propositional calculus if all the formulas deducible in the calculus are valid in $\mathfrak{M}$. Models of IPC are called *Heyting algebras* or *pseudo-Boolean algebras*. If $\mathfrak{M}$ is a Heyting algebra, then one can define a partial order $\leq$ on $\mathfrak{M}$ in the following way: if $a, b \in \mathfrak{M}$, then $a \leq b$ iff $a \rightarrow b = 1$. A detailed exposition of pseudo-Boolean algebras can be found in the book Rasiowa and Sikorski (1963). By Completeness Theorem (see e.g. Jaśkowski 1936), if a propositional formula A is not deducible in IPC, then there exists a finite Heyting algebra $\mathfrak{M}$ such that A is not valid in $\mathfrak{M}$.

Consider some notions introduced by V. A. Yankov (1963b). A finite logical matrix $\mathfrak{M}$ is called *a finite implicative structure* iff $\mathfrak{M}$ is a model of the calculus IPC. Thus, by Completeness Theorem, finite implicative structures are exactly the finite Heyting algebras. We reformulate other Yankov's definitions replacing the term 'a finite implicative structure' by 'a finite Heyting algebra'.

There are two important ways of constructing Heyting algebras. *Cartesian product* of algebras $\mathfrak{M}_1, \ldots, \mathfrak{M}_n$ is a Heyting algebra defined as the set $M_1 \times \cdots \times M_n$ equipped with the component-wise operations. Another operation Γ for any Heyting algebra $\mathfrak{M}$ gives an algebra $\Gamma(\mathfrak{M})$ obtained by adding to $\mathfrak{M}$ a new element ω with the property: $x \leq \omega$ for each $x \neq 1$ in $\mathfrak{M}$ and $\omega \leq 1$.

A finite Heyting algebra $\mathfrak{M}$ is called *Gödelean* iff for any its elements a, b, whenever $a \vee b = 1$, we have either $a = 1$ or $b = 1$. If a finite Heyting algebra $\mathfrak{M}$ contains more than one element, then it is easy to prove that $\mathfrak{M}$ is Gödelean iff there is an element ω which is the greatest among the elements different from 1. This element is called *a Gödelean element* of the algebra $\mathfrak{M}$. Note that in Completeness Theorem, one can do only with finite Gödelean algebras. Moreover, every algebra of the form $\Gamma(\mathfrak{M})$ is Gödelean.

Let $\mathfrak{M}$ be a finite Gödelean algebra, ω be its Gödelean element. To each element $a \in \mathfrak{M}$ assign in a one-to-one way a propositional variable p_a. Let K be the conjunction of all the formulas $p_{a \circ b} \equiv p_a \circ p_b$ ($\circ \in \{\&, \vee, \rightarrow\}$) and $p_{\neg a} \equiv \neg p_a$ for $a, b \in \mathfrak{M}$. Thus the formula K simulates the tables defining the operations & , $\vee$, $\rightarrow$, and $\neg$. The formula $K \to p_\omega$ is called *a characteristic formula* for $\mathfrak{M}$ and is denoted by $X_{\mathfrak{M}}$. It is obvious that the evaluation $f(p_a) = a$ refutes $X_{\mathfrak{M}}$ on $\mathfrak{M}$. Yankov has proved the following theorem (see Jankov 1963b, Theorem 2).

Theorem 8.1 *For any finite Gödelean algebra $\mathfrak{M}$ and any propositional formula A, the following conditions are equivalent:*
(1) $\mathsf{IPC} + A \vdash X_{\mathfrak{M}}$;
(2) $\mathfrak{M} \not\models A$.

A property S of propositional formulas is called intuitionistic if the fact that a formula A has the property S implies that every formula deducible in the calculus $\mathsf{IPC} + A$ has this property too. The following theorem (Jankov 1963b, Theorem 3) is a consequence of Theorem 8.1.

Theorem 8.2 *For any finite Gödelean algebra $\mathfrak{M}$ and any intuitionistic property S, the following conditions are equivalent:*
(1) each formula having the property S is valid in $\mathfrak{M}$;
(2) the formula $X_{\mathfrak{M}}$ does not have the property S.

This theorem is important in studying various semantics for intuitionistic propositional logic. Let an interpretation (a semantics) of propositional formulas be given. Suppose that IPC is sound with respect to this semantics and we are interested in completeness of IPC. If there exists a formula A valid in this semantics and nondeducible in IPC, then A is refuted on a finite Gödelean algebra $\mathfrak{M}$. In this case, the formula $X_{\mathfrak{M}}$ is also valid and nondeducible. Thus we can look for an example of this kind among the characteristic formulas.

We say that elements $a_1, \ldots, a_n$ of a finite Gödelean algebra $\mathfrak{M}$ form a base if every element in $\mathfrak{M}$ can be obtained from $a_1, \ldots, a_n$ by means of $\neg$, & , $\vee$, and $\to$. The following theorem proved by Yankov (see Jankov 1963b, Theorem 4) is useful for applications.

Theorem 8.3 *If a finite Gödelean algebra $\mathfrak{M}$ has a base consisting of k elements, then there is a formula A with k propositional variables such that A and $X_{\mathfrak{M}}$ are deductively equivalent.*

8.4 Medvedev Logic

Heyting algebras give a rather formal characterization of intuitionistic propositional logic. An interesting although informal interpretation of that logic was proposed by Kolmogorov (1932). He considers propositional formulas as schemes of problems. Let A and B be arbitrary problems. Then A & B is the problem 'To solve the problems A and B', $A \vee B$ is the problem 'To solve the problem A or to solve the problem B', $A \to B$ is the problem 'To reduce the problem B to the problem A', i.e., 'To solve the problem B assuming that a solution of the problem A is given', the problem $\neg A$ is the problem $A \to \bot$, where $\bot$ is a problem having no solution. A propositional formula $A(p_1, \ldots, p_n)$ is considered as a principle of *the logic of problems* if there exists a uniform solution for all the problems of the form $A(A_1, \ldots, A_n)$, where $A_1, \ldots, A_n$ are arbitrary problems. Kolmogorov shows that the calculus IPC is sound with respect to this interpretation: all the formulas deducible in IPC are principles of the logic of problems.

In order to state the problem of completeness of IPC relative to the interpretation by means of problems, we have to make the notion of a problem more precise from the mathematical point of view. This was done by Yu. T. Medvedev (1962). A *finite problem* is a pair $\langle F, X\rangle$, where F is a finite non-empty set, X is its subset (possibly empty). Intuitively, F is the set of *possible solutions* of the problem, X is the set of its *actual solutions*. Let $\bot$ be a fixed problem without any actual solutions, for example, $\bot = \langle\{0\}, \emptyset\rangle$. For a finite problem $\mathfrak{A} = \langle F, X\rangle$ we denote F by $\varphi(\mathfrak{A})$ and X by $\chi(\mathfrak{A})$. Logical operations on finite problems are defined in the following way:

$$\varphi(\mathfrak{A} \,\&\, \mathfrak{B}) = \varphi(\mathfrak{A}) \times \varphi(\mathfrak{B}),\ \chi(\mathfrak{A} \,\&\, \mathfrak{B}) = \chi(\mathfrak{A}) \times \chi(\mathfrak{B}),$$

where $X \times Y$ means Cartesian product;

$$\varphi(\mathfrak{A} \vee \mathfrak{B}) = \varphi(\mathfrak{A}) \oplus \varphi(\mathfrak{B}),\ \chi(\mathfrak{A} \vee \mathfrak{B}) = \chi(\mathfrak{A}) \oplus \chi(\mathfrak{B}),$$

where $X \oplus Y = (X \times \{0\}) \cup (Y \times \{1\})$;

$$\varphi(\mathfrak{A} \to \mathfrak{B}) = \varphi(\mathfrak{B})^{\varphi(\mathfrak{A})},\ \chi(\mathfrak{A} \to \mathfrak{B}) = \{f \in \varphi(\mathfrak{B})^{\varphi(\mathfrak{A})} \mid f(\chi(\mathfrak{A})) \subseteq \chi(\mathfrak{B})\},$$

where Y^X means the set of maps $f : X \to Y$;

$$\neg\mathfrak{A} = \mathfrak{A} \to \bot.$$

Let $A(p_1, \ldots, p_n)$ be a propositional formula, $\mathfrak{A}_1, \ldots, \mathfrak{A}_n$ be finite problems. Then $A(\mathfrak{A}_1, \ldots, \mathfrak{A}_n)$ is a finite problem obtained by substituting $\mathfrak{A}_1, \ldots, \mathfrak{A}_n$ for $p_1, \ldots, p_n$ in $A(p_1, \ldots, p_n)$.

Let $\mathcal{F}$ be a system of non-empty finite sets $F_1, \ldots, F_n$. We say that a propositional formula $A(p_1, \ldots, p_n)$ is *valid* over $\mathcal{F}$ if there exists an uniform actual solution for all the problems of the form $A(\mathfrak{A}, \ldots, \mathfrak{A}_n)$, where $\mathfrak{A}, \ldots, \mathfrak{A}_n$ are finite problems such that $\varphi(\mathfrak{A}_i) = F_i$ $(i = 1, \ldots, n)$. A propositional formula $A(p_1, \ldots, p_n)$ is called *finitely valid* if it is valid over every system $F_1, \ldots, F_n$. The set of finitely valid formulas is called *Medvedev Logic*; we denote it by ML.

Every formula deducible in IPC is finitely valid, but IPC is not complete relative to Medvedev's interpretation. For example, the formulas

$$(\neg p \to q \vee r) \to ((\neg p \to q) \vee (\neg p \to r)) \tag{8.1}$$

and

$$((\neg\neg p \to p) \to (\neg p \vee \neg\neg p)) \to (\neg p \vee \neg\neg p) \tag{8.2}$$

are finitely valid but are not deducible in IPC.

No axiomatization of ML is known, but it is proved that it can not be axiomatized by any system of axiom schemes with bounded number of variables (see Maksimova et al. 1979). In particular, this logic is not finitely axiomatizable.

It was proved Medvedev (1966) that ML has a finite model property: there exists a sequence of finite Heyting algebras Φ^n $(n = 1, 2, \dots)$ such that a propositional formula A is in Medvedev Logic iff $\forall n\, [\Phi^n \models A]$. The sequence Φ^n is defined as follows. Let $I_n = \{1, 2, \dots, n\}$ $(n \geq 1)$, σ^n be the family of its non-empty subsets, i.e., $\sigma^n = \{E \mid E \subseteq I_n \text{ and } E \neq \emptyset\}$. For $\sigma \subseteq \sigma^n$, its closure is defined as the family $\sigma^* = \{E \in \sigma^n \mid \exists E_0\, [E_0 \in \sigma \text{ and } E_0 \subseteq E]\}$. A family σ is called *closed* if $\sigma^* = \sigma$. In other words, σ is closed iff $E \in \sigma$ implies that all supersets of E are in σ. The operation * has the following properties: (1) $\emptyset^* = \emptyset$; (2) $\sigma^* \supseteq \sigma$; (3) $\sigma^{**} = \sigma^*$; (4) $(\sigma_1 \cup \sigma_2)^* = \sigma_1^* \cup \sigma_2^*$; (5) if σ_1 and σ_2 are closed, then $\sigma_1 \cap \sigma_2$ is closed.

Let Φ^n be the set of all closed families, i.e., $\Phi^n = \{\sigma \subseteq \sigma^n \mid \sigma^* = \sigma\}$. Define operations $\&$, $\vee$, $\rightarrow$, and $\neg$ on Φ^n as follows: $\sigma_1 \,\&\, \sigma_2 = \sigma_1 \cup \sigma_2$; $\sigma_1 \vee \sigma_2 = \sigma_1 \cap \sigma_2$; $\sigma_1 \rightarrow \sigma_2 = (\overline{\sigma_1} \cap \sigma_2)^*$; $\neg\sigma_1 = (\overline{\sigma_1})^*$, where $\overline{\sigma_1} = \sigma^n \setminus \sigma_1$. The set Φ^n with these operations and a partial order $\leq$ defined as $a \leq b \leftrightharpoons a \supseteq b$, is a Heyting algebra. Its greatest element relative to $\leq$ is $\emptyset$ and the least element is σ^n.

The following theorem was proved by Medvedev (1966, Theorem 1).

Theorem 8.4 *A propositional formula F is finitely valid iff $\Phi^n \models F$ for any $n \geq 1$.*

Note that Φ^n is a Gödelean algebra. Namely $\omega_n = \{I_n\}$ is its Gödelean element. Thus for each algebra Φ^n, we can construct its characteristic formula X_n. For example, consider the case $n = 3$. We have that $I_3 = \{1, 2, 3\}$ and σ^3 consists of the following 7 elements: $a_i = \{i\}$, where $i = 1, 2, 3$, $a_{ij} = \{i, j\}$, where $i, j = 1, 2, 3$, $i \neq j$, and $a_{123} = I_3$. There are 18 closed subfamilies of σ^3. Let us find all of them. Evidently, the closed 7-element family $\mathbf{a}_0 = \sigma^3$ is the greatest element relative to $\subseteq$ (and the least element relative to $\leq$). Removing from it one of the minimal elements a_1, a_2, a_3, we obtain the closed 6-element families $\mathbf{a}_1 = \sigma^3 \setminus \{a_1\}$, $\mathbf{a}_2 = \sigma^3 \setminus \{a_2\}$, $\mathbf{a}_3 = \sigma^3 \setminus \{a_3\}$. Note that in this case, $\mathbf{a}_0 = \mathbf{a}_1 \cup \mathbf{a}_2 = \mathbf{a}_1 \,\&\, \mathbf{a}_2$. After removing from $\mathbf{a}_0$ a pair of minimal elements, we obtain the following closed 5-element families $\mathbf{a}_4 = \mathbf{a}_1 \cap \mathbf{a}_2 = \mathbf{a}_1 \vee \mathbf{a}_2$, $\mathbf{a}_5 = \mathbf{a}_1 \cap \mathbf{a}_3 = \mathbf{a}_1 \vee \mathbf{a}_3$, $\mathbf{a}_6 = \mathbf{a}_2 \cap \mathbf{a}_3 = \mathbf{a}_2 \vee \mathbf{a}_3$. Further, removing from $\mathbf{a}_0$ all the minimal elements a_1, a_2, a_3, we obtain the closed 4-element family $\mathbf{a}_7 = \mathbf{a}_1 \cap \mathbf{a}_2 \cap \mathbf{a}_3 = \mathbf{a}_1 \vee \mathbf{a}_2 \vee \mathbf{a}_3$. Removing from $\mathbf{a}_0$ the element a_{12}, in order to obtain a closed family, we must remove also the elements a_1 and a_2. Thus we obtain the closed 4-element family $\mathbf{a}_8 = \{a_3, a_{13}, a_{23}, a_{123}\}$. Note that $\mathbf{a}_8 = \neg\mathbf{a}_3$. Indeed, $\overline{\mathbf{a}_3}$ is the family $\{a_3\}$ and $\mathbf{a}_8$ is its closure. In the same manner, we obtain the closed 4-element families $\mathbf{a}_9 = \neg\mathbf{a}_2$ and $\mathbf{a}_{10} = \neg\mathbf{a}_1$. Removing from $\mathbf{a}_7$ the minimal element a_{12}, we obtain the closed 3-element family $\mathbf{a}_{11} = \{a_{13}, a_{23}, a_{123}\} = \mathbf{a}_7 \cap \mathbf{a}_8 = \mathbf{a}_7 \vee \mathbf{a}_8$. In the same way, we obtain the closed 3-element families $\mathbf{a}_{12} = \mathbf{a}_7 \cap \mathbf{a}_9 = \mathbf{a}_7 \vee \mathbf{a}_9$, $\mathbf{a}_{13} = \mathbf{a}_7 \cap \mathbf{a}_{10} = \mathbf{a}_7 \vee \mathbf{a}_{10}$. Other closed 2-element families are $\mathbf{a}_{14} = \mathbf{a}_{11} \cap \mathbf{a}_{12} = \mathbf{a}_{11} \vee \mathbf{a}_{12}$, $\mathbf{a}_{15} = \mathbf{a}_{11} \cap \mathbf{a}_{13} = \mathbf{a}_{11} \vee \mathbf{a}_{13}$, $\mathbf{a}_{16} = \mathbf{a}_{12} \cap \mathbf{a}_{13} = \mathbf{a}_{12} \vee \mathbf{a}_{13}$. Finally, we have the Gödelean element of the algebra Φ^3, namely $\mathbf{a}_{17} = \{I_3\} = \mathbf{a}_8 \cap \mathbf{a}_9 \cap \mathbf{a}_{10} = \mathbf{a}_8 \vee \mathbf{a}_9 \vee \mathbf{a}_{10}$, and the greatest relative to $\leq$ element $\mathbf{a}_{18} = \emptyset = \neg\mathbf{a}_0$. A visual diagram of the algebra Φ^3 is provided in Citkin (1978, Fig. 2).

We see that the elements $\mathbf{a}_1$, $\mathbf{a}_2$, $\mathbf{a}_3$ form a base for the algebra Φ^3. By Yankov's Theorem 8.3, there is a propositional formula L with three variables such that L is deductively equivalent to the characteristic formula X_3. In order to obtain a formula L, we assign the variables p_1, p_2, p_3 to the elements $\mathbf{a}_1$, $\mathbf{a}_2$, $\mathbf{a}_3$ and simulate the tables defining the operations of the algebra Φ^3 by propositional formulas bearing in mind that all the elements are obtained from the basic ones by means of $\neg$, &, $\vee$, $\rightarrow$. It turned out that the elements $\mathbf{a}_8$, $\mathbf{a}_9$, $\mathbf{a}_{10}$, i.e., $\neg\mathbf{a}_1$, $\neg\mathbf{a}_2$, $\neg\mathbf{a}_3$, form a base as well. For example, the Gödelean element $\mathbf{a}_{17}$ can be presented as $\neg\mathbf{a}_1 \vee \neg\mathbf{a}_2 \vee \neg\mathbf{a}_3$. Thus we can construct the formula L from $\neg p_1$, $\neg p_2$, $\neg p_3$. A general method of constructing such a formula is described in Jankov (1969). In our case, we obtain the following formula:

$$\begin{aligned}
&(\neg\neg p_1 \equiv (\neg p_2 \,\&\neg p_3)) \,\&\, (\neg\neg p_2 \equiv (\neg p_3 \,\&\neg p_1)) \,\&\, (\neg\neg p_3 \equiv (\neg p_1 \,\&\neg p_2)) \,\&\\
&\&\, ((\neg p_1 \rightarrow (\neg p_1 \vee \neg p_3)) \rightarrow (\neg p_2 \vee \neg p_3)) \,\&\\
&\&\, ((\neg p_2 \rightarrow (\neg p_3 \vee \neg p_1)) \rightarrow (\neg p_3 \vee \neg p_1)) \,\&\\
&\&\, ((\neg p_3 \rightarrow (\neg p_1 \vee \neg p_2)) \rightarrow (\neg p_1 \vee \neg p_2)) \rightarrow (\neg p_1 \vee \neg p_2 \vee \neg p_3).
\end{aligned} \tag{8.3}$$

8.5 Propositional Logic of Realizability

Another way of a specification of the informal intuitionistic semantics is to define a mathematically precise notion of a general effective operation. There is such a notion in mathematics, namely the notion of an algorithm. A variant of intuitionistic semantics based on interpreting effective operations as algorithms was proposed by S. C. Kleene. He introduced (see Kleene 1945) the notion of *recursive realizability* for the sentences of first-order arithmetic. The main idea was to consider natural numbers as the codes of verifications and partial recursive functions as effective operations. Partial recursive functions are coded by their Gödel numbers. A code of a verification of a sentence is called *a realization* of the sentence.

The unary partial recursive function with the Gödel number x will be denoted by $\{x\}$. The relation '*a natural number e realizes an arithmetic sentence A*' ($e \mathsf{r} A$) is defined by induction on the number of logical connectives and quantifiers in A.

- If A is an atomic sentence $t_1 = t_2$, then $e \mathsf{r} A$ means that $e = 0$ and A is true.
- $e \mathsf{r} (A \,\&\, B)$ iff e is of the form $2^a \cdot 3^b$ and $a \mathsf{r} A$, $b \mathsf{r} B$.
- $e \mathsf{r} (A \vee B)$ means that either e is of the form $2^0 \cdot 3^a$ and $a \mathsf{r} A$ or e is of the form $2^1 \cdot 3^b$ and $b \mathsf{r} B$.
- $e \mathsf{r} (A \rightarrow B)$ means that for any a, if $a \mathsf{r} A$, then $\{e\}(a) \mathsf{r} B$.
- $e \mathsf{r} \neg A$ means that $e \mathsf{r} (A \rightarrow 0 = 1)$.
- $e \mathsf{r} \forall x\, A(x)$ means that for any n, $\{e\}(n) \mathsf{r} A(n)$.
- $e \mathsf{r} \exists x\, A(x)$ means that e is of the form $2^n \cdot 3^a$ and $a \mathsf{r} A(n)$.

The following theorem is proved by D. Nelson (1947).

Theorem 8.5 *Every formula deducible in intuitionistic arithmetic* HA *is realizable.*

There are various variants of the notion of realizability for propositional formulas. Let $A(p_1, \dots, p_n)$ be a propositional formula, $A_1, \dots, A_n$ be arithmetical sentences. Then $A(A_1, \dots, A_n)$ is an arithmetical sentence obtained by substituting $A_1, \dots, A_n$ for $p_1, \dots, p_n$ in $A(p_1, \dots, p_n)$. It follows from the proof of Nelson Theorem that for any propositional formula A deducible in IPC, there exists a number a such that $a \mathrel{\mathsf{r}} A(A_1, \dots, A_n)$ for any arithmetical sentences $A_1, \dots, A_n$. In this sense, deducible formulas are uniformly realizable.

The incompleteness of IPC with respect to recursive realizability was discovered by G. F. Rose (1953), who has found a propositional formula which is uniformly realizable but is not deducible in IPC. Later other examples of this kind were proposed by various authors. Deductive interrelations between them are completely studied by D. P. Skvortsov (1995). He found a list of four realizable propositional formulas which are deductively independent in IPC and any other known realizable formula is deducible from these four formulas.

Yankov made some contributions to the study of propositional realizability logic. His paper Jankov (1963a) is dedicated to this topic. In particular, Yankov proposed new examples of realizable propositional formulas which are not deducible in IPC. These formulas are:

(1) $(\neg(p \,\&\, q) \,\&\, \neg(\neg p \,\&\, \neg q) \,\&\, ((\neg\neg p \to p) \to (p \vee \neg p)) \,\&$
$\& \, ((\neg\neg q \to q) \to (q \vee \neg q))) \to (\neg p \vee \neg q)$;

(2) a series of formulas I_n $(n \geq 3)$, where I_n is the formula
$\&_{1 \leq i < j \leq n} \neg(p_i \,\&\, p_j) \,\&\, \&_{1 \leq i < n} ((\&_{1 \leq j < n, j \neq i} \neg p_j) \to (p_i \vee p_n)) \to (p_n \vee \neg p_n)$;

(3) a series of formulas K_n $(n \geq 3)$, where K_n is the formula
$\&_{1 \leq i < j \leq n} (\neg(p_i \,\&\, p_j) \,\&\, ((\&_{1 \leq k \leq n. k \neq i, k \neq j} \neg p_k) \to (p_i \vee p_j)) \to \bigvee_{1 \leq i \leq n} p_i$.

Note that the first formula is included in Skvortsov's list.

Yankov shows deductive relationships between his formulas and the Rose formula. Besides, he considers the 7-element Heyting algebra $\mathfrak{M} = \Gamma(\Gamma(I_0) \times I_0)$, where I_0 is the classical two-element algebra $\{0, 1\}$. Yankov proves that all realizable propositional formulas are valid in $\mathfrak{M}$. In this case, simulating the algebra $\mathfrak{M}$ in the language of arithmetic is used. Namely, an arithmetical formula $F_a(x)$ is assigned to every element $a \in \mathfrak{M}$ in such a way that operations in $\mathfrak{M}$ correspond to the logical operations over formulas in the realizability semantics. For example, if $a \to b = c$ in $\mathfrak{M}$, then the formula $\forall x\, ((F_a(x) \to F_b(x)) \equiv F_c(x))$ is realizable. At the same time, the formula $\forall x\, F_a(x)$ is realizable only in the case $a = 1$. This method was later used by the author in the proof of the completeness of the disjunction-free part of IPC relative to realizability (see Plisko 1973).

It turns out that the formula (8.2) is refuted on the algebra $\mathfrak{M}$. In fact, this formula is deductively equivalent to the characteristic formula of the algebra $\mathfrak{M}$. Since realizability is an intuitionistic property of propositional formulas, it follows from Yankov's Theorem 8.2 that the formula (8.2) is not realizable. This formula is of interest because the Rose formula is a substitutional instance of (8.2). Namely, the Rose formula is obtained by substituting $\neg q \vee \neg r$ for p in (8.2).

A comprehensive survey of propositional realizability logic can be found in Plisko (2009).

8.6 Realizability and Medvedev Logic

We consider relations between propositional logic of realizability and Medvedev Logic. First of all, note that there are propositional formulas in Medvedev Logic which are not realizable, for example, the formulas (8.1) and (8.2). Consider the converse inclusion. Imagine that we are looking for a propositional formula F that is realizable but is not finitely valid. We know that such a formula should be refuted on the algebra Φ^n for an appropriate n. On the other hand, realizability is an intuitionistic property. Thus by Yankov's Theorem 8.2, if the formula F is realizable and is refuted on Φ^n, then the characteristic formula X_n of the algebra Φ^n is realizable. Moreover, the formula X_n is not finitely valid. Thus we can consider only the formulas X_n. It turned out that X_3 is already the formula we are looking for. Of course, any formula which is deductively equivalent to X_3 has the desired property. We have constructed above such a formula, namely (8.3). Denote it by $\mathbf{L}(p_1, p_2, p_3)$.

Theorem 8.6 *The formula* $\mathbf{L}(p_1, p_2, p_3)$ *is uniformly realizable.*

Proof We prove that there exists a natural number e such that for any arithmetical sentences $\mathrm{A}_1, \mathrm{A}_2, \mathrm{A}_3$, e realizes the arithmetical sentence $\mathbf{L}(\mathrm{A}_1, \mathrm{A}_2, \mathrm{A}_3)$, i.e., the sentence

$$\begin{aligned}&(\neg\neg \mathrm{A}_1 \equiv (\neg \mathrm{A}_2 \,\&\neg \mathrm{A}_3)) \,\&\, (\neg\neg \mathrm{A}_2 \equiv (\neg \mathrm{A}_3 \,\&\neg \mathrm{A}_1)) \,\&\, (\neg\neg \mathrm{A}_3 \equiv (\neg \mathrm{A}_1 \,\&\neg \mathrm{A}_2)) \,\&\\ &\&\, ((\neg \mathrm{A}_1 \to (\neg \mathrm{A}_1 \vee \neg \mathrm{A}_3)) \to (\neg \mathrm{A}_2 \vee \neg \mathrm{A}_3)) \,\&\\ &\&\, ((\neg \mathrm{A}_2 \to (\neg \mathrm{A}_3 \vee \neg \mathrm{A}_1)) \to (\neg \mathrm{A}_3 \vee \neg \mathrm{A}_1)) \,\&\\ &\&\, ((\neg \mathrm{A}_3 \to (\neg \mathrm{A}_1 \vee \neg \mathrm{A}_2)) \to (\neg \mathrm{A}_1 \vee \neg \mathrm{A}_2)) \to (\neg \mathrm{A}_1 \vee \neg \mathrm{A}_2 \vee \neg \mathrm{A}_3)\end{aligned} \tag{8.4}$$

Begin with some auxiliary statements. Let $\pi_0, \pi_1, \pi_2, \ldots$ be sequential prime numbers, and $(a)_i$ be the exponent with which the prime number π_i appears in the decomposition of the number a into prime factors.

Lemma 8.1 *Let f be a total recursive function with values* 0 *and* 1. *Then there exists a number x such that either* $(\{x\}(x))_0 = 0$ *and* $f(x) = 1$ *or* $(\{x\}(x))_0 = 1$ *and* $f(x) = 0$.

Proof This is an easy exercise in the theory of recursive functions.

Lemma 8.2 *Suppose that the formula*

$$(\neg\neg A_1 \equiv (\neg A_2 \,\&\neg A_3))\&(\neg\neg A_2 \equiv (\neg A_3 \,\&\neg A_1))\&(\neg\neg A_3 \equiv (\neg A_1 \,\&\neg A_2)) \tag{8.5}$$

is realizable and numbers a_1, a_2, a_3 are such that

$$a_1 \,\mathbf{r}\, ((\neg A_1 \to (\neg A_2 \vee \neg A_3)) \to (\neg A_2 \vee \neg A_3)),$$
$$a_2 \,\mathbf{r}\, ((\neg A_2 \to (\neg A_3 \vee \neg A_1)) \to (\neg A_3 \vee \neg A_1)),$$
$$a_3 \,\mathbf{r}\, ((\neg A_3 \to (\neg A_1 \vee \neg A_2)) \to (\neg A_1 \vee \neg A_2)).$$

For any x and $i = 1, 2, 3$, let $f_i(x) = (\{a_i\}(x))_0$ and $\alpha(x)$ be the Gödel number of the constant function $g(y) = 2^{(\{x\}(x))_0} \cdot 3^0$. Then there exist x and $i \in \{1, 2, 3\}$ such that either $(\{x\}(x))_0 = 0$ and $f_i(\alpha(x)) = 1$ or $(\{x\}(x))_0 = 1$ and $f_i(\alpha(x)) = 0$.

Proof It follows from realizability of the formula (8.5) that exactly two of the formulas $\neg \mathrm{A}_1, \neg \mathrm{A}_2, \neg \mathrm{A}_3$ are realizable. As the situation is absolutely symmetric, let us consider the case when $\neg \mathrm{A}_1$ and $\neg \mathrm{A}_2$ are realizable and $\neg \mathrm{A}_3$ is not. Then the formula $\neg \mathrm{A}_3 \to (\neg \mathrm{A}_1 \vee \neg \mathrm{A}_2)$ is realized by every number, thus $\{a_3\}$ is a total recursive function. The function α is also total and recursive. Therefore the function $f(x) = f_3(\alpha(x))$ is total, and we obtain the proposition under the proof as an immediate consequence of Lemma 8.1.

Let a realization of the premise of the formula (8.4) be given. It follows that the formula (8.5) is realizable and numbers a_1, a_2, a_3 as in Lemma 8.2 can be found effectively. Then, by Lemma 8.2, there exist x and $i \in \{1, 2, 3\}$ such that either $(\{x\}(x))_0 = 0$ and $f_i(\alpha(x)) = 1$ or $(\{x\}(x))_0 = 1$ and $f_i(\alpha(x)) = 0$, where f_i and α are the same functions as in Lemma 8.2. Note that the number x can be found effectively. Now we can describe an algorithm which allows to find a realization of the conclusion of the formula (8.4). Obviously, it is enough to indicate a realizable member in the disjunction $\neg \mathrm{A}_1 \vee \neg \mathrm{A}_2 \vee \neg \mathrm{A}_3$. We state that:

- if $i = 1$ and $f_1(\alpha(x)) = 0$, then $\neg \mathrm{A}_2$ is realizable;
- if $i = 1$ and $f_1(\alpha(x)) = 1$, then $\neg \mathrm{A}_3$ is realizable;
- if $i = 2$ and $f_2(\alpha(x)) = 0$, then $\neg \mathrm{A}_3$ is realizable;
- if $i = 2$ and $f_2(\alpha(x)) = 1$, then $\neg \mathrm{A}_1$ is realizable;
- if $i = 3$ and $f_3(\alpha(x)) = 0$, then $\neg \mathrm{A}_1$ is realizable;
- if $i = 3$ and $f_3(\alpha(x)) = 1$, then $\neg \mathrm{A}_2$ is realizable.

Consider the first case, namely $f_1(\alpha(x)) = 0$ for some x such that $(\{x\}(x))_0 = 1$. Note that $\{x\}(x)$ is defined and $\alpha(x)$ is the Gödel number of the constant function whose only value is $2^{(\{x\}(x))_0} \cdot 3^0$. We have to prove that $\neg \mathrm{A}_2$ is realizable. Assume that $\neg \mathrm{A}_2$ is not realizable. Then $\neg \mathrm{A}_1$ and $\neg \mathrm{A}_3$ are realizable. In this case, the number $\alpha(x)$ is a realization of the formula $\neg \mathrm{A}_1 \to (\neg \mathrm{A}_2 \vee \neg \mathrm{A}_3)$. Indeed, every number a realizes $\neg \mathrm{A}_1$. Note that $\{\alpha(x)\}(a) = 2^{(\{x\}(x))_0} \cdot 3^0 = 2^1 \cdot 3^0$ and in fact, this number realizes the formula $\neg \mathrm{A}_2 \vee \neg \mathrm{A}_3$. Therefore $\{a_1\}(\alpha(x))$ realizes the formula $\neg \mathrm{A}_2 \vee \neg \mathrm{A}_3$ and $(\{a_1\}(\alpha(x))_0 = 1$, that is $f_1(\alpha(x)) = 0$. But this is not the case. The contradiction means that the formula $\neg \mathrm{A}_2$ is realizable. Other cases are considered in the same way.

Thus we have an algorithm for finding a realizable member in the disjunction $\neg \mathrm{A}_1 \vee \neg \mathrm{A}_2 \vee \neg \mathrm{A}_3$ if a realization of the premise of the formula (8.4) is given. This means that the formula (8.4) is realizable. As the realization does not depend on the sentences $\mathrm{A}_1, \mathrm{A}_2, \mathrm{A}_3$, the formula $\mathbf{L}$ is uniformly realizable.

Acknowledgements The reported study was funded by RFBR, project number 20-01-00670.

References

Citkin, A. I. (1978). On structurally complete superintuitionistic logics (Russian). *Doklady Akademii Nauk SSSR, 241*, 40–43. Translation *Soviet Mathematics Doklady, 19*, 816–819.

Heyting, A. (1930). Die formalen Regeln der intuitionistischen Logik. Sitzungsberichte der Preussischen Akademie der Wissenschaften. *Physikalisch-mathematische Klasse*, 42–52.

Jankov, V. A. (1963a). On realizable formulas of propositional logic (Russian). *Doklady Akademii Nauk SSSR, 151*, 1035–1037. Translation *Soviet Mathematics Doklady, 4*, 1146–1148.

Jankov, V. A. (1963b) On the connection between deducibility in the intuitionistic propositional calculus and finite implicative structures (Russian). *Doklady Akademii Nauk SSSR, 151*, 1293–1294. Translation *Soviet Mathematics Doklady, 4*, 1203–1204.

Jankov, V. A. (1969). Conjunctively indecomposable formulas in propositional calculi (Russian). *Izvestiya Akademii Nauk SSSR. Seriya Matematicheskaya, 33*, 18–38. Translation *Mathematics of the USSR-Izvestiya, 3*, 17–35.

Jaśkowski, S. (1936). Recherches sur le système de la logique intuitioniste. Actes du Congrès International de Philosophie Scientifique, VI, Philosophie des mathématique. *Actualités scientifique et industrielle, 393*, 58–61. Translation *Studia Logica, 34*, 117–120.

Kleene, S. C. (1945). On the interpretation of intuitionistic number theory. *Journal of Symbolic Logic, 10*, 109–124.

Kolmogorov, A. N. (1925). On the tertium non datur principle (Russian). *Matematicheskii sbornik, 32*, 646–667. Translation (van Heijenoort 1967), pp. 414–437.

Kolmogorov, A. N. (1932). Zur Deutung der intuitionistischen Logik. *Mathematische Zeitschrift, 35*, 58–65.

Maksimova, L. L., Skvortsov, D. P., & Shekhtman, V. B. (1979). The impossibility of a finite axiomatization of Medvedev's logic of finitary problems (Russian). *Doklady Akademii Nauk SSSR, 245*, 1051–1054. Translation *Soviet Mathematics Doklady, 20*, 394–398.

Medvedev, Y. T. (1962). Finite problems (Russian). *Doklady Akademii Nauk SSSR, 142*, 1015–1018. Translation *Soviet Mathematics Doklady, 3*, 227–230.

Medvedev, Y. T. (1966). Interpretation of logical formulae by means of finite problems (Russian). *Doklady Akademii Nauk SSSR, 169*, 20–23. Translation *Soviet Mathematics Doklady, 7*, 857–860.

Nelson, D. (1947). Recursive functions and intuitionistic number theory. *Transactions of the American Mathematical Society, 61*, 307–368.

Plisko, V. (1973). On realizable predicate formulae (Russian). *Doklady Akademii Nauk SSSR, 212*, 553–556. Translation *Soviet Mathematics Doklady, 14*, 1420–1424.

Plisko, V. (2009). A survey of propositional realizability logic. *The Bulletin of Symbolic Logic, 15*, 1–42.

Rasiowa, H., & Sikorski, R. (1963). *The mathematics of metamathematics*. Warszawa: Państwowe Wydawnictwo Naukowe.

Rose, G. F. (1953). Propositional calculus and realizability. *Transactions of the American Mathematical Society, 75*, 1–19.

Skvortsov, D. P. (1995). A comparison of the deductive power of realizable sentential formulas (Russian). *Logicheskie issledovania (Logical Investigations), 3*, 38–52.

van Heijenoort, J. (Ed.). (1967). *From Frege to Gödel: A source book in mathematical logic, 1879–1931*. Cambridge: Harvard University Press.

Chapter 9
A Note on Disjunction and Existence Properties in Predicate Extensions of Intuitionistic Logic—An Application of Jankov Formulas to Predicate Logics

Nobu-Yuki Suzuki

Abstract Predicate extensions of intuitionistic logic (PEI's) are intermediate predicate logics having the same propositional part as intuitionistic logic. Intuitively, PEI's must resemble intuitionistic logic. We discuss PEI's from the viewpoint of disjunction property (DP) and existence property (EP). Note that DP and EP are regarded as "hallmarks" of constructivity of intuitionistic logic. There are, however, uncountably many PEI's having both of DP and EP. Moreover, there are two continua of PEI's: (1) each of which lacks both of DP and EP, and (2) each of which has EP but lacks DP. Now, a natural question arises: *Do there exist uncountably many PEI's each of which has DP and lacks EP?* We answer this question affirmatively. Specifically, we construct uncountably many such PEI's by making use of modified Jankov formulas. This result suggests that although PEI's are living near to intuitionistic logic, the diversity of their nature seems rich. In other words, logics among PEI's are fascinating from the logical point of view and yet to be explored.

Keywords Disjunction property · Existence property · Intermediate logics · Jankov formula

2020 Mathematics Subject Classification: Primary: 03B55, Secondary: 03F50

9.1 Introduction

Predicate extensions of intuitionistic logic (PEI's) are intermediate predicate logics having the same propositional part as intuitionistic logic. Intuitively, PEI's must resemble intuitionistic logic. We discuss PEI's from the viewpoint of disjunction

N.-Y. Suzuki (✉)
Faculty of Science, Department of Mathematics, Shizuoka University, Ohya 836, Suruga-ku, Shizuoka 422-8529, Japan
e-mail: suzuki.nobuyuki@shizuoka.ac.jp

A. Citkin and I. M. Vandoulakis (eds.), *V. A. Yankov on Non-Classical Logics, History and Philosophy of Mathematics*, Outstanding Contributions to Logic 24,
https://doi.org/10.1007/978-3-031-06843-0_9

property (DP) and existence property (EP). Note that DP and EP are regarded as distinguishing characteristics and features of constructivity of intuitionistic logic. However, Suzuki (1999) constructed a continuum of PEI's having both of EP and DP. There exits a continuum of PEI's without both of EP and DP as well. In 1983, Minari (1986) and Nakamura (1983) independently proved that some well-known PEI's have DP and fail to have EP. Recently, in Suzuki (2021), a continuum of PEI's having EP and lacking DP was constructed.[1]

Now, a natural question arises: *Do there exist uncountably many PEI's each of which has DP and lacks EP?* We answer this question affirmatively. Specifically, we construct uncountably many such PEI's by giving a recursively enumerable sequence of concrete predicate axiom schemata. These axiom schemata are obtained by modifying the Jankov formulas (Jankov 1963, 1968, 1969).

Jankov created an invaluable research tool for the study of non-classical propositional logics[2]; the Jankov formulas provide us with a connection between algebraic property of Heyting algebras and inclusion relation among propositional logics. In this paper, we give an application of Jankov's tool to non-classical *predicate* logics. Since Jankov's method deals with propositional logics, its straightforward application to predicate logics inevitably yields logics having their propositional parts differing from intuitionistic logic. We introduce our formulas with an appropriate modification of Jankov's to keep them having intuitionistic propositional part.

Accordingly, to show our main result, we prove three Lemmata 9.2, 9.6, and 9.9; Lemma 9.2 states that our modified Jankov formulas yield PEI's lacking EP; Lemma 9.6 states that they yield PEI's having DP; from Lemma 9.9, it holds that we can generate uncountably many PEI's by using them. We show Lemma 9.2 by making use of algebraic semantics. Our idea for the proof of Lemma 9.6 comes from the above-mentioned idea of Minari (1986) and Nakamura (1983) based on Kripke frame semantics. Lemma 9.9 is proved by *algebraic Kripke sheaf semantics* introduced in Suzuki (1999).

We assume readers' some familiarity with Heyting algebras and Kripke frames. To make this paper rather self-contained, we briefly explain some notions and definitions on these semantical tools needed in this paper. Algebraic Kripke sheaves are semantical framework obtained from integrating algebraic semantics into Kripke semantics. Since general algebraic Kripke sheaves are (as of now) not so simple to handle, we introduce restricted algebraic Kripke sheaves, called Ω-*brooms*, and use them with a result in Suzuki (1999) for the proof of our main result.

In Sect. 9.2, brief explanation of intermediate (propositional and predicate) logics and some related definitions as well as DP and EP are given. In Sect. 9.3, we introduce Jankov formulas and modified Jankov formulas. Here, we prove that modified ones as axiom schemata yield PEI's without EP (Lemma 9.2) . In Sects. 9.4 and 9.5, we prove that these axiom schemata enjoy DP (Lemma 9.6), and that they generate a

[1] Thus, DP and EP in intermediate predicate logics were proved to be independent. This result contrasts with Friedman (1975) and Friedman and Sheard (1989).

[2] His tool have been, and is being, extended to many propositional logics variously. See e.g., Citkin (2018). The reader will find recent development there.

continuum of PEI's (Lemma 9.9), and we complete the proof of the main result (Theorem 9.4). In Sect. 9.6, we make some concluding remarks.

9.2 Preliminaries

Intermediate logics are logics falling intermediate between intuitionistic and classical logics. There are two types of intermediate logics: intermediate *propositional* logics and intermediate *predicate* logics. We refer readers to Ono (1987) for an information source.

We use a *pure* first-order language $\mathcal{L}$. Logical symbols of $\mathcal{L}$ are propositional connectives: $\vee$, $\wedge$, $\supset$, and $\neg$ (disjunction, conjunction, implication, and negation, respectively), and quantifiers: $\exists$ and $\forall$ (existential and universal quantifiers, respectively). $\mathcal{L}$ has a denumerable list of individual variables and a denumerable list of m-ary predicate variables for each $m < \omega$. All 0-ary predicate variables are identified with propositional variables; thus, the propositional language $\mathcal{L}_{proposition}$ is contained in $\mathcal{L}$. Note that $\mathcal{L}$ contains neither individual constants nor function symbols.

The idea of introducing intermediate logics is the identification of each logic and the set of formulas provable in it. For example, intuitionistic propositional logic $\mathbf{H}$ and intuitionistic predicate logic $\mathbf{H}_*$ are identified with the sets of formulas provable in $\mathbf{H}$ and $\mathbf{H}_*$, respectively. Also, classical propositional and predicate logics, $\mathbf{C}$ and $\mathbf{C}_*$, are treated in the same way.

Definition 9.1 A set $\mathbf{J}$ of formulas of propositional language $\mathcal{L}_{proposition}$ is said to be an *intermediate propositional logic*, if $\mathbf{J}$ satisfies the conditions: (P1) $\mathbf{H} \subseteq \mathbf{J} \subseteq \mathbf{C}$ and (P2) $\mathbf{J}$ is closed under the rule of modus ponens (from A and $A \supset B$, infer B) and uniform substitution for propositional variable.

A set $\mathbf{J}$ of formulas of $\mathcal{L}_{proposition}$ is said to be a *super-intuitionistic propositional logic*, if $\mathbf{J}$ satisfies (P1') $\mathbf{H} \subseteq \mathbf{J}$ and (P2). Let Ψ_0 be the set of all propositional formulas. The Ψ_0 is the only super-intuitionistic propositional logic that is not an intermediate propositional logic.

Definition 9.2 A set $\mathbf{L}$ of formulas of $\mathcal{L}$ is said to be an *intermediate predicate logic*, if $\mathbf{L}$ satisfies the three conditions: (Q1) $\mathbf{H}_* \subseteq \mathbf{L} \subseteq \mathbf{C}_*$ and (Q2) $\mathbf{L}$ is closed under the rule of modus ponens, the rule of generalization (from A, infer $\forall x A$), and uniform substitution[3] for predicate variable.

A set $\mathbf{L}$ of formulas of $\mathcal{L}$ is said to be a *super-intuitionistic predicate logic*, if $\mathbf{L}$ satisfies (Q1') $\mathbf{H}_* \subseteq \mathbf{L}$ and (Q2). There are uncountably many superintuitionistic predicate logics that are not intermediate predicate logics.

When $A \in \mathbf{L}$, we sometimes write $\mathbf{L} \vdash A$, and say "A is provable in $\mathbf{L}$." For a logic $\mathbf{L}$ and a set Γ of formulas, the smallest logic containing $\mathbf{L}$ and Γ (as sets) is denoted

[3] *Cf.* the operator $\check{\mathsf{S}}$ in Church (1956).

by $\mathbf{L} + \Gamma$. Let $\mathbf{L}$ be a predicate logic. Then, $\pi(\mathbf{L}) = \mathbf{L} \cap \Psi_0$ is a propositional logic. It is called the *propositional part* of $\mathbf{L}$.

For each propositional logic $\mathbf{J}$, a predicate logic $\mathbf{L}$ is called a *predicate extension* of $\mathbf{J}$, if $\pi(\mathbf{L}) = \mathbf{J}$. A predicate logic $\mathbf{L}$ is said to be a *predicate extension of intuitionistic logic* (a PEI), if $\pi(\mathbf{L}) = \mathbf{H}$.

Definition 9.3 (*cf. Church* 1956; *Sect. 32*) To each predicate variable p, we associate a unique propositional variable $\pi(p)$. For a given formula A of $\mathcal{L}$, we define the *associated formula of the propositional calculus* (afp) by (1) deleting all quantifiers $\forall x$ and $\exists x$ in A and (2) substituting $\pi(p)$ to $p(v_1, \ldots, v_n)$ in A for each predicate variable[4] p occurring in A. The afp of A is denoted by $\pi(A)$.

Proposition 9.1 *Let* $\mathbf{L}$ *be a predicate logic. It holds that* $\pi(\mathbf{H}_* + \Gamma) = \mathbf{H} + \{\pi(A)\ ;\ A \in \Gamma\}$.

Definition 9.4 A logic $\mathbf{L}$ is said to have the *disjunction property* (DP), if for every A and every B, $\mathbf{L} \vdash A \vee B$ implies either $\mathbf{L} \vdash A$ or $\mathbf{L} \vdash B$.

A formula A is said to be *congruent* to a formula B, if A is obtained from B by alphabetic change of bound variables which does not turn any free occurrences of variables newly bound (*cf.* Kleene 1952; p. 153). A predicate logic $\mathbf{L}$ is said to have the *existence property* (EP), if for every $\exists x A(x)$, $\mathbf{L} \vdash \exists x A(x)$ implies that there exist a formula $\widetilde{A}(x)$ which is congruent to $A(x)$ and an individual variable v such that v is free for x in $\widetilde{A}(x)$ and $\mathbf{L} \vdash \widetilde{A}(v)$ (*cf.* Kleene 1962).

Formulas congruent to a formula $A(x)$ are intuitionistically equivalent to each other. They are usually written by the same symbol $A(x)$ for the sake of simplicity (*cf.* Gabbay et al. 2009; Sect. 2.3).

Definition 9.5 (*cf. Jankov* 1968) A sequence $\{\mathbf{L}_i\}_{i<\omega}$ of logics is said to be *strongly independent*, if $\mathbf{L}_i \not\subseteq \bigcup_{j \neq i} \mathbf{L}_j$ for each $i < \omega$, where $\bigcup_{j \neq i} \mathbf{L}_j$ is the smallest logic containing all $\mathbf{L}_j$ $(j \neq i)$.

Proposition 9.2 *Let* $\{\mathbf{L}_i\}_{i<\omega}$ *be a strongly independent sequence of logics.*

(1) *For every* $I, J \subseteq \omega$, $I = J$ *if and only if* $\bigcup_{i \in I} \mathbf{L}_i = \bigcup_{i \in J} \mathbf{L}_i$.

(2) *The set* $\{\bigcup_{i \in I} \mathbf{L}_i\ ;\ I < \omega\}$ *has the cardinality* 2^{ω}.

Proof It suffices to show that $I \neq J$ implies $\bigcup_{i \in I} \mathbf{L}_i \neq \bigcup_{i \in J} \mathbf{L}_i$. Suppose $I \neq J$. Without loss of generality, we may assume that there exists a $k \in I \setminus J$. It is obvious that $\bigcup_{i \in J} \mathbf{L}_i \subseteq \bigcup_{i \neq k} \mathbf{L}_i$. By the assumption, we have $\mathbf{L}_k \not\subseteq \bigcup_{i \neq k} \mathbf{L}_i$. Thus, $\mathbf{L}_k \not\subseteq \bigcup_{i \in J} \mathbf{L}_i$. Therefore, we have $\bigcup_{i \in I} \mathbf{L}_i \neq \bigcup_{i \in J} \mathbf{L}_i$. □

For a sequence $\{X_i\}_{i<\omega}$ of formulas, we can define a sequence $\{\mathbf{H}_* + X_i\}_{i<\omega}$ of logics. If $\{\mathbf{H}_* + X_i\}_{i<\omega}$ is strongly independent, we say that $\{X_i\}_{i<\omega}$ is *strongly independent* .

---

[4] In Church (1956), predicate variables are called functional variables.

9.3 Modified Jankov Formulas—Learning Jankov's Technique

In this section, we briefly explain Jankov formulas of finite subdirectly irreducible Heyting algebras. Then, we introduce a variant of Jankov formulas modified to achieve our aim. We show that these modified Jankov formulas as axiom schemata generate PEI's without EP.

9.3.1 Heyting Algebras and Jankov Formulas

Let $\mathbf{A}$ be a Heyting algebra. In what follows, we denote basic operations of $\mathbf{A}$ by: $\cup_{\mathbf{A}}$ (join), $\cap_{\mathbf{A}}$ (meet), $\neg_{\mathbf{A}}$ (pseudo-complementation), and $\rightarrow_{\mathbf{A}}$ (relative pseudo-complementation). We use the same letter $\mathbf{A}$ to denote its underlying set. The partial order determined by the lattice structure of $\mathbf{A}$ is denoted by $\leq_{\mathbf{A}}$. Also, $1_{\mathbf{A}}$ and $0_{\mathbf{A}}$ are the greatest and least element of $\mathbf{A}$. We sometimes omit the subscript $\mathbf{A}$. The two-element Boolean algebra is denoted by $\mathbf{2}$ $(= \{1_{\mathbf{2}}, 0_{\mathbf{2}}\})$.

Definition 9.6 A Heyting algebra $\mathbf{A}$ is said to be *subdirectly irreducible*, if $\mathbf{A} \setminus \{1_{\mathbf{A}}\}$ has the greatest element. This element is denoted by $\star_{\mathbf{A}}$.

Example 9.1 A non-empty partially ordered set $\mathbf{M} = (\mathbf{M}, \leq_{\mathbf{M}})$ is said to be a *Kripke base*, if it has the least element $0_{\mathbf{M}}$. A subset $S \subseteq \mathbf{M}$ is said to be *open*, if S is upward-closed (i.e., for every $x \in S$ and every $y \in \mathbf{M}$, $x \leq_{\mathbf{M}} y$ implies $y \in S$). Then the set $O(\mathbf{M})$ of all open subsets of $\mathbf{M}$ is a subdirectly irreducible Heyting algebra with respect to the set-inclusion as its partial ordering. The second greatest element of $O(\mathbf{M})$ is $\mathbf{M} \setminus \{0_{\mathbf{M}}\}$.

Let $\mathbf{A}$ be a Heyting algebra, PV the set of all propositional variables. A mapping $v : PV \rightarrow \mathbf{A}$ is said to be an *assignment* on $\mathbf{A}$. By the usual induction, we extend the v to the mapping $v : \Psi_0 \rightarrow \mathbf{A}$. A propositional formula C is said to be *valid in* $\mathbf{A}$, if $v(C) = 1_{\mathbf{A}}$ for every assignment v on $\mathbf{A}$. The set of all propositional formulas valid in $\mathbf{A}$ is denoted by $E(\mathbf{A})$.

Proposition 9.3 *For every non-degenerate Heyting algebra* $\mathbf{A}$, *the set* $E(\mathbf{A})$ *is an intermediate propositional logic.*

Let $\mathbf{A}$ be a finite subdirectly irreducible Heyting algebra. For each $a \in \mathbf{A}$, we can attach a unique propositional variable $p_a \in PV$. The *diagram* $\delta(\mathbf{A})$ of $\mathbf{A}$ is the finite set of propositional formulas defined by:

$$\delta(\mathbf{A}) = \{p_{a\cup_{\mathbf{A}}b} \supset (p_a \vee p_b), (p_a \vee p_b) \supset p_{a\cup_{\mathbf{A}}b} \ ; \ a, b \in \mathbf{A}\}$$
$$\bigcup\{p_{a\cap_{\mathbf{A}}b} \supset (p_a \wedge p_b), (p_a \wedge p_b) \supset p_{a\cap_{\mathbf{A}}b} \ ; \ a, b \in \mathbf{A}\}$$
$$\bigcup\{p_{a\to_{\mathbf{A}}b} \supset (p_a \supset p_b), (p_a \supset p_b) \supset p_{a\to_{\mathbf{A}}b} \ ; \ a, b \in \mathbf{A}\}$$
$$\bigcup\{p_{a\to_{\mathbf{A}}0_{\mathbf{A}}} \supset (\neg p_a), (\neg p_a) \supset p_{a\to_{\mathbf{A}}0_{\mathbf{A}}} \ ; \ a \in \mathbf{A}\} \ .$$

The *Jankov formula* $J(\mathbf{A})$ of $\mathbf{A}$ is the propositional formula defined by:

$$J(\mathbf{A}) \ : \ \big(\bigwedge \delta(\mathbf{A})\big) \supset p_{\star_{\mathbf{A}}},$$

where $\bigwedge \delta(\mathbf{A})$ is the conjunction of all formulas in $\delta(\mathbf{A})$. Then it is easy to see that $J(\mathbf{A})$ is not valid in $\mathbf{A}$ by taking the assignment $v_{\mathbf{A}}$: $p_a \mapsto a$ for each $a \in \mathbf{A}$. Since $v_{\mathbf{A}}(\bigwedge \delta(\mathbf{A})) = 1_{\mathbf{A}}$, we have $J(\mathbf{A}) \notin E(\mathbf{A})$. Moreover, we have the following prominent result due to Jankov (1963), which provide us with a connection between validity of Jankov formula and algebraic property.

Lemma 9.1 (cf. Jankov 1963; 1968; 1969) *Let* $\mathbf{A}$ *be a finite subdirectly irreducible Heyting algebra. If* $J(\mathbf{A})$ *is not valid in a Heyting algebra* $\mathbf{B}$, *then there exists a quotient algebra* $\mathbf{B}'$ *of* $\mathbf{B}$ *such that* $\mathbf{A}$ *is embeddable into* $\mathbf{B}'$.

For further discussion, we need a denumerable sequence $\{\mathbf{A}_i\}_{i<\omega}$ satisfying;

(A1) for each $i < \omega$, $\mathbf{A}_i$ is a finite subdirectly irreducible Heyting algebra;
(A2) for every $i, j < \omega$, if $i \neq j$, then $\mathbf{A}_i$ is not embeddable into any quotient algebra of $\mathbf{A}_j$.

In fact, Jankov (1968) and Wroński (1974) constructed concrete sequences of Heyting algebras with the above properties. Let us fix one of these sequence. By the virtue of their construction, we have the following by Lemma 9.1.

Corollary 9.1 (cf. Jankov 1968 and Wroński 1974) $\{J(\mathbf{A}_i)\}_{i<\omega}$ *is strongly independent.*

Proof Define $\mathbf{L}_i = \mathbf{H}_* + J(\mathbf{A}_i)$ for each $i < \omega$. Pick an arbitrary $i_0 \in \omega$. Then, for every $j \neq i_0$, $\mathbf{A}_j$ is not embeddable into any quotient algebra of $\mathbf{A}_{i_0}$. Thus, it holds that $J(A_j) \in E(\mathbf{A}_{i_0})$ for every $j \neq i_0$. Therefore, we have $\bigcup_{j\neq i_0} \mathbf{L}_j \subseteq E(\mathbf{A}_{i_0}) \not\ni J(\mathbf{A}_{i_0}) \in \mathbf{L}_{i_0}$. Hence, $\mathbf{L}_{i_0} \nsubseteq \bigcup_{j\neq i_0} \mathbf{L}_j$. □

Thus, by putting $\mathbf{J}(I) = \bigcup_{i\in I} \mathbf{L}_i$ for each $I \subset \omega$, we have a continuum $\{\mathbf{J}(I) \ ; \ J \subset \omega\}$ of logics by Proposition 9.2. Note that no logic in this continuum has the propositional part being identical to intuitionistic logic. We must modify the original $J(\mathbf{A})$ so as to achieve our aim of this paper.

9.3.2 Modified Jankov Formulas for PEI's Without EP

In this subsection, we introduce *modified Jankov formulas*. The idea of our modification comes from observation of behavior of the sentence the sentence F: $\exists x(p(x) \supset \forall y p(y))$, where p is a unary predicate variable. Clearly, F is provable in classical predicate logic $\mathbf{C}_*$, but $p(v) \supset \forall y p(y)$ is not so for every individual variable v. Thus, this F is a typical counterexample to EP of $\mathbf{C}_*$. Note that the afp $\pi(F)$ of F is $p \supset p$, and hence the propositional part $\pi(\mathbf{H}_* + F)$ of $\mathbf{H}_* + F$ equals $\mathbf{H}$ by Proposition 9.1. This F also gives a counterexample to EP of $\mathbf{H}_* + F$ that is a PEI. Moreover, Minari (1986) and Nakamura (1983) independently proved that $\mathbf{H}_* + F$ has DP, and hence they showed that $\mathbf{H}_* + F$ is a PEI having DP and lacking EP. Our modified Jankov formulas play the same role as F and have a property similar to that of the original Jankov formula shown in Lemma 9.1 (see Lemma 9.12).

Let $\mathbf{A}$ be a finite subdirectly irreducible Heyting algebra, $J(\mathbf{A})$ the Jankov formula of $\mathbf{A}$. Pick a fresh individual variable v. Let $\Delta(\mathbf{A})$ be the finite set of sentences obtained from $\delta(\mathbf{A})$ by replacing all occurrences of $p_{\star_\mathbf{A}}$ by F. Define a formula $PJ(\mathbf{A})(v)$ and a sentence $QJ(\mathbf{A})$ by:

$$PJ(\mathbf{A})(v) \;:\; \bigwedge \Delta(\mathbf{A}) \supset (p(v) \supset \forall y p(y)) \ ,$$
$$QJ(\mathbf{A}) \;:\; \exists v PJ(\mathbf{A})(v) \ .$$

Jankov (1968) and Wroński (1974) constructed a concrete sequence $\{\mathbf{H}_i\}_{i<\omega}$ of finite subdirectly irreducible Heyting algebras satisfying the conditions (A1) and (A2) in Sect. 9.3.1 together with the following (A3):

(A3) for each $i < \omega$, there are exactly three elements in $\mathbf{H}_i$ having no incomparable element (i.e., 0, 1 and $\star$).

Let us fix one of their sequences. Then, we can construct $QJ(\mathbf{H}_i)$ $(i < \omega)$ one by one concretely and in a recursively enumerable manner. To achieve our main aim, we use this sequence of Heyting algebras and show that $\{QJ(\mathbf{H}_i)\}_{i<\omega}$ satisfies the following three conditions:

- (*cf.* Lemma 9.2) for every $I \subseteq \omega$, $\mathbf{H}_* + \{QJ(\mathbf{H}_i)\ ;\ i \in I\}$ is a PEI lacking EP,
- (*cf.* Lemma 9.6) for every $I \subseteq \omega$, $\mathbf{H}_* + \{QJ(\mathbf{H}_i)\ ;\ i \in I\}$ has DP.
- (Lemma 9.9) $\{QJ(\mathbf{H}_i)\}_{i<\omega}$ is strongly independent.

In the rest of this subsection, we show that modified Jankov formulas axiomatize PEI's lacking EP. Specifically, we show the

Lemma 9.2 *Let $\mathcal{S}$ be a set of finite non-degenerate subdirectly irreducible Heyting algebras having at least three elements. Then, $\mathbf{H}_* + \{QJ(\mathbf{A})\ ;\ \mathbf{A} \in \mathcal{S}\}$ is a PEI lacking EP.*

Note that $QJ(\mathbf{A})$ is provable in $\mathbf{C}_*$. It is clear that $\pi(QJ(\mathbf{A})) = \pi(\bigwedge \Delta(\mathbf{A})) \supset (p \supset p)$ is provable in intuitionistic logic $\mathbf{H}$. Hence, for every set $\mathcal{S}$ of finite subdirectly irreducible Heyting algebras, the intermediate predicate logic $\mathbf{H}_* + \{QJ(\mathbf{A})\ ;\ \mathbf{A} \in \mathcal{S}\}$ is a PEI. It suffices to show the

Lemma 9.3 *Let* $\mathbf{A}$ *be a finite subdirectly irreducible Heyting algebra having at least three elements. Then,* $PJ(\mathbf{A})(v)$ *is not provable in* $\mathbf{C}_*$.

We introduce algebraic semantics for predicate logics, and some definitions and propositions on Heyting algebras without proofs, and show this Lemma.

Definition 9.7 For each non-empty set U, the language obtained from $\mathcal{L}$ by adding the name $\overline{u}$ for each $u \in U$ is denoted by $\mathcal{L}[U]$. In what follows, we use the same letter u for the name $\overline{u}$ of u, when no confusion can arise. We sometimes identify $\mathcal{L}[U]$ with the set of all sentences of $\mathcal{L}[U]$.

A Heyting algebra $\mathbf{A}$ is said to be κ-*complete* for some cardinal κ, if both of the supremum $\bigcup S$ and the infimum $\bigcap S$ exist in $\mathbf{A}$ for every subset S of $\mathbf{A}$ having the cardinality at most κ. A pair $\mathcal{A} = (\mathbf{A}, U)$ of a non-degenerate $|U|$-complete Heyting algebra $\mathbf{A}$ and a non-empty set U is said to be an *algebraic frame*, where $|U|$ is the cardinality of U.

A mapping V of the set of all atomic sentences of $\mathcal{L}[U]$ to $\mathbf{A}$ is said to be an *assignment* on $\mathcal{A}$. We extend V to a mapping of $\mathcal{L}[U]$ to $\mathbf{A}$ inductively as follows:

- $V(A \wedge B) = V(A) \cap V(B)$,
- $V(A \vee B) = V(A) \cup V(B)$,
- $V(A \supset B) = V(A) \to V(B)$,
- $V(\neg A) = V(A) \to 0$,
- $V(\forall x A(x)) = \bigcap_{u \in U} V(A(\overline{u}))$,
- $V(\exists x A(x)) = \bigcup_{u \in U} V(A(\overline{u}))$.

Since $\mathbf{A}$ is κ-complete, the right hand sides of the last two equalities are well-defined. A pair $(\mathcal{A}, V)$ of an algebraic frame $\mathcal{A}$ and an assignment V is said to be an *algebraic model*. A formula A of $\mathcal{L}$ is said to be *true* in an algebraic model $(\mathcal{A}, V)$, if $V(\overline{A}) = 1$, where $\overline{A}$ is the universal closure of A. A formula of $\mathcal{L}$ is said to be *valid* in an algebraic frame $\mathcal{A}$, if it is true in $(\mathcal{A}, V)$ for every assignment V on $\mathcal{A}$. The set of formulas of $\mathcal{L}$ valid in $\mathcal{A}$ is denoted by $L(\mathcal{A})$ or $L(\mathbf{A}, U)$.

Proposition 9.4 *For each algebraic frame* $\mathcal{A}$*, the set* $L(\mathcal{A})$ *is a super-intuitionistic predicate logic.*

It is well-known that $\mathbf{C}_* \subseteq L(\mathbf{2}, \{0, 1\})$. To show Lemma 9.3, we construct an appropriate assignment V on $(\mathbf{2}, U)$ for each finite subdirectly irreducible Heyting algebra $\mathbf{A}$ having at least three elements, and show that $V(\overline{PJ(\mathbf{A})(v)}) \neq 1_{\mathbf{2}}$.

Lemma 9.4 *Let* $\mathbf{A}$ *be a finite subdirectly irreducible Heyting algebra having at least three elements. There exists a propositional assignment* μ *on* $\mathbf{2}$ *such that* $\mu(\bigwedge \delta(\mathbf{A})) = \mu(p_{\star_{\mathbf{A}}}) = 1_{\mathbf{2}}$.

Proof Take an assignment v such that $v(p_a) = a$ for every $a \in \mathbf{A}$. Then, $v(\bigwedge \delta(\mathbf{A})) = 1$ and $v(p_{\star_\mathbf{A}}) = \star_\mathbf{A}$. The set $\{a \in \mathbf{A}\ ;\ a = \neg_\mathbf{A}\neg_\mathbf{A} a\}$ forms a Boolean algebra with respect to the restriction of $\leq_\mathbf{A}$ to this set. We denote this Boolean algebra by $\mathbf{A}^{\neg\neg}$. Since $\mathbf{A}$ is non-degenerate, $\mathbf{A}^{\neg\neg}$ is non-degenerate. Let $\neg\neg$ be the mapping of $\mathbf{A}$ to $\mathbf{A}^{\neg\neg}$ defined by $\neg\neg(a) = \neg_\mathbf{A}\neg_\mathbf{A} a$ for every $a \in \mathbf{A}$. Then, $\neg\neg$ is a Heyting homomorphism. We have: $\neg\neg \circ v(\bigwedge \delta(\mathbf{A})) = \neg\neg \circ v(p_{\star_\mathbf{A}}) = 1_{\mathbf{A}^{\neg\neg}}$. Since, $\mathbf{A}^{\neg\neg}$ is non-degenerate, there exists an ultrafilter $\mathcal{U}$ on this Boolean algebra such that $\mathbf{A}^{\neg\neg}/\mathcal{U} \simeq \mathbf{2}$. Let ρ be the canonical projection of $\mathbf{A}^{\neg\neg}$ onto $\mathbf{2}$. Then, we have: $\rho \circ \neg\neg \circ v(\bigwedge \delta(\mathbf{A})) = \rho \circ \neg\neg \circ v(p_{\star_\mathbf{A}}) = 1_2$. Putting $\mu = \rho \circ \neg\neg \circ v$, we have the conclusion. □

Taking the assignment μ in Lemma 9.4, we define an assignment V on $(\mathbf{2}, \{0, 1\})$ by:

$$V(A) = \begin{cases} \mu(a) & \text{if } A \text{ is } p_a \text{ for some } a \in \mathbf{A}, \\ 1_2 & \text{if } A \text{ is } p(\overline{1}), \\ 0_2 & \text{otherwise.}, \end{cases}$$

for each atomic sentence A of $\mathcal{L}[U]$. It is easy to check that $V(F) = 1_2$. Then, we have the

Lemma 9.5 *Let X be a propositional formula having no propositional variable other than $\{p_a\ ;\ a \in \mathbf{A}\}$. By X', we denote the formula obtained from X by replacing all occurrences of $p_{\star_\mathbf{A}}$ by the sentence F. Then, we have $V(X') = \mu(X)$.*

Proof We can show this Lemma by induction on the length of X. Since $\rho \circ \neg\neg$ is a Heyting-homomorphism, it suffices to check the Basis-part. But it is obvious by the fact that $V(F) = \mu(p_{\star_\mathbf{A}}) = 1_2$. □

Note that $a = v_\mathbf{A}(p_a)$ for each $a \in \mathbf{A}$. Now, we show Lemma 9.3. By Lemma 9.5, we have $V(\bigwedge \Delta(\mathbf{A})) = \mu(\bigwedge \delta(\mathbf{A})) = 1_2$. Thus we have $V(PJ(\mathbf{A})(\overline{1})) = V(\bigwedge \Delta(\mathbf{A})) \to_2 (V(p(\overline{1})) \to_2 V(\forall y p(y))) = 1_2 \to_2 (1_2 \to_2 0_2) = 0_2$. Therefore, we have $V(PJ(\mathbf{A})(v)) = 0_2 \neq 1_2$, i.e., $PJ(\mathbf{A})(v) \notin L(\mathbf{2}, \{0, 1, \}) \supseteq \mathbf{C}_*$. This completes the proofs of Lemmas 9.3 and 9.2.

9.4 Modified Jankov Formulas Preserve DP—Learning Minari's and Nakamura's Idea

In this section, we show that the modified Jankov formulas as axiom schemata preserve DP. More specifically, we show the

Lemma 9.6 *Let $\mathcal{S}$ be a set of finite non-degenerate subdirectly irreducible Heyting algebras. Then, $\mathbf{H}_* + \{QJ(\mathbf{A})\ ;\ \mathbf{A} \in \mathcal{S}\}$ has DP.*

We show this Lemma by making use of Kripke frame semantics. In Sect. 9.4.1, we introduce Kripke frame semantics for predicate logics. Next, in Sect. 9.4.2, a technique is given in a simplified form . A part of the result in Minari (1986) and Nakamura (1983) is presented to illustrate this technique.

9.4.1 Kripke Frame Semantics

Recall that a partially ordered set $\mathbf{M} = (\mathbf{M}, \leq_{\mathbf{M}})$ with the least element $0_{\mathbf{M}}$ is said to be a *Kripke base*. For example,

Example 9.2 The set $\mathcal{P}(\mathbf{A})$ of all prime filters of a subdirectly irreducible Heyting algebra $\mathbf{A}$ together with its set-inclusion relation forms a Kripke base with the least element $\{1_{\mathbf{A}}\}$.

Definition 9.8 Let S be a non-empty set. A mapping D of a Kripke base $\mathbf{M}$ to 2^S is called a *domain* over $\mathbf{M}$, if $\varnothing \neq D(a) \subseteq D(b)$ for all $a, b \in M$ with $a \leq b$. A pair $\mathcal{K} = \langle \mathbf{M}, D \rangle$ of a Kripke base $\mathbf{M}$ and a domain D over $\mathbf{M}$ is called a *Kripke frame*.

Intuitively, each $D(a)$ is the individual domain of the world $a \in \mathbf{M}$. For each $a \in \mathbf{M}$ and each $b \in \mathbf{M}$ with $a \leq b$, every sentence of $\mathcal{L}[D(a)]$ is also a sentence of $\mathcal{L}[D(b)]$. A binary relation $\models$ between each $a \in \mathbf{M}$ and each atomic sentence of $\mathcal{L}[D(a)]$ is said to be a *valuation* on $\mathcal{K} = \langle \mathbf{M}, D \rangle$, if for every $a, b \in \mathbf{M}$ and every atomic sentence A of $\mathcal{L}[D(a)]$, $a \models A$ and $a \leq b$ imply $b \models A$. We extend $\models$ to the relation between each $a \in \mathbf{M}$ and each sentence of $\mathcal{L}[D(a)]$ inductively as follows:

- $a \models A \wedge B$ if and only if $a \models A$ and $a \models B$,
- $a \models A \vee B$ if and only if $a \models A$ or $a \models B$,
- $a \models A \supset B$ if and only if for every $b \in \mathbf{M}$ with $a \leq b$, either $b \not\models A$ or $b \models B$,
- $a \models \neg A$ if and only if for every $b \in \mathbf{M}$ with $a \leq b$, $b \not\models A$,
- $a \models \forall x A(x)$ if and only if for every $b \in \mathbf{M}$ with $a \leq b$ and every $u \in D(b)$, $b \models A(\overline{u})$,
- $a \models \exists x A(x)$ if and only if there exists $u \in D(a)$ such that $a \models A(\overline{u})$.

A pair $(\mathcal{K}, \models)$ of a Kripke frame $\mathcal{K}$ and a valuation $\models$ on $\mathcal{K}$ is said to be a *Kripke-frame model*. A formula A of $\mathcal{L}$ is said to be *true* in a Kripke-frame model $(\mathcal{K}, \models)$, if $0_{\mathbf{M}} \models \overline{A}$. A formula of $\mathcal{L}$ is said to be *valid* in a Kripke frame $\mathcal{K}$, if it is true in $(\mathcal{K}, \models)$ for every valuation $\models$ on $\mathcal{K}$. The set of formulas of $\mathcal{L}$ that are valid in $\mathcal{K}$ is denoted by $L(\mathcal{K})$. The following propositions are fundamental properties of Kripke semantics.

Proposition 9.5 *For every Kripke-frame model $(\langle \mathbf{M}, D \rangle, \models)$, every $a, b \in \mathbf{M}$, and every sentence $A \in \mathcal{L}[D(a)]$, if $a \models A$ and $a \leq b$, then $b \models A$.*

Proposition 9.6 *For each Kripke frame $\mathcal{K}$, the set $L(\mathcal{K})$ is a super-intuitionistic predicate logic.*

It is well-known that $\mathbf{H}_*$ is strongly complete with respect to Kripke frame semantics. That is,

Theorem 9.1 *Let Γ be a set of sentences of $\mathcal{L}$. If a formula $A(v_1, \ldots, v_n)$ of $\mathcal{L}$ having no free variables other than $v_1, \ldots, v_n$ is not provable from Γ in $\mathbf{H}_*$, then there exist a Kripke-frame model $(\langle \mathbf{M}, D \rangle, \models)$ and elements $d_1, \ldots, d_n \in D(0)$, where 0 is the least element of $\mathbf{M}$, such that (1) $0 \models S$ for every $S \in \Gamma$ and (2) $0 \not\models A(d_1, \ldots, d_n)$.*

9.4.2 *Pointed Joins of Kripke-Frame Models*

Let U and V be non-empty sets, $f : U \to V$ a mapping. The f induces the translation $\cdot^f$ from $\mathcal{L}[U]$ to $\mathcal{L}[V]$; for each sentence A of $\mathcal{L}[U]$, the symbol A^f denotes the sentence of $\mathcal{L}[V]$ obtained from A by replacing occurrences of $\overline{u}$ $(u \in U)$ by the name $\overline{f(u)}$ of $f(u)$.

Definition 9.9 Let $\mathcal{K}_1 = \langle \mathbf{M}_1, D_1 \rangle$ and $\mathcal{K}_2 = \langle \mathbf{M}_2, D_2 \rangle$ be Kripke frames with the least elements 0_1 and 0_2, respectively. Take a fresh element 0 and define a Kripke base $\{0\} \uparrow (\mathbf{M}_1 \oplus \mathbf{M}_2)$ as the partially ordered set obtained from the disjoint union $\mathbf{M}_1 \oplus \mathbf{M}_2$ of $\mathbf{M}_1$ and $\mathbf{M}_2$ by adding 0 as the new least element. Then, we define a Kripke frame $0 \uparrow (\mathcal{K}_1 \oplus \mathcal{K}_2)$ on $\{0\} \uparrow (\mathbf{M}_1 \oplus \mathbf{M}_2)$ by associating the domain $D^\uparrow$:

$$D^\uparrow(a) = \begin{cases} D_1(0_1) \times D_2(0_2) & \text{if } a = 0, \\ D_1(a) \times D_2(0_2) & \text{if } a \in \mathbf{M}_1, \\ D_1(0_1) \times D_2(a) & \text{if } a \in \mathbf{M}_2, \end{cases}$$

where $U \times V$ denotes the Cartesian product of U and V. The Kripke frame $0 \uparrow (\mathcal{K}_1 \oplus \mathcal{K}_2) = (\{0\} \uparrow (\mathbf{M}_1 \oplus \mathbf{M}_2), D^\uparrow)$ is called the *pointed join*[5] of $\mathcal{K}_1$ and $\mathcal{K}_2$.

Let $\pi_i := \{(\pi_i)_a : D^\uparrow(a) \to D_i(a) \,;\, a \in \{0\} \cup \mathbf{M}_i\}$ $(i = 1, 2)$ be families of mappings defined by:

$$(\pi_i)_a((d_1, d_2)) = d_i \quad \text{for } (d_1, d_2) \in D^\uparrow(a) \text{ and } a \in \{0\} \cup \mathbf{M}_i.$$

Observe that π_i induces translations of $\mathcal{L}[D^\uparrow(a)]$ to $\mathcal{L}[D_i(a)]$ $(a \in \mathbf{M})$ or of $\mathcal{L}[D^\uparrow(0)]$ to $\mathcal{L}[D_i(0_i)]$; for every sentence $A \in \mathcal{L}[D^\uparrow(a)]$ (or $A \in \mathcal{L}[D^\uparrow(0)]$), the sentence translated by π_i is denoted simply by A^{π_i}.

Let $\models_1$ and $\models_2$ be valuations on Kripke frames $\mathcal{K}_1 = \langle \mathbf{M}_1, D_1 \rangle$ and $\mathcal{K}_2 = \langle \mathbf{M}_2, D_2 \rangle$, respectively. A Kripke-frame model $(0 \uparrow (\mathcal{K}_1 \oplus \mathcal{K}_2), \models)$ is said to be the *pointed join model* of $(\mathcal{K}_1, \models_1)$ and $(\mathcal{K}_2, \models_2)$, if for each $a \in \{0\} \cup \mathbf{M}_1 \oplus \mathbf{M}_2$ and each atomic sentence $p((d_1^1, d_2^1), \ldots, (d_1^n, d_2^n)) \in \mathcal{L}[D^\uparrow(a)]$,

[5] In Suzuki (2017), a more general definition of pointed joins was introduced for Kripke sheaf models.

$$a \models p((d_1^1, d_2^1), \ldots, (d_1^n, d_2^n))$$
$$\text{if and only if} \begin{cases} a \in \mathbf{M}_1 \text{ and } a \models_1 p(d_1^1, \ldots, d_1^n), \\ \text{or} \\ a \in \mathbf{M}_2 \text{ and } a \models_2 p(d_2^1, \ldots, d_2^n). \end{cases}$$

Then, the following Lemma clearly holds.

Lemma 9.7 *Let $(0 \uparrow (\mathcal{K}_1 \oplus \mathcal{K}_2), \models)$ be the pointed join model of $(\mathcal{K}_1, \models_1)$ and $(\mathcal{K}_2, \models_2)$. For each $i = 1, 2$ and each $a \in \mathbf{M}_i$, it holds that*

$$\textit{for every } A \in \mathcal{L}[D^{\uparrow}(a)], \quad a \models A \ \textit{ if and only if } \ a \models_i A^{\pi_i}.$$

Definition 9.10 A formula A is said to be *axiomatically true* in a Kripke-frame model $(\mathcal{K}, \models)$, if all of the substitution instances of A in the language $\mathcal{L}$ are true in $(\mathcal{K}, \models)$.

A formula A is said to be *pointed-join robust*, if A is true in Kripke-frame models $(\mathcal{K}_1, \models_1)$ and $(\mathcal{K}_2, \models_2)$, then A is true in the pointed join model of them.

If the axiomatic truth of a formula A is preserved under the pointed-join construction of two Kripke models, then $\mathbf{H}_* + A$ has DP. More precisely,

Theorem 9.2 (cf. Minari 1986 and Nakamura 1983) *Let A be a formula of $\mathcal{L}$ satisfying:*

$(*)$ *every substitution instance of A is pointed-join robust.*

Then $\mathbf{H}_* + A$ has DP.

Proof Suppose that $\mathbf{H}_* + A \nvdash B_1$ and $\mathbf{H}_* + A \nvdash B_2$. We show $\mathbf{H}_* + A \nvdash B_1 \vee B_2$. Without loss of generality, we may assume that B_1 and B_2 contain no free variables other than $v_1, \ldots, v_m$, and we write B_i as $B_i(v_1, \ldots, v_m)$ $(i = 1, 2)$. By the strong completeness theorem of $\mathbf{H}_*$ with respect to Kripke-frame models (i.e., Theorem 9.1), we have two Kripke-frame models $(\langle \mathbf{M}_1, D_1 \rangle, \models_1)$ and $(\langle \mathbf{M}_2, D_2 \rangle, \models_2)$, and elements $d_i^1, \ldots, d_i^m \in D_i(0_i)$, where 0_i is the least element of $\mathbf{M}_i$ $(i = 1, 2)$, such that A is axiomatically true in both of them and $0_i \not\models_i B_i(d_i^1, \ldots, d_i^m)$ $(i = 1, 2)$. Take the pointed join model $(\mathcal{K}, \models)$ of them. By $(*)$, we have that A is axiomatically true in $(\mathcal{K}, \models)$. By Lemma 9.7, we have $0_i \not\models B_i((d_1^1, d_2^1), \ldots, (d_1^m, d_2^m))$ $(i = 1, 2)$, and hence $0 \not\models B_i((d_1^1, d_2^1), \ldots, (d_1^m, d_2^m))$ $(i = 1, 2)$, where 0 is the least element of $\mathcal{K}$. Therefore, $0 \not\models (B_1 \vee B_2)((d_1^1, d_2^1), \ldots, (d_1^m, d_2^m))$. Thus we have $\mathbf{H}_* + A \nvdash B_1 \vee B_2$. □

We can show the following in the same way as the above.

Corollary 9.2 *Let Γ be a set of formulas satisfying the condition $(*)$ in Theorem 9.2. Then $\mathbf{H}_* + \Gamma$ has DP.*

Lemma 9.8 *Let p be a unary predicate variable, S a sentence. Then, $\exists x(S \wedge p(x) \supset \forall y p(y))$ satisfies the condition $(*)$ in Theorem 9.2.*

Proof Suppose otherwise. Then, there exist a substitution instance I of $\exists x(S \wedge p(x) \supset \forall y p(y))$ and Kripke-frame models $(\langle \mathbf{M}_1, D_1 \rangle, \models_1)$ and $(\langle \mathbf{M}_2, D_2 \rangle, \models_2)$ such that I is true in both of them but I is not true in the pointed join model $(0 \uparrow (\langle \mathbf{M}_1, D_1 \rangle \oplus \langle \mathbf{M}_2, D_2 \rangle), \models)$. We may assume that the I contains no free variables other than $v_1, \ldots, v_n$, and these variables are distinct from x and y. There exist two formulas $B(v_1, \ldots, v_n)$ and $C(w, v_1, \ldots, v_n)$ of $\mathcal{L}$ having no free variables other than $v_1, \ldots, v_n$ and $w, v_1, \ldots, v_n$, respectively, such that I is obtained from $\exists x(S \wedge p(x) \supset \forall y p(y))$ by substituting $C(w, v_1, \ldots, v_n)$ to all occurrences of $p(w)$ (here w is a fresh variable) and replacing S by $B(v_1, \ldots, v_n)$. Thus, I is of the form:

$$\exists x(B(v_1, \ldots, v_n) \wedge C(x, v_1, \ldots, v_n) \supset \forall y C(y, v_1, \ldots, v_n)).$$

For the sake of simplicity, we assume $n = 1$. Since I is not true in the pointed join model, there exist an element $a \in \{0\} \uparrow \mathbf{M}_1 \oplus \mathbf{M}_2$ and a $d \in D^{\uparrow}(a)$ such that $a \not\models \exists x(B(d) \wedge C(x, d) \supset \forall y C(y, d))$. Suppose $a \in \mathbf{M}_1$. Then, by Lemma 9.7, it holds that $a \not\models_1 \exists x(B(\pi_1(d)) \wedge C(x, \pi_1(d)) \supset \forall y C(y, \pi_1(d)))$. This contradicts the assumption that I is true in $(\langle \mathbf{M}_1, D_1 \rangle, \models_1)$. Therefore, $a \notin \mathbf{M}_1$. Similarly, we have $a \notin \mathbf{M}_2$, and hence $a = 0$. Since $0_i \models_i \exists x(B(\pi_i(d)) \wedge C(x, \pi_i(d)) \supset \forall y C(y, \pi_i(d)))$ for $i = 1, 2$, there exist $s_1 \in D_1(0_1)$ and $s_2 \in D_2(0_2)$ such that $0_i \models_i B(\pi_i(d)) \wedge C(s_i, \pi_i(d)) \supset \forall y C(y, \pi_i(d))$ $(i = 1, 2)$. Therefore, by Lemma 9.7, we have $0_i \models B(d) \wedge C((s_1, s_2), d) \supset \forall y C(y, d)$ $(i = 1, 2)$.

Now we have two cases: $0 \not\models B(d) \wedge C((s_1, s_2), d)$ and $0 \models B(d) \wedge C((s_1, s_2), d)$. The former case implies $0 \models B(d) \wedge C((s_1, s_2), d) \supset \forall y C(y, d)$. That is, $0 \models \exists x(B(d) \wedge C(x, d) \supset \forall y C(y, d))$. Next, we assume the latter case where $0 \models B(d) \wedge C((s_1, s_2), d)$. Then, we have $0_i \models B(d) \wedge C((s_1, s_2), d)$, and hence $0_i \models \forall y C(y, d)$ $(i = 1, 2)$. If it holds that $0 \models C(t, d)$ for every $t \in D^{\uparrow}(0)$, then we have that $0 \models \forall y C(y, d)$, and hence we trivially have $0 \models B(d) \wedge C((s_1, s_2), d) \supset \forall y C(y, d)$. That is, we have $0 \models \exists x(B(d) \wedge C(x, d) \supset \forall y C(y, d))$. Thus we have that there exists $t \in D^{\uparrow}(0)$ such that $0 \not\models C(t, d)$. Consider the sentence $B(d) \wedge C(t, d) \supset \forall y C(y, d)$. We have that $0 \not\models B(d) \wedge C(t, d)$, and that $m \models \forall y C(y, d)$ for every $m \neq 0$. Hence, we have that $0 \models B(d) \wedge C(t, d) \supset \forall y C(y, d)$. Therefore, it holds that $0 \models \exists x(B(d) \wedge C(x, d) \supset \forall y C(y, d))$. This contradicts the assumption. □

We give here a proof of a result of Minari (1986) and Nakamura (1983) in this setting.

Corollary 9.3 (cf. Minari 1986 and Nakamura 1983) $\mathbf{H}_* + \exists x(p(x) \supset \forall y p(y))$ *has DP but lacks EP.*

Proof Take a fresh propositional variable q. Then, $\exists x(p(x) \supset \forall y p(y))$ is equivalent to $\exists x((q \supset q) \wedge p(x) \supset \forall y p(y))$ in $\mathbf{H}_*$. By Lemma 9.8, $\exists x(p(x) \supset \forall y p(y))$ satisfies the condition $(*)$ in Theorem 9.2, and hence we have the conclusion. □

Now, we prove Lemma 9.6. Let $\mathcal{S}$ be a set of finite non-degenerate subdirectly irreducible Heyting algebras. Recall that $QJ(\mathbf{A})$ $(\mathbf{A} \in \mathcal{S})$ is of the form $\exists v(S \supset$

$(p(v) \supset \forall y p(y)))$ with S being a sentence. Then, $QJ(\mathbf{A})$ is equivalent to $\exists v(S \wedge p(v) \supset \forall y p(y))$. From Lemma 9.8, it follows that $QJ(\mathbf{A})$ satisfies the condition $(*)$ in Theorem 9.2. By Corollary 9.2, it holds that $\mathbf{H}_* + \{QJ(\mathbf{A})\ ;\ \mathbf{A} \in \mathcal{S}\}$ has DP. This completes the proof of Lemma 9.6.

9.5 Strongly Independent Sequence of Modified Jankov Formulas—Jankov's Method for Predicate Logics

In this section, we show the following Lemma 9.9, and then the main Theorem (Theorem 9.4). Recall that $\{\mathbf{H}_i\}_{i<\omega}$ is the sequence of finite subdirectly irreducible Heyting algebras introduced in Sect. 9.3.2 and that $\{\mathbf{H}_i\}_{i<\omega}$ satisfies three conditions (A1), (A2) (in Sect. 9.3.1), and (A3) (in Sect. 9.3.2).

Lemma 9.9 *$\{QJ(\mathbf{H}_i)\}_{i<\omega}$ is strongly independent.*

For the proof, we use *algebraic Kripke sheaf semantics* for super-intuitionistic predicate logics. The algebraic Kripke sheaf is a framework for extended semantics obtained from a Kripke base equipped with a *domain-sheaf* and a *truth-value-sheaf*. A domain-sheaf is a covariant functor which integrates interpretations of equality into Kripke semantics for predicate logics.[6] A truth-value-sheaf is a contravariant functor which provides each possible world with an algebraic structure of "truth values" at the world.[7] In this paper, we use a simplified version of algebraic Kripke sheaves, called *Ω-brooms*, and apply a result in Suzuki (1999).

In Sect. 9.5.1, our simplified algebraic Kripke sheaf semantics is introduced. In Sect. 9.5.2, toolkit (a definition, lemmata, and notation) needed later is presented. In Sect. 9.5.3, we prove Lemma 9.9 and then Theorem 9.4.

9.5.1 *Special Algebraic Kripke Sheaves*

Definition 9.11 (*cf.* Suzuki 1999) A Kripke base $\mathbf{M}$ can be regarded as a category in the usual way. A covariant functor D from a Kripke base $\mathbf{M}$ to the category of all non-empty sets is called a *domain-sheaf* over $\mathbf{M}$, if $D(0_{\mathbf{M}})$ is non-empty. That is,

---

[6] Dragalin (1988) and Gabbay (1981) introduced Kripke frames with the equality, each of which is a Kripke frame equipped with a family of appropriate equivalence relations on the individual domains as the interpretation of equality. A pair $\mathcal{K} = \langle \mathbf{M}, D \rangle$ of a Kripke base $\mathbf{M}$ and a domain-sheaf D over $\mathbf{M}$ is called a *Kripke sheaf* (for super-intuitionistic predicate logics). Each Kripke sheaf is obtained from a Kripke frame with the equality as the quotient sets of domains by the equivalence relations together with the family of canonical projections.

[7] In the original Kripke semantics, each possible world has two possibilities for each formula: *true* or *not-true*. In this setting, from a viewpoint of algebraic semantics, each possible world has **2** as the set of truth values. Instead of **2**, we take an algebra $P(a)$ for each a ($\in \mathbf{M}$) as the set of truth values at a (*cf.* Suzuki 1999).

(DS1) $D(0_{\mathbf{M}})$ is a non-empty set,
(DS2) for every $a, b \in \mathbf{M}$ with $a \leq_{\mathbf{M}} b$, there exists a mapping $D_{ab} : D(a) \to D(b)$,
(DS3) D_{aa} is the identity mapping of $D(a)$ for every $a \in \mathbf{M}$,
(DS4) $D_{ac} = D_{bc} \circ D_{ab}$ for every $a, b, c \in \mathbf{M}$ with $a \leq_{\mathbf{M}} b \leq_{\mathbf{M}} c$.

Intuitively, $D(a)$ is the set of individuals at the world $a \in \mathbf{M}$. For each $d \in D(a)$ and each $b \in \mathbf{M}$ with $a \leq_{\mathbf{M}} b$, the element $D_{ab}(d) \in D(b)$ is said to be the *inheritor* of d at b. According to this intuition, each A ($\in \mathcal{L}[D(a)]$) with $a \leq_{\mathbf{M}} b$ has its unique *inheritor* $A^{D_{ab}}$ ($\in \mathcal{L}[D(b)]$). The $A^{D_{ab}}$ is denoted simply by $A_{a,b}$. In this paper, we deal only with domain-sheaves with the following additional condition:

(DS5) for every $a \in \mathbf{M}$, $D(a) = \begin{cases} \omega\ (= \{0, 1, \dots\}) \text{ if } a = 0_{\mathbf{M}}, \\ \{0\} \qquad\qquad\qquad \text{otherwise.} \end{cases}$

Thus, D_{ab}'s are trivially determined as follows:

$$D_{ab}(i) = \begin{cases} i \ \text{ if } a = b = 0_{\mathbf{M}}, \\ 0 \text{ otherwise,} \end{cases}$$

for every $i \in D(a)$. Then, for every $a \neq 0_{\mathbf{M}}$, the inheritor $A_{a,b}$ of $A \in \mathcal{L}[D(a)]$ at b is identical to A.

The category $\mathcal{H}$ of all non-degenerate complete Heyting algebras with arrows being complete monomorphisms between complete Heyting algebras. A contravariant functor P from a Kripke base $\mathbf{M}$ to the category $\mathcal{H}$ is called a *truth-value-sheaf* over $\mathbf{M}$. That is,

(TVS1) P_a is a non-degenerate complete Heyting algebra: $P(a) = (P(a), \cap^a, \cup^a, \to^a, 0^a, 1^a)$,
(TVS2) for every $a, b \in \mathbf{M}$ with $a \leq_{\mathbf{M}} b$, there exists a complete monomorphism $P_{ab} : P(b) \to P(a)$,
(TVS3) P_{aa} is the identity mapping of $P(a)$ for every $a \in \mathbf{M}$,
(TVS4) $P_{ac} = P_{ab} \circ P_{bc}$ for every $a, b, c \in \mathbf{M}$ with $a \leq_{\mathbf{M}} b \leq_{\mathbf{M}} c$.

A triple $\mathcal{K} = \langle \mathbf{M}, D, P \rangle$ of a Kripke base $\mathbf{M}$, a domain-sheaf D over $\mathbf{M}$, and a truth-value-sheaf P over $\mathbf{M}$ is called an *algebraic Kripke sheaf*. Intuitively, $P(a)$ is the set of *truth values* at a. If $a \leq_{\mathbf{M}} b$ (i.e., b is accessible from a), the P_{ab} sends computations of truth values in $P(b)$ into $P(a)$.

Let Ω be $\{1/n\ ;\ n = 1, 2, \dots\} \cup \{0\}$. With the natural ordering, Ω is a complete Heyting algebra having the greatest element 1 and the least element 0. The lattice order on Ω is denoted by $\leq_\Omega$ or simply $\leq$. In this paper, we deal only with algebraic Kripke sheaves with the following condition:

(TVS5) for every $a \in \mathbf{M}$, $P(a) = \begin{cases} \Omega \text{ if } a = 0_{\mathbf{M}}, \\ \mathbf{2} \ \text{ otherwise.} \end{cases}$

Thus, P_{ab}'s are essentially set-inclusions up to the identification: $1_2 = 1 = 1_\Omega$ and $0_2 = 0 = 0_\Omega$. Then, our algebraic Kripke sheaves are characterized by the Kripke base $\mathbf{M}$. Moreover, by (DS5), we may regard our algebraic Kripke sheaf as a Kripke frame for propositional logics, except at the least element $0_\mathbf{M}$ of its Kripke base. To make the difference clear, we will call an algebraic Kripke sheaf satisfying (DS5) and (TV5) an Ω-*broom*. An Ω-broom having $\mathbf{M}$ as its Kripke base is denoted by $\mathcal{B}(\mathbf{M})$.

A mapping V which assigns each pair (a, A) of an $a \in \mathbf{M}$ and an atomic sentence $A \in \mathcal{L}[D(a)]$ to an element $V(a, A)$ of $P(a)$ is said to be a *valuation* on $\langle \mathbf{M}, D, P \rangle$, if $a \leq_\mathbf{M} b$ implies $V(a, A) \leq^a P_{ab}(V(b, A_{a,b}))$, where $\leq^a$ is the lattice order of $P(a)$. In our setting, $P(a)$'s are all trivial subalgebras of $\Omega = P(0_\mathbf{M})$, and P_{ab}'s are set-inclusions. Thus, this condition can be written simply as: $a \leq_\mathbf{M} b$ implies $V(a, A) \leq_\Omega V(b, A_{a,b})$. We extend V to the mapping which assigns to each pair (a, A) of an $a \in \mathbf{M}$ and a sentence $A \in \mathcal{L}[D(a)]$ an element $V(a, A)$ of $P(a)$ as follows:

- $V(a, A \wedge B) = V(a, A) \cap V(a, B)$,
- $V(a, A \vee B) = V(a, A) \cup V(a, B)$,
- $V(a, A \supset B) = \bigcap_{b:a \leq_\mathbf{M} b}(V(b, A_{a,b}) \to V(b, B_{a,b}))$,
- $V(a, \neg A) = \bigcap_{b:a \leq_\mathbf{M} b}(V(b, A_{a,b}) \to 0)$,
- $V(a, \forall x A(x)) = \bigcap_{b:a \leq_\mathbf{M} b} \bigcap_{u \in D(b)} V(b, A_{a,b}(\overline{u}))$,
- $V(a, \exists x A(x)) = \bigcup_{u \in D(a)} V(a, A(\overline{u}))$.

Note that operations of Heyting algebra in the right hand sides are those of Ω. In the original definition in Suzuki (1999), these induction steps, especially of $\supset$, $\neg$, and $\forall$, are slightly more complicated. However, by the virtue of (TV5), these simple steps work well.[8]

A pair $(\mathcal{B}, V)$ of an Ω-broom $\mathcal{B}$ and a valuation V on it is said to be an Ω-*broom model* (in the general case, an *algebraic Kripke-sheaf model*). A formula A of $\mathcal{L}$ is said to be *true* in an Ω-broom model $(\mathcal{B}, V)$, if $V(0_\mathbf{M}, \overline{A}) = 1$. A formula of $\mathcal{L}$ is said to be *valid* in an Ω-broom $\mathcal{B}$, if it is true in $(\mathcal{B}, V)$ for every valuation V on $\mathcal{B}$. The set of formulas of $\mathcal{L}$ that are valid in $\mathcal{B}$ is denoted by $L(\mathcal{B})$. The following propositions are fundamental properties of algebraic Kripke sheaf semantics (*cf.* Suzuki 1999).

Proposition 9.7 (cf. Proposition 9.5) *For every Ω-broom model $(\mathcal{B}(\mathbf{M}), V)$, every $a, b \in \mathbf{M}$, and every sentence $A \in \mathcal{L}[D(a)]$, if $a \leq_\mathbf{M} b$, then $V(a, A) \leq_\Omega V(b, A_{a,b})$.*

Proposition 9.8 (cf. Propositions 9.4 and 9.6) *For each Ω-broom $\mathcal{B}$, the set $L(\mathcal{B})$ is a super-intuitionistic predicate logic.*

[8] In Suzuki (1999), the $\cap$, $\cup$, $\to$, 0 must have appropriate superscripts $\cdot^a$ and $\cdot^b$, and appropriate P_{ab}'s in front of V's in the right-hand sides.

9.5.2 Toolkit for Ω-Brooms

Definition 9.12 (*cf. Suzuki* 1999) Let **M** be a finite Kripke base. Take the Jankov formula $J(O(\mathbf{M}))$ and replace all occurrences of $p_{\star_{O(\mathbf{M})}}$ in $J(O(\mathbf{M}))$ by F (i.e., $\exists x(p(x) \supset \forall y p(y))$). Then, we denote the resulting sentence by $J(\mathbf{M}; F)$.

The following Lemma gives the relationship between $QJ(O(\mathbf{M}))$ and $J(\mathbf{M}; F)$ in Ω-brooms.

Lemma 9.10 *In every Ω-broom, the sentence* $(q \supset \exists x r(x)) \supset \exists x(q \supset r(x))$ *is valid, where* q *and* r *are a propositional variable and a unary predicate variable, respectively.*

Proof Let C and D be $q \supset \exists x r(x)$ and $\exists x(q \supset r(x))$, respectively. Let V be an arbitrary valuation on an Ω-broom $\mathcal{B}(\mathbf{M}) = \langle \mathbf{M}, D, P \rangle$. Note that for each $b \neq 0_{\mathbf{M}}$, we have that $V(b, \exists x r(x)) = V(b, r(0))$ and that the inheritor $r(i)_{0_{\mathbf{M}}, b}$ of $r(i)$ at b is $r(0)$ for every $i \in \omega = D(0_{\mathbf{M}})$. Then, for every $b \neq 0_{\mathbf{M}}$, we have

$$\begin{aligned} V(b, C) &= \bigcap \{V(c, q) \to V(c, \exists x r(x)) \; ; \; c \geq b\} \\ &= \bigcap \{V(c, q) \to V(c, r(0)) \; ; \; c \geq b\} \\ &= V(b, q \supset r(0)) \\ &= V(b, \exists x(q \supset r(x))) \\ &= V(b, D) \; . \end{aligned}$$

Hence, it holds that $V(0_{\mathbf{M}}, C \supset D) = \bigcap \big[\{V(0_{\mathbf{M}}, C) \to V(0_{\mathbf{M}}, D)\} \cup \{V(b, C) \to V(b, D) \; ; \; b \neq 0_{\mathbf{M}}\}\big] = V(0_{\mathbf{M}}, C) \to V(0_{\mathbf{M}}, D)$. Therefore, it suffices to show that $V(0_{\mathbf{M}}, C) \leq V(0_{\mathbf{M}}, D)$. Let us check the value $V(0_{\mathbf{M}}, C)$:

$$\begin{aligned} & V(0_{\mathbf{M}}, C) \\ &= V(0_{\mathbf{M}}, q \supset \exists x r(x)) \\ &= \bigcap \{V(a, q) \to V(a, \exists x r(x)) \; ; \; a \in \mathbf{M}\} \\ &= \bigcap \big[\{V(0_{\mathbf{M}}, q) \to V(0_{\mathbf{M}}, \exists x r(x))\} \cup \{V(b, q) \to V(b, \exists x r(x)) \; ; \; b \neq 0_{\mathbf{M}}\}\big] \\ &= \bigcap \big[\{V(0_{\mathbf{M}}, q) \to V(0_{\mathbf{M}}, \exists x r(x)) \cup \{V(b, q) \to V(b, r(0)) \; ; \; b \neq 0_{\mathbf{M}}\}\big] \; . \end{aligned}$$

We have two cases: (1) $V(b, q) \to V(b, r(0)) = 0$ for some $b \neq 0_{\mathbf{M}}$, and (2) $V(b, q) \to V(q, r(0)) = 1$ for every $b \neq 0_{\mathbf{M}}$.

Suppose that (1) holds. Since 0 belongs to $\{V(b, q) \to V(b, r(0)) \; ; \; b \neq 0_{\mathbf{M}}\}$, we have $V(0_{\mathbf{M}}, C) = 0 \leq V(0_{\mathbf{M}}, D)$. Next, suppose that (2) holds. Since $\{V(b, q) \to V(b, r(0)) \; ; \; b \neq 0_{\mathbf{M}}\} = \{1\}$, we have $V(0_{\mathbf{M}}, C) = V(0_{\mathbf{M}}, q) \to V(0_{\mathbf{M}}, \exists x r(x))$. Since $V(0_{\mathbf{M}}, \exists x r(x)) = \max_{i \in \omega}\{V(0_{\mathbf{M}}, r(i))\}$, there exists an $i_0 \in \omega$ such that $V(0_{\mathbf{M}}, \exists x r(x)) = V(0_{\mathbf{M}}, r(i_0))$. Hence, it holds that

$$V(0_{\mathbf{M}}, C) = \begin{cases} V(0_{\mathbf{M}}, r(i_0)) & \text{if } V(0_{\mathbf{M}}, r(i_0)) < V(0_{\mathbf{M}}, q), \\ 1 & \text{if } V(0_{\mathbf{M}}, q) \leq V(0_{\mathbf{M}}, r(i_0)). \end{cases}$$

To calculate $V(0_{\mathbf{M}}, D)$, we put

$$v_i = V(0_{\mathbf{M}}, q \supset r(i))$$

for each $i \in \omega$, and we have $V(0_{\mathbf{M}}, D) = \max_{i \in \omega} v_i$. Let us check the value v_i:

$$\begin{aligned} v_i &= V(0_{\mathbf{M}}, q \supset r(i)) \\ &= \bigcap \{V(a, q) \to V(a, r(i)_{0_{\mathbf{M}}, a}) \ ; \ a \in \mathbf{M}\} \\ &= \bigcap \left[\{V(0_{\mathbf{M}}, q) \to V(0_{\mathbf{M}}, r(i))\} \cup \{V(b, q) \to V(b, r(0)) \ ; \ b \neq 0_{\mathbf{M}}\}\right] . \end{aligned}$$

By the assumption (2), it holds that $\{V(b, q) \to V(b, r(0)) \ ; \ b \neq 0_{\mathbf{M}}\} = \{1\}$. Therefore, we have $v_i = V(0_{\mathbf{M}}, q) \to V(0_{\mathbf{M}}, r(i))$. If $V(0_{\mathbf{M}}, r(i_0)) < V(0_{\mathbf{M}}, q)$, we have $V(0_{\mathbf{M}}, C) = V(0_{\mathbf{M}}, r(i_0)) = v_{i_0} \leq \max_{i \in \omega} v_i = V(0_{\mathbf{M}}, D)$. If $V(0_{\mathbf{M}}, q) \leq V(0_{\mathbf{M}}, r(i_0))$, we have $v_{i_0} = 1 = V(0_{\mathbf{M}}, D) \geq V(0_{\mathbf{M}}, C)$. □

From this Lemma and Proposition 9.8, it follows that $(A \supset \exists x B(x)) \supset \exists x(A \supset B(x))$ is valid in every Ω-broom model, where A does not contain x as a free variable. Thus, we have the[9]

Lemma 9.11 *In every Ω-broom model, $J(\mathbf{M}; F) \supset QJ(O(\mathbf{M}))$ is valid.*

Next we recall a Lemma in Suzuki (1999). We describe the Lemma in the setting of the present paper.[10] This Lemma asserts that $QJ(O(\mathbf{M}))$ and $J(\mathbf{M}; F)$ have a property similar to that of the original Jankov formula shown in Lemma 9.1.

Lemma 9.12 (*cf.* Lemma 9.1 and Suzuki 1999; Lemma 4.10) *Let* $\mathbf{M}$ *be a finite Kripke base such that* $O(\mathbf{M})$ *satisfies the condition* (A3) *in Sect. 9.3.2. For each Ω-broom $\mathcal{B}(\mathbf{N})$ with* $\mathbf{N}$ *having at least two elements, if $QJ(O(\mathbf{M})) \notin L(\mathcal{B}(\mathbf{N}))$, then* $O(\mathbf{M})$ *is embeddable into a quotient algebra of* $O(\mathbf{N})$.

Proof (*Sketch*) Suppose that $QJ(O(\mathbf{M})) \notin L(\mathcal{B}(\mathbf{N}))$. Then, by Lemma 9.11, we have $J(\mathbf{M}; F) \notin L(\mathcal{B}(\mathbf{N}))$. Since $J(\mathbf{M}; F)$ is obtained from $J(O(\mathbf{M}))$ by replacing $p_{\star_{\mathbf{M}}}$ by the sentence F, the original Jankov formula $J(O(\mathbf{M}))$ is not valid in $\mathcal{B}(\mathbf{N})$. This implies that $J(O(\mathbf{M}))$ is not valid in the Heyting algebra[11] $\big(O(\mathbf{M})/\star\big) \oplus \Omega$.

---

[9] Since $\exists x(q \supset r(x)) \supset (q \supset \exists x r(x))$ is provable in $\mathbf{H}_*$, it follows that $(q \supset \exists x r(x))$ and $\exists x(q \supset r(x))$ are equivalent in every Ω-broom model. Thus, $J(\mathbf{M}; F)$ is equivalent to $QJ(O(\mathbf{M}))$ in every Ω-broom.

[10] The condition (A3) is denoted by (#) in Suzuki (1999). In Lemma 4.10 of Suzuki (1999), F is replaced by an arbitrary sentence.

[11] The algebra $\big(O(\mathbf{M})/\star\big) \oplus \Omega$ is the *sum* of $O(\mathbf{M})/\star$ and Ω. Here, $O(\mathbf{M})/\star$ is the quotient algebra of $O(\mathbf{M})$ modulo $\star = [\star_{O(\mathbf{M})})$, where $[\star_{O(\mathbf{M})})$ is the filter generated by the second greatest element $\star_{O(\mathbf{M})}$ of $O(\mathbf{M})$). Note that $\big(O(\mathbf{M})/\star\big) \oplus \Omega$ is denoted by $O(\mathbf{M}) \lhd \Omega$ in Suzuki (1999).

From Lemma 9.1, it follows that $\mathcal{O}(\mathbf{M})$ is embeddable into a quotient algebra of $(\mathcal{O}(\mathbf{M})/\star) \oplus \Omega$. By (A3), we have that $\mathcal{O}(\mathbf{M})$ is embeddable into a quotient algebra of $\mathcal{O}(\mathbf{N})$. □

Lemma 9.13 *Let* $\mathbf{M}$ *be a finite Kripke base. Then* $QJ(\mathcal{O}(\mathbf{M}))$ *is not valid in* $\mathcal{B}(\mathbf{M})$.

Proof Let $\mathcal{B}(\mathbf{M})$ be $\langle \mathbf{M}, D, P\rangle$. Define a valuation V by

$$V(a, p_O) = \begin{cases} 1 \text{ if } a \in O, \\ 0 \text{ if } a \notin O, \end{cases}$$

for every $a \in \mathbf{M}$ and every $O \in \mathcal{O}(\mathbf{M})$, and

$$V(a, p(i)) = \begin{cases} 1 & \text{if } a \neq 0_{\mathbf{M}}, \\ 1/(i+1) & \text{if } a = 0_{\mathbf{M}}, \end{cases}$$

for every $a \in \mathbf{M}$ and every $i \in D(a)$. Take an $a \neq 0_{\mathbf{M}}$. Since $V(a, \forall y p(y)) = \bigcap_{b \geq a} V(b, p(0)) = 1$, we have that $V(a, F) = V(a, p(0) \supset \forall y p(y)) = \bigcap_{b \geq a}(1 \to V(b, \forall y p(y))) = \bigcap_{b \geq a} V(b, \forall y p(y)) = 1$. Thus, it holds that $V(a, p(0) \supset \forall y p(y)) = V(a, F) = 1$ for every $a \neq 0_{\mathbf{M}}$. Next, check that $V(0_{\mathbf{M}}, \forall y p(y)) \leq \bigcap_{i \in \omega} V(0_{\mathbf{M}}, p(i)) = 0$. For a fixed $i \in \omega$, we have that $V(0_{\mathbf{M}}, p(i) \supset \forall y p(y)) = \bigcap_{a \in \mathbf{M}}\{V(a, p(i)_{0_{\mathbf{M}}, a}) \to V(a, \forall y p(y))\} \leq V(0_{\mathbf{M}}, p(i)) \to V(0_{\mathbf{M}}, \forall y p(y)) = 1/(i+1) \to 0 = 0$. Thus, we have $V(0_{\mathbf{M}}, F) = 0$.

Consider an assignment v on $\mathcal{O}(\mathbf{M})$ defined by

$$v(p_O) = \begin{cases} \{a\ ;\ V(a, p_O) = 1\} \text{ if } O \neq \mathbf{M} \setminus \{0_{\mathbf{M}}\}, \\ \{a\ ;\ V(a, F) = 1\} \ \ \text{if } O = \mathbf{M} \setminus \{0_{\mathbf{M}}\}, \end{cases}$$

for every $O \in \mathcal{O}(\mathbf{M})$. Then, v is nothing but the assignment $v_{\mathcal{O}(\mathbf{M})}$ that makes the Jankov formula not true in $\mathcal{O}(\mathbf{M})$.

Claim. *Let* X *be a propositional formula having no propositional variable other than* $\{p_O\ ;\ O \in \mathcal{O}(\mathbf{M})\}$. *By* X', *we denote the formula obtained from* X *by replacing all occurrences of* $p_{\star_{\mathcal{O}(\mathbf{M})}}$ *by the sentence* F. *Then, we have* $V(a, X') \in \{0, 1\}$ *for every* $a \in \mathbf{M}$ *and that* $v(X) = \{a\ ;\ V(a, X') = 1\}$.

This Claim can be proved by induction on the length of X. The Basis-part is already clear by the discussion just after the definition of V. We check the Induction Steps. Suppose that X is of the form $Y \supset Z$. Take an arbitrary $a \in v(Y \supset Z)$. Then for every $b \geq a$, either $b \notin v(Y)$ or $b \in v(Z)$. By the induction hypothesis, we have that $\{V(b, Y'), V(b, Z')\} \subseteq \{0, 1\}$ for every $b \in \mathbf{M}$, and $v(Y) = \{c\ ;\ V(c, Y') = 1\}$ and $v(Z) = \{c\ ;\ V(c, Z') = 1\}$. Thus, for every $b \geq a$, either $V(b, Y') = 0$ or $V(b, Z') = 1$. Therefore, we have $V(b, Y') \to V(b, Z') = 1$ for every $b \geq a$, and hence $V(a, Y' \supset Z') = 1$. Next take an arbitrary $b \notin v(Y \supset Z)$. There exists $b \geq a$ such that $b \in v(Y)$ and $b \notin v(Z)$. By the induction hypothesis, we have $V(b, Y') = 1$ and $V(b, Z') = 0$. Thus, $V(a, Y' \supset Z') = 0$. Therefore, for every $a \in \mathbf{M}$, we have

$V(a, Y' \supset Z') \in \{0, 1\}$ and $v(Y \supset Z) = \{a \; ; \; V(a, Y' \supset Z') = 1\}$. Other cases can be proved similarly. This completes the proof of the Claim.

From this Claim, it follows that $\mathbf{M} = 1_{O(\mathbf{M})} = v(\bigwedge \delta(O(\mathbf{M}))) = \{a \; ; \; V(a, \bigwedge \Delta(O(\mathbf{M}))) = 1\}$. That is, we have $V(a, \bigwedge \Delta(O(\mathbf{M}))) = 1$ for every $a \in \mathbf{M}$. Note that $V(0_\mathbf{M}, QJ(O(\mathbf{M})) = \max_{i \in \omega}\{V(0_\mathbf{M}, \bigwedge \Delta(O(\mathbf{M})) \supset (p(i) \supset \forall y p(y)))\}$. Let us check for an $i \in \omega$:

$$\begin{aligned} & V(0_\mathbf{M}, \bigwedge \Delta(O(\mathbf{M})) \supset (p(i) \supset \forall y p(y))) \\ &= \bigcap \{V(a, \bigwedge \Delta(O(\mathbf{M}))) \to V(a, (p(i) \supset \forall y p(y))_{0_\mathbf{M}, a}) \; ; \; a \in \mathbf{M}\} \\ &= \bigcap \{V(a, (p(i) \supset \forall y p(y))_{0_\mathbf{M}, a}) \; ; \; a \in \mathbf{M}\} \\ &\leq V(0_\mathbf{M}, p(i) \supset \forall y p(y)) \\ &= 0 \end{aligned}$$

Therefore, we have $V(0_\mathbf{M}, QJ(O(\mathbf{M})) = 0$. □

As we already mentioned in Examples 9.1 and 9.2, we have the correspondence between subdirectly irreducible Heyting algebras and Kripke bases. When they are finite, we have more exact correspondence:

Fact 9.3 (1) *For each finite subdirectly irreducible Heyting algebra* $\mathbf{A}$*, the* $\mathcal{OP}(\mathbf{A})$ *is isomorphic to* $\mathbf{A}$*.*
(2) *For each finite Kripke base* $\mathbf{M}$*, the* $\mathcal{PO}(\mathbf{M})$ *is isomorphic to* $\mathbf{M}$*.*

Thus, we may identify finite subdirectly irreducible Heyting algebras and finite Kripke bases. In the rest of this paper, we denote by $\mathbf{N}_i$ the Kripke base corresponding to $\mathbf{H}_i$ $(i < \omega)$. That is, $\mathbf{N}_i = \mathcal{P}(\mathbf{H}_i)$ and $\mathbf{H}_i = O(\mathbf{N}_i)$ for each $i < \omega$.

9.5.3 Proofs of Lemma 9.9 and the Main Theorem

The proof of Lemma 9.9 proceeds similarly to the proof of Corollary 9.1 by the virtue of the discussion of the previous subsection. Define $\mathbf{L}_i = \mathbf{H}_* + QJ(\mathbf{H}_i)$ for each $i < \omega$. We show that $\{\mathbf{L}_i\}_{i<\omega}$ is strongly independent. Pick an arbitrary $i_0 \in \omega$. Then, for every $j \neq i_0$, the $\mathbf{H}_j$ is not embeddable into any quotient algebra of $\mathbf{H}_{i_0}$. Thus, by Lemma 9.12, it holds that $QJ(\mathbf{H}_j) \in L(\mathcal{B}(\mathbf{N}_{i_0}))$ for every $j \neq i_0$. Therefore, $\bigcup_{j \neq i_0} \mathbf{L}_j \subseteq L(\mathcal{B}(\mathbf{N}_{i_0}))$. By Lemma 9.13, we have $QJ(\mathbf{N}_{i_0}) \notin L(\mathcal{B}(\mathbf{N}_{i_0}))$. Hence, $\mathbf{L}_{i_0} \not\subseteq \bigcup_{j \neq i_0} \mathbf{L}_j$. This completes the proof of Lemma 9.9.

Theorem 9.4 (Main Theorem) *There exits a continuum of PEI's having disjunction property but lacking existence property.*

Proof By Lemma 9.2, for every $I \subseteq \omega$, the logic $\mathbf{H}_* + \{QJ(\mathbf{H}_i) \; ; \; i \in I\}$ is a PEI lacking EP. By Lemma 9.6, for every $I \subseteq \omega$, this logic has DP. By Lemma 9.9, $\{QJ(\mathbf{H}_i)\}_{i<\omega}$ is strongly independent. Thus, we have the conclusion. □

By examining the definition of $\{QJ(\mathbf{H}_i)\ ;\ i < \omega\}$, we have shown that $\{QJ(\mathbf{H}_i)\ ;\ i < \omega\}$ is a recursively enumerable sequence of concrete predicate axioms schemata.

Corollary 9.4 *There exits a continuum of PEI's having none of DP and EP.*

Let Lin^* be $(q(x) \supset q(y)) \vee (q(y) \supset q(x))$, where q is a fresh unary predicate variable. Then, it is obvious that $\mathbf{H}_* + Lin^*$ is a PEI without DP. We have the

Lemma 9.14 *Lin^* is valid in every Ω-broom $\mathcal{B}$.*

Proof Let V be a valuation on $\mathcal{B}(\mathbf{M})$. If $a \in \mathbf{M}$ and $a \neq 0_{\mathbf{M}}$, the inheritors of $\forall y((p(i) \supset p(y)) \vee (p(y) \supset p(i))$, $(p(i) \supset p(j)) \vee (p(j) \supset p(i))$, and $p(i) \supset p(j) \in \mathcal{L}[D(0_{\mathbf{M}})]$ at a are $\forall y((p(0) \supset p(y)) \vee (p(y) \supset p(0))$, $(p(0) \supset p(0)) \vee (p(0) \supset p(0))$, and $p(0) \supset p(0)$, respectively. Clearly, $V(a, p(0) \supset p(0)) = 1$, and hence for $a \neq 0_{\mathbf{M}}$,

$$\begin{aligned} &V(a, \forall y((p(0) \supset p(y)) \vee (p(y) \supset p(0)))) \\ &= \bigcap\{V(b, (p(0) \supset p(0)) \vee (p(0) \supset p(0)))\ ;\ b \geq a\} \\ &= 1. \end{aligned}$$

And also,

$$\begin{aligned} &V(0_{\mathbf{M}}, p(i) \supset p(j)) \\ &= \bigcap[\{V(b, p(0)) \to V(b, p(0)))\ ;\ b \neq 0_{\mathbf{M}}\} \cup \{V(0_{\mathbf{M}}, p(i)) \to V(0_{\mathbf{M}}, p(j))\}] \\ &= V(0_{\mathbf{M}}, p(i)) \to V(0_{\mathbf{M}}, p(j)). \end{aligned}$$

Therefore,

$$\begin{aligned} V(0_{\mathbf{M}}, Lin^*) &= V(0_{\mathbf{M}}, (p(i) \supset p(j)) \vee (p(j) \supset p(i))) \\ &= V(0_{\mathbf{M}}, p(i) \supset p(j)) \cup V(0_{\mathbf{M}}, p(j) \supset p(i)) \\ &= (V(0_{\mathbf{M}}, p(i)) \to V(0_{\mathbf{M}}, p(j))) \cup (V(0_{\mathbf{M}}, p(j)) \to V(0_{\mathbf{M}}, p(i))) \\ &= 1. \end{aligned}$$

Thus, Lin^* is valid in $\mathcal{B}(\mathbf{M})$. □

Let us consider the sequence $\{Lin^* \wedge QJ(\mathbf{H}_i)\}_{i<\omega}$ of sentences. Then, by putting $\mathbf{K}_i = \mathbf{H}_* + Lin^* \wedge QJ(\mathbf{H}_i)$ $(i < \omega)$, we can show that $\{\mathbf{K}_i\}_{i<\omega}$ is strongly independent. It is clear that for every non-empty subset $\mathcal{S}$ of $\{\mathbf{K}_i\ ;\ i < \omega\}$, the logic $\bigcup \mathcal{S}$ fails to have DP and EP. This completes the proof of Corollary 9.4.[12]

Note that the sequence $\{Lin^* \wedge QJ(\mathbf{H}_i)\}_{i<\omega}$ is recursively generated.

---

[12] In fact, Corollary 9.4 can be shown as a corollary to the proof in Suzuki (1995; p. 184). Let $\widetilde{F}$ and $\widetilde{W_2}$ be $\exists x \exists y(p(x) \wedge p(y) \supset \forall z p(z))$ and $\bigvee_{i=1}^{3}(q(x_i) \supset \bigvee_{j\neq i} q(x_j))$, respectively. By putting $\mathbf{H}_* + \widetilde{F} + \widetilde{W_2}$ as $\mathbf{L}$ in Suzuki (1995; p. 184), we can show that there exists a continuum of logics between $\mathbf{H}_* + \widetilde{F} + \widetilde{W_2}$ and $\mathbf{H}_* + (\exists x p(x) \supset \forall x p(x)) \vee (r \vee \neg r)$. Since $\mathbf{H}_* + \widetilde{F} + \widetilde{W_2}$ fails to have DP and EP, all of these logics lack EP and DP. Note that $\mathbf{H}_* + (\exists x p(x) \supset \forall x p(x)) \vee (q \vee \neg q)$ is the greatest PEI.

9.6 Concluding Remarks

We constructed a recursively enumerable set of concrete predicate axiom schemata. By adding these schemata to $\mathbf{H}_*$, we obtained a strongly independent sequence of predicate extensions of intuitionistic; and this sequence yields a continuum of predicate extensions of intuitionistic logic each of which has DP but lacks EP.

This result suggests that although PEI's are living near to intuitionistic logic, the diversity of their nature seems rich. In other words, logics among PEI's are fascinating from the logical point of view and yet to be explored.

We have four types of continua of logics: "with EP and DP," "without EP and DP," "with DP but without EP," and "with EP but without DP." Other than the last one, three of them can be obtained by recursively enumerable construction of concrete axiom schemata. Recall that DP ad EP are regarded as "hallmarks" of constructivity of intuitionistic logic. It seems interesting that continua of logics with/without properties closely related to constructivity are constructively generated by sequences of axiom schemata. However, for the continuum: "with EP but without DP," we do not have such a sequence of axiom schemata. So we pose a

Problem. *Does there exist a recursively enumerable and strongly independent sequence of axiom schemata such that all the logics yield by this sequence are PEI's, have EP, and fail to have DP?*

We make two Remarks on the consideration of the Problem.

Remark 9.1 As shown in Suzuki (2021), if an intermediate logic $\mathbf{L}$ has EP, then $\mathbf{L}$ has DP, provided that $\mathbf{L}$ has a very weak DP: $\mathbf{L} \vdash A \vee (p(x) \supset p(y))$ implies $\mathbf{L} \vdash A$ whenever A contains no occurrence of the symbols: p, x, and y. Note that this weak DP seems natural for reasonable logics such as logics complete with respect to a class of Kripke bases or to a class of complete Heyting algebras. (Even classical logic possesses it.) Thus, it is not straightforward to create semantically a logic that has EP and does not have DP

Remark 9.2 In Suzuki (2021), we gave a method to create a PEI with EP but without DP from a given logic with EP. Let $\mathbf{H}^*$ be the superintuitionistic predicate logic $\mathbf{H} + \exists x p(x) \supset \forall x p(x)$, where p is a unary predicate variable. Then, $\mathbf{H}^*$ is the greatest superintuitionistic predicate logic having the same propositional part as $\mathbf{H}$. If $\mathbf{L}$ is an intermediate predicate logic having EP, then $\mathbf{L} \cap \mathbf{H}^*$ has EP but lacks DP, provided that $\mathbf{L}$ is NOT a PEI.

If we try to use this method to solve the problem affirmatively, we need appropriate logics with EP. Ferrari and Miglioli (1993) gave a continuum of intermediate predicate logics having both of EP and DP. These logics are all not PEI's. However, their logics are non-recursively generated. Hence, the resulting logics by this method are not recursively generated. We cannot use their logics to solve the Problem. On the other hand, (Suzuki 1999)'s strongly independent sequence are recursively generated, but we cannot apply the method to them, because these logics are PEI's (they are fixed points of the Δ-operation. *cf.* Suzuki 1996). Hence, we cannot use these logics neither.

Acknowledgements The author thanks the Japan Society for the Promotion of Science (JSPS), Core-to-Core Program (A. Advanced Research Networks) and Grand-in-Aid for Scientific Research (C) No.16K05252 and No.20K03716 for supporting the research.

References

Church, A. (1956). *Introduction to mathematical logic I*. Princeton: Princeton University Press.

Citkin, A.(2018). Characteristic formulas over intermediate logics. In S. Odintsov (Ed.), *Larisa Maksimova on implication, interpolation, and definability, outstanding contributions to logic* (Vol. 15, pp. 71–98). Cham: Springer.

Dragalin, A. G. (1988). *Mathematical intuitionism. Introduction to proof theory*. Translated from the Russian by E. Mendelson, *Translations of mathematical monographs* (Vol. 67). Providence: American Mathematical Society.

Ferrari, M., & Miglioli, P. (1993). Counting the maximal intermediate constructive logics. *Journal of Symbolic Logic, 58*, 1365–1401.

Friedman, H. (1975). The disjunction property implies the numerical existence property. *Proceedings of the National Academy of Sciences of the United States of America, 72*, 2877–2878.

Friedman, H., & Sheard, M. (1989). The equivalence of the disjunction and existence properties for modal arithmetic. *Journal of Symbolic Logic, 54*, 1456–1459.

Gabbay, D. M. (1981). *Semantical investigation of Heyting's intuitionistic logic, synthese library, studies in epistemology, logic, methodology, and philosophy of science* (Vol. 148). Dordrecht: D. Reidel Publishing Company.

Gabbay, D. M., Shehtman, V. B., & Skvortsov, D. P. (2009). *Quantification in nonclassical logic* (Vol. 1), Studies in logic and the foundations of mathematics (Vol. 153). Amsterdam: Elsevier.

Jankov, V. A. (1963). The relationship between deducibility in the intuitionistic propositional calculus and finite implicational structures. *Soviet Mathematics Doklady, 4*, 1203–1204.

Jankov, V. A. (1968). Constructing a sequence of strongly independent superintuitionistic propositional calculus. *Soviet Mathematics Doklady, 9*, 806–807.

Jankov, V. A. (1969). Conjunctively indecomposable formulas in propositional calculi. *Mathematics of the USSR-Izvestiya, 3*, 17–35.

Kleene, S. C. (1952). *Introduction to metamathematics*. New York: D. Van Nostrand.

Kleene, S. C. (1962). Disjunction and existence under implication in elementary intuitionistic formalisms. *Journal of Symbolic Logic, 27*, 11–18.

Minari, P. (1986). Disjunction and existence properties in intermediate predicate logics. In: *Atti del Congresso Logica e Filosofia della Scienza, oggi. San Gimignano, dicembre 1983. Vol.1 – Logica. CLUEB, Bologna (Italy)* (pp. 7–11).

Nakamura, T. (1983). Disjunction property for some intermediate predicate logics. *Reports on Mathematical Logic, 15*, 33–39.

Ono, H. (1987). Some problems in intermediate predicate logics. *Reports on Mathematical Logic, 21*(1987), 55–67.

Suzuki, N.-Y. (1995). Constructing a continuum of predicate extensions of each intermediate propositional logic. *Studia Logica, 54*, 173–198.

Suzuki, N.-Y. (1996). A remark on the delta operation and the Kripke sheaf semantics in superintuitionistic predicate logics. *Bulletin of the Section of Logic, University of Łódź, 25*, 21–28.

Suzuki, N.-Y. (1999). Algebraic Kripke sheaf semantics for non-classical predicate logics. *Studia Logica, 63*, 387–416.

Suzuki, N.-Y. (2017). Some weak variants of the existence and disjunction properties in intermediate predicate logics. *Bulletin of the Section of Logic, Department of Logic, University of Łódź, 46*, 93–109.

Suzuki, N.-Y. (2021). A negative solution to Ono's problem P52: Existence and disjunction properties in intermediate predicate logics. In N. Galatos & K. Terui (Eds.), *Hiroakira Ono on substructural logics. Outstanding contributions to logic* (Vol. 23). Springer.

Wroński, A. (1974). The degree of completeness of some fragments of the intuitionistic propositional logic. *Reports on Mathematical Logic, 2*, 55–62.

Part II
History and Philosophy of Mathematics

Chapter 10
On V. A. Yankov's Contribution to the History of Foundations of Mathematics

Ioannis M. Vandoulakis

Abstract The paper examines Yankov's contribution to the history of mathematical logic and the foundations of mathematics. It concerns the public communication of Markov's critical attitude towards Brouwer's intuitionistic mathematics from the point of view of his constructive mathematics and the commentary on A.S. Esenin-Vol'pin program of ultra-intuitionistic foundations of mathematics.

Keywords V.A. Yankov · A.A. Markov · A.S. Esenin-Vol'pin · L.E.J. Brouwer · A. Heyting · E.A. Bishop · B.A. Kushner · Markov's constructive mathematics · Intuitionistic mathematics · Philosophy of constructive mathematics

Mathematics Subject Classification: 03-03, 03F55, 01A60, 01A72

10.1 Introduction

Vadim A. Yankov is well-known as a mathematical logician and philosopher, a specialist in constructive logic, and a pupil of Andrei A. Markov (1903–1979), who established the Russian school of constructive mathematics and theory of algorithms. He is also known for his political activity in USSR as a political dissident and former political prisoner. However, his contribution to the history and philosophy of mathematics is not generally known. Part of his contribution was presented for the first time by the author (Vandoulakis 2015).

Yankov's contribution to the history of mathematical logic and the foundations of mathematics is connected with his participation in Markov's school and the foundational debates in the Moscow milieu of mathematicians of the 20th century.

His first contribution concerns the preservation of Markov's attitude towards Luitzen Egbertus Jan Brouwer's (1881–1966) intuitionistic mathematics that was

I. M. Vandoulakis (✉)
The Hellenic Open University, Athens, Greece and FernUniversität in Hagen, Hagen, Germany
e-mail: i.vandoulakis@gmail.com; vandoulakis.ioannis@ac.eap.gr

A. Citkin and I. M. Vandoulakis (eds.), *V. A. Yankov on Non-Classical Logics, History and Philosophy of Mathematics*, Outstanding Contributions to Logic 24,
https://doi.org/10.1007/978-3-031-06843-0_10

exposed from the viewpoint of his (constructive) mathematics. Yankov succeeded in making Markov's views on Brouwer's intuitionism public by suggesting exposing his views in the endnotes to the Russian translation of Heyting's *Intuitionism* (Heyting 1956) published in Moscow in 1965. These endnotes are the only historical record of Markov's views on Brouwer's program of intuitionistic foundations of mathematics.

The second contribution concerns Yankov's commentary on Aleksandr S. Yesenin-Vol'pin's[1] (1924–2016) works related to ultra-intuitionistic foundations of mathematics that were reprinted in a volume dedicated to Esenin-Vol'pin (Finn and Daniel' 1999).

There is a peculiar relation between Yankov, Markov and Esenin-Vol'pin. Their general relation is conceptual since all of them held constructivist views on the foundations of mathematics. However, Yankov followed Markov's constructivist viewpoint, whereas Esenin-Vol'pin formulated his ultra-intuitionistic program proceeding from the substitution of Markov's principle of *potential realizability* with his concept of *factual (practical) realizability* (Esenin-Vol'pin 1959). Thus, in some sense, Esenin-Vol'pin's program was formulated as a more strict finitistic alternative to Markov's program of constructive mathematics.

On the other hand, their relation also involves a political aspect: in 1968, Yankov and Markov co-signed the famous letter of the 99 Soviet mathematicians addressed to the Ministry of Health and the General Procurator of Moscow asking for the release of imprisoned Esenin-Vol'pin (Finn and Daniel' 1999, 328–330). As a result, Yankov lost his job at the Moscow Institute of Physics and Technology (MIPT).

10.2 Logic and Foundations of Mathematics in Russia and the Soviet Union and the Rise of Constructive Mathematics

Before proceeding to the appraisal of Yankov's contribution, we will outline the historical background against which Yankov and Esenin-Vol'pin developed their activities. Generally, we will describe the context within which foundational debates and constructive mathematics were advanced in Russia and especially during the Soviet era.

It appears to be an enigmatic question, how constructive mathematics appeared in the Soviet Union. Hostile attitudes towards logic were widespread during the Soviet era; they can be traced back to medieval Rus'. The view that logic was a product of the West alien to the Slavic mentality was propounded by Pyotr Chaadaev's (1794–1856) declaration that "the syllogism of the West is unknown to us" (Chaadaev 1991, I, 93). Nevertheless, scholastic logic was introduced in Muscovite Rus' by Sophronios

[1] Transliterated also as Esenin-Volpin, Yessenin-Volpin, and Ésénine-Volpine.

Leichoudis (1652–1730), who taught for the first time a systematic course of Aristotelian logic in the Moscow Slavo-Graeco-Latin Academy in 1690 (Vandoulakis 2014).

During the 18th century, traditional logic was included as a necessary subject in the university curricula. The founding decree of the Moscow University explicitly states that the professor of philosophy must also teach logic (Vernadsky 1972, 389). Further, traditional Aristotelian logic was also introduced in the Corps of Noble Cadets, established in 1731. During the second half of the 18th century, textbooks in traditional logic were compiled by Russian authors, such as prefect of the Moscow Slavo-Graeco-Latin Academy Makary (Mark) Petrovich (1734–1765), Aleksandr Nikolsky (1755–1834) and others (Anelis 1992, 30–32).

Beginning from the Petrine era (1682–1725), we observe a shift of interest from the traditional logic to research in the field of logic current in the West, namely in the algebra of logic. Platon S. Poretsky (1846–1907) is possibly the most notable representative of Russian researchers in the algebra of logic with several significant contributions (Bazhanov 2007, 147–163). Euler's best-known contribution to logic is the logic diagrams, called *Eulerian circles*, invented in Berlin. They are part of his famous *Lettres à une Princesse d'Allemagne*, written between 1760 and 1762 and addressed to Friederike Charlotte of Brandenburg-Schwedt and her younger sister Louise. However, their first publication (Euler 1768) occurred while he was in Saint-Petersburg. Research beyond the sphere of traditional logic was initiated by Alexander N. Radishchev (1749–1802), who advanced a logic of relations that is absent in traditional Aristotelian logic (Silakov and Stiazhkin 1962, 15).

During the 19th century, foundational studies were advanced at Kazan University. They concern Lobachevsky's studies in the foundations of geometry, notably the advance of an 'imaginary' geometry based on a negation of the axiom of parallels. Following a similar way (Béziau 2017), and working within the framework of Aristotelian logic, but simultaneously criticising traditional Aristotelian logic and the contemporary algebra of logic, Nikolai A Vasiliev (1880–1940) advanced an 'imaginary' logic without the laws of contradiction and excluded middle (Vasiliev 1989). His contribution marks the beginning of non-Aristotelian logic in Russia (Bazhanov 2001, 2016). However, it also marks the beginning of foundational studies in logic. Nikolai N. Luzin (1883–1950) viewed Vasiliev's logic as logic without Law of Excluded Middle and thereby as anticipation, in a certain sense, of Brouwer's (1881–1966) intuitionistic logic (Vandoulakis and Denisova 2020, 21–22).

Besides, an intellectual contact between the Russian logicians with Western developments in the field of foundational studies was established by many translations. Among them, the debate between Louis Couturat (1868–1914) and Henri Poincaré (1854–1912) on logicism that took place in the *Revue de Métaphysique et des Morales* (Poincaré 1905, 1906) became known in Russia in 1915 (Poincaré and Couturat 2007). Couturat's *Les Principes des Mathématiques* (1905a) was translated and published in Saint Petersburg in 1913, and his *L'Algèbre de la logique* (1905b) in Odessa in 1909. Bernard Bolzano's (1781–1848) *Paradoxien des Unendlichen* was published in Odessa in 1911 (Bolzano 1851).

After 1918, several crucial developments in logic remained unknown for a long time, such as the work of Gottlob Frege (1848–1925) and the functional approach to logic. Frege's works became a subject of study quite later, in 1959 by Boris V. Birjukov (1922–2014), on the suggestion of Sofia A. Yanovskaya (1896–1966). Hence, during the early Soviet era, there was no immediate experience of the sensational situation with the discovery of logical paradoxes in Frege's system and the subsequent formulation of competent foundational programmes for the reconstruction of mathematics. Instead, logic and foundational studies had a relatively independent development and were centred around two central axes:

(1) The challenge of the universal validity of the laws of logic.
(2) The doubts against the abstraction of the actual infinite, which included the concept of the actual infinite of the series of natural numbers and the involvement of the axiom of choice in the continuum.

The questions around the first axis appear mainly in logic by challenging the laws of contradiction and excluded middle initiated by N.A. Vasiliev. The second group of problems appear in the circles of the Moscow school of theory of functions, notably in the works of N.N. Luzin (1883–1950).

N.N. Luzin adopted a decisive view against the actual infinite, including the infinity of the set of natural numbers. His views were formed under the influence of the French function theorists, notably Émile Borel (1871–1956), in 1905–06 (Demidov 2018a, b). In a letter to Pavel A. Florensky (1882–1937), dated August 1915, Luzin writes

> "There is no actual infinite! When we dare to talk about it, in fact, we always talk about the finite and about the fact that after *n* there is *n*+1 …; that's all!" (Demidov 2018a, 342 (my translation)).

The same idea is repeated in his *Leçons sur les ensembles analytiques et leurs applications*

> "Ce que nous appelons *l'infini actuel, ce n'est (ou ne serait) que le fini fixe et très grand*" (Lusin 1930, 322).

Further, to Luzin is ascribed the following bold statement

> "The series of natural numbers does not seem to be an absolutely objective structure. It seems to be an artifact of the brain of the mathematician who happens to be speaking about the natural numbers" (Demidov, S. S.; Levshin 2016 (1999), 12).

What is meant is the actual infinite of the series of natural numbers, which Luzin views as a creation of a mathematician's mind.

By referring directly to Luzin, Pyotr Konstantinovich Rashevsky (or Rashevskii) (1907–1983) states some original reflections on the nature of the series of natural numbers. In his view, the mathematical idea of the actual infinite of natural numbers is unnecessary in natural science. Physicists need a mathematical theory of natural numbers, in which the numbers could acquire a "blurred meaning" when they become too great. For a mathematician, adding a unit changes the number. However, what changes for a physicist if a molecule is added into a container with gas? If we adopt this idea, we must abandon our standard concept that any number of the natural series can be obtained by successive counting of units. Rashevsky notes that this conception challenges the principle of mathematical induction, and the "numbers" of such a hypothetical natural series would be objects of another nature (Rashevsky 1973, 244). These reflections anticipate in some sense Esenin-Vol'pin's ideas on the natural series and his criticism of the principle of mathematical induction, but also ideas advanced within the strict finitist approach concerning the sorites paradox and vague predicates (Magidor 2012; Dean 2018).

Ideological hostility towards logic and the philosophy of mathematics did not favour the development of mathematical logic and the discussion on foundational problems during the early Soviet period. How then a whole trend in the foundations of mathematics suddenly appeared, notably Andrei A. Markov's school of constructive mathematics and theory of algorithms? Nagorny points out that Markov's switch to constructivism happened at the end of World War II, when ideological pressure was intense, leading to denouncing several sciences, such as cybernetic and mathematical logic (Nagorny 1994, 469).

Kolmogorov is commonly considered as starting point in the history of mathematical logic and foundational studies in the Soviet Union (Mints 1991), marked by his seminal paper on the principle of excluded middle (Kolmogorov 1925), in which the classical propositional calculus is embedded into an intuitionistic system, called later "the minimal (propositional and predicate) calculus". However, this is not the whole historical picture. We showed above that there was a constant interest in logic and the foundations of mathematics since the 19th century. This current continues during the early Soviet era. However, it was underground, unofficial, and has not been adequately described in histories of mathematical logic in the Soviet Union, which focus on the demonstration of the commitment of Soviet logicians working in the field of the foundations to the 'orthodoxy' of Marxist ideology (Küng 1961; Bocheński 1961, 1973; Bazhanov 1995). Information about this period is very scarce. Much evidence was communicated in personal memories and very often in oral narratives. For instance, Izabella G. Bashmakova (1921–2005) recalled that A.N. Kolmogorov hosted an informal "seminar" or gathering at his home before the Department of Mathematical Logic at the Moscow Lomonosov University was established in 1959. Markov was a regular participant of it. Alexandr S. Kuzichev, a specialist in combinatory logic, also confirmed the work of this seminar. However, it is unknown what topics were discussed in these gatherings, for how long it functioned and how regular it was.

Nevertheless, it is historically confirmed that there was a standard practice during the Soviet period to discuss topics of common interest, be it mathematics, science, or literature, in an informal environment of unofficial gatherings of people who get together, called 'kruzhki' (literally, little circles). In Moscow, there were circles led by Kolmogorov, Alexandrov, Markov, Gelfand and others (Gerovitch 2013, 2016). Kushner mentions a gathering in the mid-1960s, in a *dacha* near Moscow, where I.G. Bashmakova, A.S. Kuzichev, S.A. Yanovskaya and himself were attending Markov's reading of his poems (Kushner 1993, 183). In St. Petersburg (Leningrad), the history of mathematics circles was described by Fomin et al. (1996). These circles functioned outside the official mainstream and served as underground forums of open discussion and dissemination of knowledge.[2] Therefore, an underground current of an unrecorded oral tradition of mathematical logic and the foundations of mathematics remained alive during the Soviet era, which was a significant factor in creating the most remarkable generation of mathematical logicians.

On the other hand, the official environment remained hostile to mathematical logic and the major foundational programmes for years. Philosophers of mathematics, like Vladimir Nikolaevich Molodshij (1906–1986), were directly opposite to Platonism underlying George Cantor's set-theoretic "paradise", but also critical to the ideas of the French "effectivists," like Émile Borel (1871–1956), Henri Lebesgue (1875–1941), René-Louis Baire (1874–1932), which were close to some of Brouwer's ideas. Their ideas were perceived as "subjective idealism" and a variation of Ernst Mach's (1838–1916) philosophy of science (Molodshij 1938, 53, 78). The hostility was strengthened by the fact that Nikolai Nikolaevich Luzin (1883–1950), who was viewed as an adherent to the ideas of the French "effectivists" was accused as "active counter-revolutionary" and persecuted (Molodshij 1938, 78–84; Demidov and Levshin 1999). Thus, David Hilbert's (1862–1943) formalistic philosophy and Brouwer's intuitionism were officially considered trends of idealistic philosophy (Molodshij 1938, 35). Even such a notable mathematician as Aleksandr Yakovlevich Khinchin (1894–1959) did not escape from a subjective exposition of Brouwer's ideas (Khinchin 1926).

The change of the official attitude against mathematical logic in the Soviet Union is substantially associated with the activities of two scholars working outside the field of logic, namely in the history and methodology of mathematics, notably Sofya Aleksandrovna Yanovskaya (1896–1966) and Adolf Pavlovich Yushkevich (1906–1993). Sofya Yanovskaya was one of Yushkevich's teachers in the gymnasium when the Yushkevich family returned to Odessa after 1917.

Yushkevich translated Herman Weyl's (1885–1955) book *Philosophie der Mathematik und Naturwissenschaft* under the concise title *On Philosophy of Mathematics* (Weyl 1934) with an introduction by Yanovskaya that for long remained in the Soviet Union a primary source of information about the Western developments in the philosophy of mathematics. Thus, Brouwer's ideas became initially known to the Soviet scholars through Weyl's perception of them. Two years later, Yushkevich translated Heyting's *Survey of Research on Foundations of Mathematics* (Heyting 1936) with an introduction by Kolmogorov.

Yanovskaya was the editor in Russian of classical logical Western books, such as David Hilbert and Wilhelm Ackermann's (1896–1962) *Grundzuge der Theoretischen Logik* (Hilbert and Ackermann 1928) in 1947, Alfred Tarski's (1901–1983) *Intro-*

[2] It seems that this practice was also popular in other countries of the Soviet block. For instance, the Hungarian mathematician Dénes Nagy remembers a similar kind of "seminar" or gathering on history of mathematics, which he regularly visited, hosted by classical scholar and historian of mathematics Árpád Szabó (1913–2001) at his home. Despite his international recognition, Szabó was persecuted by the Hungarian authorities because of his support to the Hungarian Uprising of 1956.

duction to the Logic and Methodology of Deductive Sciences (Tarski 1940) in 1948,[3] George Pólya's (1887–1985) *Mathematics and Plausible Reasoning* (Pólya 1954) in 1957, Rudolf Carnap's (1891–1970) *Meaning and Necessity* (Carnap 1947) in 1959, Alan Turing's (1912–1954) "Computing Machinery and Intelligence," (Turing 1950) translated as *Can Machines Think?* in 1960. Moreover, Yanovskaya delivered for the first time a course on mathematical logic in the Faculty of Mathematics and Mechanics at the Moscow Lomonosov University in 1936 and established a research seminar there in cooperation with Ivan Ivanovich Zhegalkin (1869–1947) and Pyotr Sergeyevich Novikov (1901–1975) in 1943. In 1957, Novikov became the head of the newly established department of mathematical logic at the Steklov Mathematical Institute of the USSR Academy of Science. In 1959, Markov became the head of the newly established department of mathematical logic at the Moscow Lomonosov University and joined the leadership of the seminar. These developments mark the institutionalisation of mathematical logic as a recognised mathematical discipline in the Soviet Union.

Gradually, the anti-Platonic orientation of Brouwer's philosophy of mathematics started to be attractive to Soviet philosophers because it was considered compatible with the principles of dialectical materialism. During the thaw, severe criticism gave its place to efforts to release the mathematical content of Brouwer's intuitionism from his general philosophical viewpoint. In this line, Brouwer's concept of *intuition* was studied, particularly by the Soviet philosopher Valentin Ferdinandovich Asmus (1894–1975) (Asmus 1963) who distinguished intuition as used in mathematical problems and mathematical creative imagination from intuition used in philosophical contexts, which is irrelevant to mathematics.

10.3 Yankov's Contribution to the History of Constructive Mathematics

During the 1950s, Markov's constructive mathematics was flourished and started to be perceived as an approach on foundations of mathematics alternative to intuitionism. However, in the absence of rival foundational programmes in the Soviet Union, such as logicism, formalism, and intuitionism in the West, Markov's constructive mathematics cannot be viewed as a direct reaction to the logical paradoxes. It originates from the debates on the concept of actual infinite, which were developed in the Moscow school of the theory of functions that go back to Luzin and his circle (Demidov 2018a, b). Although Markov's constructive mathematics can be considered as a variation of intuitionistic mathematics (Troelstra and van Dalen 1988, 1, 3–4), the source out of which it came was Markov's research in applied mathematics, notably on the theory of *normal algorithms* which he mainly advanced while he was

[3] Yanovskaya was attacked by V.P. Tugarinov and L.E. Majstrov (1950) for these two publications, and was compelled to reply to the critics (Yanovskaya 1950). See also Yanovskaya's reply to Iwasaki (1977) on a number of related questions.

in Leningrad in the course of his research on the identity problem for semigroups (Markov 1954a).

Markov's constructivism has never acquired the status of a trend in Soviet philosophy of mathematics. No 'constructivist philosophy of mathematics' was ever formulated neither by Markov or his followers nor by the Soviet philosophers of mathematics. This fact was partly caused by the reluctance of the mathematicians of Markov's school, including Markov himself, from publicly expressing their philosophical views. They preferred to remain on the solid ground of mathematical proving activity. This stance was reasonable, provided that Markov became a target of attacks by philosophers and mathematicians, ardent champions of Marxist philosophy, as Nagorny reports (Nagorny 1994, 469).

Moreover, the school included repressed and persecuted mathematicians, such as Nikolai M. Nagorny (1928–2007), who was sent to a GULAG camp with his wife (as they revealed to me in 1990), Vadim A. Yankov, and others. None of Markov's pupils became a professor at the department of mathematical logic of the Moscow Lomonosov University or a fellow at the Steklov Mathematical Institute of the Academy of Sciences (Nagorny 1994, 469). Vadim Yankov was involved in the dissident movement, arrested in 1982 and sentenced to four years in prison and three years in exile. He was given amnesty and released in January 1987, and rehabilitated on 30th October 1991.

Boris Abramovich Kushner (1941–2019) indicates that Markov had a conception of constructive mathematics. Kushner witnessed a curious encounter between Markov and Errett Albert Bishop (1928–1983) in 1966 during the International Mathematical Congress held in Moscow. He recalls that when he was ready to enter the office of the department of mathematical logic on the sixteenth floor of the main building of the Moscow Lomonosov University, he heard voices inside, and Bishop rushed out, followed by Markov with an enigmatic smile and one of his closest associates (whom Kushner does not name) who repeated excitedly "But he has no standpoint!" Kushner implies that a debate between Markov and Bishop was taken place during which it was revealed that Markov had a philosophical or methodological standpoint, whereas Bishop was concerned with "live" mathematical activity without following any elaborate conception (Kushner 1993, 188–190).[4] However, if Markov had some philosophical or methodological viewpoint, as Kushner implies, what this 'constructivist' viewpoint was?

---

[4] That Bishop lacked any methodological standpoint is a rather exaggerated claim. Kushner does not ascribe this view to Markov, but to his anonymous associate. The first chapter of his book (Bishop 1967, 1–10), entitled "A Constructivist Manifesto" contains what Kushner calls "an inspired" standpoint (Kushner 1993, 190). Further, Bishop summarizes the principles of his methodological viewpoint in (Bishop 1983). Billinge argues that Bishop

> "did his mathematics in a constructive manner for explicitly philosophical reasons."

Nevertheless, Bishop's philosophical ideas "cannot be rounded out into an adequate philosophy of constructive mathematics" (Billinge 2003, 177).

By the time of the International Mathematical Congress in Moscow, Markov's "standpoint" underlying his constructive mathematics was sketched in 1950 (Markov 1950) and outlined in 1962 (Markov 1962). These works can be characterised primarily as methodological; they briefly expose the methodological principles of constructive mathematics but avoid discussing philosophical problems. Markov's philosophical views are nowhere formulated. In lack of explicit formulation of Markov's philosophical standpoint, Nikolai Nagorny, for instance, believed that Markov was possibly a positivist, as said to me in 1990.

Fortunately, we have evidence about Markov's views on other foundational programs because of Yankov, who had the brilliant idea to suggest to Markov add in the Russian edition of Heyting's *Intuitionism* a new person—"Con,"—representing the constructivist, in Markov's sense. Thus, "Con" enters in a peculiar one-sided dialogue with Heyting's fictional persons "Class," "Form," "Int," "Pragm," and "Sign," which represent classical mathematics, formalism, intuitionism, pragmatism and significism, respectively. "Con" entertains the extraordinary privilege that he can criticise all other representatives of the team without himself being liable to criticism! (Vandoulakis 2015).

"Con" focuses his criticism predominantly against Brouwer, whom he perceives as his principal antagonist. He essentially ignores the other representatives, except David Hilbert, to which he devotes a sarcastic comment about his program to "save" the "valuable" mathematical results that lacked content:

> "what to save and why?" (Heyting (Markov) 1956, 162 (my translation)).

This comment should be taken to mean that Markov, like Brouwer, rejected classical mathematics, specifically mathematical results about the abstract existence of mathematical objects, obtained by indirect proof that cannot be found by an 'algorithmic' procedure.

10.4 Markov's Philosophy of Constructive Mathematics

Because of Yankov's initiative, we have today evidence about Markov's view primarily on intuitionistic mathematics from the point of view of what could be called "Markov's philosophy (or methodology) of constructive mathematics." This evidence is given in Markov's endnotes to Heyting's *Intuitionism*, which includes "Con"'s criticism. This philosophy of constructive mathematics evolves around certain principal points outlined below that Markov advances while criticising Brouwer's views.

10.4.1 Mathematical Objects

Markov admits only *constructive objects* in mathematics. By constructive objects, he means not mentally constructed objects, like in intuitionism, but concrete objects, like the letters of an alphabet, that is a (finite or infinite) collection of discernible signs. Thus, mathematical objects are real objects obtained as an outcome of a process executable in a computer. Hence, constructive mathematics studies *constructive processes* and *constructive objects* generated by them. By constructive processes are meant step-by-step processes which from certain initial configurations of signs, new configurations are formed following defined formation rules. It should be noted that the concepts of "constructive process" and "constructive object" remain undefined.

Markov agrees with Brouwer that the study of constructive objects requires a new form of logic.[5] However, on the question of what logic is required, they diverge: for the latter, it is intuitionistic logic, but for the former, it is the constructive (in Markov's sense) mathematical logic. Nevertheless, both these logics are free from the Law of the Excluded Middle.

10.4.2 The Infinite

Markov considers that the infinite is introduced in mathematics by abstraction (idealisation). He distinguishes between the "unclear" (in Markov's view) *abstraction of the actual infinite*, which is used to introduce (unintuitable) complete infinite totalities, and the *abstraction of potential realizability*[6] that abstracts from any practical spatial, temporal or material limitations on our capacity of constructing (concrete or abstract) mathematical objects. This abstraction enables us to conduct reasoning on as lengthy constructive processes and as large constructive objects as required. Therefore, constructive objects are only those, which are *not* generated by abstractions more powerful than the abstraction of potential realizability.

Markov assumes a philosophical stand about abstractions in the late 50s:

> "Abstractions are necessary for mathematics; however, they must not be devised for their own sake and lead where there is no return down to "earth". We should always remember to pass from abstract thinking to practice as a necessary step of human cognition of objective reality. In case that the possibility of such a passage is turned out to be too doubtful, it is necessary to reconsider the abstractions applied and try to modify them." (Markov 1958, 315–316 (my translation)).

[5] On this point there is a fundamental divergence between both Markov and Brouwer, on the one hand, and Bishop, on the other hand, who does not assumes any constructive logic for his constructivisation of mathematics.

[6] This term was introduced by Markov during the second half of 1940s.

In line with this thesis, Markov understands Brouwer's mental constructions as potentially realisable since they have (practically) realisable material constructions as archetypes. In this way, he reinterprets Brouwer's idea of *potential infinite* in terms of his abstraction of potential realizability in an attempt to "return down to earth".

10.4.3 Mathematical Existence

Markov identifies mathematical existence not with constructability, like Brouwer, but with the potential realizability of a construction. As stated already in 1962,

> "in constructive mathematics, the existence of an object with given properties is considered proved only when a way of potentially realisable construction of the object with these properties is indicated." (Markov 1962, 9 (my translation)).

However, it should be emphasised that this construction is not perceived as a process evolving in time like in Brouwer's concept of creative subject or Kripke's scheme. An object exists whenever it can be indicated as a complete finite word in an alphabet or given by a pair (letter, algorithm), and it is known that the algorithm applies to the letter. If such a pair cannot be constructed, or the algorithm's applicability cannot be established, this does not mean that the object does not exist. An object does not exist only when the impossibility for the object to be constructed is proved. For instance, if the inapplicability of the corresponding algorithm is proved. In this case, the object under consideration does not exist eternally.

10.4.4 Normal Algorithms

Markov rejected Brouwer's concept of *infinitely proceeding sequences*, which is not defined by a definite law but can be an object of ever-creating mental construction. He regarded it as non-evident and possibly non-constructive. A possible explanation for Markov's view might be the indeterminacy or eventual impossibility of practical realisation of the acts of determination of the successive components of the sequence. The following comment suggests this explanation:

> I cannot but feel sorry for the man whom you are ready to force to do so many [acts of] "free choice" or "dice drops." My understanding of the infinitely proceeding sequence is more humane [than yours] since a computer can efficiently execute algorithms. Moreover, what is most important is that my understanding is constructive because the concept of the algorithm can be standardised, which makes possible the coding of an algorithm and its recording by "letters" in a fixed alphabet. In turn, algorithms themselves can become constructive objects. It is possible to apply other algorithms to them, which is very important in constructive analysis.Your infinitely proceeding sequences are not constructive objects, and I cannot manage them. (Heyting (Markov) 1956, 166 (my translation)).

Thus, Markov suggests his concept of the *normal algorithm* that he discovered in his study on the identity problem for semigroups.

Accordingly, Markov reinterprets Heyting's concept of *spread* (given in terms of the spread-law and the complementary law) by using the concept of normal algorithm instead of the concept of law. However, under the new understanding of the concept of spread, in Markov's sense, the fan theorem is no longer true (Zaslavsky and Tseitin 1962).

10.4.5 Church Thesis

Church Thesis was stated independently by Alonso Church (1903–1995), Emil Leon Post (1897–1954) and Alan Turing in 1936 (Olszewski et al. 2006, 7) and expressed the fact that certain refinements of the concept of the algorithm (such as, for instance, the concepts of the recursive function, λ-definable function, Turing machines, etc.) are adequate explications of the broad intuitive concept of algorithm.

In Markov's constructive mathematics, the Church Thesis assumes the form of the *principle of normalisation of algorithms*, which states that every verbal algorithm in an alphabet V is equivalent with respect to V to some normal algorithm in V, or, concisely, every verbal algorithm is normalisable.

Church Thesis is a point of fundamental divergence between Markov's constructive mathematics and intuitionism. Heyting considers this thesis in two subsequent papers (Heyting 1962, 1969), arguing against its adoption.

10.4.6 The Concept of Number and the Continuum

Markov's concept of natural number is essentially the same as the intuitionistic understanding presented in Heyting's *Intuitionism*. For Markov, natural numbers are defined as words of the form |, | |, | | |, and so forth, over the alphabet |. The abstraction of potential realizability does not allow the formation of "infinite" words or the collection of "all" words over a given alphabet taken for a completed totality.

By adding to the natural numbers all the words of the form $-N$, where "$-$" is a new letter and N is a natural number, we get the integers.

Further, rational numbers are understood as words of a certain type over the alphabet {|, –, /} ("–" is the sign of minus, "/" is the sign of fraction). A constructive sequence of rational numbers is a normal algorithm that maps every natural number into a rational number.

A pair of normal algorithms (encoded appropriately by a word) is a constructive real number if the first algorithm is a constructive sequence of rational numbers and the second effectively estimates the rate of convergence of this sequence.

This kind of constructive continuum has essential properties that do not occur in the classical continuum. For instance, all constructive real functions are continuous, i.e., no real function can have a constructive discontinuity at any point (Tseitin 1962).

10.4.7 Constructive Mathematics is a Technological Science

This idea is not fully developed by "Con" in Heyting's *Intuitionism*. "Con" criticises Heyting's view that mathematics studies certain functions of the human mind and, therefore, it is more akin to philosophy, history, and the social sciences (Heyting 1956, 3rd ed. 1971, 10). Markov's objection is based on the argument that a human, together with his mental constructions, is part of nature. Mental constructions, such as the construction of greater and greater natural numbers, have material archetypes in reality. Moreover, mental constructions, such as complex algorithms, are initially conceived as mental constructions but are implemented afterwards as computer programs. Consequently, mental constructions are not considered by Markov as falling under social sciences.

This line of argumentation is advanced in an unfinished manuscript written during the last months of Markov's life that N. Nagorny published in 1987. In this manuscript, it is stated that constructive mathematics is essentially engineering

> "because [constructive mathematics] investigate and supply instruments, applied in various spheres of human activity. In this respect, it is like engineering" (Markov 1987, 212 (my translation)).

Thus, Markov's view on constructive mathematics is a viewpoint of a specialist primarily interested in the applications of mathematics. At this point, it seems that Markov meets Bishop. Thus, Markov's viewpoint can be possibly more adequately characterised as a metamathematical (methodological) viewpoint with naturalistic philosophical underpinnings.

It is noteworthy that Heyting examined Markov's comments expressed in the endnotes in the third edition of his *Intuitionism* in 1971. Thus, because of Yankov's contribution, these endnotes served as a starting point of a historical "dialogue" initiated between the mathematical schools of Brouwer's intuitionism and Markov's constructivism.

It should be clarified that Markov's notes do not offer an overall comparison between his version of constructive mathematics and intuitionism as exposed in Heyting's *Intuitionism.* Certain fundamental divergences are not examined here, for instance, the so-called *Markov's principle of constructive selection* [Markov 1954b, 1956, 1962]. According to this principle, if a constructive process, given by some prescription, is not potentially infinite, then the process terminates. The principle is essential for the proof of certain theorems in Markov's mathematical analysis. However, the intuitionists do not accept it because it involves an *ad hoc* use of an indirect argument. Moreover, it remained controversial and insufficiently evident even among some Markovian constructivists (Kushner 1973, 45).

Markov does not provide a detailed account of his constructive logic in comparison to Brouwer's intuitionistic logic. The semantics for Markov's constructive logic has profound differences from that of intuitionistic logic and was a later development; it is based on the idea of a hierarchy of (formal) languages (Markov 1974a, b, c, d, e, f, g).

10.5 Yankov on Esenin-Vol'pin's Ultra-Intuitionism

In contradistinction to Markov's achievements, which are presented in due detail in the Soviet bibliography in the history of mathematics, Esenin-Vol'pin's work on foundations of mathematics was not presented or analysed in the early Soviet historiography of mathematics. For instance, his name is absent in (Kurosh et al. 1948).

The first references to Esenin-Vol'pin can be found in (Kurosh et al. vol. 1, 1959). S.A. Yanovskaya, who wrote the section on mathematical logic and foundations of mathematics, gives due merit to Esenin-Vol'pin's achievements. According to A.S. Kuzichev's oral evidence, Yanovskaya had made substantial efforts by exerting her influence on the political hierarchy of the Communist Party to protect Esenin-Vol'pin, release him from the forced hospitalisation, and secure his right to publish his works. Esenin-Vol'pin's results on axiomatic set-theory (Esenin-Vol'pin 1954, 1957) are extensively presented in the above volume, namely his version of axiomatic set theory without the axiom of choice is mentioned, in which the continuum hypothesis and Suslin hypothesis are not derivable (Kurosh et al. vol. 1, 1959, 20–22, 90–91, 99). Special emphasis is given to his criticism of the abstraction of potential realizability and the principle of mathematical induction. Further, Esenin-Vol'pin's concept of *factual realizability* is discussed. Yanovskaya notes that this concept is not evident from the classical point of view and is not formalisable in a traditional logical system.[7]

[7] An attempt to formalize Essenin-Vol'pin's ultra-intuitionistic viewpoint was undertaken by Geiser (1974) by means of nonstandard analysis. Nonstandard analysis provides a possible interpretation of some aspects of Esenin-Vol'pin's proof-theoretic concepts (Esenin-Vol'pin 1961, 1970; Yessenin-Volpin 1981) that can give a proof of consistency for Zermelo-Fraenkel set theory. Geiser's proof

Yanovskaya outlines Esenin-Vol'pin's foundational program to prove the consistency of classical set theory by ultra-finitistic means (Kurosh et al. vol. 1, 1959, 24–25).[8]

In the section on topology, P.S. Aleksandrov and V.G. Boltyansky mention Esenin-Vol'pin's early contribution to abstract topology (Esenin-Vol'pin 1949a, b), notably the theory of dyadic bicompacta (Kurosh et al. vol. 1, 1959, 232).

The exposition of Esenin-Vol'pin's results in (Shtokalo et al. 1966-1970, 1968, vol. 3, 457; vol. 4.2, 439–440) is more concise and adds nothing essentially new compared to Yanovskaya's survey. Four significant contributions are mentioned (Esenin-Vol'pin 1954, 1959, 1960, 1967); the second and the fourth of them were reprinted in a volume dedicated to Esenin-Vol'pin (Finn and Daniel' 1999). The first two sections of the last volume are devoted to philosophy and logic. The first section contains Esenin-Vol'pin's major work on the ultra-intuitionistic program of foundations of mathematics (Esenin-Vol'pin 1993) and Finn's commentary on it. The second section contains four works (Esenin-Vol'pin 1959, 1967, 1971, 1999) and Yankov's concise commentary.

10.5.1 On the Concept of Natural Numbers and "Factual (Practical) Realizability"

Yankov associates Esenin-Vol'pin with the trend of Brouwer's intuitionism, which reduced mathematics from the clouds of set theory to the earth of the (potentially infinite) sequence of natural numbers and the sequences isomorphic to it. However, Esenin-Vol'pin challenged the concept of the (potentially infinite) sequence of natural numbers. He believed that the assumption that one can reach *any* number by successive steps, beginning from the unit, contains a powerful idealisation.[9] Esenin-Vol'pin claims that neither a human nor a machine can reach a number if it is too great, e.g. numbers of the form 10^{12}.[10] Therefore, he suggests abandoning the idealisation of the (potentially infinite) sequence of natural numbers in an analogous way that Brouwer abandoned the idealisation of actual infinite. This thesis is the core of Esenin-Vol'pin's viewpoint.

Consequently, Essenin-Vol'pin considers the human mind capable of grasping and handling only the small finite, although also able to understand the potentially infinite. Thus, the large finite may also be understood in a potential or modal sense. Accordingly, Essenin-Vol'pin challenges Markov's concept of abstraction of potential realizability, which he substitutes with his assumption of *factual (practical) realizability* and the related principle of mathematical induction. In sum, he suggests substituting the concept of the infinite with the concept of *unrealisable*. This assump-

concerns a weak system of set theory with a Dedekind infinity axiom. To this effect, Geiser is compelled to define a non-classical concept of proof.

[8] A reconstruction of Esenin-Vol'pin's consistency proof was given by Gandy (1982).

[9] Cf. Rashevsky's reflections presented earlier.

[10] Cf (Isles 1992).

tion leads to another picture, in which, as Yankov notes, there exist many "sequences of natural numbers" in this case, which are not isomorphic.[11] One of them might be "shorter" or "lengthier" than the other. Although this situation might appear to be paradoxical, there is no contradiction in it. Thus, Yankov concludes that this kind of mathematics is legitimate.

10.5.2 On the Ultra-Intuitionistic Program of Foundations of Mathematics

Esenin-Vol'pin interprets the classical Zermelo-Frenkel set-theoretic mathematics in ultra-intuitionistic terms. This interpretation is constructed on the assumption that different sequences of natural numbers exist so that some of them are closed under certain operations and cannot reach other sequences. In this way, classical mathematics is proved consistent from the standpoint of ultra-intuitionism. However, the appeal to assumptions diminishes the value of the proof, according to Yankov. To support his view on this point, Yankov also communicates the opinion of Pyotr Novikov expressed personally to him. Yankov's cautious attitude towards Esenin-Vol'pin's result is not groundless. Pavel Pudlák's view is more vividly expressed

> "He claimed that he proved the consistency of a set theory up to some large number *n* using a finitistic argument. Unfortunately his arguments were obscure and I have not met anybody who could explain to me, what actually Esenin-Volpin proved" (Pudlák 1996, 66).

Esenin-Vol'pin's program attracted the attention of Paul Bernays (1888–1977) and Kurt Gödel (1906–1978). Bellotti presents an exchange of letters between Bernays and Gödel in 1962–1963 with an extensive discussion on Esenin-Vol'pin's program and the possibility of proof of the consistency of classical mathematics in the ultra-intuitionistic framework (Gödel 2003, 204–233). Gödel shows a sceptical attitude, but Bernays is more sympathetic to Esenin-Vol'pin, although ultimately unconvinced by his program (Bellotti 2008, 14–17).

Yankov adopts a moderate view and poses certain open questions.

[11] See also (Isles 1981).

> "The mathematicians have to deal with them [Esenin-Vol'pin's works] and give an answer to the questions: can ultra-intuitionism be interpreted in classical mathematics or an appropriate extension of it? In this case, what do the ultra-intuitionistic theorems mean when translated into the classical language? Or, maybe this is not possible? However, if, as Esenin-Vol'pin demanded, all classical mathematics can be embedded into ultra-intuitionism, what enrichment is attained compared to classical mathematics?
> When this is done, we will be able to evaluate the content and significance of the titanic work to which A.S. [Esenin-Vol'pin] devoted most of his energy and life" (Finn and Daniel' 1999, 117–118 (my translation)).

Consequently, Yankov considers Esenin-Vol'pin's ultra-intuitionistic program an unfinished project, possibly not unified, but admitting various approaches.

Research on Esenin-Vol'pin's program continues, and many interesting results have been obtained. Nevertheless, no satisfactory reduction of the infinite to the ("hyper")-finite, in some sense compatible with Esenin-Vol'pin's insight, was attained. A survey of these attempts is given by Bellotti (2008). An interpretation of ultra-finitism in terms of *feasibility* was proposed by Parikh (1971). However, his interpretation is not inspired by the view of reduction of the infinite to the finite. Instead, it is motivated by an "anthropomorphic" view of mathematics. Another interpretation was advanced by Engeler (1981), who is guided by explaining why finite minds can perceive infinite totalities (Engeler 1981, 347).

10.5.3 Esenin-Vol'pin's Works on Modal and Deontic Logics

A last noticeable contribution of Yankov is the inclusion in the volume (Finn and Daniel' 1999) of three works of Esenin-Vol'pin (1967, 1971, 1999) that remained in the shadow of his ultra-intuitionistic program. The last work was published during the 1970s; however, as Elena Lysanyuk informed me, it is impossible to find an original copy. Thus, Yankov has also contributed in this way to the preservation of Esenin-Vol'pin's legacy.

Although Yankov does not touch these works in his commentary, they have become a subject of modern research. S.M. Kuskova's (2012) study on deontic logic is devoted to the first two works. She pays attention to Yesenin-Vol'pin's concept of "tactics" as the set of orders correlated with suitable situations and his hierarchy of tactics which, in her view, enables us to explain the principles of Kant's autonomic ethics and legislative systems. On the other hand, Olga S. Kovalevich and Elena N. Lisanyuk (2015) associate the works mentioned above with agent-oriented and indeterministic approaches to deontic logic.

10.6 Conclusion

As shown, Russia and USSR followed a different way from Western development in the history of foundational studies. Russia passed through the Western stages of the development of logic, i.e., established intellectual contacts with scholastic logic and the algebra of logic up to the 19th century. However, the Soviet logicians did not experience the sensational situation with the logical paradoxes in set theory that gave rise to foundational studies in the West. Frege's works were translated into Russian much later, and the philosophies of the great foundational programs—logicism, formalism, and intuitionism—were not widespread during the early Soviet era because of the official hostile attitude towards them.

Reflexion over the foundations of mathematics appeared in Russia and the USSR concerning the fundamental concepts of mathematical analysis and the theory of functions. These reflexions are connected with the concept of actual infinite, challenged by N.N. Luzin, under the influence of the French school of theory of functions, notably Émile Borel. The controversies about the actual infinite were widespread in the milieu of the Moscow mathematical school at the beginning of the 20th century, and the idea of cautious or negative attitude towards the actual infinite remained alive in the underground mathematical circles of Moscow. The rejection of the actual infinite is the starting point of both trends developed in Moscow during the Soviet period—Markov's constructivism and Esenin-Vol'pin's ultra-intuitionism. Nevertheless, none of these trends had ever become official, despite their evident anti-platonic orientations. On the contrary, several of their members were persecuted.

Yankov's place in the history of foundational studies in the Soviet Union is crucial because of his association with these schools. His significant contribution is the preservation of Markov's views about Brouwer's intuitionism and, to a lesser extent, to Hilbert's formalism. He achieved to get a historical "instant shot" of Markov's views advanced by 1965 by including Markov as a peculiar interlocutor of Brouwer and Hilbert in the Russian translation of Heyting's *Intuitionism*. Markov's dialogue with Heyting's fictional persons initiated a real dialogue between Markov and the intuitionists since Heyting responded to it in the third edition of his book in 1971. Moreover, it offers the necessary historical material to reconstruct the principles underlying Markov's "philosophy of constructivism."

References

Anellis, I. H. (1992). Theology against logic: The origins of logic in old Russia. *History and Philosophy of Logic, 13*, 15–42.

Asmus, V. F. (1963). *Problema intuicii v filosofii i matematike, The problem of intuition in the philosophy of mathematics*. 2-e izdanie, Moscow, 1965. Available online [in Russian]. http://psylib.ukrweb.net/books/asmus01/index.htm (Accessed 10-11-2020).

Bazhanov, V. A. (1995). *Prervannyj polet: istorija "universitetskoj" filosofii i logiki v Rossii, The interrupted flight. The history of "University" philosophy and logic in Russia*. Moscow: Moscow University Press [in Russian].

Bazhanov, V. A. (2001). *The origins and emergence of non-classical logic in Russia (nineteenth century until the turn of the twentieth century), Zwischen traditioneller und moderner Logik* (pp. 205–217). Nichtklassiche Ansatze: Mentis-Verlag, Paderborn.
Bazhanov, V. A. (2007). *Istorija logiki v Rossii i SSSR*. Moscow: History of logic in Russian and the USSR. [in Russian].
Bazhanov, V. A. (2016). Russian origins of non-classical logics. In F. F. Abeles & M. E. Fuller (Eds.), *Modern logic 1850–1950* (pp. 197–203). East and West: Studies in universal logic.
Bellotti, L. (2008). Some attempts at a direct reduction of the infinite to the (large) finite. *Logique et Analyse, 201*, 3–27.
Béziau, J. -Y. (2017). Is modern logic non-Aristotelian? In: V. Markin & D. Zaitsev (Eds.), *The logical legacy of Nikolai Vasiliev and modern logic*. Synthese library (Studies in epistemology, logic, methodology, and philosophy of science) (Vol. 387). Cham: Springer. https://doi.org/10.1007/978-3-319-66162-9-3.
Billinge, H. (2003). Did Bishop have a philosophy of mathematics? *Philosophia Mathematica, 3*, 176–194.
Bishop, E. A. (1983). Schizophrenia in contemporary mathematics. In *Errett Bishop: reflections on him and his research*. San Diego, California, 1983, Contemp. Math. 39 (Prov-idence, RI, 1985), 1–32.
Bishop, E. A. (1967). *Foundations of constructive analysis*. New York: McGraw-Hill.
Bocheński, J. M. (1961). Soviet logic. *Studies in Soviet Thought, 1*, 29–38.
Bocheński, J. M., & Janovskaja, S. A. (1973). *Studies in Soviet Thought, 13*, No. 1/2, 1–10.
Bolzano, B. (1851). *Paradoxien des Unendlichen*. Leipzig, C.H. Reclam. English translation: *Paradoxes of the infinite*. Translated by D. A. Steele. London: Routledge, 1950. Russian translation: *Paradoksy biezkoniečnaho*. Odessa: Mathesis, 1911 (2nd ed.) 1999.
Carnap, R. (1947). *Meaning and necessity: A study in semantics and modal logic*. Enlarged edition 1956. Russian edition *Znacheniye i neobkhodimost'*. Introduction by S. A. Yanovskaya. Moscow, Gos. Izd-vo inostrannoj literatury, 1959.
Chaadaev, P. (1991). *Pollnoe sobranie sochinenij i izbrannye pis'ma [Collected works and selected correspondence]*. 2 vols. Moscow: Nauka [in Russian].
Couturat, L. (1906). Pour la Logistique (réponse à M. Poincaré), *Revue de Métaphysique et des Morales*, XIV, 208–250.
Couturat, L. (1905a). *L'Algèbre de la logique*. Russian translation: *Algebra logiki [Algebra of logic]*. Odessa [1909].
Couturat, L. (1905b). *Les Principes des Mathématiques: avec un appendice sur la philosophie des mathématiques de Kant*. Republished 1965, Georg Olms. Russian translation: *Filosofskiye printsipy matematiki*. SPb., Izd. Karabasnikova, 1913.
Dean, W. (2018). Strict finitism, feasibility, and the sorites. *Review of Symbolic Logic, 11*, 295–346.
Demidov, S. S. (2018a). N. N. Luzin i struktura chislovogo kontinuuma [N.N.Luzin and the structure of the numeric continuum], *Algebra, teoriya chisel i diskretnaya geometriya: Sovremennyye problemy i prilozheniya*. Materialy XV Mezhdunarodnoy konferentsii, posvyashchennoy stoletiyu so dnya rozhdeniya professora Nikolaya Mikhaylovicha Korobova. Tula: Izdatel'stvo: Tul'skiy gosudarstvennyy pedagogicheskiy universitet im. L.N. Tolstogo, 341–343 [in Russian].
Demidov, S. S. (2018b). N. N. Luzin i problema stroyeniya kontinuuma [N. N.Luzin and the problem of the structure of the continuum], *Funktsional'nye prostranstva. Differentsial'nye operatory. Problemy matematicheskogo obrazovanija*. Tezisy dokladov Pyatoj Mezhdunarodnoj konferentsii, posvjashchonnoy 95-letiju so dnja rozhdenija chlena-korrespondenta RAN, akademika Evropeyskoj akademii nauk L. D. Kudrjavtseva. Moscow: Rossijskij universitet druzhby narodov 422 [in Russian].
Demidov, S. S., & Levshin, B. V. (Eds.) (1999). *Delo akademika Nikolaya Nikolaevicha Luzina*. [*The case of Academician Nikolai Nikolaevich Luzin*]. Institut istorii estestvoznanija i tehniki im. S. I. Vavilova RAN, Arhiv Rossijskoj Akademii Nauk. St. Petersburg: Russkii Khristianskii Gumanitarnyi Institut. Available online [in Russian] http://www.ihst.ru/projects/sohist/books/

luzin.pdf. (Accessed 10-11-2020). English translation: *The case of Academician Nikolai Nikolaevich Luzin*, translated from the Russian by Roger Cooke. American Mathematical Society, 2016.

Engeler, E. (1981). An algorithmic model of strict finitism. B. Dömölki & T. Gergely (Eds.), *Mathematical logic in computer science*. Colloquia Mathematica Societas János Bolyai (Vol. 26, pp. 345–357). Amsterdam: North-Holland.

Esenin-Vol'pin, A. S. (1949a). O zavisimosti mezhdu lokal'nym i integral'nym vesom v diadicheskikh vikompaktakh [On the relation between the local and integral weight in dyadic bicompacta]. *Doklady Akademii Nauk SSSR, 68*, 441–444. [in Russian].

Esenin-Vol'pin, A. S. (1949b). O syschestvovanii universalnogo bikompakta ljubogo vesa [On the existence of universal bicompact of any weight]. *Doklady Akademii Nauk SSSR, 68*, 649–652. [in Russian].

Esenin-Vol'pin, A. S. (1954). Nedokazuemost' gipotezy Suslina bez pomoschi aksiomy vybora v sisteme aksiom Bernajsa-Mostovskogo [Non-provability of Suslin's hypothesis without the help of the axiom of choice in the Bernays-Mostowski system]. *Doklady Akademii Nauk SSSR, 96*, 9–12. [in Russian].

Esenin-Vol'pin, A. S. (1957). Dokazatel'stvo neprotivorechivosti klassicheskoj arifmetiki s pomosch'ju induksii do ε_0 (po Shjute). Proof of inconsistency of classical arithmetic with the help of induction (according to Schütte), Appendix VII in the Russian translation of S. C. Kleene *Introduction to metamathematics* (1952), 1957, translated by A. S. Esenin-Vol'pin, 485–492 [in Russian].

Esenin-Vol'pin, A. S. (1959). Analiz potentsial'noj osuschestvimosti [Analysis of the potential relizability]. *Logicheskie issledovanija*, 218–262 [in Russian].

Esenin-Vol'pin, A. S. (1960). K obosnovaniju teorii mnozhestv [On the foundation of set theory]. *Primenenie logiki v nauke i tekhnike*, 22–118 [in Russian].

Esénine-Volpine, A. S. (1961). Le programme ultra-intuitionniste des fondements des mathématiques, *Infinitistic methods* (*Proceedings of Symposium. Foundations of Mathematics*, Warsaw, 1959), Oxford: Pergamon, 201–223. See also: MR 0147389 Reviewed by Kreisel, G. and Ehrenfeucht, A. 1967, Le Programme Ultra-Intuitionniste des Fondements des Mathematiques by A. S. Ésénine-Volpine, *The Journal of Symbolic Logic*, Association for symbolic logic, *32*, 517.

Esenin-Vol'pin, A. S. (1967). O teorii modal'nostej [On the theory of modalities]. *Logika i metodologija nauki. IV Vsesojuznyj simpozium. Kiev, ijun', 1965* (pp. 56–67) [in Russian].

Esenin-Vol'pin, A. S. (1971). O logike nravstvennykh nauk [On the logic of moral sciences], Obshchestvennyye problemy: Samizdat-zhurnal / pod red. V. Chalidze. No 12.

Esenin-Vol'pin, A. S. (1993). Ob antitraditsionnoj (ul'traintuitsionistskoj) programme osnovanij matematiki i estestvenno-nauchnom myshlenii [On the anti-traditional (ultra-intuitionistic) program of foundations of mathematics and natural-scientific thinking]. *Semiotika i informatika, 33*, 13–67. [in Russian].

Esenin-Vol'pin, A. S. (1999). O teorii disputov i logike doveriya [On the theory of disputes and the logic of trust], [Finn, Daniel' 1999, 178–192].

Euler, L. (1768). *Lettres à une Princesse d'Allemagne*, St. Petersburg; l'Academie Imperiale des Sciences.

Finn, V. K., & Daniel', AYu. (Eds.). (1999). *A.S. Esenin-Vol'pin. Filosofija, Logika, Poezija, Zaschita prav cheloveka (A.S. Esenin-Vol'pin. Philosophy, logic, poetry, human rights)*. Moscow: Russian State University for Humanities.

Fomin, D., Genkin, S., & Itenberg, I. (1996). *Mathematical circles: (Russian experience)*. Imprint Providence: American Mathematical Society.

Gandy, R. O. (1982). Limitations to mathematical knowledge. In: D. Van Dalen, D. Lascar & T. J. Smiley (Eds.), *Studies in logic and the foundations of mathematics* (Vol. 108, pp. 129–146). Logic Colloquium '80, Papers intended for the European Summer Meeting of the Association for Symbolic Logic. Amsterdam.

Geiser, J. R. (1974). A formalisation of Essenin-Volpin's proof theoretical studies by means of nonstandard analysis. *The Journal of Symbolic Logic, 39*(1), 81–87.

Gerovitch, S. (2013). Parallel worlds: Formal structures and informal mechanisms of postwar Soviet mathematics. *Historia Scientiarum, 22–3*, 181–200.

Gerovitch, S. (2016). Creative discomfort: The culture of the Gelfand seminar at Moscow University. In B. Larvor (Ed.), *Mathematical cultures: The London meetings* (pp. 2012–2014). Basel: Birkhäuser.

Gödel, K. (2003). *Collected works* (Vol. 4). In: F. Solomon (Ed.), Correspondence, A–G. New York: Oxford University Press.

Heyting, A. (1936). *Obzor issledovanij po osnovanijam matematiki* [A survey of research in the foundations of mathematics]. German original:*Mathematische Grundlagenforschung, Intuitionismus*, Beweistheorie. Berlin. 1934.

Heyting, A. (1956). *Intuitionism. An introduction*. Amsterdam: North-Holland. 2nd revised edition, 1966. 3rd revised edition, 1971. Russian translation by Vadim A. Yankov and edited by Andrei A. Markov from the first edition: Moscow: Mir 1965.

Heyting, A. (1962). After thirty years. *Logic, methodology and philosophy of science, proceedings of 1960 international congress* (pp. 194–197). Stanford: Stanford University Press.

Heyting, A. (1969). Wat is berekenbaar? *Nieuw Archiv van Wiskunde, 17*, 5–6.

Hilbert, D., & Ackermann W. (1928). *Grundzüge der theoretischen Logik*. English translation: *Principles of mathematical logic*. Chelsea. Translation of 1938 German edition. Russian translation: *Osnovy teoreticheskoy logiki*, Moscow, 1947.

Isles, D. (1981). On the notion of standard non-isomorphic natural number series. Lecture Notes in MathematicsIn F. Richman (Ed.), *Constructive mathematics* (Vol. 873, pp. 111–134). Berlin-Heidelberg-New York: Springer.

Isles, D. (1992). What evidence is there that 2power65536 is a natural number?. *Notre Dame Journal of Formal Logic*, *33*(4), 465–480.

Iwasaki, C. (1977). O pis'me S. A. Yanovskoy ot 29-go dek. 1955 g. po nekotorym filosofskim voprosam matematicheskoy logiki [On S. A. Yanovskaya's Letter dated 29th December 1955 on some philosophical questions of mathematical logic]. *Hitotsubashi Journal of Social Studies*, *9*(1 (9)), 23–31. Available online http://www.jstor.org/stable/43294241, (Accessed 10-11-2020) [in Russian].

Khinchin, A. Ja. (1926). Idei intuicionizma i spor o predmete sovremennoj matematiki [The ideas of intuitionism and the dispute on the subject of modern mathematics]. *Vestnik kommunisticheskoj akademii, 16*, 184–192 [in Russian]. Translated into English in Lukas M. Verburg, Olga Hoppe-Kondrikova (co-translator) On A. Ya. Khinchin's paper Ideas of intuitionism and the struggle for a subject matter in contemporary mathematics (1926): A translation with introduction and commentary. *Historia Mathematica*, *43*(2016), 369–398.

Kolmogorov, A. N. (1925). On the principle of *tertium non datur*. *Matematicheski Sbornik*, vol. 32, 646–667. Available online [in Russian] http://www.mathnet.ru/links/ffe370e498d5a44fddaf9ee09d0b27c8/sm7425.pdf (Accessed 10-11-2020). English translation, On the law of the excluded middle, From J. van Heijenoort (Ed.), *Frege to Godel. A source-book in mathematical logic, 1879-1931* (pp. 414–437). Cambridge: Harvard University Press, 1967.

Kovalevich, O. S., & Lisanyuk, E. N. (2015). Printsipy deonticheskoj logiki dejstvij A. S. Esenina-Vol'pina [The principles of A. Essenin-Volpin's logic as founded on the logic of action]. *Logiko-filosofskie shtudii, 12*, 142–156. [in Russian].

Küng, G. (1961). Mathematical logic in the Soviet union (1917–1947 and 1947–1957). *Studies in Soviet Thought, 1*, 39–43.

Kurosh, A. G. (Ed.). (1959). *Matematika v SSSR za 40 let 1917-1947 [Mathematics in USSR during 40 years 1917-1957]*. Moscow: Gosudarstvennoe izd-vo fiziko-matematicheskoj literatury [in Russian].

Kurosh, A. G., Markushevich, A. I., & Rashevsky, P. K. (Eds.). (1948). *Matematika v SSSR za 30 let 1917-1947 [Mathematics in USSR during 30 years 1917-1947]*. Moscow: OGIZ [in Russian].

Kushner, B. A. (1973). *Lektsii po konstruktivnomu matematicheskomu analyzu* [*Lectures on constructive mathematical analysis*]. Moscow: Nauka. English translation by E. Mendelson, *Lectures on constructive mathematical analysis. American Mathematical Society*, 1984.

Kushner, B. A. (1993). A. A. Markov and Errett Bishop, Smilka Zdravkovska, Peter L. Duren (Eds.), *Golden years of Moscow mathematics*. American Mathematical Society, Russian original: A.A. Markov i Errett Bishop. *Voprosy Istorii Estestvoznananija i Tekhniki* 1 (1992), 70–81.

Kuskova, S. M. (2012). 'Teoriya modal'nostey A. S. Yesenina-Vol'pina [A. S. Yesenin-Vol'pin's theory of modalities]. *Logiko-filosofskie shtudii, 10*, 68–78. [in Russian].

Lusin, N. (1930). *Leçons sur les ensembles analytiques et leurs applications*. Paris: Gauthier-Villars.

Magidor, O. (2012). Strict finitism and the happy sorites. *Journal of Philosophical Logic, 41*, 471–491.

Markov, A. A. (1950). Konstruktivnaja logika [Constructive logic]. *Uspekhi matematicheskikh nauk, 5*(3), 187–188. [in Russian].

Markov, A. A. (1954a). Teorija algorifmov [Theory of algorithms]. *Trudy matematicheskogo instituta im. V. A. Steklova, 42*, 3–375. Available online http://www.mathnet.ru/links/13a476d238e313f7a66ec80f1a961123/tm1178.pdf. (Accessed 10-11-2020) [in Russian].

Markov, A. A. (1954b). O nepreryvnosti konstruktivnyh funkcij [On the continuity of constructive functions]. *Uspehi matematicheskikh nauk, 9*(3), 226–229. [in Russian].

Markov, A. A. (1956). Ob odnom principe konstruktivnoj matematicheskoj logiki [On a constructive principle of mathematical logic]. *Trudy tret'ego Vsesojuznogo matematicheskogo s"ezda*, t. 2, Moscow, 146–47 [in Russian].

Markov, A. A. (1958). O konstruktivnyh funkcijah [On constructive functions]. *Trudy matematicheskogo instituta im. V.A. Steklova, 52*, 315–348. Available online http://www.mathnet.ru/links/b2dc4d7cdef827fa07f160497a2a6a78/tm1320.pdf. (Accessed 10-11-2020) [in Russian].

Markov, A. A. (1962). O konstruktivnoj matematike [On constructive mathematics]. *Trudy matematicheskogo instituta im. V. A. Steklova, 67*, 8–14. Available online http://www.mathnet.ru/links/80837f80d53ef16c770dd46a2757389c/tm1756.pdf. (Accessed 10-11-2020) [in Russian].

Markov, A. A. (1974a). O jazyke Я_0 [On the language Я_0]. *Doklady Akademii Nauk SSSR, 214*, 40–43. [in Russian].

Markov, A. A. (1974b). O jazyke Я_1 [On the language Я_1]. *Doklady Akademii Nauk SSSR, 214*, 279–282. [in Russian].

Markov, A. A. (1974c). O jazyke Я_2 [On the language Я_2]. *Doklady Akademii Nauk SSSR, 214*, 513–516. [in Russian].

Markov, A. A. (1974d). O jazyke Я_3 [On the language Я_3]. *Doklady Akademii Nauk SSSR*, 765–768 [in Russian].

Markov, A. A. (1974e). O jazykah $\text{Я}_4, \text{Я}_5, \ldots$ [On the languages $\text{Я}_4, \text{Я}_5, \ldots$]. *Doklady Akademii Nauk SSSR, 214*, 1031–1034. [in Russian].

Markov, A. A. (1974f). O jazyke Я_ω On the language Я_ω]. *Doklady Akademii Nauk SSSR, 214*(1–6), 1262–1264. [in Russian].

Markov, A. A. (1974g). O jazyke $\text{Я}_{\omega|}$ [On the language $\text{Я}_{\omega|}$]. *Doklady Akademii Nauk SSSR, 215*, 57–60. [in Russian].

Markov, A. A. (1987). Chto takoe konstruktivnaja matematika? (Vvedenie, Publikacija i predislovie N.M. Nagornogo) [What is constructive mathematics? (Introduction, Publication and preface by N. M. Nagorny)]. In: Panov M. I. (red.) *Zakonomernosti razvitija sovremennoj matematiki*. Moskva: Nauka, 209–212 [in Russian].

Mints, G. (1991). Proof theory in the USSR 1925–1969. *The Journal of Symbolic Logic, 56*(2), 385–424.

Molodshij, V. N. (1938). *Jeffektivizm v matematike*. [*Effectivism in mathematics*]. Moscow [in Russian].

Nagorny, N. M. (1994). Andrei Markov and Mathematical Constructivism. In: P. Dag, S. Brian, W. Dag (Eds.), *Logic, methodology and philosophy of science IX. Proceedings of the ninth international congress of logic, methodology and philosophy of science*. Uppsala, Sweden, August 7–14, 1991, (pp. 467–479). Elsevier: The Netherlands.

Olszewski, A., Woleński, J., & Janusz, R. (Eds.). (2006). *Church's thesis after 70 years*. Heusenstamm: Ontos.

Parikh, R. (1971). Existence and feasibility in arithmetic. *Journal of Symbolic Logic, 36*, 494–508.

Poincaré, H. (1905, 1906). Les mathématiques et la logique, *Revue de Métaphysique et des Morales*, XIII (1905), 815–835 ; XIV (1906), 17–34.

Poincaré H., Couturat L. (2007). *Matematika i logika* [*Mathematics and Logic*]. Moscow, Izd-vo LKI. 2nd ed. 1st ed. 1915.

Pólya, G. (1954). *Mathematics and plausible reasoning*. Volume I: *Induction and analogy in mathematics*. Volume II: *Patterns of plausible inference*. Princeton University Press. Russian Translation *Matematika i pravdopodobnyye rassuzhdeniya*, Edited and Introduction by S. A. Yanovskaya. Moscow, 1957, 2nd edn. 1975.

Pudlák, P. (1996). On the lengths of proofs of consistency. In: *Collegium logicum* (Annals of the Kurt-Gödel-Society) (Vol. 2). Vienna: Springer.

Rashevsky P. K. (1973). O dogmate natural'nogo rjada [On the dogma of the natural series]. *Uspekhi matematicheskikh nauk* 28(4)172, 243–246. English translation: Rashevskii P.K. On the dogma of the natural numbers, *Russian Math. Surveys*, 28(4), 1973, 277–279.

Shtokalo, I. Z., Yushkevich, A. P., & Bogoljubov, A. N. (Eds.). (1966-1970). *Istorija otechestvennoj matematiki [History of Russian mathematics]*. Kiev: Naukova Dumka, Vol.1: 1966; Vol. 2: 1967; Vol. 3: 1968; Vol. 4, Books 1 & 2: 1970 [in Russian].

Silakov, V. D., & Stiazhkin, N. I. (1962). *Kratkiij ocherk istorii obshcheij i matematicheskoij logiki v Rossii [A brief survey of the history of general and mathematical logic in Russia]*. Moscow [in Russian].

Tarski, A. (1940). *Introduction to logic and to the methodology of deductive sciences*, 1st ed., Oxford University Press (New York) (and many subsequent editions). Russian translation *Vvedeniye v logiku i metodologiyu deduktivnykh nauk*. Edited and Introduction by S. A. Yanovskaya. Moscow, Gos. Izd-vo inostrannoj literatury, 1948.

Troelstra, A. S., & van Dalen, D. (1988). *Constructivism in mathematics: An introduction (two volumes)*. Amsterdam: North-Holland.

Tseitin, G. S. (1962). Algorithmic operators in constructive metric spaces. Trudy matematicheskogo instituta im. V.A. *Steklova, 67*, 295–361. [in Russian].

Tugarinov, V. P., & Majstrov, L. E. (1950). Protiv idealizma v matematiceskoj logike [Against idealism in mathematical logic]. *Voprosy filosofii, 3*, 331–339.

Turing, A. (1950). Computing machinery and intelligence. *Mind*, *59*, 433–60. Russian translation Mozhet li mashina myslit'? [Can a machine think?] Edited and Introduction by S. A. Yanovskaya. Moscow, Gos. Izd-vo inostrannoj literatury, 1960.

Vandoulakis, I. (2014). Materials on Lichoudae's grammar and logic courses in the Moscow Slavo-Graeco-Latin academy. *Rossija i Khristianskij Vostok, 4–5*, 350–356. [in Russian].

Vandoulakis, I. (2015). On A. A. Markov's attitude towards Brouwer's intuitionism. In: E. B. Pierre, H. Gerhard, H. Wilfrid & S.H. Peter (Eds.). Selected contributed papers from the 14th international congress of logic, methodology and philosophy of science. *Philosophia Scientiæ, 19*, 143–158. https://doi.org/10.4000/philosophiascientiae.1054.

Vandoulakis, I., & Tatiana, D. (2020). On the historical transformations of the square of opposition as semiotic object. *Logica Universalis, 14*(2020), 7–26. https://doi.org/10.1007/s11787-020-00248-z

Vasiliev, N. A. (1989). *Voobrazhaemaja logika [Imaginary Logic]*. In: V. A. Smirnov (Ed.). Moscow: Nauka [in Russian].

Vernadsky, G. (Ed.). (1972). *A source book for Russian history from early times to 1917 (Vol. 2). Peter the Great to Nicholas I*. New Haven and London: Yale University Press.

Weyl, H. (1934). O filosofii matematiki [On the philosophy of mathematics]. M.-L. [in Russian] Reprinted 2005. German original: Philosophie der Mathematik und Naturwissenschaft, 1927 (2nd ed.) (1949). *English translation: Philosophy of mathematics and natural science* (p. 2009). Princeton: Princeton University Press.

Yanovskaya, S. A. (1950). Pis'mo v redakciju [Letter to the editors]. *Voprosy filosofii, 3*, 339–342.

Yessenin-Volpin, A. S. (1970). The ultra-intuitionistic criticism and the anti-traditional program for foundations of mathematics. In: A. Kino, J. Myhill & R. E. Vesley (Eds.), *Intuitionism and proof theory* (*Proceedings of the Conference*, Buffalo, N.Y., 1968), Studies in logic and the

foundations of mathematics (pp. 3–45). Amsterdam: North-Holland. See also: MR 0295876 Reviewed by Geiser, James R. (1975), The ultra-intuitionistic criticism and the antitraditional program for foundations of mathematics by A. S. Yessenin-Volpin. *The Journal of Symbolic Logic*, Association for symbolic logic, *40*, 95–97.
Yessenin-Volpin, A. S. (1981). About infinity, finiteness and finitization (in connection with the foundations of mathematics). In: F. Richman (Ed.), *Constructive mathematics* (Las Cruces, N.M., 1980). Lecture Notes in Mathematics (Vol. 873, pp. 274–313). Berlin-New York: Springer.
Zaslavsky, I. D., & Tseitin, G. S. (1962). O singuljarnyh pokrytijah i svjazannyh s nimi svojstvah konstruktivnyh funkcij [Singular coverings and properties of constructive functions connected with them]. *Problemy konstruktivnogo napravlenija v matematike. 2. Konstruktivnyj matematicheskij analiz*, Sbornik rabot, Tr. MIAN SSSR, 67, Izd-vo AN SSSR, M.–L., 458-502. Available online [in Russian] http://www.mathnet.ru/links/c205e0b68376378fef3887294d31ae06/tm1761.pdf. (Accessed 23-11-2014).

Chapter 11
On V. A. Yankov's Existential Interpretation of the Early Greek Philosophy. The Case of Heraclitus

Tatiana Yu. Denisova

Abstract In this paper, we examine Yankov's interpretation of the early Greek philosophy and particularly the idea of anthropological (and even existential) problematic in it. Examining his broad research program in the case of Heraclitus, we show that Yankov's contribution is that he transformed an earlier vague hypothesis about the existential problematic in the early Greek philosophy into a research problem. He challenged the interpretation of the early Greek philosophy as natural philosophy and rejected its reduction to the model of the genesis and structure of the world, in which man stands as a neutral observer.

Keywords V.A. Yankov · Early Greek philosophy · Heraclitus · Existential problematic · Ontology

Mathematics Subject Classification: 01A20, 01A99, 03-03

11.1 Introduction

The plurality of interpretations of philosophical texts and even the conflict of interpretations (P. Ricoeur) is considered the norm and indicates their depth and multilayeredness. Philosophical ideas, concepts and teachings live in a particular way; they are problematised anew and interpreted within new cultural contexts and remain open to subsequent interpretations. However, in reality, any attempt at an unconventional interpretation of a classical text, different from its "classical" reading, is often met with envious rejection by the professional community, accusations against the author for amateurism, or it is just ignored.

The history of interpretations of the texts of the early Greek philosophers starts from the very beginning of their appearance. These interpretations are divided into "standard" or "classical," i.e., established interpretations, and "unconventional," i.e.,

T. Yu. Denisova (✉)
Department of Philosophy and Law, Surgut State University, Surgut, Russia
e-mail: tatiana.denisova.1209@gmail.com

A. Citkin and I. M. Vandoulakis (eds.), *V. A. Yankov on Non-Classical Logics, History and Philosophy of Mathematics*, Outstanding Contributions to Logic 24,
https://doi.org/10.1007/978-3-031-06843-0_11

not recognised by the scientific community. Any attempt to say a new word in the interpretation of classical texts always has the risk of either repetition or excessive liberty. V.A. Yankov's *An Interpretation of Early Greek Philosophy* is just such an attempt undertaken by a leading logician and philosopher of mathematics that calls for another look at familiar texts.

The research novelty and value of Yankov's analysis of early Greek philosophy is the following: firstly, his examination proceeding from the ontological and metaphysical nature of the early Greek philosophical constructions and, secondly, that he reveals their existential meaning.

11.2 A General Outline of V.A. Yankov's Interpretation of Early Greek Philosophy

The idea to study the history of human thought through the prism of existential problems is stated by V.A. Yankov for the first time in his article "A Sketch of an Existential History," published in 1998. The idea has met then harsh criticism, which to a great extent is justified.

As per its title, "Sketch," the article gives the idea of a sketch that needs improvement and is needed for the author only to catch the main idea in mind and fix it on paper. It is pretty chaotically built, suffers from stylistic negligence, contains several controversial historical interpretations and even factual errors, the references do not support its statements, and the conclusions are not justified. However, Yankov's aim to link different cultures into a single semantic whole and track their development in time, taking the existential idea as the central axis, seems new and promising.

By the existential idea, Yankov means[1]

> "the deep self-understanding of man, which usually does not reach conscious manifestation, but is expressed in the whole way of his actions, as their basis, as a truly existential force" (Yankov 1998, 3).

Self-understanding of man is a part of his entire picture of the world, but, as Yankov emphasises, it is a particular part, the basis of its constitutive power, with which all the other parts are correlated. Yankov formulates in the article a research program aiming to

> "reveal and describe the existential types of man, a hidden or semi-hidden understanding of oneself and the world in the main historical cultures" (Yankov 1998, 3)

The expected outcome of this program would be an "existential history", that is, a history of human development and his entire picture of the world. Yankov neither fulfils nor even set before himself the task of creating an integral "existential history," but his undoubted merit is the very idea of viewing human history from this point of view.

---

[1] All translations of quotations from Yankov's works are mine.

A few years later, he implements this program in his book *An Interpretation of Early Greek Philosophy* (Yankov 2011) with Greek natural philosophy, significantly narrowing the field of his analysis. However, the original project is impressive in its scale. In his research, he gives a panorama of the Greek picture of the world of the 5th century BC, an epoch-making milestone in the history of humanity, which required a radically new way of understanding the world and man in it.

11.3 On the Ontological Essence of Early Greek Philosophy

Based on extensive material, Yankov convincingly shows that the Greek natural philosophers, who are often presented as naive materialists who were looking for a common fundamental principle of the world in certain tangible, visible substances, actually tried to comprehend the a priori metaphysical form of the world, which predetermined the ontology of its objective spheres and intelligible meanings.

The early Greek philosophies are primarily metaphysical because they are related to the a priori foundations of the world, its ontological structure. Yankov claims that the Greeks transformed the idea of knowing the world inherited from the East, replacing the Eastern "vision of the concrete" with the Greek contemplation of the Whole. Creating metaphysical models, discovering universal ontological laws and formulating categories, the Greeks realise a unique view of the world as an ordered whole, consistent in all its parts. This striving for integral contemplation and metaphysics thinking is manifested in all spheres of intellectual and artistic creativity of the Greeks.

As Yankov notes, Greek geometry is not just a body of knowledge necessary for utilitarian purposes. The theory of regular polyhedra is not just the result of observations over specific three-dimensional objects but a movement towards the classical understanding of the geometric concepts of point, line, and surface (Yankov 2011, 9). Yankov points out that the intention for integral contemplation is reflected in the very word ϑεωρία (*theoria* which means viewing, speculation, contemplation (Peters 1967, 194)), created by the Greek philosophers, as well as in the representation of knowledge about the world in abstract concepts, logical and rhetorical schemes, and artistic canons.

11.4 On the Existential Ideas in the Early Greek Philosophy

The ontological concepts of the early Greek philosophers also include an existential aspect of the general scheme of the world order. This aspect is neither an anthropomorphisation of natural forces (like in the mythological models) nor a projection of human relations, values and connections onto the world (like in Eastern philosophy). It is an attempt to look at the world objectively—what it is in itself, and not through the eyes of man, and what is man's place in this world, how can he live in it, how to embed the meanings of his finite existence into the objective order and rhythm of the infinite world.

Yankov's critics perceived the existential line in the analysis of early Greek philosophy as an accidental and optional element of a generally quite "faithful" interpretation, as a hypothetical idea that was a "side effect" of his amateurism and unprofessionalism for which he was accused.

However, Yankov talks about it as a principal thesis. He says that the traditional interpretation of the early Greek philosophers as materialists and hylozoists never satisfied him:

> "The vital nerve of their thinking was too attached to the Greek tragic worldview to allow them, rising over the man and the universe, to engage in cold-blooded calculation from which to build the world" (Yankov 2011, 7).

He admits that he was fascinated by the opportunity to reinterpret the Milesians in terms of the Chinese *Qi*. Greece felt, he notes, the strong influence of Eastern culture, in particular, Asia Minor, and knew about its scientific achievements and religious ideas, but

> "the Greek man, perceiving this theme, plays it in the Greek way, so that in interaction with the Eastern ideas he appears as a man who creates his world from within, by free deployment" (Yankov 2011, 8).

This is the specific Greek way of thinking: the free creation of man and the world. Yankov affirms the freedom of the Greek mind as its most important characteristic. He argues that the Greeks perceive the world as a rationally organised and rationally explicable system and a place for the free play of various forces. The man in the Greek outlook is by no means only a *part* of the cosmos and a *part* of the *polis*, subordinate to the Whole, determined in his actions by the rigid, that is, the indestructible *Ananke* (Yankov 2011, 57). The primary quality of man is freedom. The main content of his life is the 'agon' (competition or contest) (Burckhardt 1999). These two qualities are the conditions for the emergence of philosophy and the creative nature of the entire Greek culture.

Yankov believes that the first who felt liberation from the power of *Ananke* was Pherecydes of Syros (fl. 6th century BC). Syros is an island belonging to the Cyclades, and this is important because, as Yankov says, the entire Cycladic culture was influenced by the Cretan civilisation and the "Minoan sense of a freely floating world" (Yankov 2011, 59). Starting with Pherecydes,

> "*Ananke* disappears from the history of the origin of the gods and men. The story itself is placed not in the context of the "megalithic" necessity, as it was the case with Hesiod, but in the context of infinite space-time, where everything which was created and born feels free" (Yankov 2011, 59).

Ananke, or, more precisely, the theme of fate in general, does not disappear from Greek thought. It simply plays a definite role in the life of the Greeks, different from the role of the blind and omnipotent Roman *Fatum*. *Ananke*

> "goes through the deeds of kings and heroes who *act from within*, realising the human meaning. Not by avoiding *Ananke*, but by participating in it, as a spontaneous and free creature, the man realises his *own* meaning" (our emphasis) (Yankov 2011, 58).

Ananke equates the man and the gods in everything, except immortality and mortality; that is, both are not only equally subordinate to *Ananke*, but both, as Yankov emphasises, are *equally free* in self-realisation, in the "free struggle of their powers."

The most crucial idea of Yankov, which became the core of his interpretation of early Greek philosophy, is his statement about individualism as a characteristic feature of Greek culture, and thereby about the personal responsibility for man's existence, about man's loneliness before the world, about man's personality principle and the possibility and value of personal choice. Individualism does not mean renouncing man's involvement in the everyday affairs of the *polis* and rejecting the perception of the world as a shared home. Yankov points to an utterly harmonious combination of these characteristics:

> "Individualism appears on stage and pervades the entire history of Greece, although at its very beginning it is combined with a vivid sense of the community of the *polis* and even the whole of Hellas" (Yankov 2011, 11).

Further, it will be shown how both these ideas are justified by V.A. Yankov on the example of his interpretation of Heraclitus' philosophy.

11.5 On the History of Existential Interpretations of the Early Greek Philosophy

The idea of the anthropological nature (anthropologically) of Greek philosophy, especially of the early Greek philosophy, and even more the idea of the presence of existential problematics in it, is not simply generally unacceptable in the modern historiography of philosophy. It is perceived as an amateurish assault on the academic tradition and meets sharp criticism. Moreover, this tradition has been preserved for many decades and supported by highly authoritative historians of philosophy.

Since the end of the 19th century, textual studies have tended towards hypercriticism against the doxography of the pre-Socratics. The interpretations of the pre-Socratics by the Stoics, who viewed the ethical components of their teachings, were declared doxographic aberrations (John Burnet (1930); Harold Cherniss (1951); Walter Burkert (1972)). Among the interpretations of Heraclitus prevailed the physicalist interpretation that goes back to Aristotle, and the relativistic interpretation, originating from Plato.

Andrei V. Lebedev, one of the most profound contemporary researchers of Heraclitus, notes with bitterness that the physicalist interpretation of Heraclitus, which prevailed because of the works of G.S. Kirk (1954) and M. Marcovich (1967), led to the belief that Heraclitus' ethical fragments from the Stobaeus collection were forgeries. Any references to ἐκπύρωσις (*ekpyrosis*—conflagration), reality as a flow, or the concept of fate, were considered uncritical. What seemed unimportant or senseless from the standpoint of analytical philosophy was removed from the history of Greek philosophy (Lebedev 2014, 97–98). Furthermore, stereotypes associated with the concept of "pre-Socratics" continue to exist even in our modern times. Since the boundary between the early Greek and classical philosophy is associated with the anthropological revolution of Socrates, it was implied that there was not and could not be any anthropological problematics before Socrates.

Thus, W.K.C. Guthrie (1906–1981), in his *History of Greek Philosophy* (1962), confines himself to the observation that Socrates' turn to anthropological issues was the beginning of the change of the interest of philosophers from the Universe to man and, accordingly, this is the boundary between the pre-Socratics and the classical period of Greek philosophy. He connects the theme of the soul (*psyche*) and fate in Heraclitus with his religious views:

> "Whatever his views about the soul and its fate, we may be sure that they will not be purely rational or without religious overtones" (Guthrie 1962, 7).

Catherine Osborne, in her survey of the teachings of the pre-Socratic philosophers, notes that the pre-Socratics (including Heraclitus) present only different descriptions of the physical world, which differ between them in the first principles that underlie the world and ensure its unity (the One, the Fire, the four elements, or other principles). In the chapter devoted to Heraclitus, she does not even mention his doctrine of the soul (Osborne 2004, 35).

In *The Oxford Handbook of Pre-Socratic Philosophy* (Curd and Graham 2009), we find the enduring tradition of division of the pre-Socratics again into two groups: the physiologists, who were looking for the material principle of the world, and the Pythagoreans, focused on a mathematical, formal description of reality. The collection presents studies on the pre-Socratics that give a new, thematically more complex picture of their views. Specifically, the relationship between theology, epistemology, cosmology, metaphysics is examined, and a fresh look at the formation of the pre-Socratic philosophy is advanced. However, the anthropological problematics is not touched in any of them, even indirectly.

Perhaps the exception among the researchers of Greek philosophy is the Irish scholar Eric R. Dodds (1893–1979). In his work, *The Greeks and the Irrational* (2004), he claims that anthropological problematics was present among the Greeks. He justifies his view by appealing to Homer and finding evidence of manifestations of *shame* (Dodds 2004, 26). It is essential that the manifestations concern shame and not *guilt*. Shame differs from guilt. Shame is determined by one's ideas about what must be done; guilt is defined by social conventions. Consequently, Homeric man is already characterised by a behaviour peculiar to a personality. In the history of Greek thought, there is convincing evidence about reflection over this behaviour.

Several authoritative researchers (Charles H. Kahn 1979; Roderick T. Long) recognise the ethical components of Heraclitus' teachings in the interpretations of the Stoics. However, as A.V. Lebedev puts it[2]

> "the positivist Heraclitus created by Burnet is not going to concede his place" (Lebedev 2014, 74).

The situation in Russian historiography of Greek philosophy is similar. The Russian philosopher and publicist of the 19th century Sergei N. Trubetskoy (1862–1905) categorically states that

> "Man as a person did not exist for the ancient consciousness" (Trubetskoy 1890 (2010), 235 (my translation)).

Outstanding Russian researcher of antiquity Aleksei F. Losev (1893–1988) noticed that in early antiquity, there was no term denoting the personality and, in general, the Greeks

> "had a poor idea of the exclusivity and peculiarity of the human personality" (Losev 1975, 538 (my translation)).

In his view, the widespread, starting with Homer, Heraclitus and Democritus, term *psyche* (soul) indicated not the uniqueness of man, but something material (breath) that connects man with the Cosmos. Hence, he concludes that

> "the Greeks simply did not have a sense of personality, as most of the peoples of the Middle East did not have" (Losev 1975, 538 (my translation)).

On the other hand, he notes, contradicting to himself, that

> "a nation that reached a certain level of culture and civilisation cannot but have a sense of personality" (Losev 1975, 537–538 (my translation)).

The Greek tragedians Aeschylus, Sophocles, Euripides, did not know the term "personality," but their tragedies are built on the tragedy of the personalities' fate, not just of their soul, notes Losev. As a result, he comes to a strange conclusion:

> "As for early classical Greece, *with the tremendous development of the sense of personality here*, the personality itself remained only an appendage of the Cosmos, a kind of its emanation, its offspring, not always even obligatory and necessary" (Losev 1975, 538 (my translation)).

[2] We use the author's unpublished translations into English available at his site https://varetis.academia.edu/AndreiLebedev. However, Yankov used Lebedev's earlier Russian translations of the Fragments of the Pre-Socratics (Lebedev 1989).

In other words, there is a sense of personality, but at the same time, there is no freedom of decisions and actions because this personality is just a dispensable, accidental emanation of the Cosmos.

The authoritative Russian and Soviet classical philologist Olga M. Freidenberg (1890–1965) emphatically asserts:

> "The only form of ancient philosophy is cosmology and ontology" (Freidenberg 1998, 330 (my translation)).

Georgian philosopher Merab Mamardashvili (1930–1990) is categorical on this point:

> "Greek thought, starting with Parmenides and ending with Aristotle, is non-anthropological, non-psychological, unethical and maximally non-humanistic" (Mamardashvili 2009, 67–68 (my translation)).

At the same time, Mamardashvili points out that in classical Greek culture (in myth, in tragedy), there is a figure of a hero whom he called "the foundation of himself" (Mamardashvili 2009, 33). The hero, in Mamardashvili's understanding, is the one who

> "becomes the beginning of the reasons for his actions (…), who does not participate in the cohesion of natural causes and actions" (Mamardashvili 2009, 35 (my translation)).

The condition of this phenomenon of grounding himself Mamardashvili calls *conscience*.

A contemporary Russian specialist in the history of ancient Greek philosophy, Gennady Drach, devotes to this problem a specialised monograph entitled *The birth of ancient philosophy and the beginning of anthropological problematics*, where, along with the general statement that anthropological problematics is crucial for philosophy as a whole, notes that

> "the anthropological status of early Greek philosophy still needs a justification" (Drach 2003, 13 (my translation)).

In Drach's interpretation, anthropology pervades all Greek natural philosophy, cosmogony turns out to be connected with the worldview problematic, the ideas of justice and fate, and the conceptions of Parmenides, Thales, Anaximander, Anaximenes are characterised by an anthropomorphic vision of the world (Drach 2003, 289).

In our view, the following idea of a modern philosopher of science, Anatoly V. Akhutin, concerning the place of man in the cosmos in the early Greek thought, is well-grounded and, besides, brilliantly resolves the contradictions inherent in the viewpoints discussed above.

> "Greek philosophy, as a product of the Greek mind, says that man here has *already* gained autonomy in the world, but he has not yet opposed himself to it. He is no longer lost in the world (and in himself), in the immensity and overwhelming power of the natural forms," which is characteristic of the East, is not entirely woven into the variegated fabric of myths, is not subject to the obsessive givenness of the custom. He stands face to face (mind to mind) with the world but does not leave the world for an autonomous Being" (Akhutin 2007, 143 (my translation)).

Concerning Heraclitus in particular, A.V. Lebedev's standpoint is convincing and deeply grounded on the study of texts and the linguistic and cultural contexts. On the grounds of his new translation of Heraclitus's texts, he declares that

> "The "Stoic" Heraclitus, i.e. the ethical, political and theological thinker, is much more authentic and closer to his Ephesian prototype than the physicalist Heraclitus of Aristotle or the relativist epistemologist of Plato" (Lebedev 2014, 98).

Heraclitus was neither an abstract metaphysician, much less a "physicist," writes Lebedev. Otherwise, why would he condemn the "polymathy" of his contemporaries who studied the nature of individual things? The work of Heraclitus was ethical and political and was undertaken in order

> "to show the similarities and differences in the organisation of the world and the norms of behaviour of gods and men (that is, nature and society), and to demand from co-citizens to bring the political, legal, moral and religious standards adopted by the Greeks of his time in line with the "divine" eternal standards; to bring local human forms of "justice," based on the subjective opinion (δόξα), in line with the universal and shared-by-all (ξυνόν) Justice (Δίκη), corresponding to the objective and natural order of things (κατὰ φύσιν)" (Lebedev 2014, 99).

Besides that, Lebedev emphasises that Heraclitus' cosmos

> "consists not of elements or corpuscles, but of "mortals and immortals," "gods and men," that is, of living wills" (Lebedev 2014, 100).

However, for what reason, despite the obvious facts and the non-obvious conclusions, Greek philosophy is traditionally denied of the "anthropological" aspect and the presence of existential problematics?

Probably, this is due to a superficial interpretation of the texts, or, perhaps, interpretive inertia to follow an established tradition, the fear of imposing on the Greeks retrospectively ideas generated in another time, a different mentality. However, a more serious reason is the unjustified contraposition of ontology and anthropology. When it is argued that Greek philosophy was "non-anthropological" because it focuses exclusively on ontology, it appears to be a preference for the object of study. The ontology of Greek philosophy means focusing on the essence of things. Greek philosophy does not at all ignore the man and does not perceive him as a "dispensable emanation of the Cosmos," but tries to understand him proceeding from the general foundations of the world, its fundamental principles, that is, it turns to the beginnings, starts from the beginnings.

11.6 The Complexity of the Interpretation of Heraclitus

This section will focus on V.A. Yankov's analysis of existential ideas in Heraclitus of Ephesus, who is recognised as one of the most profound and most controversial thinkers in the history of philosophy.

To an inexperienced mind, the reason for the mysteriousness of Heraclitus is quite apparent. Firstly, twenty-five centuries separate us from him, and, secondly, his teachings have come down to us in a few fragments (about one hundred in total), commentaries to them and retellings-interpretations of thinkers of subsequent eras Leonardo Tarán writes about this:

> "The interpretation of Heraclitus will always remain controversial. The main reasons are his peculiar mode of expression, the fragmentary character of the evidence, and the very way ancient authors quoted his sayings" (Tarán 1986, 1).

Therefore, it may seem that we could understand him unmistakably if his teaching reached us in full. However, as M. Heidegger rightly notes in his *Lectures on Heraclitus*, it was his contemporaries, and not the descendants of Heraclitus, who, despite the availability of the philosopher's work in its entirety, called him σκοτεινός ("The Dark"). Thus, the ancients had no advantage in understanding Heraclitus since it was too new and unfamiliar for them and beyond their life and intellectual experience.

E. Fink, in the seminar on Heraclitus, carried out by him together with M. Heidegger, talks about the primary intention of the interpretation of Heraclitus:

> "Confronted with his texts, left to us only as fragments, we are not so much concerned with the philological problematic, as important as it might be,' as with advancing into the matter itself, that is, toward the matter that must have stood before Heraclitus' spiritual view" (Heidegger and Fink 1979), 3).

Any interpretation is *dialogical*, and the interpretation of the teachings of one philosopher by another philosopher is always a meeting of two texts, and the result of this meeting is unpredictable. It is possible to understand a thinker only by thinking, and if genuine thinking is audacity, according to Heidegger, then any attempt to interpret it is inevitably an audacious and risky response to a challenge. As Heidegger puts it, the reason for the "darkness" of Heraclitus is not in the (intentionally or accidentally) obscure way of expressing his thoughts but in the thought itself, which is unusually profound and new.

> "Heraclitus is thus ὁ Σκοτεινός, 'The Obscure,' not because he intentionally or unintentionally expresses himself in a manner that is incomprehensible, but rather because every merely reasonable thinking excludes itself from the thinking of the thinker (i.e., from essential thinking)" (Heidegger 1994, 24)

In general, it must be said that interpreting a philosophical text is always extremely difficult and risky. However, this risk is justified. As Heidegger notes,

> "Discerning minds understand that Heraclitus speaks in one way to Plato, in another to Aristotle, in another to a Church Father, and in others to Hegel and to Nietzsche. The respective difference of each dialogical interpretation of thought is a sign of the unspoken fullness to which even Heraclitus himself could only speak by following the path of the insights afforded him. Wishing to pursue the 'objectively correct' teaching of Heraclitus means refusing to run the salutary risk of being confounded by the truth of a thinking" (Heidegger 1985, 105–106).

Is it possible, in principle, to be sure of the correctness of an interpretation? Are there methodologies or criteria enabling us to establish the correctness of the interpretation of a philosophical text?

The problem lies in the *limits of freedom* of interpretation, the right to one's interpretation. Vladimir V. Bibikhin (1938–2004), a prominent Soviet and Russian philosopher, best known for his translations of Martin Heidegger, notes in his *Reading Philosophy*:

> "We always knew how to read, *extracting* what we need, but to let the other be oneself, this kind of reading would be worthy of learning, if it is possible at all" (Bibikhin 2009, 30 (my translation)).

The distortion of someone else's meaning and the subordination of someone else's thought to our own goals are typical troubles of an interpreter. However, if we wish to "let the other be oneself," how do we do it? To follow pedantically the letter of someone else's text or to try to be inspired by the spirit of its creator? V.V. Bibikhin says that

> "A philosophical thing does not scare of movement, reinterpretation, even distortion; in this sense, it differs from a natural thing that a rude attitude can break it, but a rude attitude cannot break a firmly worked philosophical thing" (Bibikhin 2009, 31 (my translation)).

Moreover, V.V. Bibikhin claims that

> "a fantastic, extreme interpretation sometimes works better than a correct one" (Bibikhin 2009, 31 (my translation)).

Why is it better? A correct interpretation follows the letter of the text, allowing at any moment to rely on what is said directly and unambiguously. A "fantastic" or imaginary interpretation is a more free reading; it results from the meeting of two texts (two minds, two intellectual experiences), one of the author, the other of the interpreter, and their dialogue.

There is a beautiful Greek word, ἀμηχανία, which aptly describes a situation of understanding and interpretation. ἀμηχανία means confusion, perplexity, difficulty, inability to see the way out in essential questions. This term denotes the inapplicability of a mechanical approach to explaining the world using technical methods. The conceptions of the pre-Socratics are clear and smooth, complete and logical only in the expositions for the first-year students and even for non-philosophical faculties. The texts of the pre-Socratics are not smooth, not only because they are poorly preserved, in fragments, but first of all, because the pre-Socratic philosophers had a sense of the uncertainly and unreliability of their guesses, the constructive insufficiency of their theoretical constructions, the artificiality of their visions of the world. Philosophy does not lead out of the state of confusion but introduces it; it does not answer but poses questions; it does not teach ready-made truths but encourages the birth of thought.

Eugene Fink, in the seminar with Heidegger on Heraclitus mentioned above, makes a critical statement:

> "We remain restless and are unable to rely on a sure interpretation of the Greeks. For us, the Greeks signify an enormous challenge" (Heidegger and Fink 1979), 3).

At the end of the seminar, he is still full of doubts about the legitimacy of free interpretation:

> "Our question is whether, not in a new turn toward what the Greeks have thought, we can encounter the Greek world with our new experience of Being" (Heidegger and Fink 1979), 161).

By claiming that existential problematics is present in early Greek philosophy, that is, attributing to the ancient man a sense of loneliness, the tragedy of personal responsibility for his choices, we have the risk of being criticised that we look at him from our "distance," imposing our vision upon him. In contrast, the Greeks might have had no idea of this concept. However, as Mikhail M. Bakhtin (1895–1975) notes,

> "the ancient Greeks did not know the most important thing about themselves, that they were the ancient Greeks and never called themselves so" (Bakhtin 1986, 506 (my translation)).

Modern man can read in the actions of the Greeks more than they could read themselves. It is necessary to have formed a sense of personality, the experience of awareness of the personality principles of modern European man to formulate the emerging intuitions of personality, self-awareness and personality values.

11.7 V.A. Yankov on the Traditional Interpretation of Heraclitus

V.A. Yankov warns against being overly enthusiastic about the idea of seeking direct embodiment (confirmation) of one's ideas in early Greek philosophy. Thus, he says that one should not follow Hegel,

> "who saw in the ancient embodiment almost his central thought" (Yankov 2011, 222)

Interpretation should be creative but must not be arbitrary. It is no coincidence that V.A. Yankov's monograph is entitled *An Interpretation of Early Greek Philosophy*. By the title of his work, the author indicates the possibility of various readings of Greek thinkers, and in no case claims that his work is the only proper understanding of them.

In his analysis of the teachings of Heraclitus, Yankov points out that the thinker is usually perceived one-sidedly by reducing the wealth of his ideas to the following:

(a) the idea of the change of all existing things,
(b) the idea of admissibility of opposite statements about the same thing, and
(c) the idea of seeking for an initial material substance common of all things like the natural philosophers.

Yankov speaks about this with a sense of bitterness

> "When one tries to imagine in full what all subsequent European thought owes to Heraclitus, becomes inevitably amazed and seems very strange that people can recall in their memory only his famous *dictum* πάντα ῥεῖ (Yankov 2011, 222).

Unfortunately, this saying became not only the hallmark of Heraclitus but, for many people, a synopsis of his entire philosophy.

In Yankov's view, the problem is not only the wrongness of the reduction but the illegitimacy of Heraclitus' qualification as the author of precisely these ideas; the misinterpretation of his main philosophical achievements. Yankov notes that the idea of the change of the material world is one of the most important in Heraclitus' teaching, but he is not its creator. This doctrine belongs to Pythagoras; it was restated by Epicharmus and became known before Heraclitus. Heraclitus only included it in his teaching, making it one of his starting points (Yankov 2011, 222).

His idea to associate fire with the primordial substance of the Universe is also not original, claims Yankov. Heraclitus could have accepted it under the influence of the Pythagoreans, particularly Hippasus, who considered fire to be the primordial substance (Yankov 2011, 222–223).

Besides, Yankov considers the credit to Heraclitus for inventing dialectics[3] as "pure misunderstanding." He shows that the combination of opposites in pairs in

[3] This view was held by G. W. F. Hegel and Friedrich Engels.

Heraclitus (clean-dirty, life-death, and the like) does not mean compatibility of opposite characteristics in one object, but different contexts of understanding this object, that is, different points of view: water is not clean and dirty at the same time, but clean for the fish and dirty for the man. There is no contradiction or dialectic in this, and the qualification of an object depends not on the object itself but the evaluating subject.

Yankov declares that he intends to penetrate the core of Heraclitus's views, leaving aside his physics and cosmology (Yankov 2011, 224) and focusing on the doctrine of the Logos and the doctrine of the soul,

> "The two main complexes of Heraclitus's ideas that laid the foundation for many branches of European thought" (Yankov 2011, 249).

The problems of the Logos and the human soul are ontologically related, in Yankov's view. They are permeated with existential meaning because the closeness to the Logos sets and determines the degree of a man's closeness to his destiny and purpose the degree of meaningfulness of his existence.

11.8 Yankov's Predecessors About Heraclitus' Existential Ideas

Yankov is not the first scholar who attempts to connect Heraclitus's ideas with existential problematics.

Plutarch talks about Heraclitus as a philosopher who teaches about the loneliness of the human "I", locked in his own body, like in an island, and also holds the idea that a man, like everything existing, is born and dies daily, without possessing a one and forever given to him essence. It is easy to see that the idea of the inescapable loneliness of every human being and the idea of the realisation of the essence in the process of existence are fundamental, characteristic existential ideas.

Friedrich Nietzsche (1844–1900) recognised himself in the tragic fate of the misunderstood loner Heraclitus. He talks about the unparalleled loneliness of Heraclitus, implying not so much his solitary existence as the impossibility of sharing thoughts. Although, as Nietzsche admits, it is common for any philosopher to "pave his way alone,' the loneliness of Heraclitus was unique. Nietzsche writes about him,

> "The feeling of solitude, however, that pierced the Ephesian hermit of the temple of Artemis, we can intuit only when we are freezing on wild desolate mountains of our own. No all-powerful feeling of compassionate emotions, no desire to help, to heal, to save, stream forth from Heraclitus. He is a star devoid of atmosphere" (Nietzsche 1962, 67)

A convincing interpretation of the human dimension of Heraclitus teachings belongs to the Greek philosopher Kostas Axelos (1924–2010). In his view, for Heraclitus, man is a whole that is a part of a larger whole—the Cosmos. The formation of the human is included in the formation of the Cosmos, participates in it. We call this dimension of Heraclitean thought, writes Axelos, *anthropological* in lack of a better term. What is important is that

> "this anthropological dimension does not surpass the others since, like all of them, it originates from the same central core of his thought. In the 'bays' of this anthropological dimension, we meet physiological, psychological and existential themes" (Axelos 1974, 223 (my translation)).

Thus, Axelos does not question the anthropological dimension of Heraclitus' teachings. He considers it to be on a par with the rest aspects, the cosmological, epistemological, political, and, notably, include the existential issues in the anthropological dimension of Heraclitus's teachings. Hence, Axelos makes a critical remark, which, most probably, Yankov would agree with, since he follows a similar position

> "All these topics, sorted by us, but not by Heraclitus, are not gathered together in some special place, since anthropology does not yet have its special separate place" (i.e. in knowledge, philosophy – T.D.) (Axelos 1974, 224 (my translation)).

Thus, the rejection of the anthropological aspect in Heraclitus, because he has no special section devoted to man, is untenable. Heraclitus does not have a systematic scheme of some practical anthropology, as Axelos emphasises. The essence of his anthropological aspect is that in his teaching, for the first time in the history of thought, something happens that "paves the way that can lead man to the heart of the universe," and this 'something' is man's search of himself, his attempt to understand himself and through this understanding to come closer to the understanding of the Universe. The point is not to view the world by human eyes and thus see a world model constructed by the man by his ability to view and understand something.

The fact is that Heraclitus was the first who speculated that there is a common code for both man's understanding of himself and the world. While thinking, the man tries to comprehend himself, but the concept of thinking and thinking ability itself has a universal meaning. Axelos perfectly grasped the essence of Heraclitus' reasoning about the Logos and the soul, saying that the search for oneself is not entirely psychological since man is an organic part of the Universe. The Ego and the Universe are interdependent, he claims. A mystical path leads from one to the other (Axelos 1974, 225). Heraclitus's anthropology is not psychological, but existential, since it is not about what man feels while existing in the world, but about the meaning of his existence within the Whole, how this meaning of private existence is related to the Being and meanings of the Whole.

Similar ideas, independently of K. Axelos, are presented in his book by Yankov, although Yankov is not aware of Axelos' work. Let us consider this in detail below.

11.9 The Existential Dimension of the Doctrine of Logos

Yankov's analysis of Heraclitus's texts proceeds from the recognition of two equally essential axes of his teaching:

(i) the problem of the unity of Logos, and
(ii) the problem of man.

The division in these two problems is only conventional, apparently, for the convenience of interpreters. Heraclitus presents them in profound unity: reflections on the Logos are permeated with the anthropological problematics, and the theme of man and human existence acquires meaning only in the context of its inclusion in the meanings of the Whole, and accord with the rhythm and order governed by the Logos.

The relation between the Logos and man sounds full of indignation and despair in the famous sentence of Heraclitus about the inability of most men to hear the voice of the Logos and follow it:

> "Of this Truth, real as it is, men always prove to be uncomprehending, both before they have heard it and when once they have/heard it" (DK 22 B1; M 1) (Marcovich 1967, 6).

However, what is the Logos? What does it mean to hear its voice? Why is it so crucial for the man to hear the voice of the Logos? Moreover, why men do not hear it?

The concept of the Logos is one of the most complex and ambiguous and is rendered, depending on the context, as "word," "speech" (collection of words), "discourse," "reason," "ground," "account," "plea," "proportion," "order." Since Heraclitus, in the teaching of which this is a fundamental concept, it occurs pretty often in the history of Greek thought.

So, according to David Hoffman,

> "Logos, the noun, occurs only twice in Homer, (*Iliad* 15.393; *Odyssey* 1.56). When it does occur, it signifies speech. Used just 0.10 times per 10,000 words in Homer, and 3.09 times per 10,000 words in Hesiod, the frequency of logos jumps to 24.06 in Herodotus and 28.92 per 10,000 in Aeschylus, and reaches its single-author high with 45.79 uses per 10,000 words in Isocrates" (Hoffman 2003, 30).

Heraclitus called *logos* both the generating and governing principle of the Universe, as well as his teaching because he was confident that he had succeeded in deciphering the grammar of the Cosmos, translating it into human language and expounding it in his treatise: οὐκ ἐμός ὁ λόγος, which means "this Logos is not mine," or, in Marcovich's rendering,

> "If you have heard [and understood] not me but the Logos" (DK 22 B50; M 26) (Marcovich 1967, 113).

According to Heraclitus, the principle of the existence of the Logos is ἓν διαφέρον ἑαυτῷ This expression can be rendered as "the one which is differing within itself", that is, the principle of unity. Although the world exists and moves because of the struggle of opposites, a proper view presupposes their finding within a single whole, their inclusion into the universal interconnection of everything with everything. Yankov puts it this way:

> "The parts find their true meaning when they are understood as rooted in a whole. Logos generates this self-consistent in all parts multiplicity" (Yankov 2011, 226).

That is why

> "it is wise to agree that all things are one" (DK 22 B50; M26) (Marcovich 1967, 113).

"To know everything as one" provides the original, the root meaning of the word *logos*. According to Hoffman's assumption, the concept of logos *comes* from the word *legein*, which means "to gather":

> "The noun *logos*, according to Liddel, Scott and Jones, is derived from the verb *legein*. Therefore an inquiry into the root meaning of *logos* must begin with an inquiry into the root meaning of *legein*.The root meaning of *legein* comes from the Indo-European stem *leg*which means "To collect; with derivative meaning 'to speak'." (Hoffman 2003, 28–29).

In Homer, one of the first who made use of this concept, its meaning is closer to "to gather",

> "all the uses of *legein* in Homer have a unifying sense, it must be closer to 'to gather' than to 'to speak'" (Hoffman 2003, 29).

According to Yankov, the Heraclitean Logos not only comprehends the Universe as a self-consistent whole, not only unites disparate parts into one meaningful picture but also *generates* this semantic unity, *produces the world* in such a way that

> "all parts have their semantic place within a single meaning" (Yankov 2011, 226).

All parts, *including man*. That is why Heraclitus says,

> "Those who will speak [i.e. act | with sense must rely on what is common to atlas a city relics on its law, and much more firmly: for all human laws are nourished by one law, the divine law: for it extends its power as far as it will and is sufficient for all [human laws] and still is left over" (DK 22 B114; M 23) Marcovich 1967, 91].

For human life to have meaning in the Whole, and, thereby, to have meaning in general, man must hear the voice of the Logos.

What does it mean to hear the voice of the Logos?

To hear the Logos means to think about the Being, the Whole, the Infinite, which are inaccessible to empirical experience, i.e., to pass over a fundamentally different intellectual space, rise above the tradition, mythological discourse, the habit of object-oriented thinking.

To hear the voice of the Logos means to perceive the world as a whole, to feel its rhythm and order, to view the structure, purposefulness, and laws, in virtue of which the Whole remains a Whole and includes man as a necessary part of it. Nonetheless, in ordinary, physical and psychological everyday life, man views the empirical connections of concrete things without exceeding the framework of empirical observations, mental habits, stereotypes of perception of the cognisable and easily describable world with the help of generally understood verbal cliches.

Heraclitus' call to hear the voice of the Logos is a call to try to view things as they are, the connections between them as they are, to discover for itself the world of beings in their logic of cause and effect. In Mamardashvili's words,

> "The logic of things as they are is what is called the Logos by the Greeks" (Mamardashvili 2009, 60 (my translation)).

Does this mean that to establish contact with the Logos, one must leave the visible, everyday world and enter another higher, invisible world? No, the Greeks did not set such a task, and no other world existed for them. Everything was this-worldly for the Greeks; everything was here; everything was revealed; it was open in certain aspects of existence. The task is not to view something else but to view this only one thing differently. The task that Heraclitus sets before man concerns the way of thinking, the way of understanding.

Why is it important to hear the voice of the Logos?

The Logos in Heraclitus is a "generating term". To hear the Logos is to be in contact with the world so that these contacts give rise to meanings in man's mind, to be able to view the coherence, unity, completeness and harmony of the world behind the finite individual things. There is no other world besides this one, and the task is not to find another world but in a deep, meaningful view. Hearing the voice of the Logos makes human life meaningful and, thereby, genuinely human.

It may seem that Heraclitus was trying to convey his teaching to people, but they either resisted or were incapable of learning, and this didactic helplessness drove him to despair, for which he became known as "the weeping philosopher." However, what did Heraclitus want to convey, whom to teach? What is about his teaching?

Mamardashvili (2009) notes that all commentators of Heraclitus face the same question: of which doctrine Heraclitus could be considered the author? Of the doctrine of the flow and change of things, which is commonly called "becoming?" Of the doctrine of the harmony of the world? Of the teaching of the combination of opposites? or of something else? Mamardashvili concludes that there is no teaching of Heraclitus but an attempt to create a special syntax, a distinctive semantic figurative structure, with the help of which it would be possible to talk about the Being. Heraclitus "invents" a new syntax because the object language, by which the world of things is described and understood, is not suitable for the interpretation of the

Being that does not have object expression. To hear the voice of the Logos requires to abandon the visual objectness of the vision of the world and the language that names visible things and indicates the visible connections between them.

The "Logoi" of Heraclitus generate a sense of ἀμηχανία. In this way, they awaken thought. Their value is not that they explain something and answer questions; on the contrary, they induce questioning where everything is "clear," they are puzzling, triggering some mechanisms of generating thought.

Heraclitus did not intend to convey to his listeners a set of guaranteed true ready-made theses that constitute a "theory of Being." His task was to provoke the birth of meanings in every individual consciousness.

There is no "theory of Being," just as there is no ready-made plan of Being. Mamardashvili explains that peculiar to the Greek thought is the understanding of what happens and exists in the world as a "present cohesion," a spontaneous folding of events and things, not given beforehand, without guarantees. There is no structure of Being once forever; everything comes to be and becomes in agreement with each other, not with some original plan. To get closer to the truth of Being, man needs to learn to view this universal coherence and harmony:

> "If you have heard [and understood] not me but the Logos, it is wise to agree that all things are one" (DK 22 B 50; M 26) (Marcovich 1967, 113).

The delusion of men is that they do not feel the one world, the one common Logos, i.e. the unity of the meanings of the Logos for all. Thus, they live according to their understanding as if they were sleeping:

> "The waking share one common world, whereas the sleeping turn aside each man into a world of his own" (DK 22 B 89; M24) (Marcovich 1967, 99).

Men form their knowledge about the world by trusting their ears and eyes, but what they view in the world is only their 'barbaric' souls are ready for that:

> "Evil witnesses are eyes and ears for men, if they have souls that do not understand their language" (DK 22 B 107; M13) (Marcovich 1967, 47).

According to Heraclitus, outside the Logos, human existence is meaningless. Hoffman formulates Heraclitus' idea of the place of man in the world in the following way:

> "In logos human Being gathers itself out of the world. In logos human Being gathers itself into conflict with the world. In this conflict the human comes to know itself as a being in a world of beings" (Hoffman 2003, 42).

Why do men not hear the voice of the Logos?

Complaining about the barbaric souls of men, Heraclitus nevertheless understands why they are deaf to the voice of the Logos:

> "It is because of want of (human) confidence that it [the Logos?] escapes men's knowledge" (DK 22 B 86; M12) (Marcovich 1967, 43).

Occasionally, we can see what Heidegger called "the lumen of Being," that is, what is inconsistent with everyday life, is not commensurate with it, the Other relative to its dimensions, tasks and values. However, in "privileged moments," this seldom happens because it is impossible to be in this ultimate state all the time.

An encounter with something great, genuine requires compatibility, commensurateness and personal response; for this reason, it is frightening. It is easy to be religious, but it is hard to be a believer. It is easy to be law-abiding, but it is challenging to have a genuinely civic consciousness. It is easy to rush into battle on command, but it is difficult to display true valour. Mamardashvili writes about this:

> "Since Being is not a thing, not a substance, then there are no guarantees for man: by discovering the Being, man does not lean against something solid, immobile, high, does not sit on it as if it were a solid foundation of his existence" (Mamardashvili 2009, 64).

Therefore, the cognition of the Being, the ability to hear it, is tragic and complicated in itself.

The Logos and the human mind ("soul")

The second "centre of Heraclitus's thinking," as Yankov calls it, that is the second problem, which, in his view, is not inferior in importance to the problem of the unity of the Logos, is the problem of man, formulated as the problem of the mind (ψυχή—"soul") (Yankov 2011, 246). However, both these "centres" are closely interrelated and are examined by Yankov exactly in their interrelationship.

In his teaching, Heraclitus does not just touch the question of the immortality/mortality of the soul, thus including the discussion started by the early Greek philosophers and continued later in the classical and Hellenistic eras. He turns the question of the immortality of the soul into a subject of special examination. He considers it in connection with the most significant ontological and existential issues: the principles of the world order and man's place in the world. The "soul" for Heraclitus is not just material evaporation, part of his physiology, as many of his contemporaries thought. The "soul" is that by which the man occupies a special place in the Whole of the Cosmos. It establishes its communion with the meaning of the Whole, its structure, its articulated coherence. The "soul" gives a man separateness and draws him into the unity of the Logos. The unity with the Logos testifies to the "ontological perfection of the soul." Therefore, according to Heraclitus, "soul" is the meaning of life, as Yankov points out (Yankov 2011, 247).

Men view the world differently, not because their sensory abilities are different or even false, but because physical senses are not sufficient to know the world. The world is cognised not by eyes and ears but by the mind ("soul"). A man participates in the immortal Logos in virtue of the immortality of his soul and man's ability to cognise it. The more fire-mind in his soul, the more he can comprehend the Fire-Logos.

Cognition of the Logos makes man's life meaningful since following the voice of the Logos takes part in the Whole and gives a place and purpose to his particular existence in this common Whole, making his existence not accidental. The more meaningful life a man lives, the more chances his soul has for immortality.

Convinced in the immortality of the soul, Heraclitus is far from the Pythagorean idea of the transmigration of souls. He is closer to the idea of Anaximenes, who explained the immortality of the soul by the immortality of its airy substrate. However, Heraclitus goes further than these views and points out that the soul, in its eternal existence, does not remain the same; it changes. Heraclitus conveys its properties through the characteristics of the physical body (dry–wet) or through the elements that permeate space (fire–moisture), using them, of course, metaphorically.

The fire of the Logos permeates the human soul, and the drier (wise) it is, the more is filled with the fire of the Logos, which is not enough for the "wet" barbarian souls. Yankov writes that "to indulge in the "water" component for the soul is disastrous" (Yankov 2011, 246) because it is *a fortiori* disconnected from the Logos, separated from the Whole, and as a result, it is mortal. In other words, the after-death destiny of souls is different and depends on what a man was during his lifetime. Since the Cosmos exists cyclically, like "a regularly igniting and dying fire," when the cosmos flares up, the souls are tested for their strength:

> "Fire, having come suddenly upon them, will judge and convict all (living beings)" (DK 22 B 66; M 82) (Marcovich 1967, 435).

According to Neoplatonist philosopher Olympiodorus the Younger (c. 495–570), Heraclitus believed that an uneducated soul dies immediately upon leaving the body,

> "and the educated soul, tempered by virtues, persists <until> the conflagration (ἐκπύρωσις) of the whole world" (DK 22 B 116a) (Lebedev 2014, 204).

Obviously, in Heraclitus, the physical composition of every individual soul is unique, different from others. This is a fundamental difference of his idea of the immortality of the soul from all other authors. The soul is not just immortal, like other souls, it is not just included in the Whole of the common to all cosmos, like other souls, it can, according to Heraclitus, contain more or less of the divine fire, that is, it can be to a more or less extent participant to the Logos. Yankov highlights this peculiarity of Heraclitus's conception, distinguishing it from the Anaximenes's viewpoint, which is the closest one:

> "If in Anaximenes, every soul participates in the divine as part of the boundless, active and divine air, in Heraclitus, the participation of the soul in the divine is determined by the portion of the fire, which carries in itself" (Yankov 2011, 241)

Consequently, not every soul becomes immortal, but only the one that participates in the Logos-Fire with its fiery dry part. In any case, this is logical. The more essential the fiery part of the soul, the more it can continue to exist as a separate part of the common fire, concludes Yankov. The souls of the heroes are immortal, unlike the souls of men living with senses, desires and emotions, but they acquired immortality not by chance, not as a gift, but because they *deserve* it. A perfect, dry soul is the result of a man's efforts, and, accordingly, the immortality of his soul is his merit.

Every soul has, to a certain degree, grains of fire. However, what is most important, is that every soul has to a different degree the possibility (although different) of cognition and communion with the Logos. The Logos must be sought in one's soul, as Yankov emphasises, but this is not easy since the Logos is outside of the ordinary, familiar things and needs efforts to establish contact with it. What was accessible to the fiery soul of Heraclitus himself was inaccessible to most lazy wet souls.

Yankov notes an additional probable reason for man's inability to hear the voice of the Logos: the *epithymia* (desire) and *thymos* (anger) that fill the soul. In this way, Heraclitus anticipates Plato's idea of the three-partite soul, consisting of the rational (*logistikon*), the "spirited" (*thymoeides*), and the appetitive (*epithymetikon*), concludes Yankov (Yankov 2011, 243). In Heraclitus, the rational corresponds to the fire and the *thymos* and *epithymia* to the moisture of the soul. In support of his conclusion, he cites two Heraclitean passages:

> "It is not better for men to get all they want" (DK 22 B 110; M71) (Marcovich 1967, 390),

and

> "It is hard to fight with the heart's desire; for whatever it wishes it buys at the price of soul" (DK 22 B 85; M70) (Marcovich 1967, 386).

It is crucial that the soul of one man does not only differ from the soul of another man in its composition, and thereby in its properties, but it is capable of growing, changing:

> "Soul has a (numerical) ratio which increases itself" (DK 22 B115; M112) (Marcovich 1967, 569).

Heraclitus makes clear that *thymos* and *reason* are choices of man. Thus, none and nothing (the deity, the circumstances, the chance, the complexity of a task) is responsible for the incomprehensibility of the Logos, but only man himself. Man is responsible for the degree of his participation in the Logos of the Cosmos. He determines to which extent he participates in the life of the Whole. Thus, he defines the degree of meaningfulness of his existence.

11.10 Conclusion

Yankov's most significant contribution is that he highlighted the idea of anthropological problematics in the early Greek philosophy, which has been stated before him as a vague hypothesis, of some deviant interpretive trend, into a research problem. He challenged the interpretation of the early Greek philosophy as natural philosophy exclusively and rejected its reduction to the modelling of the genesis and structure of the world, in which man stands as a neutral observer.

The reason existential and anthropological problematics are rejected in the early Greek philosophy should be sought in the narrow interpretation of the concept of "existence." The latter is reduced to the experience of the tragic and inescapable loneliness, which the Greeks did not have. The Greeks were aware of their separateness in this world, assumed responsibility for existence, felt guilt and shame for their actions without blaming fate or the gods. Nevertheless, they never felt abandoned into the world, did not suffer from orphanhood or eventuality. They were aware that they participate in the world, and the meaning of their life co-participates in the meanings of the Cosmos depends on them, influences them, is consonant with them.

In his research, Yankov does not examine any earlier unknown texts of the early Greek philosophers; he works with the same sources of Heraclitus' ideas as most other researchers, i.e., with the fragments and the comments published by H. Diels and the translation of the fragments into Russian and comments by A.V. Lebedev. However, the idea of programmatic research on the existential aspects of the early Greek philosophy (in our study, confined to the case of Heraclitus) is achieved by Yankov because of the novel perspective he adopted and his hermeneutic boldness. Heraclitus's known texts are interweaving with each other in a new way and, considered from a new standpoint, give at the end a convincing, uncommon reconstruction of the teachings of Heraclitus. Yankov's book is not just another systematic account of the history of early Greek philosophy, well-founded with texts, but an original interpretation of the early Greek worldview, explaining (among other things) the cause of the "Greek miracle."

References

Akhutin, A. V. (2007). *Antichnye nachala filosofii. [Ancient origins of philosophy].* St. Petersburg, Nauka (in Russian).

Axelos, K. (1974). *Ο Ηράκλειτος και η φιλοσοφία.* Athens: Exandas Publ. 1974 (in Greek). French original: *Héraclite et la philosophie: La première saisie de l'être en devenir de la Totalité*, 1962.

Bakhtin, M. M. (1986) Otvet na vopros redaktsii 'Novogo mira' [Reply to the editors of 'Novyj Mir']. In: Bakhtin M. M. *Literaturno-kriticheskiye stat'i [Literary-critical articles].* Moscow: Khudozhestvenaja literatura, 501–508 (in Russian).

Bibikhin, V. V. (2009). *Chtenie filosofii [Reading philosophy].* Series *Slovo o suschem [Discourse about the existent].* St. Petersburg, Nauka (in Russian).

Burckhardt, J. (1999). *The Greeks and Greek civilization.* O. Murray (Ed.), New York: St Martin's Griffin. German original: *Griechische Kulturgeschichte*, 1898–1902.

Burkert, W. (1972). *Lore and Science in Ancient Pythagorenism*, translated by Edwin L. Minar Jr. Cambridge, Mass.: Harvard University Press. German original: *Weisheit und Wissenschaft: Studien zu Pythagoras, Philolaos, und Platon.* Nuremberg: Hans Carl, 1962.

Burnet, J. (1930). *Early Greek philosophy* (4th ed.; 1st ed. 1892). London.

Cherniss, H. (1951). The characteristics and effects of Pre-Socratic philosophy. *Journal of the History of Ideas, 12*(3), 319–345.

Curd, P., & Graham, D. W. (Eds.). (2009). *The Oxford handbook of Pre-Socratic philosophy.* Oxford: Oxford University Press.

Diels, H., & Kranz, W. (1972-1974). (DK) *Die Fragmente der Vorsokratiker griechisch und deutsch von Hermann Diels, hrsg. von Walther Kranz.* Band I-III. / H. Diels, W. Kranz. Berlin: Weidmannsche Verlag Buchhandlung.

Dodds, E. R. (2004). *The Greeks and the irrational.* University of California Press.

Drach, G. V. (2003). *Rozhdenie antichnoj filosofii i nachalo antropologicheskoj problematiki. [The birth of ancient philosophy and the beginning of anthropological problem]*. Moscow. Gardariki (in Russian).

Freidenberg, O. M. (1998). *Mif i literatura drevnosti [Ancient myth and literature]*. Moscow: Vostochnaya literatura RAN (in Russian).

Guthrie, W. K. C. (1962). *A history of Greek philosophy* (Vol. 1). *The earlier pre-Socratics and the Pythagoreans*. Cambridge: Cambridge University Press.

Heidegger, M. (1985). *Early Greek thinking*. Harper San-Francisco: Translated by David Farrell Krell and Frank A. Capuzzi.

Heidegger, M. (1994). *The inception of occidental thinking logic: Heraclitus's doctrine of the logos*. Translated by Julia Goesser Assaiante and S. Montgomery Ewegen (3rd ed.). USA: Bloomsbury Publishing Plc.

Heidegger, M., & Fink, E. (1979). *Heraclitus seminar 1966/1967*. Trans. by Charles H. Seibert. The University of Alabama Press.

Hoffman, D. (2003). Logos as composition. *Rhetoric Society Quarterly, 33*(3), 27–53.

Kahn, C. (1979). *The art and thought of Heraclitus. An edition of the fragments with translation and commentary*. Cambridge: Cambridge University Press.

Kirk, G. S. (1954). *Heraclitus. The cosmic fragments*. Edited with an Introduction and Commentary by G. S. Kirk. Cambridge: Cambridge University Press (reprinted 1970).

Lebedev, A. V. (Ed.) (1989). *Fragmenty rannikh grecheskikh filosofov*. Chast' 1. *Ot epicheskikh teokosmogonii do vozniknoveniya atomistiki. [Fragments of the early Greek philosophers.* Part 1. *From the epical Theo-cosmogonies to the birth of atomism]*. Moscow, Nauka (in Russian).

Lebedev, A. V. (2014). *Logos Geraklita. Rekonstruktsiya mysli i slova [The Logos of Heraclitus. Reconstruction of thought and text]*. Saint Petersburg: Nauka (in Russian) Parts of this work is translated into English by the author and is available online at https://varetis.academia.edu/AndreiLebedev.

Losev, A. F. (1975). *Istoriya antichnoj estetiki* (Vol. 4). *Aristotel' i pozdnyaya klassika. [History of ancient aesthetics]* (Vol. 4). *[Aristotle and the late classical [period]]*. Moscow. Iskusstvo (in Russian).

Mamardashvili, M. K. (2009). *Lektsii po antichnoj filosofii. [Lectures on ancient philosophy]*. Moscow: Progress-Traditsija (in Russian).

Marcovich, M. (1967). *Heraclitus*. Greek text with a short commentary. Editio maior. Merida: The Los Andes University Press.

Nietzsche, F. (1962). *Philosophy in the tragic age of Greeks translated by Marianne Cowan*. Washington: Regnery Gateway.

Osborne, C. (2004). *Pre-Socratic philosophy. A very short introduction*. New York: Oxford University Press.

Peters, F. E. (1967). *Greek philosophical terms. A historical lexicon*. New York: New York University Press.

Taran, L. (1986). The first fragment of Heraclitus. *Illinois Classical Studies, 11*(1/2), 1–15.

Trubetskoy, S. N. (2010). *Metafizika Drevnei Gretsii [Metaphysics of ancient Greece]*. Moscow (1890); Reprinted by Mysl' Publ. (in Russian).

Yankov, V. A. (1998). Eskiz ekzistencial'noj istorii [A sketch of an existential history]. *Voprosy filosofii, 6*, 3–28. (in Russian).

Yankov, V. A. (2011). *Istolkovanie rannej grecheskoj filosofii [An interpretation of early Greek philosophy]*. Moscow: Publication of the Russian State University for Humanities (in Russian).

Chapter 12
On V. A. Yankov's Hypothesis of the Rise of Greek Mathematics

Ioannis M. Vandoulakis

Abstract The paper examines the main points of Yankov's hypothesis on the rise of Greek mathematics. The novelty of Yankov's interpretation is that the rise of mathematics is examined within the context of the rise of ontological theories of the early Greek philosophers, which mark the beginning of rational thinking, as understood in the Western tradition.

Keywords V.A. Yankov · Early Greek mathematics · Early Greek philosophy · Mathematical proof · Hippocrates of Chios · Euclid · Proclus · *ratio* · Infinite divisibility of magnitudes · Analysis and synthesis · Reasoning by *reductio ad absurdum* · Árpád Szabó

Mathematics Subject Classification: 03-03 · 03F55 · 01A60 · 01A72

12.1 On Yankov's Motivation to Study the Rise of Rational Thinking

It may seem strange that Vadim A. Yankov, a known mathematical logician and philosopher, a specialist in constructive logic, suddenly decided to study early Greek philosophy and early Greek mathematics. The outcome of his enormous work is his book *An Interpretation of Early Greek Philosophy* (Yankov 2011) that numbers over 850 pages! It is neither a textbook nor an encyclopedia on the early Greek philosophy but a profound study on the beginnings of mathematics, philosophy, and generally rational thinking.

Yankov is not the first scientist who became interested in the history of mathematics or the history of philosophy. In the first half of the twentieth century, several mathematicians showed interest in the history of Greek mathematics. Bartel

I. M. Vandoulakis (✉)
The Hellenic Open University, Athens, Greece and FernUniversität in Hagen, Hagen, Germany
e-mail: i.vandoulakis@gmail.com; vandoulakis.ioannis@ac.eap.gr

A. Citkin and I. M. Vandoulakis (eds.), *V. A. Yankov on Non-Classical Logics, History and Philosophy of Mathematics*, Outstanding Contributions to Logic 24,
https://doi.org/10.1007/978-3-031-06843-0_12

Leendert van der Waerden (1903–1996) turned to the history of mathematics in his later years (van der Waerden 1954, 1967, 1983, 1985). However, van der Waerden does not touch philosophy in any of his works on Greek mathematics. Hans Freudenthal (1905–1990) devoted some works on the history of Greek mathematics [1953, 1966, 1976], but the topics studied focus on rather specific questions. Logician and historian of mathematics Oscar Becker's (1889–1964) work in the history of Greek mathematics is primarily connected with the pre-Euclidean theory of ratios, which he identified in Aristotle's *Topics*, and attributes to Theaetetus, and the foundational "crisis" in Greek mathematics (Hasse and Scholz 1928) occasioned by the discovery of incommensurability of the side and the diagonal of a square attributed to Hippasus of Metapontum (c. 530–c. 450 BC) (Becker 1933–1936). Becker is also the first who noticed that most of Euclid's proofs do not require the use of the principle of excluded middle. On the other hand, Yankov's project to study the rise of mathematics in combination with the appearance of Greek philosophy has no precedent in the historiography of science.

In the second half of the twentieth century, several philosophers of science turned to Greek science and philosophy. First, Karl Popper (1902–1994) proclaimed to go back to the beginning of philosophy, notably the pre-Socratics [1958–1959, 1998]. Popper interprets the pre-Socratic philosophers not as creators of "natural philosophical" conceptions but as the discoverers of first theories or intuitions and fascinating cosmological explanations, which were brilliant insights but not the result of observation. Popper rejects the Baconian viewpoint, according to which science starts from observation and then slowly and cautiously proceeds to theories. He claims that this was not the way of thinking of the pre-Socratic philosophers, who conceived bold, revolutionary, and most portentous ideas in the whole history of human thought, not by observation, but by reasoning (Popper 1958–1959, 4). According to Popper, their way of thinking shows how actually science advances.

Although Yankov is unaware of Popper's appeal to return to the pre-Socratics, his project has many similarities. Denisova, in her paper in this volume, soundly emphasises that

> "Yankov convincingly shows that the Greek natural philosophers, who are often presented as naive materialists who were looking for a common fundamental principle of the world in certain tangible, visible substances, actually tried to comprehend the *a priori* metaphysical form of the world, which predetermined the ontology of its objective spheres and intelligible meanings" [Denisova 2022, 273].

Furthermore, Thomas S. Kuhn (1922–1996) points that

> "Only the civilisations that descend from Hellenic Greece have possessed more than the most rudimentary science" (Kuhn 1962/69, 168).

Thus, the problem of the rise of mathematics and philosophy, i.e., the question of the appearance of rational thinking that became the pillar of European civilisation, is central in modern philosophy of science because the idea of Greek origins of science is in the borderline of the distinction between science and other, earlier modes of thought that are generally understood as inferior. This question of demarcation between science and non-science is central to the modern philosophy of science. In this respect, Yankov's attempt to approach this problem, undertaking a comparative analysis with the non-European philosophical thinking, primarily Chinese philosophy, is fully justified.

An apparent direct motivation for Yankov to study the question of the rise of Greek mathematical and philosophical thinking was my PhD Thesis *On the Formation of Mathematical Science in Ancient Greece* (Vandoulakis 1991) that was prepared at the Department of History and Methodology of Mathematics and Mechanics and submitted to the Faculty of Mathematics of the Moscow M.V. Lomonosov in fulfilment of the requirements for the Degree of "Candidate of Science." Earlier, Yankov did not publish anything on the rise of Greek mathematics.

Yankov was a referee of my Thesis. In his speech before the Committee on 4th October 1991, he states explicitly the reason that motivated him to develop his subsequent project.

> "Despite the highly abstract conceptual level of modern mathematics and its "ideological" ramifications, its fundamental concepts are rooted in intuition. Consequently, when mathematics reaches a new level, it always makes sense to wonder whether these roots can be traced back to that mathematics which, apparently, for the first time exceeded the limits of the technique of counting and made a decisive step towards both proof and the intellectual intuition associated with it." [Minutes of the PhD Session dated 4th October 1991].

In his interpretation of early Greek philosophy, Yankov focuses on several issues highlighted in my dissertation: the problem of the finite and the infinite, and thereby of the finitary and infinitary methods of handling the infinite and the modes of reasoning about it; the problem of truth and falsity in its close connection with the problem of meaningfulness and meaninglessness; the problem concerning linguistic expressions, their meaning and reference, and others.

In the sequel, we will focus on Yankov's hypothesis of the rise of Greek mathematics.

12.2 Outline of Yankov's Hypothesis of the Rise of Greek Mathematics

Yankov focuses on the appearance of proof in ancient Greece. The Greek concept of proof is a distinctive feature of European civilisation since it does not appear, for instance, in Hindu, Chinese or Japanese mathematics. We will outline the main points of Yankov's hypothesis in due detail since it is nowhere exposed in the English-language literature. Yankov's hypothesis was exposed in [Yankov 1997, 2000, 2001, 2003] before appearing in his major monograph (Yankov 2011). In his research,

Yankov relies on Lebedev's Russian edition of the *Fragments of the pre-Socratic philosophers* (Lebedev 1989).

(1) As first textual evidence of proof, Yankov considers Hippocrates' of Chios (c. 470–c. 410 BC) proofs on lunules, reported by Simplicius. These proofs are not strict since Hippocrates lacked the concept of the ratio of magnitudes (lines, angles, figures, solids), although Simplicius does not comment on that (van der Waerden 1954, 132). Yankov observes that two possible proofs preceded Hippocrates's propositions:

i The proof that the sum of angles of a triangle is two right angles mentioned by Proclus, which according to Aristotle (384–322 BC) and Geminus (fl. 1st century BC), was proved for "all three kinds" of triangles (equilateral, isosceles and scalene triangles) (Hankel 1874).

ii The proof of the infinite divisibility of magnitudes found in the Scholium 1 to Book X of Euclid's *Elements* (Heiberg 1883–1916, vol. V, 414–417). Although it is doubtful that the Scholium contains the exact text of the Pythagoreans, the main idea of the proof is conveyed.

(2) Proclus reports a classification of proofs, ascribed to Porphyry (c. 234–c. 305 AD) (Morrow 1970, 198–199):

(a) arguments that proceed *from* starting points that are subdivided into

 (a_1) those which proceed from *common notions*, that is, from self-evident propositions.

 (a_2) those which proceed from previously demonstrated propositions.

(b) arguments that proceed *to* the starting points, which are either affirmative of them or destructive and are subdivided into

 (b_1) those that affirm first principles are called "analyses" and their reverse procedures "syntheses".

 (b_2) those that are destructive are called *reductio ad absurdum* (ἡ εἰς τὸ ἀδύνατον ἀπόδειξις).

Yankov notes that the method of analysis and synthesis is rather a heuristic discovery method that may lead to a result whenever all the steps of the arguments are invertible, since it assumes proof from a possible state of affairs, not from a fact. Thus, Yankov claims that this kind of reasoning was possibly advanced later.

The mode of reasoning by *reductio ad absurdum* is close in logical structure to the reasoning by analysis. Both concern an inference of a proposition B from a proposition A. However, in the case of analysis, A is assumed as possible, and B is the starting point, whereas, in the case of the *reductio ad absurdum*, B is assumed to be false, and A is the starting point.

Yankov argues against Árpád Szabó's (1913–2001) view that the reasoning by *reductio ad absurdum* comes from the Eleatic philosophers, notably Parmenides. He notes that such an argument can be found earlier in Anaximander's proof that the Earth floats very still in the centre of the infinite (*apeiron*), not supported by anything. In any case, the mode of reasoning by *reductio ad absurdum* is a common method of reasoning in everyday life when somebody states, for instance, "John did not come; otherwise, I would have seen him." This is because, according to Yankov, no denial can be proved by any other method, except by *reductio ad absurdum*; this is part of the meaning of negation. Thus, the method of *reductio ad absurdum* could not have appeared without the prior development of methods of proving affirmative statements; the latter is primary. Accordingly, the proof by *reductio ad absurdum* could have appeared when a problem, the solution of which is negative, made its appearance.

Yankov conjectures that such a problem could occur for the first time in Pythagorean arithmetic: from a given oblong (ἑτερομήκης) number the one side of which has double units than the other to construct a square number equal to the former. This statement of the Pythagorean theory of even and odd numbers is exposed in Propositions 21–29 of Book IX of Euclid's *Elements*. This statement can be demonstrated using the methods of pebble arithmetic and could be discovered by the early Pythagoreans or even by Pythagoras himself.

Another problem that requires reasoning by *reductio ad absurdum* could be the division of the octave in music theory. Although this is equivalent to the previous arithmetical statement, this equivalence cannot be perceived immediately.

Yankov assumes that Porphyry's meta-methodological observations are based upon an already existing practice of mathematical proving. In Yankov's view, the idea of proof by reduction to common notions, i.e., in the sense of (a_1), comes possibly from the practice of the mathematicians when they noticed that in various proofs, they had to repeat the same reasoning.

Further, Yankov claims that the first proofs concerned evident statements proved by visual reasoning over a figure, as discussed by Wilbur Richard Knorr (1945–1997) (Knorr 1975, 59–74). He notes that in the pre-Platonic period, the proofs were closely associated with the figures they referred to because the objects of mathematics were still perceived as real, not abstract, objects. Non-intuitive concepts and reasoning, i.e., not associated with some figure, could appear at this stage in the theory of music.

(3) Yankov distinguishes a primary stage of development of proof when reasoning is predominantly visual in character and concerns real objects. To this stage, he associates Thales' geometry, which concerned real objects (points, lines, etc.) drawn on some medium, and the proofs were conducted by visual reasoning based on mirror symmetry, rotation or folding. This is a kind of "empirical" geometry, as called by Proclus

> "attacking some problems in a general way and others more empirically (αἰσθητικώτερον)" [Proclus (Morrow) 1970, 52].

(4) Between Thales's geometry and Pythagoras's mathematics, Yankov picks Mamercus, mentioned immediately after Thales in Proclus's list of geometers and before Pythagoras. Of him, nothing is known beyond Proclus's poor information:

> "brother of the poet Stesichorus, is remembered as having applied himself to the study of geometry" [Proclus (Morrow) 1970, 52].

Nevertheless, Yankov conventionally calls "Mamercovian" the pre-Pythagorean geometry to which he ascribes the use of the first three Postulates exposed in Euclid's *Elements*, Book I, namely,

1. To draw a straight line from any point to any point.
2. To produce a finite straight line continuously in a straight line.
3. To describe a circle with any centre and radius.

These three postulates go beyond the geometry of figures drawn on some medium; they define geometry not as a science of real objects drawn and immediately perceivable but about line segments and figures wherever they are found on a plane or in space.

He also ascribes to this period the appearance of the method of proof by *indication* (δεῖξις) and *application of areas* (ἐφαρμογή τῶν χωρίων). Nevertheless, this kind of "Mamercovian" geometry concerns also real objects; however, a new "ideal" element appears in it, which is the notion of the "direction to the infinite."[1]

(5) The concept of the infinite (ἄπειρον) appears in Anaximander's (c. 610–c. 546 BC) philosophy to designate an original principle (ἀρχή) of the world. The *apeiron* is understood as a spatially indefinite, unbounded, limitless substance

> "from which arise all the heavens and the worlds within them" (Theophrastus *Physic Opinion*, fr. 2, Diels 1879. *Dox.* 476) (Burnet 1930, 52).

In Yankov's view, Anaximander's concept of *apeiron* has a dynamic aspect, involving the notion of tending to the infinite. Consequently, according to Yankov, this philosophical concept is naturally correlated with the concept of the indefinite extension of a line.

(6) In the Pythagorean philosophy, the *apeiron* is complemented by a limiting principle: the *boundary* (πέρας). The One or the unity (τὸ ἕν) is the product of the imposition of the *peras* upon the *apeiron*. Yankov adopts the view that the Pythagorean unit was understood as an atomic physical entity; however, the question of divisibility of the unit was not stood yet at that time. Thus, Pythagorean arithmetic, as it appears in the expositions of the Neo-Pythagoreans, such as Nicomachus of Gerasa (c. 60–c. 120 AD) (D'Ooge 1926), concerns various configurations of units in the above sense. Here, Yankov heavily rests on the results of my research initially exposed in my PhD Thesis and later in (Vandoulakis 2009).

Concerning the discrepancy between Becker (1936) and Knorr (1975, 147–148) on the representation of numbers using pebbles (διά ψήφων), Yankov puts under question the notion of "representation" of numbers by the Pythagoreans. The Pythagoreans identified the numbers with the configurations of units. Hence, there was no need for the numbers to be "represented." We can talk about the representation

---

[1] The lack of historical evidence about Mamercus had met some skepticism by Izabella G. Bashmakova that she expressed in our discussions. Nevertheless, she recognized that for the major "heroes" of Greek mathematics, historical evidence is inadequate or completely absent (e.g. Thales, Pythagoras, Euclid, Diophantus). In our view, Mamercus appears in Yankov's conception not as a "hero" (to whom one could ascribe certain notable discoveries) but as "boundary point" that marks two distinct periods of Greek mathematics characterised by Yankov in logical terms.

of numbers when the latter are understood abstractly, like, for instance, in Euclid's *Elements*. However, the Pythagorean numbers configurations are not necessary to be unique, and, thereby, different variants could be used, including those of Becker and Knorr.

Consequently, the "Mamercovian" period is characterised by reasoning over indefinitely extended configurations, both in arithmetic and geometry, which Yankov contrasts with Thales's geometry of finite drawings. The pebbles used by the early Pythagoreans are real physical entities that represent the unit; the units are the real objects of Pythagorean arithmetic. In the early geometry, a concrete drawing could represent some indefinite configuration of geometric objects; however, this configuration consisted of points, lines, angles, and other objects, drawn on a medium, so that reasoning over them did not involve any ideal element.

(7) Rigor in reasoning over ideal entities appears when they are constituted in our mind and preserve in operations what is embedded in them by our mind. This aspect of the "constitution" of objects is essential in both Pythagorean mathematics and the associated mystical generation of the world out of numbers. Anaximander's spontaneously generated *apeiron* is replaced by the Pythagoreans by the vision of an orderly generation of the world by the successive addition of units and the construction of arithmetic configurations. This insight explains the mystical experience related to the cosmogonic construction, highlighted by Walter Burkert (1931–2015) (Burkert 1962). The cosmogonic construction generates arithmetic construction.

(8) According to Proclus,

> "Pythagoras transformed mathematical philosophy into a scheme of liberal education, surveying its principles from the highest downwards and investigating its theorems in an immaterial (ἀΰλως) and intellectual (νοερῶς) manner" [Proclus (Morrow) 1970, 52–53].

Proclus's assessment is commonly considered unjustified. However, Yankov focuses on Proclus's characterisation of Pythagorean investigations as "immaterial" and "intellectual" and suggests the following interpretation. The passage expresses Pythagoras intention to release mathematics from its "material" model of concrete physical objects, such as the pebbles and concrete geometric drawings, but possibly did not complete his project. This was realised by the later Pythagoreans. Therefore, the passage fuses Pythagoras's program with its subsequent realisation by his pupils, who ascribed their achievements to their teacher.

(9) The Pythagoreans probably attempted to construct a geometry modelled upon the arithmetic of units (*monads*). In this kind of geometry, a line segment would have had consisted of a great, yet finite, number of units (*monads*). However, this kind of discrete "monadic" geometry could not be advanced for internal reasons—it could not yield important affine and metric concepts. Besides, it could not be developed in virtue of the proposition of the infinite divisibility of magnitudes talked about in the Scholium 1 to Book X of Euclid's *Elements* mentioned above.

This impasse paved the way for the introduction of ideal geometrical entities, namely a point understood as "that which has no part," a line as "that which has no breadth," a surface as that "which has length and breadth only." Accordingly, the geometric magnitudes were divided into three kinds—lines, figures, and solids—and the comparison could be made among magnitudes of the same kind.

This marks a turning point in the development of geometry. In place of Thales' "empirical" geometry or the geometry of the "Mamercovian" period or the possible "monadic" geometry of the early Pythagoreans appeared the geometry of magnitudes, advanced by the Pythagoreans of the 5th century. This new geometry was later exposed in Euclid's *Elements*, Book I and II, in a version possibly reworked by Euclid.

The appearance of the geometry of magnitudes also means that the real physical objects of geometry were replaced by ideal initial objects, such as those described in Euclid's Definitions 1, 2, and 5. This ultimately realises Pythagoras's intention to release mathematics from its "material" content of concrete physical objects and investigate theorems in an "immaterial" (ἀΰλως) and "intellectual" (νοερῶς) manner, as Proclus states. However, Yankov ascribes not to Pythagoras but Hippasus the development of the geometry of magnitudes, i.e., the geometry, in which segments, figures, angles, and solids are treated as magnitudes.

(10) Hippasus is usually associated with the discovery of incommensurability (of the side and diagonal of a square), although the latter is not explicitly ascribed to Hippasus by any ancient writer. Further, a foundational "crisis" (Hasse and Scholz 1928) occasioned by the discovery of incommensurability is emphasised in the traditional historiography of mathematics (Becker, van der Waerden, and others).

Yankov adopts Szabó's criticism of this interpretation (Szabó 1969, 123–127), remarkably, that no traces of this "crisis" have been survived. This interpretation was challenged later also by Knorr (2003, 39–42) and Fowler (1999, 289–302), but Yankov does not refer to them.

Yankov believes that the discovery of incommensurability had no tragic consequences for the Pythagoreans. Nevertheless, a revolution took place, and evidence about this revolution has been survived. This revolution is the discovery of the infinite divisibility and lack of common measure for all magnitudes of the same genus (segments or plane figures). The proof of the discovery of the infinite divisibility, as it appears in the Scholium 1, is not proof "within geometry" as understood in the *Elements*. Yankov associates it with the "Mamercovian" style of geometry. In Yankov's view, the discovery of the infinite divisibility paved the way for creating the geometry of magnitudes, which he considers a "revolution" in mathematics. Without the awareness of the infinite divisibility, the well-known method of *anthyphairesis* that enables discovering new kinds of incommensurables would be impossible.

Nevertheless, Yankov seems sceptical about several historical questions:

- Did the discovery of infinite divisibility precede the creation of the theory of magnitudes and their ratios? Or
- The theory of magnitudes was advanced earlier within the "empirical" geometry?

He notes that there is no trace of the notion of infinite divisibility of the world in Parmenides' fragments. The first evidence about the notion of infinite divisibility of the world is found in Zeno of Elea. In any case, we cannot categorically assert that the beginning of the development of the geometry of magnitudes is the discovery of infinite divisibility, although it is very plausible.

(11) The ontology associated with the last version of geometry faced some difficulties. The continuum became a conglomeration of an infinite number of points, each of which lacked magnitude. This concept is challenged in Zeno's of Elea (c. 495–c. 430 BC) paradoxes of plurality. The idea disputed by Zeno is that the world consists of "points."

In reaction to Zeno's challenge, Anaxagoras (c. 500–c. 428 BC) conceived the idea of a physical world consisting of actual infinitesimals. Such magnitudes were known to Greek geometers. Proclus mentions the *horn-like* (κερατοειδής) angle (between the circumference of a circle and a tangent), which is less than a right angle since it is less than an acute angle but is not an acute angle. However, there is no evidence that Anaxagoras used horn-like angles to illustrate his physical theory.

(12) Moreover, this kind of continuum has not a standard order, like the order in Pythagorean arithmetic. The objects of geometry (points, lines, segments, figures) had to be considered as "given," not "generated," although generation is not excluded yet from something already given. The primary givenness of an entity could not be overcome.

Thus, although Pythagorean arithmetic and the new geometry now dealt with ideal objects, they were opposite in spirit. The numbers (linear, plane, solid) were understood as generated and potentially infinite (Vandoulakis 2009), whereas geometry had to do with the "ocean" of the *apeiron*, i.e., faced with the concept of the actual infinite. This development determined the route of the axiomatisation of mathematics in ancient Greece.

Arithmetic, in virtue of its constructive nature, did not need identification of certain primary principles (ἀρχαί) to be developed. The (full) axiom of mathematical induction was never formulated in Greek arithmetic (see also (Vandoulakis 2009)). To talk about "principles" or "beginnings" is needed when the objects of mathematics are given beforehand and not generated. In this case, we must guess or imagine the relationship between the given objects to conduct reasoning over them. In geometry, the initial objects—points, lines, surfaces—are understood as given, i.e., some ideal givenness.

(13) Yankov claims that the first abstract mathematical concept introduced to mathematics by the Pythagoreans was the *ratio* of numbers. Yankov agrees with Szabó that this concept might have appeared in music due to experiments on the string length of a monochord. These experiments led to the discovery of the concordant musical intervals (the octave, fifth and fourth) that correspond to the number ratios 2:1, 3:2 and 4:3, respectively. In this case, the concept of ratio might have appeared outside mathematics out of experimentation.

12.3 An appreciation of Yankov's Hypothesis

Yankov's hypothesis is a profound conception of the rise of Greek mathematics, and primarily of the Greek concept of proof, which is systematically examined in relationship to the development of ontology in early Greek philosophy. It is impossible to review and comment on all aspects of Yankov's hypothesis in this paper, so we will focus on certain points that seem to us the most important.

1. Yankov, as a logician, views parallelism between the ontology underlying the early Greek mathematical theories with ontological conceptions developed in early Greek philosophy. Thus, he associates Anaximander's *apeiron*, particularly its aspect of tending to the infinite, with the concept of the indefinite extension of a line, i.e., the "Mamercovian" pre-Pythagorean geometry.

Further, the arithmetic of finite pebbles configurations is associated with the Pythagorean doctrine of generation of things by the imposition of the *peras* upon the *apeiron*. The use of this principle is mentioned in the Neo-Pythagorean expositions of arithmetic (Vandoulakis 1991, 2009). However, Yankov uses this parallelism to explain the mystical experience related to the cosmogonic orderly construction highlighted by Burkert.

The appearance of the geometry of magnitudes is associated with Zeno's logical paradoxes, which involve a concept of the world perceived as a conglomeration of an infinite number of points.

These are not the only possible parallelisms between early Greek mathematics and pre-Socratic philosophy. I have claimed that there is a relationship between the Parmenidean semantic conception and the Pythagorean arithmetic. This relation can be expressed as follows:

> "*Parmenides' theory of truth may have been reached through a process of reflection on Pythagorean arithmetic in which truth was identified with genetic constructability.*"

In other words, Pythagorean arithmetic could have served as a model for Parmenidean semantics (Vandoulakis 2020).

Therefore, there is a relationship between early mathematical practice and certain philosophical conceptions of the early Greek philosophers. However, this does not mean that these conceptions are straightforward generalisations from the mathematical practice of the time; they are not philosophies of mathematics of those times. They are speculative philosophical conceptions concerning man and the world, in the shaping of which we can trace the role that early mathematical practices have probably played.

2. Yankov's view that the objects of Hippocrates's geometry are perceived as real objects is textually supported, although Yankov does not seek textual evidence. In Simplicius's fragment on the calculation of the area of some Hippocratic *lunules,* we meet a kind of reference to geometric objects which is not commonly used by other Greek mathematicians, namely the following locutions are used (Fig. 12.1):

(a) "The point on which A stands" (or, "is marked by A") (τὸ ἐφ' ᾧ̃ A) refers to a point.

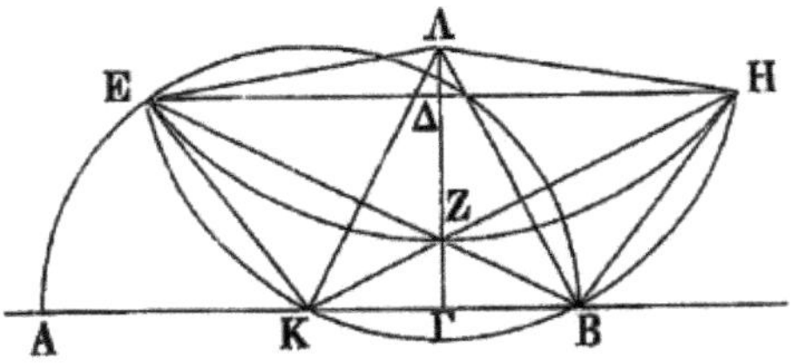

Fig. 12.1 This figure is taken from (Bulmer-Thomas, 1939, I, 242)

(b) "The line on which AB stand" (or, "is marked by AB") (ἡ ἐφ’ ᾗ AB) refers to a line segment.

(c) "The trapezium on which EKBH stands" (or, "is marked by EKBH") (τὸ τραπέζιον ἐφ’ οὗ EKBH) refers to a figure, notably a trapezium.

Thus, the letters used in Fig. 12.1, do not *name* geometrical objects (the point, the line segment or the trapezium, respectively), and the locution used—τὸ ἐφ’ ᾧ, ἡ ἐφ’ ᾗ, τὸ ἐφ’ οὗ—mean "stand on." This is not the way that Euclid uses for referring to mathematical objects designated by line segments. In Hippocrates, the letters serve as *markers* or *indicators* to label or indicate *concrete* geometrical objects in the figure. Thus, for Hippocrates, AB is not the name of diameter, as in modern mathematical symbolism. AB is a visible sign pattern that points to the diameter, i.e., it is a "label", pointing to the diameter in Fig. 12.1. Its function is to help the reader to identify the diameter in the figure. Therefore, letters in Hippocrates' text are signs that provide (spatial) evidence of the object being designated (Papadopetrakis 1990; Vandoulakis 2017).

3. Yankov is right when he claims that the development of proving affirmative statements precedes the methods of proving negative statements. Negative statements are not found in the texts of Hippocrates of Chios on the quadrature of the lunules, quoted by Simplicius. Hippocrates's visual mode of notation described above does not name geometrical objects but serves as an index or marker indicating concrete geometrical objects; this is not compatible with an abstract concept of negation.

Negative statements are not also found in Book II of Euclid's *Elements*, which is considered of early origin. Negation is alien to Parmenides. His ontological universe is a positive, true Being, lacking negative facts (Vandoulakis 2015).

There is no trace either of any concept of negation in the Pythagorean version of arithmetic, even in its elaborate expositions by Neo-Pythagorean authors (Nicomachus, Theon, and others). This kind of arithmetic concerns affirmative sentences stating something 'positive' that the construction of some configuration can confirm. No statements postulating the existence of a number identified by a negative property (or lack of a property) or statements asserting the impossibility of construction is ever formulated in the extant sources (Vandoulakis 2009).

It was Plato who first examined the structure of simple statements and defined negation in a way close to the concept of logical negation, irrespectively of their linguistic expression by the two Greek words: μή or οὐ. Euclid, in his *Elements*, deliberately uses both these forms as logical negations in the style of Plato (Vandoulakis 2015).

As regards the mode of reasoning by *reductio ad absurdum*, it also appears in dialectical reasoning of refutation (ἔλεγχος), where the truth of a proposition *P* is checked. If the proposition leads to absurd, then we conclude that the negation of *P* is true, i.e., that not-*P* holds. This kind of refutation is "intuitive" and compatible with the Pythagorean style of visual arithmetic. The need to employ reasoning of this kind appears from the problem of examination of hypotheses that seem "intuitively" to be true but turn out to be false. One example of such an "intuitive" hypothesis is the Pythagorean notion that all segments are commensurable (Waszkiewicz and Wojciechowska 1990, 88; Cf. Gardies 1991).

The Eleatics may have taken over this method of reasoning from the mathematicians, as the Kneales have conjectured [Kneales 1962, 8], notably from the Pythagorean arithmeticians, or they might have invented it independently in the process of developing dialectics. However, the need for such a mode of reasoning in dialectics is slightly different from that in arithmetic. In dialectics, it stems from the necessity to formulate and reject choices or alternatives, rejecting incompatible (yet, not necessarily contradictory) alternatives.

Wherever reasoning by *reductio ad absurdum* was invented, be it in Pythagorean arithmetic, as a hypothesis-checking method, or in Eleatic dialectics, as a method of refutation of one of the alternative options, the event was closely connected with the logical problems of negation and the definition of truth for negative facts both solved by Plato. Thus, the problem of negation is a connecting link between Greek philosophy and mathematics (Vandoulakis 2015).

4. Yankov adopts the view that Greek arithmetic was never axiomatised. As I have demonstrated, Euclid's arithmetical Books are not constructed using axiomatic assumptions. Euclid's arithmetic lacks the concept of absolute number or any elaborated concept of equality. It is constructed as 'formal' theory of numbers, designated by segments, which serve as a kind of 'formalisation' of the metalinguistic concept of 'multitude' (πλῆθος). The only "principle" or "starting point" of Euclid's arithmetic system is the *unit*. Assuming the existence of units in nature, that is, in an ontological sense that is never applied in arithmetical reasoning itself, Euclid proceeds to introduce new kinds of numbers through effective procedures.

Even when reasoning is conducted over infinite processes, where a form of the principle of infinite descent, the least number principle and mathematical induction are used, no use of the abstraction of actual infinite is involved. These uses do not go beyond finitary arithmetic (Vandoulakis 1998). As Yankov emphasises, there was not any internal reason for the axiomatisation of arithmetic by the Greeks. Although the Greek mathematicians applied the axiomatic approach to geometry, arithmetic seems to have been developed based on effective procedures in a way described in (Vandoulakis 1998). This approach also concerns the (Neo-)Pythagorean arithmetic, as shown in (Vandoulakis 2009).

Consequently, the underlying ontology of Greek arithmetic is an expanding universe of numbers increasing (potentially) *ad infinitum*, beginning from a designated object called the unit (*monas*). In the second chapter of his first Prologue, where he considers the Limit and the Unlimited as common ontological principles of mathematics, Proclus explains that although the number can increase *ad infinitum*, any given number is finite:

> "For number, beginning with the unity, is capable of indefinite increase, yet any [individual] number you choose is finite." [Proclus (Morrow), 1970, 5].

This cannot be an ontology for geometry, which is associated with the question of infinite divisibility, specific to the domain of the continuous. In this way, Yankov refocuses historical research from the general question of the axiomatisation of mathematics to the more specific question: How did the axiomatisation of geometry occur in ancient Greece?

Moreover, as Yankov clarifies, the objects of geometry (points, lines, segments, figures) had to be considered as "given," not "generated" (like the objects of arithmetic) although generation is not excluded, yet from something already given. In this context, Yankov paves the way for a reexamination from a new perspective of Euclid's *Data* (Δεδομένα), which deals precisely with the question of what magnitude can be considered as "given" in geometrical problems. In this context, the *Data* can be viewed as a work of meta-mathematical nature.

5. As we stated above, Yankov distinguishes between two traditions of pre-Euclidean geometry: Thales's "empirical" geometry, which concerns real objects (points, lines, etc.) drawn on some medium, and proofs conducted by visual reasoning based on intuitive principles (mirror symmetry, rotation or folding) and the "Mamercovian," pre-Pythagorean geometry which also deals with real objects but a new "ideal" element makes its appearance in reasoning: the notion of "tending to the infinite," which goes beyond the geometry of the (bounded) drawn figures.

It is noteworthy that a similar distinction, although not identical and based on different grounds, was suggested by W.R. Knorr. Namely, he states:

> "I find it possible to distinguish two quite different styles in the Euclidean plane geometry:
> (i) One approach, designatable as 'topological,' predominates in Books I, III, and VI. The primary objective is the examination of the topological relations of point, line, plane, plane figure; the angle is of particular interest; the triangle is the principal plane studied, ... inequalities are often introduced.
> (ii) A second approach. which we may call 'metrical', is central to Books II, IV, X and XIII and parts of Book VI; there is an overlap with the content, but not the mode of presentation of the latter parts of Book I. The principal problem is the measurement of area; ... the basic figure of study is the rectangle, with the special case of the square; equalities are always at issue, never inequalities. ...
> Tradition (i) represents the older material. the characteristically Ionic geometry, and its most brilliant representative and first systematiser was Hippocrates of Chios (c. 410 BC). ...
> Tradition (ii) material is much younger. Only in the time of Theodorus did the formalisation of the more elementary portions even begin (cf. Book II). (Knorr 1975, 6–7).

Knorr defines two styles of geometric thinking: a "topological" and a metrical one. On the other hand, Yankov's criterion is the reasoning conducted over mathematical objects. He claims that the objects are real in both traditions, but the geometry on a drawing medium is studied in the Ionian tradition, whereas the "Mamercovian" geometry goes beyond the "topological" features of figures and studies figures anywhere on the plane or in space.

12.4 In Lieu of a Conclusion

Yankov's monograph is the first systematic attempt to study the rise of Greek mathematics in relation to the appearance of early Greek philosophy. Both phenomena concern the rise of rational thinking, but the two fields were studied by historians of science and historians of philosophy separately. Yankov focuses on questions of ontology in philosophy and mathematical theories and their possible relationship.

Although Yankov characterised his hypothesis of the rise of mathematics as speculative (hypothetical, philosophical), his insights are profound and reasonable and provide a new consistent narrative of the historical development of logical thinking in mathematics and philosophy and their interrelationship. Unfortunately, this paper cannot touch the wide variety of facets elaborated in scrupulous detail in his seminal monograph. Nevertheless, we hope that it contributes to the acquaintance of the scholarly community with a work that deserves due attention.

References

Becker, O. (1933–1936). „Eudoxus-Studien: I: Eine voreudoxische Proportionenlehre und ihre Spuren bei Aristoteles und Euklid," *Quellen und Studien zur Geschichte der Mathematik, Astronomie und Phyik* B. II (1933), 311–330. II: Warum haben die Griechen die Existenz der vierten Proportionale angenommen," 369–387. "III: Spuren eines Stetigkeitsaxioms in der Art des Dedekindschen zur Zeir des Eudoxos," vol. 3 (1936), 236–244. IV: Das Prinzip des ausgeschlossenen Dritten in der griechischen Mathematik, 370–388. V: Die eudoxische Lehre von den Ideen und den Farben", 3 (1936), 389–410.

Becker, O. (1936). "Lehre vom Geraden und Ungeraden im Neunten Buch der euklidischen *Elemente*," *Quellen und Studien zur Geschichte der Mathematik, Astronomie und Physik* **3**, 533–553. Reprinted in Becker, Oskar (ed.) 1965. *Zur Geschichte der griechischen Mathematik*. Darmstadt. 125–145.

Bulmer-Thomas, I. (1939). *Selections illustrating the history of Greek mathematics*. Vol. 1. *From Thales to Euclid*. Cambridge, Mass.: Harvard University Press. Reprint as *Greek Mathematical Works*. Loeb Classical Library.

Burkert, W. (1962). *Lore and Science in Ancient Pythagoreanism*, translated by Edwin L. Minar Jr. Cambridge, Mass.: Harvard University Press.German original: *Weisheit und Wissenschaft: Studien zu Pythagoras, Philolaos und Platon*. Nuremberg, Hans Carl.

Burnet, J. (1930). *Early greek philosophy*. Great Britain: A. & C. Black Ltd.

Denisova, T. Y. (2021). "On V.A. Yankov's existential interpretation of early Greek Philosophy. The case of Heraclitus." A. Citkin, I.M. Vandoulakis *V.A. Yankov on Non-Classical Logics, History and Philosophy of Mathematics*. Springer: Outstanding Contributions to Logic, 271–294.

Diels-Kranz (1969–1971). *Die Fragmente der Vorsokratiker*. 1st ed., ed. Hermann Diels. Berlin, 1903. 6th ed., Walther Kranz (ed.) Berlin, 1951–1952. 3 vols. Repr. Dublin/Zürich, 1969–1971.

D'Ooge, M. L. (Ed. and Tr.). (1926). *Nicomachus of Gerasa: Introduction to Arithmetic translated into English*. State New York.

Fowler, D. H. (1999). *The mathematics of Plato's academy*. A New Reconstruction (2nd ed.) Oxford: Clarendon.

Freudenthal, H. (1953). Zur Geschichte der vollständigen Induktion. *Archives Internationales d'Histoire des Sciences, 22*, 17–37.

Freudenthal, H. (1966). Y avait-il une crise des fondements des mathématiques dans l'Antiquité. *Bulletin de la Société Mathématique de Belgique, 18*, 43–55.

Freudenthal, H. (1976). What is algebra and what has been its history? *Archive for History of Exact Sciences, 16*, 189–200.

Gardies, J.-L. (1991). *Le raisonnement par l'absurde*. Presses Universitaires de France.

Hankel, H. (1874). „*Zur Geschichte der Mathematik in Altertum und Mittelalter*". Leipzig

Hasse, H., & Scholz, H. (1928). „Die Grundlagenkrisis in der griechischen Mathematik". *Kant-Studien, 33*(1–2), 4–34.

Heiberg, J. L. (Ed.). (1883–1916). *Euclides opera omnia*. Leipzig. Vol. 1 (*Elements* i-iv), 1883. Vol. 2 (*Elements* v-ix), 1884. Vol. 3 (*Elements* x), 1886. Vol. 4 (*Elements* xi-xiii), 1885. Vol. 5 (*Elements* xiv-xv – scholia to the *Elements* with prolegomena critica), 1888. Vol. 6 (*Data* with the commentary of Marinus and scholia), 1896. Vol. 7 (*Optics*, Theon's recension of the *Optics*, and *Catoptrics* with scholia, 1895). Vol. 8 (*Phenomena* and musical writings), 1916.

Kneale, W., & Kneale, M. (1962). *The development of logic*. Oxford: Clarendon Press.

Knorr, W. R. (1975). *The evolution of the Euclidean elements: A study of the theory of incommensurable magnitudes and its significance for early greek geometry*. Dordrecht: D. Reidel Publishing Co.

Kuhn, T. S. (1962/69). *The structure of scientific revolutions*. Chicago: University of Chicago Press.

Lebedev A. V. (1989). *Fragmenty rannykh grecheskikh filosofov* [*Fragments of Early Greek Philosophers*]. Part 1: *Ot epicheskikh teokosmogonij do vozniknovenija atomistiki* [*From the epic theocosmogonies to the appearance of the atomists*]. Moscow: Nauka (in Russian).

Morrow, G. R. (1970). *Proclus: A commentary on the first book of Euclid's elements*. Translated with introduction and notes. Princeton.

Papadopetrakis, E. (1990). *Quantification in the Greek Mathematical Discourse* [*Η ποσόδειξη στον ελληνικό μαθηματικό λόγο*]. PhD. Thesis. Patras, 1990 [in Greek].

Popper, K. (1958–1959). Back to the Pre-socratics. *Proceeding of the Aristotelian Society*, 1–24.

Popper, K. (1998). *The World of Parmenides. Essays on the Presocratic Enlightenment*. 1998, (Edited by Arne F. Petersen with the assistance of Jørgen Mejer). Routledge.

Szabó, Á. (1969). *Anfänge der griechischen Mathematik*. Munich, Vienna. English translation: *The Beginnings of Greek Mathematics*. Dordrecht, Boston, Mass. 1978.

van der Waerden, B. L. (1954). *Science awakening*. English translation by Arnold Dresden. Groningen: P. Noordhoff Ltd.

van der Waerden, B. L. (Ed.). (1967). *Sources of quantum mechanics*. Amsterdam: North-Holland.

van der Waerden, B. L. (1983). *Geometry and algebra in ancient civilizations*. Berlin, Heidelberg: Springer.

van der Waerden, B. L. (1985). *A history of algebra. From al-Khwārizmī to Emmy Noether*. Berlin, Heidelberg: Springer.

Vandoulakis, I. M. (1991). *On the formation of mathematical science in ancient Greece*. PhD Thesis, Moscow M.V. Lomonosov University.

Vandoulakis, I. M. (1998). Was Euclid's approach to arithmetic axiomatic? *Oriens - Occidens Cahiers du Centre d'histoire des Sciences et des philosophies arabes et Médiévales, 2*, 141–81.

Vandoulakis, I. M. (2009). A genetic interpretation of neo-pythagorean arithmetic. *Oriens - Occidens Cahiers du Centre d'histoire des Sciences et des philosophies arabes et Médiévales, 7*, 113–154.
Vandoulakis I. M. (2015). Negation and truth in Greek mathematics and philosophy. *Book of Abstracts of the 15th Congress of Logic, Methodology and Philosophy of Science*, Helsinki, 3–8 August 2015.
Vandoulakis, I. M. (2017). Sign and reference in greek mathematics. *Gaṇita Bhāratī, 39* (2), (July-2017 to December-2017), 125–145.
Vandoulakis, I. M. (2020). On the origins of mathematics and scientific reasoning. *Filosofia Maravilhosa*, Journal of the Brazilian Academy of Philosophy. Inaugural Issue.
Waszkiewicz, J., & Wojciechowska, A. (1990). On the origin of reductio ad absurdum. In Ewa Żarnecka-Biały (Ed.), *Logic counts*. Dordrecht, Boston, London: Kluwer.
Yankov, V. A. (1997). "Stanovlenie dokazatel'stva v rannej grecheskoj matematike (gipoteticheskaja rekonstruktsija" [The appearance of proof in early Greek mathematics (a hypothetical reconstruction)]. *Istoriko-Matematicheskie Issledovanija*, Series 2, *2(37)*, 200–236 (in Russian).
Yankov, V. A. (2000). "Gippas i rozhdenie geometrii velichin" [Hippasus and the rise of the geometry of magnitudes]. *Istoriko-Matematicheskie Issledovanija*, Series 2, *5(40)*, 192–221 (in Russian).
Yankov, V. A. (2001). "Geometrija posledovatelej Gippasa" [The geometry of Hippasus's successors]. *Istoriko-Matematicheskie Issledovanija*, Series 2, *6(41)*, 285–366 (in Russian).
Yankov, V. A. (2003). "Geometrija Anaksagora" [Anaxagoras's geometry]. *Istoriko-Matematicheskie Issledovanija*, Series 2, *8(43)*, 241–318 (in Russian).
Yankov, V. A. (2011). *Istolkovanie rannej grecheskoj filosofii [An Interpretation of Early Greek Philosophy]*. Moscow: Publication of the Russian State University for Humanities (in Russian).

Index

A. Citkin and I. M. Vandoulakis (eds.), *V. A. Yankov on Non-Classical Logics, History and Philosophy of Mathematics*, Outstanding Contributions to Logic 24,
https://doi.org/10.1007/978-3-031-06843-0

Milton Keynes UK
Ingram Content Group UK Ltd.
UKHW020607151123
432609UK00002B/11

9 783031 068454